3.

POWER SEMICONDUCTOR DEVICES

ABOUT THE AUTHOR

B. Jayant Baliga
(Ph.D., Rensselaer Polytechnic Institute) is a professor in the Department of Electrical and Computer Engineering at North Carolina State University, where he founded the international Power Semiconductor Research Center in 1991. He pioneered the concept of MOS-bipolar functional integration and invented the insulated gate bipolar transistor; he currently holds over 70 patents in the field. Professor Baliga has published more than 350 articles and five books prior to this text. He has won numerous awards, including the IEEE William E. Newell Award, the highest honor given by the Power Electronics Society. He has been elected to the National Academy of Engineering and named a Fellow of the IEEE. Currently, Dr. Baliga is serving as a member of the Presidential Task Force on Higher Education.

POWER SEMICONDUCTOR DEVICES

B. JAYANT BALIGA

North Carolina State University

PWS PUBLISHING COMPANY

I(T)P   An International Thomson Publishing Company

Boston • Albany • Bonn • Cincinnati • Detroit
London • Melbourne • Mexico City • New York
Paris • San Francisco • Singapore • Tokyo • Toronto

PWS PUBLISHING COMPANY
20 Park Plaza, Boston, MA 02116-4324

Copyright ©1996 by PWS Publishing Company, a division of International Thomson Publishing Inc.

All rights reserved. No part of this book may be reproduced, stored in a retrieval system, or transmitted, in any form or by any means -- electronic, mechanical, photocopying, recording, or otherwise -- without prior written permission of PWS Publishing Company.

I(T)P™
International Thomson Publishing
The trademark ITP is used under license.

 This book is printed on recycled, acid-free paper

For more information, contact:

PWS Publishing Co.
20 Park Plaza
Boston, MA 02116

International Thomson Editores
Campos Eliseos 385, Piso 7
Col. Polanco
11560 Mexico D.F., Mexico

International Thomson Publishing Europe
Berkshire House 168-173
High Holborn
London WC1V 7AA
England

International Thomson Publishing GmbH
Konigswinterer Strasse 418
53227 Bonn, Germany

Thomas Nelson Australia
102 Dodds Street
South Melbourne, 3205
Victoria, Australia

International Thomson Publishing Asia
221 Henderson Road
#05-10 Henderson Building
Singapore 0315

Nelson Canada
1120 Birchmont Road
Scarborough, Ontario
Canada M1K 5G4

International Thomson Publishing Japan
Hirakawacho Kyowa Building, 31
2-2-1 Hirakawacho
Chiyoda Ku, Tokyo 102
Japan

Library of Congress Cataloging-in-Publication Data

Baliga, B. Jayant
    Power semiconductor devices / B. Jayant Baliga.
       p. cm.
    Includes bibliographical references and index.
    ISBN 0-534-94098-6
    1. Power semiconductors. I. Title
TK7871.85.B35   1995   94–42308
621.317—dc20   CIP

Sponsoring Editor: *Jonathan Plant*
Developmental Editor: *Mary Thomas*
Production Editor: *Monique A. Calello*
Editorial Assistant: *Lai Wong*
Marketing Manager: *Nathan Wilbur*
Manufacturing Coordinator: *Wendy Kilborn*
Cover Designer: *Elise S. Kaiser*
Cover Printer: *New England Book Components*
Text Printer: *Quebecor Printing/Martinsburg*

Printed and bound in the United States of America
96 97 98 99 — 10 9 8 7 6 5 4 3 2

To

my parents, Vittal and Sanjivi

my wife, Pratima

and

my sons, Avinash and Vinay

PREFACE

The microelectronics revolution has been called the *First Electronic Revolution*. It has had an enormous impact upon society because it has provided inexpensive information processing capabilities. The access to superior computation and communication services at the global level has altered our society, bringing changes in business, industrial, and commercial enterprises. The semiconductor business already exceeds $100 billion with significant growth projected into the foreseeable future. More importantly, the microelectronic chips are the core of all electronic systems because they dictate the performance. The economic impact of the microelectronics revolution therefore far exceeds the semiconductor business.

The development of power electronics systems based on semiconductor devices began concurrently with the advent of the microelectronics technology. The first power devices, capable of handling significant currents and voltages, became commercially available in the 1950s with the introduction of the power bipolar transistors and power thyristors. Due to the many advantages of semiconductor devices when compared with vacuum tubes, there was a constant demand for increasing the power ratings of the devices. During the next 30 years, the technology for these bipolar devices reached a high degree of maturity. By the 1970s, bipolar power transistors capable of handling hundreds of amperes at over 500 volts became available. More remarkably, technology was developed that enabled the manufacturing of an individual power thyristor from an entire 4-inch wafer with a voltage rating of over 5000 volts.

While the developments in the power semiconductor technology during this period were very impressive, these developments digressed from those taking place in the microelectronics industry where the emphasis shifted from bipolar devices to the *complementary metal oxide semiconductor (CMOS)* technology. Since the semiconductor market was soon dominated by the microelectronics technology for memories and microprocessors, the equipment manufacturers did not view the power semiconductor technology with the same degree of commercial interest. In addition, many short comings of the bipolar power devices were becoming evident from the application point of view. The most important among these short comings was that the bipolar devices were essentially current controlled with a relatively poor current gain. This resulted in complex discrete control circuits that were expensive to manufacture and maintain in the field. It was recognized in the 1970s that the development of power semiconductor devices based upon MOS gate technology would provide a solution to these problems.

The first power MOSFETs were introduced in the 1970s. In addition to their high input impedance, these devices offered superior switching speed when compared with the bipolar power transistors. However, due to the maturity of the bipolar technology and the higher-resolution lithography requirements for the power MOSFETs, the cost of these devices was significantly higher than that for the bipolar devices for the same power rating. This difference was initially offset by the simplification of the control circuit, which reduced the overall system cost. However, over the last 20 years, the power MOS technology has also moved down a fast learning curve, making the manufacturing of the discrete MOS-gated devices competitive with bipolar devices. In this regard, the power MOS technology has the advantage that advances being continually made in the CMOS technology can be utilized whenever appropriate. This has brought about a convergence of power semiconductor technology with mainstream microelectronics technology.

The high input impedance and faster switching speed advantages of the power MOSFET were initially considered sufficient for the complete displacement of the bipolar power transistor in applications. However, this did not take place because it was found that the current-handling capability of the power MOSFET degrades rapidly when the structure is designed to support high voltages. Its greatest impact was therefore felt in low-power systems operating at voltages below 200 volts. Some of the important applications in which the power MOSFET has dominated over the bipolar power transistor are *computer power supplies/peripheral drives, telecommunications, and automotive electronics*.

In order to overcome the poor current-handling capability of power MOSFETs at high operating voltages, the concept of merging MOS gate technology with bipolar current conduction was introduced in the 1980s. Initially, attempts were made to use discrete power MOSFETs and discrete bipolar power transistors to create improved composite switches with high input impedance and smaller net silicon area for satisfying the power ratings. A much more innovative and fruitful approach was next developed, when the physics of MOS gate control was merged with bipolar current flow within the same device. The most commercially successful structure that emerged from this effort was the *insulated gate bipolar transistor (IGBT)*. This device enabled the efficient control of very high currents and voltages together with a high input impedance. As a consequence, it has rapidly displaced the bipolar power transistor in medium power applications. Some of the important applications in which the IGBT has dominated over the bipolar power transistor are *heating/ventilating/air-conditioning (HVAC) systems, factory automation/numerical controls/robotics, lighting ballasts, appliance controls, and electric vehicle drives*.

The availability of power semiconductor devices with high input impedance for a broad range of power control has encouraged the development of integrated control circuits. In general, integrated control circuits are preferred over their discrete counterparts because of the reduced manufacturing cost at high volumes and the improved reliability because of reduced interconnections. An example of an application where the volume is sufficient to justify manufacturing of integrated control circuits is the H-bridge circuit used for motor control. These integrated circuits provide the required gate drive voltage and perform the level-shifting function. Since the cost of including additional circuitry in an integrated circuit is relatively small, the incorporation of protective features such as *overtemperature, overcurrent, and overvoltage* has become cost effective. In addition, these chips contain the encode/decode CMOS circuitry required to interface with a central microprocessor or computer in the system.

This technology has been labeled *smart power technology*.

The advent of smart power technology portends a *Second Electronic Revolution*. In contrast to integrated circuits used for information processing, this technology enables efficient control of energy. These technologies can therefore be regarded as complementary, similar to the brain and muscles in the human body. As in the case of microelectronic chips, smart power technology is also expected to have major social impact. It has been documented that over 70% of all electricity used in the world flows through power semiconductor devices. With the more efficient control of this power flow, enormous savings in losses can be projected, resulting in conservation of fossil fuels and less environmental pollution. In addition, the deployment of electric cars, in which the power electronics plays a critical role in determination of efficiency, is a major step toward providing a healthier environment in urban areas.

With these developments, it is anticipated that there will be an increasing need for technologists trained in the discipline of designing and manufacturing of power semiconductor devices. Although many universities are offering courses at the graduate level on power devices, the lack of a suitable textbook on this topic has been a hinderance. Two earlier books, *Semiconductor Power Devices* by S.K. Ghandhi and *Modern Power Devices* by B.J. Baliga, are out of print. Further, many new developments have taken place since these books were written. Consequently, this book is written at the tutorial level to fill this need. In preparing the book, it has been assumed that the reader is familiar with the fundamental concepts of current transport in semiconductors and has a basic knowledge of process technology. The book can be used for self-study by professional engineers practicing the art of power device design and fabrication. It is particularly suitable as a textbook for a graduate-level course on power devices. In fact, the material in the book is based upon notes from a two-semester course taught at *North Carolina State University*.

In this book, a chapter has been devoted to each of the following power devices: *power rectifiers, power bipolar transistors, power thyristors, power MOSFETs, IGBTs, and MOS-gated thyristors*. Each of these chapters is organized to first introduce the basic structure for the device and its electrical output characteristics. This material can be utilized to provide an introduction to power devices at the undergraduate level in a power electronics course. A detailed discussion of the physics of device operation in the static blocking and current conduction states is then provided. This is followed by analysis of the switching behavior including the safe-operating-area. A set of references is provided at the end of each chapter for further reading on some special topics. In addition, a set of problems has been provided at the end of each chapter. The reader is encouraged to solve these problems to obtain a deeper insight and grasp of the material discussed in each chapter.

Throughout the book, emphasis is placed on deriving simple analytical expressions that describe the underlying physics and enable representation of the device electrical characteristics. This treatment is invaluable for teaching a course on power devices because it allows the operating principles and concepts to be conveyed with quantitative analysis. The availability of user-friendly software for numerical simulation of semiconductor devices has greatly simplified the analysis and design of devices. However, it has also created a tendency for the loss of insight into the device physics that is an essential component for innovation. The analytical approach used in this book based on physical insight will provide a good foundation for the reader even if numerical simulation is subsequently performed to obtain a more accurate assessment of the device characteristics. In support of the analytical solutions used to describe the physics of operation of the power devices, this book contains a chapter where the fundamental transport properties of silicon have been provided. Once again, an effort has been made to formulate analytical expressions that describe the empirical data available in the literature. The information in this chapter has been restricted to silicon because, at present, all power devices are manufactured from this material.

Power semiconductor devices share a common and distinguishing feature in being able to withstand relatively high voltages applied to their terminals. Consequently, no book on power devices would be complete without a thorough discussion of the physics of avalanche breakdown and leakage current flow within the device structures. In addition, since all devices have a finite size, it becomes extremely important to provide a suitable edge termination that does not degrade the voltage-blocking capability. In general, it has been found that the breakdown voltage within the device structure can be made quite close to the intrinsic capability of the silicon drift region. Consequently, the maximum voltage-blocking capability of a power device is usually limited by the edge termination. Considerable effort has been expended by the power semiconductor industry to develop suitable edge terminations that not only allow proper utilization of the intrinsic capability of the drift region but also ensure that the device can be passivated to produce stable operation through its operating life. Since these techniques are applicable to all the power devices, the study of the chapter on breakdown voltage is essential to the design of all the devices discussed in this book.

This textbook is an outgrowth of a two-semester course for graduate students at North Carolina State University. In the first semester, the basic silicon carrier transport physics and the analysis of avalanche breakdown/edge terminations are initially treated to provide a sound basis for analysis of the power devices. The operating physics and electrical characteristics of bipolar power devices are then discussed. In the second semester, after a short review of the silicon carrier transport physics, the physics of MOSFETs, IGBTs, and MOS-gated thyristors are discussed. A brief introduction to smart power technology is also provided, but this material is taken from the literature and an IEEE Press book titled *High Voltage Integrated Circuits*, edited by B.J. Baliga.

ACKNOWLEDGMENTS

I would like to thank *John Wiley and Sons, New York,* for permission to include edited portions of my textbook *Modern Power Devices* in this book. I would also like to thank the following reviewers for their comments which helped improve upon the manuscript: Dr. Richard M. Bass, Georgia Institute of Technology; Dr. Allan G. Potter, Iowa State University; and Dr. Ziyad Salameh, University of Massachusetts at Lowell. I am grateful to the many people at PWS Publishing Company that were responsible in the production of the book. In particular, I want to express my appreciation for the careful scrutiny of the manuscript by the copyeditor, Jean Peck, who was invaluable in correcting many typographical and idiomatic errors in the original manuscript. It was a pleasure working with the engineering editor, Jonathan Plant, who approached me to consider this endeavor, and the production editor, Monique Calello, for her very professional handling of the manuscript through the production process.

In closing, I would like to take this opportunity to thank Dr. Ralph Cavin, III, Head of the Electrical and Computer Engineering Department, North Carolina State University, for encouraging me to undertake the writing of this book. I would also like to thank the sponsors of the *Power Semiconductor Research Center* at North Carolina State University for their loyal support over the years. This support has been invaluable for performing the research that has led to a deeper understanding of the physics of operation of the power devices discussed in the book. I also wish to acknowledge the many interesting discussions with my graduate students over the last five years that have assisted me in presenting the information in a tutorial format. The preparation of all the illustrations and equations interspersed and properly formatted to create a camera ready manuscript for publication has been an enormously time-consuming undertaking over a period of 18 months. I hope that this effort will be appreciated by the widespread use of the book in the teaching of courses on power devices throughout the world.

B. JAYANT BALIGA

CONTENTS

Chapter 1 INTRODUCTION   1

    1.1 **Basic Power Device Characteristics** 3
          1.1.1 Basic Power Rectifier Characteristics 4
          1.1.2 Basic Power Switch Characteristics 5
    1.2 **Historical Perspective** 7
    References *8*

Chapter 2 TRANSPORT PHYSICS   9

    2.1 **Silicon Mobility** 9
          2.1.1 Temperature Dependence 10
          2.1.2 Dopant Concentration Dependence 11
          2.1.3 Electric Field Dependence 13
          2.1.4 Injection Level Dependence 16
          2.1.5 Surface Scattering Effects 18
    2.2 **Silicon Resistivity** 27
          2.2.1 Intrinsic Resistivity 28
          2.2.2 Band Gap Narrowing 29
          2.2.3 Extrinsic Resistivity 33
          2.2.4 Neutron Transmutation Doping 36
    2.3 **Lifetime** 41
          2.3.1 Shockley-Read-Hall Recombination 42
          2.3.2 Space Charge Generation 45
          2.3.3 Recombination-Level Optimization 47
          2.3.4 Lifetime Control 55
          2.3.5 Auger Recombination 59
    References *61*
    Problems *65*

Chapter 3 BREAKDOWN VOLTAGE 66

    3.1 **Avalanche Breakdown** 67
        3.1.1 Ionization Coefficients 67
        3.1.2 Multiplication Coefficient 68
    3.2 **Abrupt Junction Diode** 71
    3.3 **Punch-Through Diode** 75
    3.4 **Linearly Graded Diode** 77
    3.5 **Diffused Junction Diode** 79
    3.6 **Edge Terminations** 81
        3.6.1 Planar Junction Terminations 82
        3.6.2 Floating Field Ring Terminations 91
        3.6.3 Multiple Floating Field Ring Terminations 98
        3.6.4 Field Plates 100
        3.6.5 Bevel Edge Terminations 103
        3.6.6 Etch Contour Terminations 110
        3.6.7 Surface Implantation Termination 111
    3.7 **Open Base Transistor Breakdown** 113
        3.7.1 Negative/Positive Bevel Combination 116
        3.7.2 Double Positive Bevel Contours 117
    3.8 **High Voltage Surface Passivation** 119
    3.9 **Comparison of Terminations** 121
References 124
Problems 126

Chapter 4 POWER RECTIFIERS 128

    4.1 **Schottky Barrier Rectifiers** 129
        4.1.1 Metal-Semiconductor Contact 129
        4.1.2 Forward Conduction 131
        4.1.3 Reverse Blocking 137
        4.1.4 Trade-Off Curves 142
        4.1.5 Power Dissipation 144
        4.1.6 Switching Behavior 145
        4.1.7 Device Technology 146
        4.1.8 Edge Terminations 148
        4.1.9 High Voltage Schottky Barrier Rectifiers 150
    4.2 **P-i-N Rectifiers** 153
        4.2.1 Forward Conduction 155
        4.2.2 Reverse Blocking 169
        4.2.3 Reverse Recovery 171
        4.2.4 Lifetime Control 175
        4.2.5 Doping Profile 177

Contents   xv

    4.2.6 Ideal Ohmic Contact 178
    4.2.7 Maximum Operating Temperature 181
  4.3 **JBS Rectifiers 182**
    4.3.1 Forward Conduction Characteristics 184
    4.3.2 Reverse Blocking Characteristics 185
    4.3.3 Device Characteristics 186
  4.4 **MPS Rectifiers 187**
    4.4.1 Forward Conduction Characteristics 188
    4.4.2 Stored Charge 189
    4.4.3 Reverse Recovery 192
    4.4.4 Reverse Blocking 192
  4.5 **Trends 192**
References 193
Problems 196

Chapter 5 BIPOLAR TRANSISTORS 198

  5.1 **Device Operation 199**
  5.2 **Current Transport 201**
    5.2.1 Emitter Injection Efficiency 203
    5.2.2 Emitter Efficiency including High Level Injection in Base 206
    5.2.3 Base Transport Factor 211
    5.2.4 Collector Bias Effects 212
    5.2.5 Voltage Saturation Region 215
    5.2.6 Base Widening at High Current Densities 222
    5.2.7 Emitter Current Crowding 228
  5.3 **Static Blocking Characteristics 232**
    5.3.1 Open Emitter Breakdown Voltage 233
    5.3.2 Shorted Emitter-Base Breakdown Characteristics 233
    5.3.3 Open-Base Breakdown Characteristics 235
  5.4 **Dynamic Switching Characteristics 236**
    5.4.1 Turn-on Transient 237
    5.4.2 Turn-off Transient 240
  5.5 **Second Breakdown Characteristics 243**
    5.5.1 Forward Biased Second Breakdown 244
    5.5.2 Reverse Biased Second Breakdown 247
  5.6 **Darlington Power Transistor 252**
  5.7 **Trends 255**
References 255
Problems 256

Chapter 6 POWER THYRISTORS 258

- 6.1 Thyristor Structure and Operation 260
- 6.2 Static Blocking Characteristics 262
  - 6.2.1 Reverse Blocking 263
  - 6.2.2 Forward Blocking 264
- 6.3 Forward Conduction Characteristics 272
  - 6.3.1 On-state Operation 272
  - 6.3.2 Gated Turn-on 273
  - 6.3.3 Holding Current 277
- 6.4 Switching Characteristics 279
  - 6.4.1 dV/dt Capability 280
  - 6.4.2 Turn-on Transient 282
  - 6.4.3 Turn-off Transient 291
- 6.5 Light Triggered Thyristors 295
- 6.6 Thyristor Self-protection 298
  - 6.6.1 Breakdown Protection 298
  - 6.6.2 [dV/dt] Turn-on Protection 300
- 6.7 Gate Turn-off Thyristors 302
  - 6.7.1 Turn-off Criterion 303
  - 6.7.2 Turn-off Time 305
  - 6.7.3 Anode Shorted GTO Structure 316
  - 6.7.4 Maximum Controllable Current 317
  - 6.7.5 GTO Layout 321
  - 6.7.6 GTO Ratings 322
- 6.8 Triacs 322
  - 6.8.1 Triac Structure and Operation 325
  - 6.8.2 [dV/dt] Limitations 329
- 6.9 Trends 331

References 331
Problems 333

Chapter 7 POWER MOSFET 335

- 7.1 Basic Structure and Operation 336
- 7.2 Output Characteristics 340
- 7.3 Static Blocking Characteristics 342
  - 7.3.1 Parasitic Bipolar Transistor 342
  - 7.3.2 Doping Profile 343
  - 7.3.3 Cell Structure 345
- 7.4 Forward Conduction Characteristics 349
  - 7.4.1 MOS Surface Physics 350
  - 7.4.2 Threshold Voltage 357

        7.4.3 Channel Resistance 362
        7.4.4 DMOSFET Specific On-resistance 367
        7.4.5 Ideal Specific On-resistance 373
        7.4.6 MOS Cell Optimization 373
        7.4.7 UMOSFET On-resistance 377
        7.4.8 Optimum Doping Profile 380
7.5 **Frequency Response 381**
        7.5.1 MOS Capacitance 382
        7.5.2 Input Capacitance 384
        7.5.3 Gate Series Resistance 386
        7.5.4 Maximum Operating Frequency 387
7.6 **Switching Performance 387**
        7.6.1 Turn-on Transient Analysis 388
        7.6.2 Turn-off Transient Analysis 392
        7.6.3 [dV/dt] Capability 395
7.7 **Safe-Operating-Area 397**
        7.7.1 Bipolar Second Breakdown 398
        7.7.2 MOS Second Breakdown 400
7.8 **Integral Diode 402**
        7.8.1 Switching Speed 404
        7.8.2 Parasitic Bipolar Transistor 406
7.9 **High Temperature Performance 406**
        7.9.1 On-resistance 407
        7.9.2 Transconductance 408
        7.9.3 Threshold Voltage 409
7.10 **Device Structures and Technology 410**
        7.10.1 DMOS (Planar) Structure 410
        7.10.2 UMOS Structure 412
        7.10.3 Gate Oxide Fabrication 414
        7.10.4 Cell Topology 415
7.11 **Silicon Carbide Power MOSFETs 417**
7.12 **Trends 420**
References 421
Problems 424

Chapter 8 INSULATED GATE BIPOLAR TRANSISTOR   426

8.1 **Device Structure and Operation 428**
8.2 **Static Blocking Characteristics 431**
        8.2.1 Reverse Blocking Capability 431
        8.2.2 Forward Blocking Capability 432
8.3 **Forward Conduction Characteristics 434**
        8.3.1 P-i-N Rectifier/MOSFET Model 436

8.3.2 Bipolar Transistor/MOSFET Model 440
8.3.3 On-state Carrier Distribution 446
8.3.4 On-state Voltage Drop 448
8.4 **Parasitic Thyristor Latch-up 451**
8.4.1 Parasitic Thyristor Latch-up Suppression via P-N-P Transistor 452
8.4.2 Parasitic Thyristor Latch-up Suppression via N-P-N Transistor 453
8.5 **Safe-Operating-Area 468**
8.5.1 Forward-Biased-Safe-Operating-Area 470
8.5.2 Reverse-Biased-Safe-Operating-Area 472
8.5.3 p-Channel versus n-Channel IGBT 472
8.5.4 DMOS Cell Design 474
8.5.5 Switching Locus 476
8.6 **Switching Characteristics 476**
8.6.1 Turn-off Time 478
8.6.2 Lifetime Control 479
8.7 **Complementary Devices 485**
8.8 **High Voltage Devices 488**
8.9 **High Temperature Characteristics 490**
8.9.1 On-state Characteristics 491
8.9.2 Switching Characteristics 493
8.9.3 Latching Current Density 494
8.10 **Trench Gate IGBT Structure 496**
8.11 **Trends 498**
References 498
Problems 501

Chapter 9 MOS-GATED THYRISTORS 503

9.1 **MOS-Gated Thyristor Turn-on 504**
9.2 **MOS-Controlled Thyristor or MOS-GTO 506**
9.2.1 Device Structure and Operation 507
9.2.2 Maximum Controllable Current Density 510
9.2.3 Complementary MCT 514
9.2.4 On-state Voltage Drop 515
9.2.5 Multicellular Devices 519
9.2.6 Turn-off Time 523
9.2.7 Temperature Dependence 524
9.2.8 MCT Characteristics 525
9.3 **Base Resistance Controlled Thyristor 526**
9.3.1 Device Structure and Operation 526
9.3.2 Maximum Controllable Current Density 530
9.3.3 On-state Voltage Drop 534
9.3.4 Turn-off Time 538

9.3.5 Temperature Dependence 541
9.3.6 BRT Characteristics 543
9.4 Emitter Switched Thyristor 543
9.4.1 Device Structure and Operation 544
9.4.2 Dual Channel Structure 548
9.4.3 Maximum Controllable Current Density 550
9.4.4 Latching Current Density 554
9.4.5 On-state Voltage Drop 557
9.4.6 Current Saturation 560
9.4.7 Turn-off Time 563
9.4.8 Temperature Dependence 564
9.4.9 EST Characteristics 567
9.5 Trends 568
References 569
Problems 571

Chapter 10 SYNOPSIS 572

10.1 Comparison of Gate Controlled Devices 572
10.1.1 Gate Control Requirements 573
10.1.2 Simplified Power Loss Analysis 573
10.2 Basic Variable Speed Motor Control Circuit 575
10.3 IGBT Power Loss Components 578
10.3.1 IGBT with PiN Rectifier 578
10.3.2 IGBT with MPS Rectifier 580
10.4 MOS-Gated Thyristor Power Loss Components 580
10.4.1 BRT with PiN Rectifier 581
10.4.2 BRT with MPS Rectifier 582
10.4.3 EST with PiN Rectifier 583
10.4.4 EST with MPS Rectifier 584
10.5 Silicon Carbide Devices 584
10.6 Comparison of Devices 586
10.7 Smart Power Integrated Circuits 587
10.8 Trends 588
References 588

INDEX 589

Chapter 1

INTRODUCTION

The applications for power semiconductor devices are quite diverse as shown in Fig. 1.1, where the power ratings of the system are shown as a function of system operating frequency.

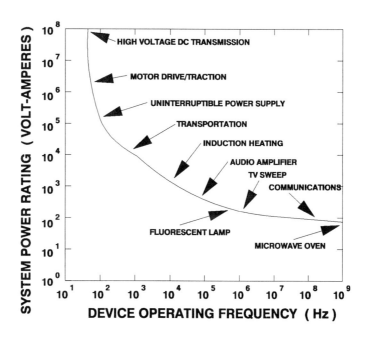

Fig. 1.1 Applications for power semiconductor devices provided as a function of system operating frequency and power handling capability.

It can be seen that the power ratings extend over a tremendous range from the level of 100 watts at microwave frequencies to 100 megawatts at low frequencies. An ideal power switch used for power conditioning should therefore be capable of handling high currents and voltages, and be able to switch at a high speed. No single semiconductor device has as yet been developed that can satisfy all three requirements simultaneously.

Another approach to classification of the applications for power semiconductor devices is in terms of the voltage and current ratings that the device must satisfy for each application.

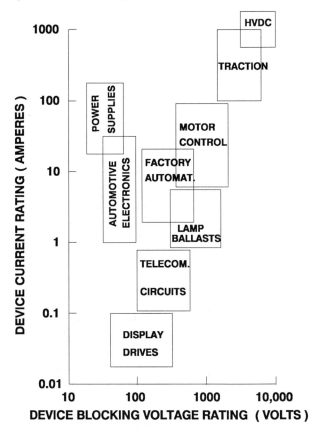

Fig. 1.2 Applications for power devices in relation to their voltage and current ratings.

This classification is shown in Fig. 1.2. From this figure, it can be seen that there is a need for devices that are capable of operating at over 6000 volts and 1000 amperes. The applications can also be noted to fall into two broad groups. The first category requires relatively low breakdown voltage (less than 100 volts) but requires high current handling capability. Two examples shown in the figure are automotive electronics and switch mode power supplies. The second group of applications lies along a trajectory of increasing breakdown voltage and current handling capability, i.e. a substantial increase in power handling capability. At the lowest power levels are applications such as display drives. These applications are being served by monolithic power integrated circuits containing multiple high voltage drive transistors. At higher power levels, smart power technology is being developed to provide monolithic chips for lamp ballasts and fractional horse power motor control. When the breakdown voltages exceed 500 volts and the applications require control of currents above 1 ampere, it becomes essential to partition the system into a control chip that drives a set of discrete power semiconductor devices. Examples of these applications are motor drives for heating, ventilating, and air-conditioning. At even higher power levels, such as those encountered in traction and high voltage DC transmission systems, the systems are implemented using discrete components.

1.1 BASIC POWER DEVICE CHARACTERISTICS

An ideal power device must be able to control the flow of power to loads with zero power dissipation. The loads may be inductive (such as motors, solenoids, etc), resistive (such as heaters and lamp filaments), or capacitive (such as transducers and display elements). Most often, the power to the load is controlled by switching the device on a periodic basis to generate pulses of current flowing through the device. The current and voltage waveforms are shown in

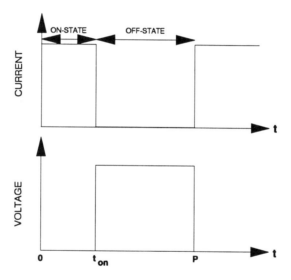

Fig. 1.3 Switching waveforms for an ideal power switch.

Fig. 1.3 for an ideal switch. In this case, there must be no voltage drop across the device when it is conducting current in the "on-state" and there must be no current flowing through the device when it is supporting voltage in its "off-state". An ideal device must also be able to switch between these states with no power dissipation. This requires an instantaneous change in the voltage and current during the switching transients.

1.1.1 Basic Power Rectifier Characteristics

Power circuits require both switches and rectifiers. In order to satisfy the requirements

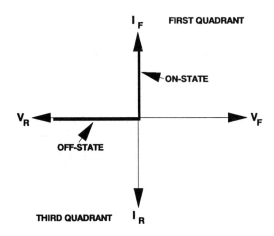

Fig. 1.4 I-V characteristics of an ideal rectifier.

for an ideal device, an ideal power rectifier must have the characteristics shown in Fig. 1.4. The ideal rectifier exhibits zero on-state voltage drop and zero reverse bias leakage current. In addition, it must be capable of switching between these states instantaneously. As an ideal device, it should also be capable of conducting infinite current in the forward direction and supporting infinite voltage in the reverse direction.

It is well known that actual power devices do not exhibit the characteristics shown for the ideal device. A typical power rectifier characteristic is shown in Fig. 1.5. It should be noted that the device has a finite on-state voltage drop (V_{on}) when conducting the on-state current (I_{ON}). This results in a significant power dissipation which limits the maximum current handling capability of the rectifier. The on-state voltage drop is a function of many design parameters including the breakdown voltage and the switching speed. In the reverse blocking mode of operation, the typical rectifier exhibits a finite "leakage" current flow and is capable of supporting a finite maximum reverse voltage, which is defined as its *breakdown voltage (BV_R)*. This results in power dissipation during the "off-state". Although this power loss is small at room temperature, it can become comparable to the "on-state" power dissipation when the device temperature increases either due to operation in a high ambient temperature or by self-heating. The device can then go into a destructive thermal runaway. The finite breakdown voltage must also be taken into account because the device can be destroyed if the operating voltage exceeds this value even under transient conditions. A typical power rectifier also exhibits finite switching times during turn-on and turn-off leading to further power losses in the circuits.

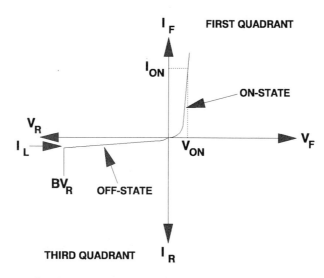

Fig. 1.5 Typical power rectifier I-V characteristics.

1.1.2 Basic Power Switch Characteristics

Power switches are used to control the power flow from the power supply to the load, which can be resistive, inductive, or capacitive in nature. In general, the power switch and its associate control electronics can be considered to be black-box between the input and output

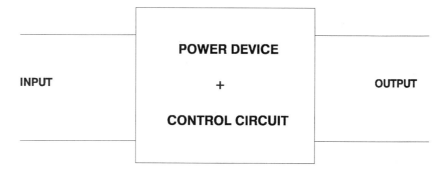

Fig. 1.6 Black-box representation of a power switch and its control circuit.

circuits as shown in Fig. 1.6. In the ideal case with 100 percent efficiency, the output power must be controlled with zero input power. In addition, the device should be able to control an arbitrarily high power level, which implies high voltage and current handling capability.

The output characteristics of an ideal three-terminal power switch are shown in Fig. 1.7. For general applicability, the ideal switch must be able to block and conduct current in both

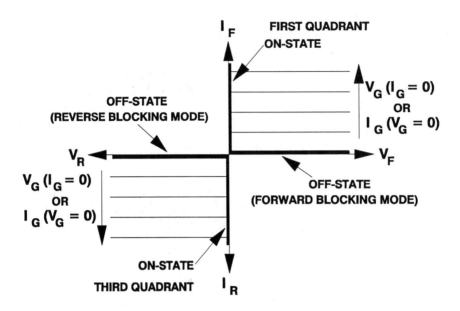

Fig. 1.7 Characteristics of an ideal power switch.

directions of current flow under the control of a signal applied to its gate electrode. This requires operation of the device in both the first quadrant and third quadrant of the I-V characteristics. In either quadrant, the device must be capable to supporting a very high voltage with zero current flow. These states are referred to as *Forward Blocking Mode* and the *Reverse Blocking Mode*. When switched to the conducting state, the device must be capable of carrying a very high current with zero voltage drop. This is referred to as the *On-State*. In addition, it is desirable for the device to regulate the output current under the guidance of the gate signal. This requires a gate controlled current saturation characteristic as shown in the figure. In order to maintain zero power losses in the input circuit, the gate control signal can be a current applied with zero voltage drop in the control circuit or a voltage signal with zero current flow in the control circuit. It has been found that it is easier to approach the ideal control case with the latter option.

It should be noted that the ideal device characteristics shown in the figure are for a *normally-off device*. This term indicates that no current flows through the device in the absence of the gate control signal. This is preferable to a device that is *normally-on*, which conducts current in the absence of a gate drive signal, because it is important to stop power flow into the load whenever the gate control circuit fails to operate. An example of this situation is during the start up of a system. In this case, it is difficult to ensure that the gate control signal will be available to control the device properly before the power is applied to the load.

The output characteristics of actual power switches that are commercially available vary greatly. The characteristics of specific devices are discussed in detail in the subsequent chapters of the book. In general, all the devices can support only a limited voltage in either direction and they exhibit a finite *leakage current* when blocking current flow. This results in power dissipation during the blocking mode that increases rapidly with increasing temperature and can lead to a thermal runaway problem unless the heat sink is designed to maintain the temperature below defined limits. Further, all the devices exhibit a finite voltage drop during the on-state current flow, leading to power loss. Due to the finite switching times for the devices, substantial power losses can occur especially at higher operating frequencies. Moreover, there can be substantial power dissipation in the input control circuits, which must be accounted for when optimizing the system performance.

1.2 HISTORICAL PERSPECTIVE

From a historical perspective, power semiconductor devices have played an increasingly important role in the development of power electronic systems over the last 50 years. The introduction of power thyristors in the 1950s led to a replacement of thyratrons for power conditioning. Due to the demand for devices that could handle higher power levels, the ratings of power thyristors have been increasing over the years. It is impressive to note that devices with blocking voltages above 6000 volts are commercially available with the ability to control several thousand amperes. These devices have relatively slow switching speed, which makes them suitable for systems with low operating frequency. Examples of such applications are high voltage direct current (HVDC) power transmission networks and high power motor drives used in steel mills.

The bipolar power transistor was developed to address the problems with the relatively slow switching frequency of thyristors. Devices with substantial power handling capability were developed for medium power applications in the 1960s. With these devices, it is possible to operate at frequencies of up to 50 kHz. The blocking voltage rating of these devices has been extended to 1200 volts for TV deflection circuits, while devices with higher current handling capability have become commercially available with breakdown voltages of about 500 volts for motor control.

In the 1970s, the power semiconductor industry assimilated the metal-oxide-semiconductor (MOS) technology developed for integrated circuits. This led to the introduction of power MOS field effect transistors (MOSFETs), which provided the user with a much faster switching capability and simplicity of control. These devices have displaced the bipolar transistor in low power conditioning systems operating at frequencies above 10 kHz particularly when the blocking voltages are less than 200 volts. Examples of applications are power supplies and computer peripherals. Although these devices were initially expected to impact the medium power conditioning area, this has not transpired due to the poor current handling capability of power MOSFETs when designed to operate at higher voltages (above 300 volts).

A further advancement in power semiconductor device capability occurred in the 1980's by the commercial introduction of the insulated gate bipolar transistor (IGBT). This device provided the user with the high input impedance feature of the power MOSFET and the high current handling capability of the bipolar devices within an integrated structure. As the ratings of these devices have grown to exceed 1500 volts and several hundred amperes, they have displaced the bipolar transistor in medium power conditioning applications. Examples of these applications are motor control for heating, ventilating, and air-conditioning, and numerical controls for robotics and factory automation systems. The device has also been selected for the drive train in electric vehicles.

In spite of these developments, a need exists for improving the power losses within the semiconductor devices used for power conditioning. One approach that is promising is the development of MOS-gated power thyristor structures. These devices offer the lower on-state voltage drop of thyristors combined with the ease of control available with MOS-gated devices. The first commercial device of this type became available in the 1990s. It remains to be seen whether it will find good market acceptance. The most promising market segment for these devices is for traction drives that are now served by gate turn-off thyristors (GTOs). However, the voltage and current ratings of these devices are as yet insufficient for use in these applications.

Looking into a longer time frame, it is anticipated that the silicon power devices may be replaced by devices fabricated using silicon carbide. The larger breakdown electric field strength in silicon carbide is projected to allow the fabrication of power devices (Schottky rectifiers and power MOSFETs) with superior characteristics for breakdown voltages as high as 5000 volts. Schottky barrier rectifiers have already been experimentally demonstrated with excellent on-state and switching characteristics with breakdown voltages of up to 500 volts. However, the commercial availability of these devices is unlikely to occur until a substantial improvement in the quality of the substrate material and a large reduction in its cost are achieved.

REFERENCES

1. B. J. Baliga, "High voltage integrated circuits", IEEE Press, New York (1988).

2. P. A. Thollot, "Power electronics technology and applications 1993", IEEE Technology Update Series, New York (1993).

3. B. J. Baliga, "An overview of smart power technology", IEEE Trans. Electron Devices, Vol. ED-38, pp. 1568-1575 (1991).

Chapter 2

TRANSPORT PHYSICS

Although the development of power semiconductor devices began in the 1950s by using germanium, most commercial power devices are fabricated by using silicon as the base material. In the 1990s, high voltage power Schottky rectifiers fabricated using gallium arsenide became commercially available for the first time. In addition, it has been theoretically demonstrated [1] that power devices fabricated using silicon carbide have superior characteristics to those made from silicon and gallium arsenide. The design of power devices and the understanding of their operating characteristics require a sound knowledge of the fundamental properties of the semiconductor material used to fabricate the devices. This chapter reviews those properties of semiconductor materials, with particular emphasis on silicon, that are relevant to the operation of the power devices discussed in this book. In addition to providing data on the various properties of silicon, such as intrinsic carrier concentration and mobility, that can be used during numerical simulation of devices, simple analytical expressions are provided wherever possible to facilitate device analysis.

Although it is the fundamental material characteristics of the semiconductor that govern the operating characteristics of power devices, the processing techniques that are used to control these properties and the technological constraints imposed by the available processes are equally important in obtaining the desired device characteristics. For this reason, the chapter includes a discussion of current technologies such as neutron transmutation doping for controlling the resistivity and electron irradiation for controlling minority carrier lifetime. These technologies have been specifically developed for power devices and are almost exclusively used for their fabrication.

2.1 SILICON MOBILITY

The *mobility* is defined as the average velocity of free carriers in a semiconductor due to an impressed electric field. In the presence of an electric field in a semiconductor, the free electrons and holes are accelerated in opposite directions. Silicon is an indirect band gap semiconductor in which electron transport under an applied electric field occurs in six equivalent conduction band minima located along the <100> crystallographic directions while the transport of holes occurs at two degenerate subbands located at the zero in k-space. The inequality between the mobility of electrons and holes in silicon arises from differences between the shapes of the conduction and valency band minima.

As the free carriers are transported along the direction of the electric field, their velocity increases until they undergo scattering. In the bulk, the scattering can occur either by interaction with the lattice or at ionized donor and acceptor atoms. As a result, the mobility is dependent upon the lattice temperature and the ionized impurity concentration. When the free carrier transport occurs near the semiconductor surface, additional scattering is observed that decreases the mobility to below the bulk value. These scattering mechanisms are dominant when the concentration of holes and electrons is never simultaneously large. However, in bipolar power devices, a high concentration of holes and electrons is simultaneously injected into the base region during forward current conduction. The probability for mutual scattering between electrons and holes is high under these conditions, and the effective mobility of the free carriers decreases with increasing injection level.

In analyzing the influence of the above parameters upon the mobility, the electric field strength is assumed to be small. The *mobility* is then defined as the proportionality constant relating the average carrier velocity to the electric field. However, at high electric fields such as those commonly encountered in power devices, the velocity is no longer found to increase in proportion to the electric field and in fact attains a saturation value. These effects have important implications to current flow in power devices. This section discusses the dependence of the mobility for holes and electrons upon the above parameters. In addition to providing data that can be used during numerical simulation of power devices, analytical expressions have been derived whenever possible in order to simplify device analysis.

2.1.1 Temperature Dependence

At low doping concentrations, the scattering of free carriers in the bulk occurs predominantly by interaction with lattice vibrations. Lattice scattering can occur by means of either optical phonons or acoustical phonons. Optical phonon scattering is important at high temperatures while acoustical phonon scattering is dominant at low temperatures. In addition, at around room temperature, intervalley scattering mechanisms become important. For high purity silicon, it has been experimentally determined that at temperatures below 50 K acoustical phonon scattering is dominant and the mobility varies as $T^{-3/2}$, where T is the absolute temperature. At around room temperature, intervalley scattering comes into effect and the mobility can be determined by using [2, 3]:

$$\mu_n = 1360 \left(\frac{T}{300} \right)^{-2.42} \quad (2.1)$$

and

$$\mu_p = 495 \left(\frac{T}{300} \right)^{-2.20} \tag{2.2}$$

where μ_n and μ_p are the electron and hole mobilities, respectively, in cm² per volt second, and T is the absolute temperature in degrees Kelvin. The variation of the mobility of electrons and holes with temperature for lightly doped silicon at low electric fields is shown by the uppermost curves in Fig. 2.1, for the case of low doping levels. Note the rapid reduction in both the electron and hole mobilities with increasing temperature. Since doping levels of below 10^{15} per

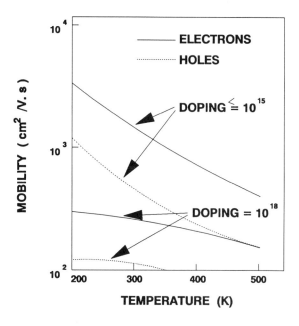

Fig. 2.1 Mobility for electrons and holes in silicon.

cm³ are necessary to achieve high breakdown voltages in silicon devices, the reduction in carrier mobility with increasing temperature is an important characteristic which must be accounted for during the design and analysis of unipolar power devices.

2.1.2 Dopant Concentration Dependence

The presence of ionized donor or acceptor atoms in the silicon lattice results in a reduction in the mobility due to the additional Coulombic scattering of the free carriers. The effect of ionized impurity scattering is the largest at low temperatures because the effect of lattice scattering becomes small. An example of the effect of ionized impurity scattering upon the mobility of electrons and holes is provided in Fig. 2.1 for the case of a doping concentration of 1 x 10^{18} per cm^3. The mobility as determined by ionized impurity scattering exhibits a positive temperature coefficient. Consequently, as the ionized impurity concentration increases, not only does the absolute mobility decrease but the temperature coefficient of the mobility also decreases [4].

At room temperature, ionized impurity scattering effects are small for doping levels below 10^{16} atoms per cm^3 and the mobility is independent of doping level as described by Eqs. (2.1) and (2.2). At higher doping concentrations, the mobility of electrons and holes decreases with increasing dopant concentration until a doping level of 10^{19} atoms per cm^3 is reached. For dopant concentrations above 10^{19} atoms per cm^3, the mobility of electrons becomes independent of donor concentration at a value of about 90 cm^2/Volt second and that of holes becomes independent of acceptor concentration at a value of about 48 cm^2/Volt second. The variation of electron and hole mobility with doping concentration is provided in Fig. 2.2.

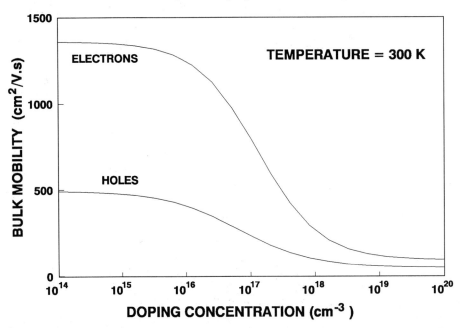

Fig. 2.2 Electron and hole mobilities for silicon as a function of doping concentration.

From the measured data [5], the empirical relationships that relate the mobility to the doping concentrations can be derived. For the case of electrons in N-type silicon:

$$\mu_n(N_D) = \frac{5.10 \times 10^{18} + 92 \, N_D^{0.91}}{3.75 \times 10^{15} + N_D^{0.91}} \qquad (2.3)$$

while for the case of holes in P-type silicon:

$$\mu_p(N_A) = \frac{2.90 \times 10^{15} + 47.7 \; N_A^{0.76}}{5.86 \times 10^{12} + N_A^{0.76}} \qquad (2.4)$$

where N_D and N_A are the donor and acceptor concentrations per cm^3, respectively. These equations are useful for computing the mobility of N- and P-type silicon from the doping concentration and can be directly used for the analysis of device characteristics. They can also be used to derive an expression for the resistivity of N- and P-type silicon as a function of doping concentration.

2.1.3 Electric Field Dependence

In previous sections, the mobility was assumed to be independent of the magnitude of the applied electric field. This is found to be true only when the electric field strength is small. In the range of electric fields where the mobility is constant, the carrier velocity will increase linearly in proportion to the electric field. For silicon, at electric field strengths above 1×10^3 V/cm, it has been found that the velocity of both electrons and holes increases sublinearly with electric fields. The variation of the velocity of electrons and holes as a function of electric field has been measured at various temperatures by using the time of flight technique [6]. Based upon these measured data, analytical expressions relating the drift velocity to the electric field and the ambient temperature have been derived.

For purposes of power device analysis, it is useful to define an average mobility as the ratio of the drift velocity to the applied electric field. At low doping levels, the average mobility (μ^{av}) for electrons can be related to the electric field (E) by:

$$\mu_n^{av}(E) = \frac{9.85 \times 10^6}{(1.04 \times 10^5 + E^{1.3})^{0.77}} \qquad (2.5)$$

while for holes:

$$\mu_p^{av}(E) = \frac{8.91 \times 10^6}{(1.41 \times 10^5 + E^{1.2})^{0.83}} \qquad (2.6)$$

These expressions provide an analytical description of the decrease in the average mobility for electrons and holes with increasing electric field strength. The variation of the average mobility of electrons and holes with electric field is shown in Fig. 2.3. It can be seen that the average mobility is essentially equal to the low field mobility as long as the electric field remains below 10^3 V/cm and drops very rapidly at fields above 10^4 V/cm. In fact, at electric fields above 10^5 V/cm, the average mobility decreases inversely with the electric field strength indicating that the drift velocity of the free carriers remains essentially constant. This phenomenon is called *drift velocity saturation*.

The saturated drift velocity is an important parameter which is required for the analysis of the characteristics of power devices operating in the presence of very high fields. At room

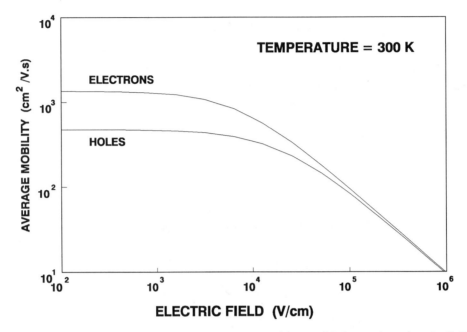

Fig. 2.3 Average mobility for electrons and holes in silicon with increasing electric field.

temperature, the saturated drift velocity for electrons in silicon is 9.9 x 10^6 cm/s and for holes in silicon is 8.4 x 10^6 cm/s. Just as the mobility of the electrons and holes changes with temperature, the saturated drift velocity is also a function of temperature. The drift velocity of both electrons and holes in silicon has been measured over a broad range of electric field strengths and ambient temperatures [6]. From these data, empirical expressions relating the drift velocity of both electrons (v_{dn}) and holes (v_{dp}) to the electric field strength (E) and the temperature (T) can be derived. For the case of electrons in silicon:

$$v_{dn} = \frac{1.42 \times 10^9 \ T^{-2.42} \ E}{[1 + (E/1.01 \ T^{1.55})^{0.0257 \ T^{0.66}}]^{1/(0.0257 \ T^{0.66})}} \quad (2.7)$$

For the case of holes in silicon:

$$v_{dp} = \frac{1.31 \times 10^8 \ T^{-2.2} \ E}{[1 + (E/1.24 \ T^{1.68})^{0.46 \ T^{0.17}}]^{1/(0.46 \ T^{0.17})}} \quad (2.8)$$

The saturated drift velocity for electrons and holes in silicon can be obtained from the above expressions by extrapolation to large values of the electric field. For the case of electrons:

$$v_{sat,n} = 1.434 \times 10^9 \ T^{-0.87} \quad (2.9)$$

Chapter 2 : TRANSPORT PHYSICS

while for the case of holes:

$$V_{sat,p} = 1.624 \times 10^8 \ T^{-0.52} \qquad (2.10)$$

Since power devices are usually rated to operate between -25°C and 150°C, the variation of the saturated drift velocity of electrons and holes in silicon over a slightly broader range of temperatures is provided in Fig. 2.4. It is worth pointing out that the saturated drift velocity

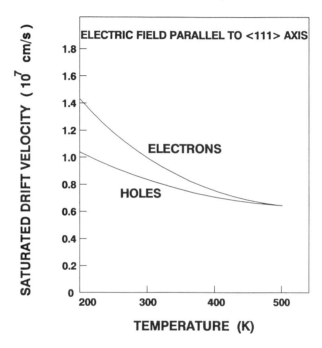

Fig. 2.4 Saturated drift velocity for electrons and holes in silicon.

of electrons is higher than that of holes in this temperature range. However, this difference decreases with increasing temperature until at 500°K the saturated drift velocity of electrons becomes equal to the saturated drift velocity of holes. It should be noted that these results are only valid for lightly doped silicon. Monte Carlo calculations of the effect of ionized impurity atoms upon high field transport in silicon at 300°K indicate that, in addition to the decrease in the mobility, the saturated drift velocity of electrons and holes also decreases when the doping concentration exceeds 10^{18} atoms per cc. Since the active base regions of power devices, where the high electric fields are present, are doped well below this level, the use of the saturated drift velocity data for electrons and holes as provided in Fig. 2.4 is usually adequate for the analysis of their characteristics.

These results apply to the transport of free carriers when the electric field is applied along the <111> crystallographic direction. Although the low field mobility of electrons and holes has been found to be independent of the orientation of the electric field with respect to the crystal, anisotropic carrier transport is observed at high electric fields [5]. This anisotropy is caused by the ellipsoidal shape of the six conduction band valleys and the warped nature of the valency sub-bands. The anisotropic transport of free carriers in silicon has generally not been taken into account during the analysis of current flow in power devices.

2.1.4 Injection level Dependence

In general, current transport in semiconductors occurs via both drift of the free carriers due to the presence of an electric field and diffusion due to the presence of a concentration gradient. In highly doped regions, majority carrier transport is dominant because of the very low minority carrier density. In lightly doped regions, the transport of both electrons and holes (i.e., both majority and minority carriers) must be taken into account in computing the current flow.

Minority carrier transport can be expressed in the form:

$$\frac{\partial p'}{\partial t} = -\frac{p'}{\tau_a} + D_a \frac{\partial^2 p'}{\partial x^2} \qquad (2.11)$$

for an n-type semiconductor and

$$\frac{\partial n'}{\partial t} = -\frac{n'}{\tau_a} + D_a \frac{\partial^2 n'}{\partial x^2} \qquad (2.12)$$

for a p-type semiconductor. In these expressions, n' and p' are the excess electron and hole concentrations, and D_a and τ_a are defined as the *ambipolar diffusion coefficient* and the *ambipolar lifetime*, respectively. The ambipolar diffusion coefficient is a function of the free carrier concentration as given by:

$$D_a = \frac{(n + p) \, D_n D_p}{(n D_n + p D_p)} \qquad (2.13)$$

where n and p are the electron and hole concentrations and D_n and D_p are the electron and hole diffusion constants, respectively. These diffusion coefficients can be obtained from the mobility by using the Einstein relationship valid for non-degenerate semiconductors:

$$D = \frac{kT}{q} \mu \qquad (2.14)$$

At low injection levels and for doping levels of several orders of magnitude above the intrinsic carrier concentration, the ambipolar diffusion coefficient is equal to the minority carrier diffusion constant. This is no longer true at high injection levels that occur in bipolar power devices when carrying high current densities in the on-state.

Chapter 2 : TRANSPORT PHYSICS

At low injection levels, where the density of the minority carriers is far less than that of the majority carriers, the carrier transport is controlled by either phonon or ionized impurity scattering. At high injection levels, which occur, for instance, in the lightly doped base regions of bipolar power devices during forward conduction, the densities of electrons and holes become approximately equal in order to satisfy charge neutrality. When the densities of both carriers become large simultaneously, the probability for the mutual Coulombic interaction of the mobile carriers about a common center of mass becomes significant, resulting in a decrease in the mobility. With the inclusion of carrier-carrier scattering, the diffusion coefficient and the mobility decrease as the injected carrier density increases. By treating carrier-carrier scattering in a manner similar to ionized impurity scattering, the mobility can be shown to vary inversely with the injected carrier concentration when the injection level exceeds 10^{17} per cm^3. The variation of the mobility with injected carrier density (n'=p') is shown in Fig. 2.5 and takes the form:

$$\frac{1}{\mu} = \frac{1}{\mu_0} + \frac{n' \ln(1 + 4.54 \times 10^{11} \, n^{-0.667})}{1.428 \times 10^{20}} \qquad (2.15)$$

where n' is the excess carrier density and μ_0 (either μ_n or μ_p) is the majority carrier mobility.

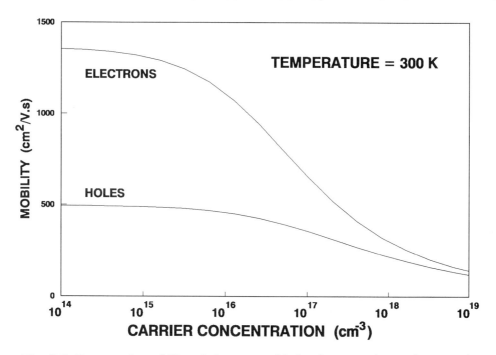

Fig. 2.5 Decrease in mobility of electrons and holes due to carrier-carrier scattering.

The decrease in the mobility at high injection levels has important implications in determining

2.1.5 Surface Scattering Effects

The transport of free carriers near the surface is of importance to power MOS field effect transistors. The forward current conduction mode of the n-channel power MOSFET is induced by the inversion of a P-type layer in the device in such a manner as to create an electron channel region connecting the source to the drain. This inversion region is created by applying an electric field normal to the semiconductor surface by using a metal gate electrode separated from the semiconductor by a silicon dioxide layer. Current transport in these devices also occurs via an accumulation layer, where a higher concentration of electrons is induced by the gate bias applied to the surface of an N-type region. The conductivity of the inversion and accumulation layers is dependent upon the total number of free carriers in the layer and their transport velocity along the surface under an applied transverse electric field. The computation of the number of free carriers in the inversion and accumulation layers available for current conduction will be discussed in the chapter on the MOS gated field effect transistor. The influence of carrier transport near the surface upon their velocity-electric field relationship is discussed in this section.

At the high electric fields applied across the oxide to create the inversion layer, the surface region within which current transport occurs is very thin. As a result, at high electric fields normal to the surface and at low temperatures, the motion of carriers within the inversion layer is quantized in a direction perpendicular to the surface. At room temperature, the kinetic energy of the electrons spreads their wave function among the quantum levels and electron transport can be treated using classical statistics.

A pictorial representation of a semiconductor surface containing an inversion layer is shown in Fig. 2.6. The electrons in the inversion layer have a Gaussian distribution located very close to the surface within a region which is typically in the range of 100 °A in thickness. In addition to Coulombic scattering due to ionized acceptors, the electrons in the inversion layer undergo several additional scattering processes: (a) surface phonon scattering due to lattice vibration, (b) additional Coulomb scattering due to surface charge in interface states and the fixed oxide charge, and (c) surface roughness scattering due to deviation of the surface from a specular interface. As in the case of the bulk, phonon scattering is important at higher temperatures when the thermal energy in the lattice is large. The influence of Coulomb scattering due to interface state and fixed charge is important for lightly inverted surfaces and when processing conditions produce high interface state or fixed charge densities. The surface roughness scattering is predominant under strong inversion conditions because the electric field normal to the surface must be increased for achieving higher inversion layer concentrations. This has the effect of increasing the velocity of the carriers towards the surface and bringing the inversion layer charge distribution closer to the surface. Both these factors enhance scattering at a rough interface such as that illustrated in Fig. 2.6. As expected, the surface roughness scattering is sensitive to processes that influence surface smoothness. The quantitative dependence of the carrier mobility in inversion and accumulation layers are of vital importance

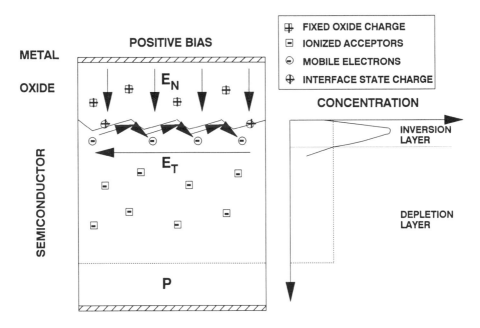

Fig. 2.6 Metal-Oxide-Semiconductor interface with current transport within inversion layer.

to the modeling and design of power MOS devices. The dependence of these mobilities upon various parameters is discussed below.

Before discussing these results it is useful to define an effective mobility for carriers in the inversion layer:

$$\mu_e = \frac{\int_0^{x_i} \mu(x) \, n(x) \, dx}{\int_0^{x_i} n(x) \, dx} \tag{2.16}$$

where $\mu(x)$ and $n(x)$ are the local mobility and free carrier concentrations in the inversion layer and x_i is the inversion layer thickness. It must be emphasized that, defined in this manner, the effective mobility is a measure of the conductance of the inversion layer and is, therefore, eminently suitable for the calculation of the characteristics of MOS field effect devices. In fact, determination of the effective mobility is performed by using MOSFET devices specially designed for these measurements. In doing these measurements, special care must be taken with regard to surface orientation, the direction of the current flow vector on the surface, as well as surface charge density, since all of these parameters have been found to influence the effective mobility.

The surface mobility is primarily a function of the electric field normal (E_N) to the surface rather than the inversion layer concentration which is dependent upon the background

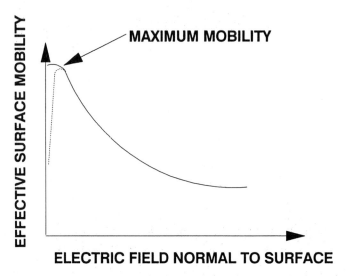

Fig. 2.7 Decrease in the effective inversion layer mobility with electric field normal to the surface.

doping level and the applied gate bias. A typical variation of the effective mobility with increasing electric field normal to the surface (E_N) is shown in Fig. 2.7. In general, it has been found that, under weak inversion conditions (i.e., at low electric fields normal to the surface), the effective mobility increases very rapidly with increasing inversion layer carrier concentration. As the inversion layer concentration increases, the effective mobility has been found to peak and then gradually decrease [7]. The increase in the effective mobility at low inversion layer concentrations has been related to Coulombic scattering at charged surface states. In fact, if the surface state density is adequately lowered by suitable processing, the effective mobility remains independent of the inversion layer concentration in the weak inversion region [8]. When the electric field normal to the surface exceeds 10^4 volts per cm, the effective mobility begins to decrease with increasing inversion layer concentration. The enhancement of surface scattering by the higher electric fields applied normal to the surface is responsible for this phenomenon. This has an important bearing upon the transconductance of power MOS devices.

2.1.5.1 Substrate Doping Dependence. The inversion layer mobility for electrons in silicon has been measured as a function of substrate doping [9]. The observed change in the maximum effective mobility with increasing substrate doping is shown in Fig. 2.8 for two cases of fixed charge density. It can be seen that the effective mobility decreases logarithmically with increasing substrate doping. This empirical observation will be incorporated in an analytical expression relating the maximum surface mobility to the substrate doping (N_A) and the fixed charge (Q_f).

It should be noted that the decrease in maximum effective mobility with increasing substrate doping is not due to enhanced ionized impurity scattering. Since the inversion layer thickness is on the order of 100 °A, the number of ionized impurities is less than 10^{10} per cm^2

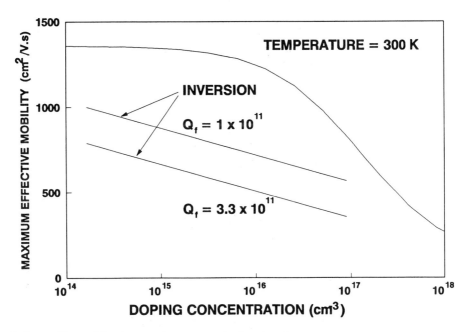

Fig. 2.8 Effect of background acceptor doping level upon electron mobility in inversion layers.

for doping levels up to 10^{16} per cm^3. This is an order of magnitude smaller than the typical fixed charge density which makes the influence of ionized impurity scattering negligible. The observed reduction in the maximum effective mobility with increasing substrate doping arises primarily from the higher electric fields which must be applied normal to the surface to produce surface inversion.

The data in Fig. 2.8 also indicate that the maximum inversion layer mobility varies from between 75 to 85 percent of the bulk value for substrate doping levels ranging from 10^{14} to 10^{17} per cm^3 when low ($<10^{11}$ per cm^2) fixed charge density is achieved. If the fixed charge density is high, the maximum effective mobility will be substantially reduced.

2.1.5.2 Oxide Charge Dependence. The inversion layer mobility for electrons in silicon has been experimentally found to decreases inversely with increasing surface charge [9]. The hyperbolic variation of the maximum effective mobility with fixed charge (Q_f) is also theoretically predicted from calculation of Coulombic scattering by charged centers [10]. Based upon these findings, an empirical relationship between the maximum effective mobility and the fixed charge has been derived :

$$\mu_{em} = \frac{\mu_0}{1 + \alpha Q_f} \qquad (2.17)$$

where the parameters μ_0 and α have been found to be a function of the background doping level as given by:

$$\mu_0 = 3490 - 164 \log(N_A) \quad (2.18)$$

and

$$\alpha = -0.104 + 0.0193 \log(N_A) \quad (2.19)$$

where N_A is the background acceptor concentration per cm³. The dependence of the maximum effective mobility upon the background doping level described in Section 2.1.5.1 is incorporated in the last two equations. It should be noted that if the interface state charge is significant, it should be added to the fixed charge (Q_f) in equation (2.17) when calculating the maximum effective mobility. However, it has been found that charges in the bulk of the oxide at distances of over 50 °A from the interface have negligible effect upon the mobility.

2.1.5.3 Electric Field Dependence As illustrated in Fig. 2.6, the inversion layer charge is distributed near the surface with a Gaussian profile. The electric field experienced by the free carriers in the inversion layer varies from the peak electric field at the oxide-semiconductor interface to that at the depletion layer interface. For the same peak surface electric field, the average electric field in the inversion layer will vary with background doping level. To describe the effect of electric field strength normal to the surface upon the effective mobility it is appropriate to use the average electric field rather than the peak surface electric field. This effective field is given by the relationship:

$$E_{eff} = \frac{1}{e_s}\left(\frac{1}{2}Q_{inv} + Q_B\right) \quad (2.20)$$

where Q_{inv} is the inversion layer charge density per cm² and Q_B is the depletion layer charge per cm². For the same peak surface electric field, the effective electric field will become smaller as the background doping level decreases.

As pointed out with the aid of Fig. 2.7, the effective mobility in an inversion layer decreases with increasing electric field. For electrons in silicon inversion layers, the effective mobility can be described as a function of the effective electric field by the following expression:

$$\mu_e = \mu_{e,\max}\left(\frac{E_c}{E_{eff}}\right)^\beta \quad (2.21)$$

where $\mu_{e,\max}$ is the maximum effective mobility for a given background doping level and fixed charge density as described by Eq. (2.17). The parameters E_c and β are empirical constants which are dependent upon the fixed charge density and the background doping level. These constants have been measured for the commonly used oxidation techniques during silicon device fabrication employing steam and dry oxygen [9]. From these measurements:

$$\beta_{wet} = 0.341 - 5.44 \times 10^{-3} \log(N_A) \qquad (2.22)$$

and

$$\beta_{dry} = 0.313 - 6.05 \times 10^{-3} \log(N_A) \qquad (2.23)$$

The parameter β is a measure of the rate of degradation of the effective mobility with increasing electric field normal to the surface. It has a strong influence upon the transconductance of power MOS devices. Under typical device processing conditions $\beta = 0.25$, i.e., the effective mobility decreases inversely as the fourth power of the effective electric field strength applied normal to the surface. The difference in β between wet and dry oxidation is believed to arise from the dissimilar surface roughness resulting from these processes. The rate of oxide growth during dry oxidation is lower than for wet oxidation resulting in a smoother interface. Since the surface roughness scattering is dominant at high surface fields, dry oxidation produces superior effective mobilities under strong inversion conditions.

The constant E_c is a function of the fixed charge density and background doping level. A general expression relating these parameters has been derived [9]:

$$E_c = 2 \times 10^{-4} \, N_A^{0.25} \, A \, e^{BQ_f} \qquad (2.24)$$

where N_A is the background doping level and Q_f is the fixed charge density. The parameters A and B are process dependent. For wet oxidation $A = 2.79 \times 10^4$ V/cm and $B = 8.96 \times 10^{-2}$, while for dry oxidation $A = 2.61 \times 10^4$ V/cm and $B = 0.13$. In Eq. (2.24), Q_f is in units of 10^{11} per cm^2. Thus, the influence of substrate doping is small whereas the fixed charge density has a strong effect on the effective mobility over a broad range of electric field strengths.

The mobility of holes in inversion layers has also been measured in several studies [7, 11] but in less detail compared with electrons. These early studies provide the data for the variation of the effective mobility for holes as a function of the inversion layer charge shown in Fig. 2.9. As in the case of electrons, the effective mobility for holes goes through a peak. The variation of the effective mobility for holes in inversion layers as a function of the electric field strength normal to the surface can also be described by equation (2.21) for the region where the inversion layer concentration exceeds 5×10^{11} per cm^2. Using the data shown in Fig. 2.9, the parameters E_c and β in Eq. (2.21) can be determined to be: $E_c = 3 \times 10^4$ V/cm and $\beta = 0.28$. The dependence of these parameters upon the substrate doping (N_D) and fixed charge (Q_f) has not been studied in detail.

2.1.5.4 Orientation Dependence The effective mobility in inversion layers has been found to be anisotropic. Its value not only depends upon the orientation of the plane of the surface on which the inversion layer is formed but also upon the direction of the current flow vector along the surface being considered. Detailed measurements [12, 13] of the effective mobility on many surface orientations for both N- and P-type wafers have shown that

$$\mu_e(100) > \mu_e(111) > \mu_e(110) \qquad (2.25)$$

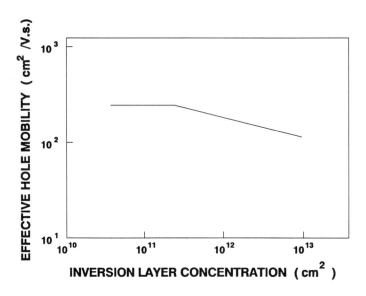

Fig. 2.9 Effective mobility for holes in silicon inversion layers.

for conduction of electrons in inversion layers and that

$$\mu_p(110) > \mu_p(111) > \mu_p(100) \qquad (2.26)$$

for conduction of holes in inversion layers. This dependence of the effective mobility upon orientation has been correlated with the anisotropy of the conductivity effective mass in silicon.

Intraplanar anisotropy of the effective electron mobility in inversion layers has also been observed [9, 12]. For the same substrate doping and fixed charge density, the effective electron mobility along the [100] direction is considerably higher than along the [110] direction, as shown in Fig. 2.10. It is worth noting that the ratio of the mobilities for the two surface orientations remains constant for the entire range of electric fields indicating that the surface roughness is isotropic and that the difference in absolute magnitude between the mobilities arises from the anisotropy of the conductivity effective mass in silicon.

The results described above apply to polished surfaces. Lower inversion layer mobilities have been observed on etched surfaces. One of the processes developed for the fabrication of power MOSFET devices relies upon the use of anisotropic etches to form V-shaped grooves on the top surface of (100) oriented wafers. Due to the higher atomic density of (111) planes in silicon, the etch rate of the anisotropic etches is slowest for the (111) plane. Consequently, these etches expose (111) surfaces creating the V-shaped grooves. The results of measurements of the effective mobility of electrons on the etched (111) surface are compared with those obtained on the unetched (i.e., polished) surface in Fig. 2.11. In this figure, two methods of anisotropic etching, based upon a solution containing either potassium hydroxide (KOH) or ethylene diamine (ED), have been examined. In addition, the mobility on the sidewalls of

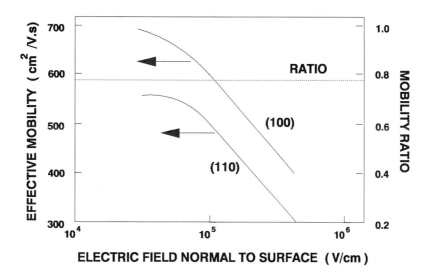

Fig. 2.10 Crystallographic orientation dependence of effective mobility for electrons in inversion layers.

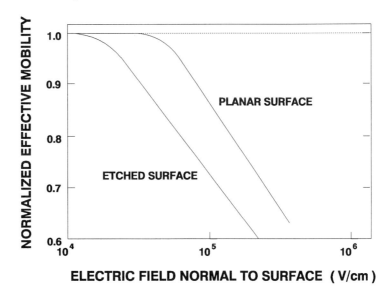

Fig. 2.11 Effect of etching on the electron mobility in inversion layers.

trenches formed by reactive ion etching has been studied. For all three cases, the maximum effective mobility was measured to be about 600 cm² per V-s. The lower effective mobilities for the etched surfaces are believed to arise from greater surface roughness scattering because the anisotropically etched silicon surface is rougher than a polished surface. These measurements indicate that planar DMOS devices fabricated on (100) wafers would have a 30 percent higher inversion layer mobility for electrons at high gate bias voltages when compared with VMOS devices which require anisotropic etching.

2.1.5.5 Accumulation versus Inversion When the direction of the electric field normal to the surface is such as to attract the majority carriers to the surface, an accumulation layer forms. The carriers in accumulation layers are distributed further from the surface than in the case of inversion layers. The effective mobility in an accumulation layer can be expected to be higher than in an inversion layer because surface roughness scattering is not as severe. The measured effective mobility for electrons in silicon accumulation layers [9] are about 80 percent of the bulk as shown in Fig. 2.12. The variation of the effective electron mobility in an accumulation layer with increasing electric field normal to the surface is similar to than in an inversion layer. The effective mobility for electrons in accumulation layers has also been found to exhibit the

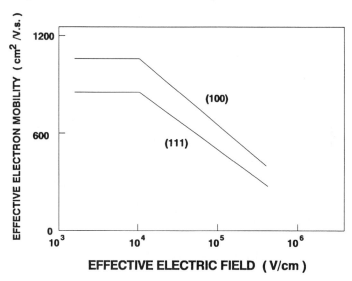

Fig. 2.12 Effective mobility for electrons in accumulation layers.

anisotropy observed in inversion layers [9]. At high electric fields normal to the surface, the mobility ratio for the [001] and [110] surface directions approaches that observed in inversion layers. This anisotropy of the effective mobility in accumulation layers is ascribed to the anisotropy in the effective mass for electrons in silicon.

Accumulation layers exist between device cells during current conduction in power MOSFET's. Current flow within the device is distributed via conduction through the accumulation layer. The conductivity of the accumulation layer plays an important part in determining the spreading resistance of these devices. The effective mobility data provided here is required for performing an accurate analysis.

2.1.5.6 Velocity Saturation. The above results apply at low electric field strengths applied parallel to the surface where the proportionality between the velocity of the free carriers and the electric field is maintained. At high field strengths, carrier velocity saturation is observed similar to that discussed for bulk silicon. The saturated drift velocity in inversion layers is a function of the surface orientation and is lower than that observed in bulk silicon. In the case of holes, the saturated drift velocity is the highest for the (100) orientation [14]. In the case of electron transport, the saturated drift velocity has been measured as a function of temperature for the (111), (110) and (100) surfaces [15]. An approximately linear decrease in the saturated drift velocity with increasing temperature is observed:

$$v_{sat,n}^i = v_0 - \delta T \qquad (2.27)$$

where v_0 is the saturated drift velocity at $0°$ K. Using the measured data, $v_0(100) = 8.9 \times 10^6$ cm/sec, $v_0(111) = 7.4 \times 10^6$ cm/sec, and $v_0(110) = 6.2 \times 10^6$ cm/sec. The constant $\delta = 7 \times 10^3$ cm/sec°K forms a good fit to the measured data for all three of these orientations.

2.2 SILICON RESISTIVITY

The resistivity of the base material used for the fabrication of power devices is an important parameter because it controls the maximum achievable breakdown voltage. In the case of unipolar devices, it also impacts the maximum current handling capability. The resistivity in silicon is controlled by the transport of holes and electrons under the applied field. Silicon is an indirect band gap semiconductor in which electron transport under applied electric fields occurs in six equivalent conduction band minima located along the <100> crystallographic directions while the transport of holes occurs at two degenerate sub-bands located at the zero in k-space. The resistivity is controlled by the concentration of free electrons in the conduction band and the concentration of holes in the valency band as well as the mobility of these carriers which relates their average velocity to the applied electric field. The mobility of free carriers was treated in the previous section.

In high purity material, the free carrier density is determined by the generation of hole-electron pairs created by the thermal excitation of electrons from the valency band into the conduction band. These carriers determine the intrinsic resistivity. The resistivity can be controlled by the addition of dopants into the silicon lattice that, when ionized, contribute either electrons or holes to the conduction process. The dopants that are commonly used for power device fabrication are boron, gallium, and aluminum for P-type regions and phosphorus, arsenic, and antimony for N-type regions. Since most of the resistance controlling the current flow in power devices during current conduction arises in that portion of the device that supports the high voltages during the blocking state of operation, power devices are generally fabricated from N-type starting material because the mobility of electrons is higher than that of holes.

2.2.1 Intrinsic Resistivity

In the absence of dopants, the resistivity is controlled by the creation of electrons and holes in the conduction and valency bands due to the thermal generation process which allows the transfer of electrons from the valency band into the conduction band. This process produces both a free electron and a free hole which can take part in current conduction. The density of these intrinsically created carriers is dependent upon the density of states in the conduction (N_c) and valency (N_v) bands and upon the energy gap (E_g):

$$n_i = \sqrt{n\,p} = \sqrt{N_c N_v}\ e^{-(E_G/kT)} \qquad (2.28)$$

where k is Boltzmann's constant and T is the absolute temperature. In the case of silicon

$$n_i = 3.87 \times 10^{16}\ T^{3/2}\ e^{-(7.02 \times 10^3)/T} \qquad (2.29)$$

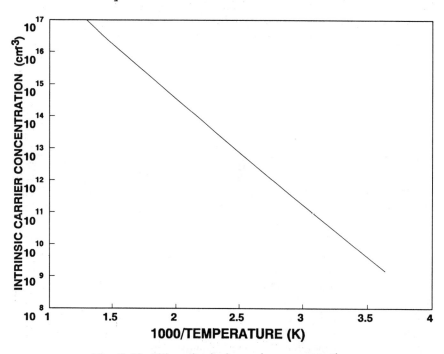

Fig. 2.13 Silicon intrinsic carrier concentration.

as long as the temperature is below 700°K [4, 16]. The increase of the intrinsic concentration with temperature is shown in Fig. 2.13. The intrinsic carrier concentration is an important parameter which determines the leakage current in power devices at elevated temperatures. It has also been found that the formation of current filaments (mesoplasmas) can be related to thermal runaway when the intrinsic concentration becomes comparable to the background concentration.

The intrinsic resistivity in the absence of dopants is related to the intrinsic carrier concentration and the mobility for holes and electrons:

$$\rho_i = \frac{1}{q \, n_i (\mu_n + \mu_p)} \qquad (2.30)$$

Using equation (2.29) for the intrinsic concentration and the mobility values at low doping levels:

$$\rho_i = 1.75 \times 10^{-7} \, T^{0.8} \, e^{(7.02 \times 10^3)/T} \qquad (2.31)$$

At room temperature, the intrinsic resistivity for silicon is 2.5×10^5 ohm-cm. This resistivity decreases rapidly with increasing temperature.

The intrinsic concentration shown in Fig. 2.13 is purely a function of temperature as long as the doping level is below 1×10^{17} per cm^3. At higher doping concentrations, the interaction between the dopant atoms becomes significant and the energy gap is no longer independent of the doping level. This phenomenon is called *band gap narrowing*. In the presence of band gap narrowing, an effective intrinsic concentration can be defined that is a function of doping level as discussed in the next section.

2.2.2 Band Gap Narrowing

At low doping levels, the energy band diagram takes the form shown in Fig. 2.14(a), where the density of states varies as the square root of energy. The donor and acceptor levels have discrete positions in the band gap that are separated from the conduction and valency bands. The well defined separation between the conduction and valency band edge is called the *energy band gap (E_g)*.

At high doping levels, three effects cause an alteration of the band structure. First, as the impurity density becomes high, the spacing between individual impurity atoms becomes small. The interaction between adjacent impurity atoms leads to a splitting of the impurity levels into an impurity band as shown in Fig. 2.14(b). Second, the conduction and valency band edges no longer exhibit a parabolic shape. The statistical distribution of the dopant atoms introduces point-by-point differences in local doping and lattice potential, leading to disorder. As illustrated in Fig. 2.14, this results in the formation of band tails due to the presence of disorder. In addition, the interaction between the free carriers and more than one impurity atom leads to a modification of the density of states at the band edges. This phenomenon is called *rigid band gap narrowing*.

The formation of the impurity band, the band tails, and the rigid band gap narrowing are majority carrier effects related to the reduced dopant impurity atom spacing. In addition to this, the electrostatic interaction of the minority carriers and the high concentration of majority carriers leads to a reduction in the thermal energy required to create an electron-hole pair. This is also a rigid band gap reduction which does not distort the energy dependence of the density of states. The rigid band gap reduction arising from the screening of the minority carriers by

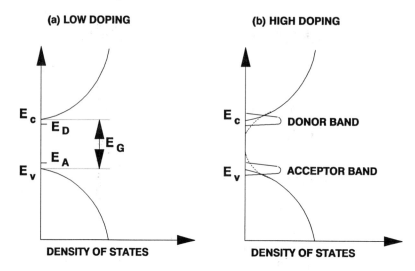

Fig. 2.14 Energy band diagram at (a) low doping concentrations and (b) high doping concentrations.

the high concentration of majority carriers can be derived by using Poisson's equation [17]:

$$\frac{1}{r^2}\frac{d}{dr}\left(r^2\frac{dV}{dr}\right) = -\frac{q}{e_s}\Delta n \qquad (2.32)$$

where Δn is the excess electron concentration given by:

$$\Delta n = n_o(e^{qV/kT} - 1) \qquad (2.33)$$

For low excess electron concentrations

$$\Delta n = \frac{q\, n_o\, V}{k\, T} \qquad (2.34)$$

Solution of Eq. (2.32) using (2.34) gives

$$V(r) = \frac{q}{4\pi e_s}\left(\frac{e^{-r/r_s}}{r}\right) \qquad (2.35)$$

where the screening radius r_s is given by:

$$r_s = \sqrt{\frac{e_s\, k\, T}{q^2\, n_o}} \qquad (2.36)$$

Chapter 2 : TRANSPORT PHYSICS

The field distribution for the screened Coulombic potential is given by:

$$E(r) = \frac{q}{4 \pi e_s r} \left(\frac{1}{r} - \frac{1}{r_s} \right) e^{-r/r_s} \qquad (2.37)$$

In comparison, at low doping concentrations, the unscreened field distribution is given by

$$E_o(r) = \frac{q}{4 \pi e_s r^2} \qquad (2.38)$$

The band gap reduction is the difference in electrostatic energy between the screened and unscreened cases:

$$\Delta E_g = \frac{e_s}{2} \int \left[E_o^2(r) - E^2(r) \right] dr \qquad (2.39)$$

Solving this equation:

$$\Delta E_g = \left(\frac{3 q^2}{16 \pi e_s} \right) \sqrt{\frac{q^2 n_o}{e_s k T}} \qquad (2.40)$$

For low injection levels and assuming that all the dopant atoms are ionized:

$$n_o = N_D^+ = N_D \qquad (2.41)$$

and

$$\Delta E_g = \left(\frac{3 q^2}{16 \pi e_s} \right) \sqrt{\frac{q^2 N_D}{e_s k T}} \qquad (2.42)$$

This expression, which is valid for small changes in band gap, indicates that the band gap narrowing will be proportional to the square root of the doping level. Using the appropriate parameters for silicon:

$$\Delta E_g = 22.5 \times 10^{-3} \sqrt{\frac{N_I}{10^{18}}} \qquad (2.43)$$

where N_I is the impurity (donor or acceptor) concentration.

The variation of the energy band gap of silicon due to high doping according to the above theory is shown in Fig. 2.15 at room temperature. The theoretical curve derived by using screening of the minority carriers by the majority carriers provides a good fit with the measured data [18-21] up to a doping level of 10^{19} per cm^3 indicating that this is the dominant effect. In terms of modeling the electrical characteristics of devices, the influence of band gap narrowing

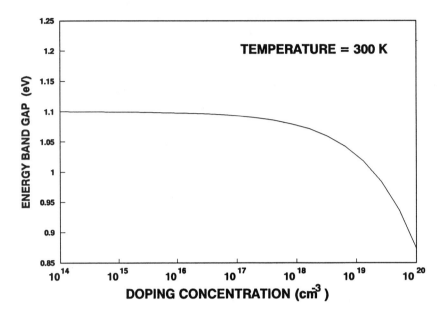

Fig. 2.15 Band gap narrowing in silicon at high doping levels.

upon the product of the equilibrium electron and hole concentrations is required. The p-n product can be represented by an effective intrinsic concentration as in the case of lightly doped semiconductors:

$$n_{ie}^2 = n\,p \tag{2.44}$$

The increase in the p-n product arising from band gap narrowing is then given by:

$$n_{ie}^2 = n_i^2\, e^{q\,\Delta E_g/kT} \tag{2.45}$$

where ΔE_g is the band gap narrowing due to the combined effects of impurity band formation, band tailing and screening. From Eq. (2.43), an expression for the effective intrinsic carrier concentration at room temperature can be derived:

$$n_{ie} = 1.4 \times 10^{10}\, \exp\left(0.433\,\sqrt{\frac{N_I}{10^{18}}}\right) \tag{2.46}$$

where N_I is the doping concentration per cm^3. The increase in the intrinsic carrier concentration with doping is shown in Fig. 2.16. It is worth pointing out that there is an order of magnitude increase in the effective intrinsic carrier concentration when the doping level exceeds 2×10^{19}

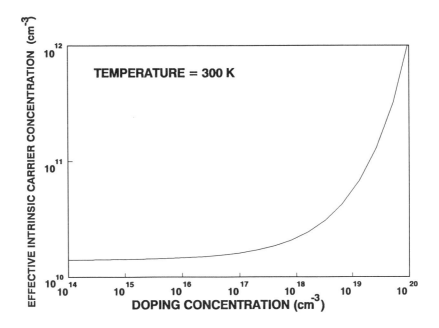

Fig. 2.16 Increase in the intrinsic carrier concentration in silicon at high doping levels.

per cm^3. These high doping levels are commonly encountered in the diffused regions of power devices. Since current conduction in bipolar power devices occurs in the presence of minority carrier injection into the diffused regions, their performance is affected by the band gap narrowing phenomena.

2.2.3 Extrinsic Resistivity

The resistivity of a semiconductor region can be altered by the addition of controlled amounts of impurities. Elements from the fifth column of the periodic table produce N-type material while elements from the third column produce P-type material. For device fabrication, the most commonly used N-type dopant is phosphorus. In addition, arsenic and antimony are used whenever their slower diffusion coefficient in silicon can be taken into advantage during device processing.

To create P-type regions, boron is the most commonly used impurity. For very high voltage, large area devices, gallium and aluminum are also utilized. These elements have a much faster diffusion rate in silicon compared with boron. The diffusion coefficients of aluminum and gallium can be compared with that for boron with the aid of Fig. 2.17. It has been experimentally found that the diffusion coefficient of aluminum is a factor of five times larger than that of boron at typical diffusion temperatures. Thus, a deep diffusion depth of 100 microns can be achieved with aluminum diffusions at 1200°C in 98 hours (4 days), while the

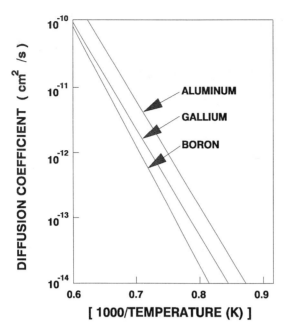

Fig. 2.17 Diffusion coefficients for P-type impurities in silicon.

same depth would require 556 hours (23 days) of diffusion time in the case of boron. It is obvious that a considerable savings in process time can be achieved by performing the diffusions with aluminum instead of boron. Other advantages of using aluminum instead of boron are that its ionic radius is comparable to that of silicon, thus, minimizing misfit strain in the lattice. Further, it does not form intermetallic compounds with silicon and superior junction breakdown characteristics are observed with aluminum diffusions because of the ability to obtain low surface concentrations by using sealed tube diffusion techniques.

The relationship between the phosphorus concentration and the resulting resistivity of silicon has been recently measured over a very wide range of doping densities [22]. The results of these measurements are provided in Fig. 2.18 for reference. Similar measurements of the resistivity of boron doped, P-type, silicon have been made [23]. This data is also provided in Fig. 2.18. The difference in the resistivity between N- and P-type silicon arises from a difference in the mobility for electrons and holes. Using the relationship between mobility and doping concentration provided by Eq. (2.3) and (2.4), the resistivity (ρ_n) can be related to the doping concentration (N_D):

$$\rho_n = \frac{3.75 \times 10^{15} + N_D^{0.91}}{1.47 \times 10^{-17} N_D^{1.91} + 8.15 \times 10^{-1} N_D} \quad (2.47)$$

Similarly for the case of P-type silicon, the resistivity (ρ_p) can be related to the doping concentration (N_A) by:

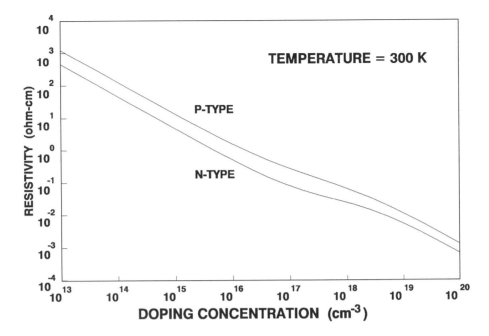

Fig. 2.18 Resistivity of N- and P-type silicon.

$$\rho_p = \frac{5.86 \times 10^{12} + N_A^{0.76}}{7.63 \times 10^{-18} N_D^{1.76} + 4.64 \times 10^{-4} N_A} \quad (2.48)$$

It is worth pointing out that, although these equations indicate a rather complex relationship between the resistivity and the doping concentration, the resistivity of starting material used for the fabrication of power devices with breakdown voltages above 100 volts exceeds 1 ohm-cm in N-type silicon and 3 ohm-cm in P-type silicon. At these higher resistivities, the mobility becomes independent of the doping concentration and the resistivity can be obtained from the following expressions:

$$\rho_n = 4.596 \times 10^{15} N_D^{-1} \quad (2.49)$$

and

$$\rho_p = 1.263 \times 10^{16} N_A^{-1} \quad (2.50)$$

The starting material for power device fabrication is preferably bulk material because of its low defect density. In most very high voltage power devices, N-type starting material is used. Since a low concentration of oxygen is required to avoid the formation of precipitates that adversely affect the breakdown characteristics, float zone (FZ) silicon is preferred over Czochralski (CZ) silicon. Power devices are generally required to carry large currents. The area of a single device is much larger than that of equivalent devices in integrated circuits. In fact, for very high power devices, a single device can be fabricated from a three or four inch diameter wafer of silicon in the 1990s. As a result, variations in the resistivity across the diameter of a wafer as well as its thickness can have a strong influence upon the device characteristics. For example, in devices containing back-to-back P-N junctions, the breakdown voltage is limited to that of an open-base transistor. As a result, if a large variation in the resistivity occurs in the wafer about its nominal value, the device breakdown voltage may be limited either by reach-through effects at the higher resistivity locations or by premature avalanche breakdown at the lower resistivity locations, as discussed in Chapter 3. Since an allowance must be made for both extremes in the resistivity distribution, a non-optimum device design is unavoidable.

The inhomogeneous resistivity in silicon wafers grown by the float zone process occurs in the form of striations that have been attributed to local fluctuations in the growth rate caused by temperature variations. These changes in the growth rate cause a variation in the local dopant incorporation in the growing crystal because the effective segregation coefficient is determined by the growth rate. The fluctuations in the resistivity can be decreased by improving the homogeneity of the temperature distribution. Advanced crystal growth methods have been successful in reducing the resistivity variations from the +15% to -15% observed in the conventional techniques to about +7% to -7%. Although this has resulted in a significant improvement in device characteristics, even superior devices can be obtained by using extremely homogeneously doped silicon produced by the neutron transmutation doping process.

2.2.4 Neutron Transmutation Doping

This technique for doping silicon is based upon the conversion of the Si^{30} isotope into phosphorus atoms by the absorption of thermal neutrons [24]. In the last few years, the control over the transmutation process has been improved to the point where it is extensively used for the production of high resistivity silicon for large area power devices. In addition, the annealing kinetics of the silicon after the neutron transmutation process are now well understood, allowing its widespread usage in the power device industry.

The neutron transmutation doping process occurs by the absorption of thermal neutrons. The thermal neutron capture by all three of the naturally occurring stable silicon isotopes takes place in proportion to their natural abundance and the capture crosssections of the respective atoms. The abundance of the three stable isotopes has been found to be 92.27 percent for the Si^{28} isotope, 4.68 percent for the Si^{29} isotope, and 3.05 percent for the Si^{30} isotope. The absorption of a neutron by these isotopes leads to the following reactions:

$$Si^{28}(n,\gamma)\ Si^{29} \tag{2.51}$$

$$Si^{29}(n,\gamma)\ Si^{30} \tag{2.52}$$

and

$$Si^{30}(n,\gamma)\ Si^{31} \rightarrow P^{31} + \beta \qquad (2.53)$$

The thermal neutron absorption cross-sections for these cases are 0.08, 0.28 and 0.13 barns, respectively (one barn equals 10^{-24} cm^2). The first two reactions merely alter the isotope ratios. The third reaction is the one that can be utilized to create phosphorus atoms. Since the naturally occurring Si^{30} isotope can be expected to be homogeneously distributed in the silicon wafer, a homogeneous distribution of phosphorus can also be created if a uniform neutron flux can be ensured within all points of the wafer being irradiated. In addition to the neutron flux distribution inside the reactor during irradiation, two other factors can influence the uniformity of the resistivity. The first is the inhomogeneous neutron flux distribution created by the absorption of the neutrons in the silicon itself and the second is the presence of a high inhomogeneous distribution of dopants in the starting material.

The influence of the absorption of the thermal neutrons has been treated for the case of a silicon ingot surrounded by a medium with the same absorption properties as silicon [25]. In this case, if a uniform neutron flux is assumed to emanate from the reactor core, the neutron flux in the silicon takes the form

$$\phi = \phi_o\, e^{-x/b} \qquad (2.54)$$

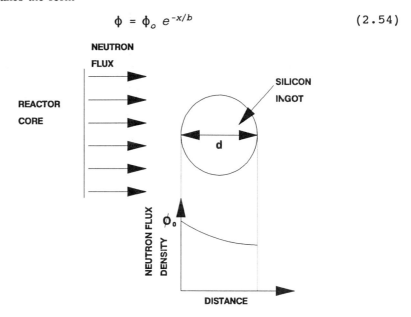

Fig. 2.19 Neutron transmutation doping of silicon ingot.

where b is the decay length for neutron absorption in silicon. For an ingot of diameter d, a phosphorus concentration proportional to ϕ_0 is created on one side of the ingot while a phosphorus concentration proportional to $\phi_0 \exp(-d/b)$ is created on the other side as illustrated in Fig. 2.19. The resulting inhomogeneous distribution of the neutron flux can be shown to be equivalent to a resistivity variation of +/- tanh (d/2b).

The decay length (b) of neutrons in silicon has been calculated to be 19 cm. Using this value, the effect of increasing the ingot diameter upon the resistivity variation is shown in Fig. 2.20. It can be seen that neutron absorption in the silicon can lead to a significant

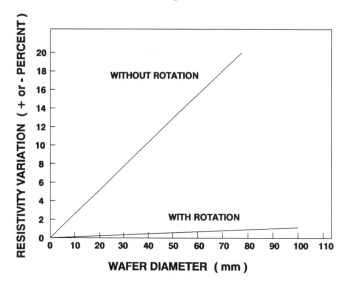

Fig. 2.20 Resistivity variation produced during neutron transmutation doping with and without rotation of ingot.

resistivity variation when the ingot diameter exceeds 40 mm. To overcome this problem, the silicon ingots are rotated in the neutron flux during irradiation. The maximum resistivity now occurs at the axis of the ingot. The ratio of the neutron dose at the periphery to that at the axis of a cylindrical ingot is given by [25]:

$$\frac{\phi(d)}{\phi(0)} = 1 + \frac{1}{16}\left(\frac{d}{b}\right)^2 \qquad (2.55)$$

The resistivity variation within the crystal will then be $+/-[32(b/d)^2 - 1]^{-1}$. This variation is also plotted in Fig. 2.20 to contrast it with the variation in the resistivity without rotation. It can be seen that a significant improvement in the doping homogeneity results from ingot rotation and that a resistivity variation of less than two percent can be achieved for ingot diameters of up to 100 mm. In actual practice, the ingot is surrounded by heavy water or graphite which tends to increase the decay length and results in even superior homogeneity than predicted by the above expressions.

The second important consideration in obtaining a homogeneous resistivity is the influence of the distribution of dopants in the starting material prior to irradiation. Due to the high segregation coefficient of boron in silicon, more homogeneous doping can be achieved in

P-type starting material than in N-type material of the same resistivity. However, the lower mobility for holes results in a greater background doping level in the P-type starting material as compared with N-type starting material. Thus, both P- and N-type starting material have been used in the preparation of neutron transmutation doped silicon. In both cases, the concentration of phosphorus created by neutron transmutation must be significantly larger than the dopant concentration in the starting material.

If the starting material has a resistivity variation of +/- α percent, then it can be shown that after irradiation, the percentage variation in the resistivity will be +/- $\alpha(\rho_f/\rho_i)$ for N-type starting material and +/- $\alpha(\mu_n/\mu_p)(\rho_f/\rho_i)$ for P-type starting material, where ρ_f is the post-radiation resistivity, ρ_i is the pre-irradiation resistivity, μ_n is the mobility of electrons and μ_p is the mobility of holes. The influence of resistivity fluctuation in the starting material upon the resistivity variation after irradiation is shown in Fig. 2.21 for various post- to pre-irradiation resistivity ratios. Since the pre-irradiation resistivity variation typically lies to the right of the dotted line in Fig. 2.21, the minimum initial resistivity required for achieving the desired

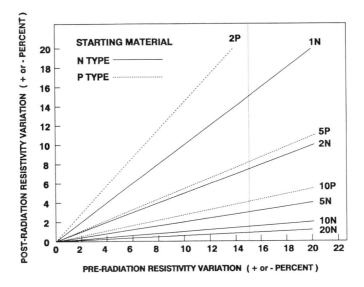

Fig. 2.21 Effect of starting material resistivity on the resistivity variation after neutron transmutation doping.

uniformity in the final resistivity can be determined from Fig. 2.21. As an example, if the starting material has a resistivity fluctuation of +/- 15 percent and the resistivity variation after irradiation must be less than +/- 3 percent, N-type starting material must have a resistivity of greater than five times the post-irradiation value, and P-type starting material must have a resistivity of greater than 15 times the post-irradiation value. Since starting material can be grown today with P-type resistivities of over 5,000 ohm-cm, homogeneous neutron doped silicon can be readily produced with resistivities of up to 200 ohm-cm.

In addition to these uniformity considerations, the occurrence of other nuclear reactions during irradiation and the creation of lattice damage must also be considered. The absorption of thermal neutrons by the phosphorus atoms created during the transmutation process can produce sulfur. Further, the irradiation invariably takes place in the presence of fast (high energy) neutrons. Fast neutron absorption by the silicon leads to the production of aluminum which would compensate the donors created by the transmutation reaction described by Eq. (2.53), and to the creation of magnesium, which is a recombination center in silicon. Recent studies [26] have shown that neither of these reactions has a significant effect upon the transmutation process. Consequently, the net donor concentration in the irradiated material can be computed solely on the basis of phosphorus atoms created by the transmutation reaction described in Eq. (2.53) and is given by:

$$C_{donor} = 2.06 \times 10^{-4} \, \phi \, t \qquad (2.56)$$

where ϕ is the neutron flux density and t is the irradiation time.

The neutron transmutation process is accompanied by severe lattice damage. The lattice damage arises from the displacement of the silicon atoms due to (a) gamma recoil during the transmutation reaction, (b) crystal irradiation from the high energy β particles emitted during the decay of Si^{31} to P^{31}, and (c) collisions of fast neutrons with silicon atoms. Although the effect of fast neutron damage can be minimized by performing the irradiation at locations in the reactor where the thermal to fast neutron ratio is high, it has been shown that the displacement effects due to fast neutrons are a thousand times greater than in the other processes [26]. Further, the damage due to the fast neutrons has been found to be more difficult to anneal out because it occurs in the form of clusters with diameters of less than 100 °A. Due to these damage mechanisms, the resistivity of the silicon is over 10^5 ohm-cm after irradiation and the minority carrier lifetime is low. In order to recover the desired resistivity due to the phosphorus doping and to obtain a high minority carrier lifetime, it is necessary to anneal the silicon after irradiation. The effect of annealing upon the resistivity and the minority carrier lifetime has been correlated with the dissociation of crystal defects in the temperature range of 400°C to 900°C [26]. These studies have shown that, although the resistivity can be recovered by annealing at 600°C for one hour, the minority carrier lifetime is still low and requires an annealing temperature above 750°C to reach its equilibrium value. Annealing the silicon at temperatures between 750°C and 800°C is therefore commonly used to obtain material from which devices can be fabricated.

In summary, neutron transmutation doping can be used to achieve superior homogeneity in resistivity as compared with conventionally doped silicon. The advantages of neutron transmutation doped silicon for power devices can be briefly stated as (a) a greater precision in the control of the avalanche breakdown voltage; (b) a more uniform avalanche breakdown distribution across the wafer, which leads to a greater capacity to withstand voltage surges; (c) a narrower base region width for achieving the desired breakdown voltage which lowers the forward voltage drop during current conduction; and (d) a more uniform current flow during forward conduction, which improves the surge current handling capability of the device.

2.3 LIFETIME

Under thermal equilibrium conditions, a continuous balance between the generation and recombination of electron-hole pairs occurs in semiconductors. Any creation of excess carriers by an external stimulus disturbs this equilibrium. Upon removal of the external excitation, the excess carrier density decays and the carrier concentration returns to the equilibrium value. The lifetime is a measure of the duration of this recovery. The recovery to equilibrium conditions can occur via several processes: (1) recombination occurring due to an electron dropping directly from the conduction band into the valency band, (2) recombination occurring due to an electron dropping from the conduction band and a hole dropping from the valency band into a recombination center, and (3) recombination occurring due to electrons from the conduction band and holes from the valency band dropping into surface traps. During these recombination processes, the energy of the carriers must be dissipated by one of several mechanisms: (1) the emission of a photon (radiative recombination), (2) the dissipation of energy in the lattice in the form of phonons (multiphonon recombination), and (3) the transmission of the energy to a third particle which can be either an electron or a hole (Auger recombination). The transitions that occur during these recombination processes are schematically illustrated in Fig. 2.22. All of these processes simultaneously assist in the recovery of the carrier density to its equilibrium value.

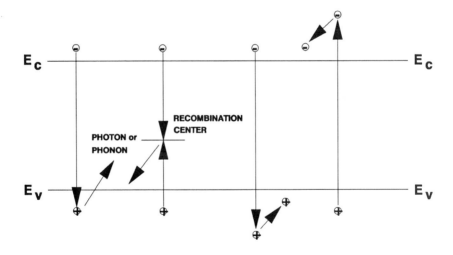

Fig. 2.22 Recombination processes that determine minority carrier lifetime in semiconductors.

Since silicon is a semiconductor with an indirect band gap structure, the probability of direct transitions from the conduction band to the valency band is small. Consequently, the direct radiative recombination lifetime for silicon is on the order of one second. In comparison to this, the density of recombination centers, even in the high purity silicon used to fabricate power devices, is sufficiently high so as to reduce the lifetime associated with recombination via deep centers in the energy gap to less than 100 microseconds. This recombination process, therefore, proceeds much more rapidly and is the predominant one under most device operating conditions.

2.3.1 Shockley-Read-Hall Recombination

The statistics of the recombination of electrons and holes in semiconductors via recombination centers were treated for the first time by Shockley, Read, and Hall [27, 28]. Their theory shows that the rate of recombination (U) in the steady state via a single level recombination center is given by:

$$U = \frac{\delta n\, p_o + \delta p\, n_o + \delta n\, \delta p}{\tau_{po}\,(n_o + \delta n + n_1) + \tau_{no}\,(p_o + \delta p + p_1)} \qquad (2.57)$$

where δn and δp are the excess electron and hole concentrations, n_o and p_o are the equilibrium concentrations of electrons and holes, τ_{no} and τ_{po} are the electron and hole minority carrier lifetimes in heavily doped P- and N-type silicon, respectively, and n_1 and p_1 are the equilibrium electron and hole density when the Fermi level position coincides with the recombination level position in the band gap. These are given by the expressions:

$$n_1 = N_c\, e^{(E_r - E_c)/kT} \qquad (2.58)$$

and

$$p_1 = N_v\, e^{(E_v - E_r)/kT} \qquad (2.59)$$

where N_c and N_v are the density of states in the conduction and valency bands, E_c, E_r, and E_v are the conduction band, recombination level, and valency band locations; k is Boltzmann's constant; and T is the absolute temperature. Under conditions of space charge neutrality, the excess electron (δn) and hole (δp) concentrations are equal and the lifetime can be defined as:

$$\tau = \frac{\delta n}{U} = \tau_{po}\left(\frac{n_o + n_1 + \delta n}{n_o + p_o + \delta n}\right) + \tau_{no}\left(\frac{p_o + p_1 + \delta n}{n_o + p_o + \delta n}\right) \qquad (2.60)$$

Since silicon power devices are generally made from N-type material due to the higher mobility of electrons, the rest of the treatment in this section will be confined to N-type silicon. Analogous equations can be derived for P-type silicon.

In the case of N-type silicon, the electron density (n_o) is much larger than the hole density (p_o). Combining this factor with Eqs. (2.58), (2.59) and (2.60) and defining a normalized injection level as $h = (\delta n/n_o)$, it can be shown that:

Chapter 2 : TRANSPORT PHYSICS 43

$$\frac{\tau}{\tau_{po}} = \left[1 + \frac{1}{(1+h)} e^{(E_r - E_F)/kT}\right] + \zeta\left[\frac{h}{(1+h)} + \frac{1}{(1+h)} e^{(2E_i - E_r - E_F)/kT}\right] \quad (2.61)$$

where E_i and E_F are the positions of the intrinsic and Fermi levels, and ζ is the ratio of the minority carrier lifetimes in heavily doped P- and N-type silicon, respectively:

$$\zeta = \frac{\tau_{no}}{\tau_{po}} \quad (2.62)$$

In deriving equation (2.61) it has been assumed that the density of states in the conduction and valency band are equal.

The minority carrier lifetime in heavily doped N- and P-type material used in the above equations is dependent upon the capture rate for holes (C_p) and electrons (C_n) at the recombination center:

$$\tau_{no} = \frac{1}{C_n N_r} = \frac{1}{v_{Tn} \sigma_{cn} N_r} \quad (2.63)$$

and

$$\tau_{po} = \frac{1}{C_p N_r} = \frac{1}{v_{Tp} \sigma_{cp} N_r} \quad (2.64)$$

where v_{Tn} and v_{Tp} are the thermal velocities of electrons and holes, σ_{cn} and σ_{cp} are the capture cross-sections for electrons and holes at the recombination center, and N_r is the density of the recombination center. Using these expressions, the ratio

$$\zeta = \frac{\tau_{no}}{\tau_{po}} = \frac{v_{Tp} \sigma_{cp}}{v_{Tn} \sigma_{cn}} = 0.827 \frac{\sigma_{cp}}{\sigma_{cn}} \quad (2.65)$$

Thus, σ is independent of the recombination center density and is a weak function of temperature.

Since well developed techniques exist for the measurement of the capture cross-sections and the position of the recombination center in the energy gap, the lifetime can be computed from these data. As an example, the variation in the lifetime with injection level is shown in Fig. 2.23 for N-type silicon for a doping level of 5×10^{13} atoms per cm^3. In Fig. 2.23, a range of values of the parameter ζ and the recombination center position (E_r) are considered. Several important observations can be made from these plots. Firstly, at very low injection levels ($h << 1$) the lifetime is observed to be independent of the injected carrier density. This low level lifetime (τ_{LL}) is given by the expression:

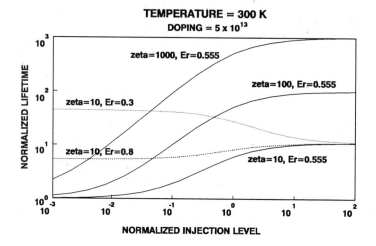

Fig. 2.23 Change in minority carrier lifetime with injection level.

$$\frac{\tau_{LL}}{\tau_{po}} = \left[1 + e^{(E_r - E_F)/kT}\right] + \zeta \, e^{(2E_i - E_r - E_F)/kT} \tag{2.66}$$

It can be seen from Fig. 2.23 that this low level lifetime is dependent upon the recombination level position in the energy gap and the capture cross-section ratio (ζ).

The variation of the low level lifetime with recombination center position (E_r) and capture cross-section for holes and electrons can be seen more clearly in the plots provided in Fig. 2.24 for N-type silicon with a doping level of 5×10^{13} cm^{-3}. The low level lifetime is observed to have its smallest value when the recombination center lies close to midgap. As the center is shifted towards either the conduction or the valency band edges, the low level lifetime increases due to the decreasing probability of capturing either holes or electrons, respectively.

Referring to Fig. 2.23, it can be seen that when the injection level becomes very large (h >> 1), the lifetime asymptotically approaches a constant value. This high level lifetime is given by expression:

$$\tau_{HL} = \tau_{po} + \tau_{no} = \tau_{po} (1 + \zeta) \tag{2.67}$$

The high level lifetime is not dependent upon the position of the recombination center and is weakly dependent on the temperature. However, it is directly dependent upon the capture cross-section ratio (ζ).

The variation in the lifetime with injection level is important to bipolar devices operating over a very large range of injection levels. The plots of Fig. 2.23 demonstrate that the lifetime can either increase or decrease with injection level depending upon the position of the recombination center in the energy gap. The magnitude of the change in lifetime with injection

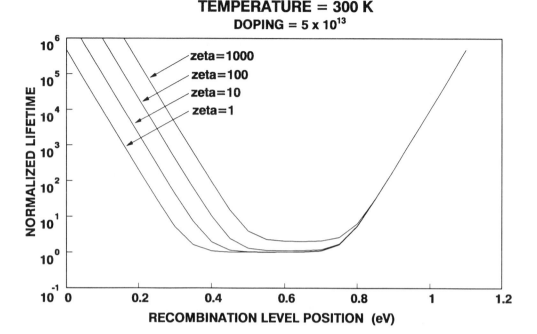

Fig. 2.24 Influence of recombination level position up on low-level lifetime.

level is controlled by the relative capture cross-sections at the recombination center, i.e., the capture cross-section parameter (ζ). In most practical cases, the lifetime is observed to increase with injection level by approximately a factor of ten.

2.3.2 Space Charge Generation

When a power device is blocking current flow, the voltage is supported across the depletion layer of a P-N junction. Under these conditions, an ideal device should exhibit no current flow. In actual reverse biased P-N junctions, a finite current flow is always observed. This 'leakage' current arises from a diffusion of free carriers which are generated within a minority carrier diffusion length from the edge of the depletion layer (*diffusion current*), and from carriers generated within the depletion layer itself (*space charge generation current*). In the case of a large energy gap semiconductor like silicon, the generation current is much larger than the diffusion current at around room temperature. At higher temperatures, the diffusion current becomes comparable to the generation current.

The contribution to the leakage current from space charge generation can be derived under the assumption of a uniform generation rate within the depletion region [29]. From this analysis, the leakage current is given by:

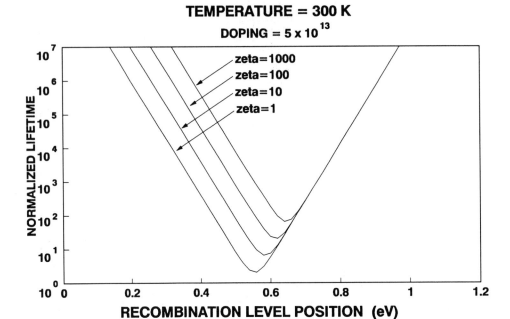

Fig. 2.25 Influence of recombination center position on space-charge generation lifetime.

$$I_{SC} = \frac{q A W n_i}{\tau_{SC}} \qquad (2.68)$$

where q is the electron charge, A is the junction area, W is the width of the depletion region, n_i is the intrinsic carrier concentration and τ_{sc} is the space-charge-generation lifetime. This lifetime can be derived from Eq. (2.57) for a single level recombination center:

$$\tau_{SC} = \tau_{po} e^{(E_r - E_i)/kT} + \tau_{no} e^{(E_i - E_r)/kT} \qquad (2.69)$$

The variation of the space charge generation lifetime with the recombination center position is shown in Fig. 2.25. The lowest space-charge generation lifetimes, and consequently the highest leakage currents, occur when the recombination center is located close to midgap. Thus, mid-gap recombination centers are not only most effective in reducing the minority carrier (low level) lifetime as shown by Fig. 2.24 but they also control the rate of generation of leakage current in power devices.

The above analysis predicts that the leakage current is proportional to the width of depletion layer and, consequently, the product of the leakage current and the junction capacitance should be independent of the junction reverse bias. Experimental measurements on silicon junctions indicate that the uniform generation of carriers occurs over only a fraction of the depletion region width [44]. The variation of this generation layer width is a function of the

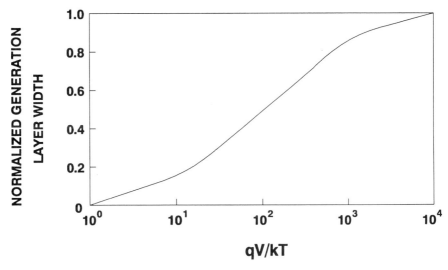

Fig. 2.26 Space-charge generation layer width as a function of reverse bias voltage applied to an abrupt junction.

applied junction bias and the substrate resistivity as shown in Fig. 2.26 for an abrupt junction. This effect is usually neglected during the analysis of power devices.

2.3.3 Recombination Level Optimization

In the previous sections, it was shown that the recombination and generation lifetimes are a strong function of the deep level position within the band gap and its capture cross-section for electrons and holes. For obtaining the best device performance, it is necessary to develop criteria for the selection of the optimum location of the deep level and its capture cross-sections. In the case of unipolar devices operating in the complete absence of minority carrier injection, the only criterion is that the space-charge generation lifetime must be large in order to minimize the leakage current. This can be achieved by maintaining a clean process which prevents the introduction of deep level impurities into the active area of the device.

In the case of devices where some low level minority carrier injection occurs, it becomes necessary to introduce deep level recombination centers in order to speed up or control the device characteristics. A small low level lifetime is highly desirable here for achieving fast switching speed. As indicated by the curves in Fig. 2.24, the recombination level should be located close to midgap to achieve the smallest low level lifetime. It is also desirable to have a large ratio (ζ) for the capture cross-sections of holes and electrons. At the same time, it is important to minimize the leakage current in the device during reverse blocking. For low leakage current, it is necessary to locate the recombination center away from midgap as indicated by the curves in Fig. 2.25. These two requirements impose conflicting demands upon the recombination center characteristics. To resolve this conflict and find an optimum location for the deep level it is necessary to consider the ratio of the space-charge generation lifetime to the low level lifetime [30]. Using equations (2.66) and (2.69) for the case of N-type silicon:

$$\frac{\tau_{SC}}{\tau_{LL}} = \frac{e^{(E_r - E_i)/kT} + \zeta e^{(E_i - E_r)/kT}}{1 + e^{(E_r - E_F)/kT} + \zeta e^{(2E_i - E_F - E_r)/kT}} \quad (2.70)$$

To achieve the best performance, it is necessary to maximize the ratio (τ_{sc}/τ_{LL}). Consider a typical example of silicon doped at a carrier concentration of 5×10^{13} per cm^3. The variation of the ratio (τ_{sc}/τ_{LL}) with recombination level position relative to the valence band is shown in Fig. 2.27 for 300°K and 400°K. It can be seen that the ratio has its highest value when the deep

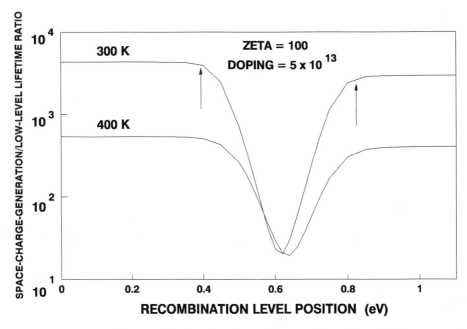

Fig. 2.27 Influence of recombination level position on the ratio of the space-charge generation to the low-level lifetime.

level is located near either the conduction band or valence band edges. The ratio can also be seen to rapidly decrease with increasing temperature. The best trade-off between low level lifetime reduction and minimizing leakage current due to space charge generation is achieved when the deep level is located near the band edges and the device operating temperature is low.

The ratio (τ_{sc}/τ_{LL}) remains high over a range of recombination level positions close to the band edges. This range is a function of the carrier concentration and the capture cross-section ratio. If the range is defined to extend to ten percent below the maximum ratio as indicated by the arrows in Fig. 2.27, the influence of doping upon the optimum position of the deep level can be quantified. For the case of N-type silicon, the preferred locations of recombination levels

Chapter 2 : TRANSPORT PHYSICS

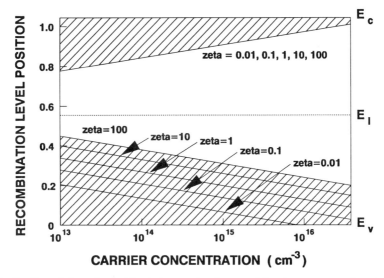

Fig. 2.28 Preferred locations (shaded areas) of recombination centers for silicon bipolar devices operating under low-level injection conditions.

within the band gap are indicated by the shaded areas in Fig. 2.28. This figure covers a spectrum of doping levels from 10^{13} to 10^{17} per cm^3 and capture cross-section ratios from 0.01 to 100. The heavily shaded area located above midgap applies to all capture cross-section ratios. Recombination levels with larger cross-section ratios (ζ) are also preferable because they can be located over a wider range within the band gap. For example, at a doping level of 1 x 10^{14} per cm^3, the recombination level must be located within 0.2 eV of the valence band for $\zeta = 0.01$ but can be located up to 0.4 eV from the valence band if $\zeta = 100$. Further, note that a broader range of recombination levels is available for lightly doped material making the selection of the lifetime controlling impurity easier.

The maximum value of the ratio (τ_{sc}/τ_{LL}) occurs when the recombination level is located close to the band edges. The maximum value is a function of the doping level but independent of the capture cross-section. It can be derived from Eq. (2.70):

$$\left(\frac{\tau_{SC}}{\tau_{LL}}\right)_{max} = e^{(E_F - E_i)/kT} = \frac{n_o}{n_i} \qquad (2.71)$$

The variation in the maximum achievable ratio (τ_{sc}/τ_{LL}) with carrier concentration is shown in Fig. 2.29 for 300°K and 400°K. It can be seen that the ratio (τ_{sc}/τ_{LL}) decreases with decreasing doping level and with increasing temperature. Consequently, it is more difficult to achieve a reduction of the low level lifetime in high resistivity silicon without substantial increase in leakage current when compared with low resistivity silicon. The problem becomes aggravated as temperature increases.

The above discussion was relevant to devices operating with low level injection during current conduction. In power devices, it is highly desirable to operate at high level injection

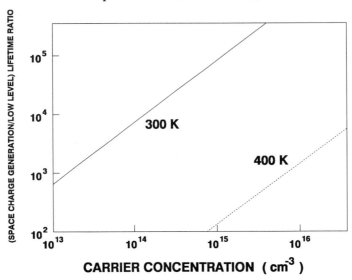

Fig. 2.29 Maximum achievable ratio of the space-charge generation and low-level lifetime.

during current conduction. The high level injection increases the conductivity of the lightly doped drift regions and greatly enhances the current handling capability. In these devices, it is desirable to have a large high level lifetime to maximize the conductivity modulation. At the same time, it is necessary to try to minimize the low level lifetime in order to speed up the switching process by rapid recombination of the minority carriers during device turn-off. The optimization of the location of the deep level for these devices can be performed by considering the ratio of the high level lifetime to the low level lifetime [31]. Using Eqs. (2.66) and (2.67):

$$\frac{\tau_{HL}}{\tau_{LL}} = \frac{1 + \zeta}{1 + e^{(E_r - E_F)/kT} + \zeta e^{(2E_i - E_F - E_r)/kT}} \qquad (2.72)$$

The ratio (τ_{HL}/τ_{LL}) can be maximized with respect to recombination level position (E_r) by setting

$$\frac{d}{dE_r}\left(\frac{\tau_{HL}}{\tau_{LL}}\right) = 0 \qquad (2.73)$$

and solving for E_r. This produces the first optimization criterion:

$$E_r = E_i + \frac{kT}{2} \ln(\zeta) \qquad (2.74)$$

The optimum deep level location is found to be independent of doping level and is purely a function of the capture cross-section ratio (ζ) and temperature.

In order to achieve the highest degree of conductivity modulation while maintaining a fast switching speed, it is necessary to achieve a large absolute value for (τ_{HL}/τ_{LL}). For N-type

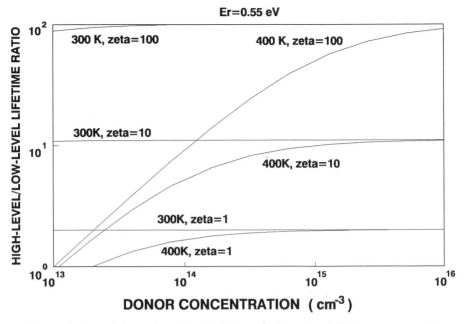

Fig. 2.30 Variation of the ratio of the high-level and low-level lifetime as a function of the background doping level.

silicon, this will occur when the capture cross-section ration (ζ) is large. It should be noted that achieving large values for the ratio (τ_{HL}/τ_{LL}) becomes more difficult at lower doping levels and at higher temperatures. This can be illustrated by plotting the ratio (τ_{HL}/τ_{LL}) as a function of the doping concentration. In Fig. 2.30, calculated curves are provided for several values of ζ at two operating temperatures. Since the high level lifetime is weakly dependent on temperature, the temperature dependence of the ratio (τ_{HL}/τ_{LL}) in Fig. 2.30 arises mainly from the variation of the low level lifetime with temperature.

In devices operating with high level injection, it is desirable to achieve a rapid transition from low level to high level lifetime with increasing injection level. This will maximize the region of injection levels over which the high level lifetime is large and provide good current conduction characteristics while simultaneously providing a small lifetime at low injection levels. The variation of the lifetime with injection level is described by Eq. (2.61). The transition from low to high level injection can be maximized by choosing a recombination level position such that

$$\frac{d}{dE_r}\left(\frac{d\tau}{dh}\right) = 0 \qquad (2.75)$$

The solution to this equation is also given by Eq. (2.74). The optimum recombination level positions obtained by using this equation are found to lie close to the center of the band gap.

It was shown in Section 2.3.2 that when the recombination center lies close to midgap, the space charge leakage current will be high. Consequently, the optimum level location defined for achieving a large (τ_{HL}/τ_{LL}) ratio conflicts with the need to maintain low leakage currents in power devices.

To resolve this conflict, it is necessary to examine the variation in the (τ_{HL}/τ_{LL}) ratio with recombination level location. Consider two examples of background donor concentrations of 5×10^{15} per cm³ and 5×10^{13} per cm³. The variation of the ratio (τ_{HL}/τ_{LL}) with recombination level position is shown in Fig. 2.31 for three values of ζ at different ambient temperatures. At the higher doping level, the ratio (τ_{HL}/τ_{LL}) exhibits a very broad maxima for all cases. Thus, the recombination level can be located away from the absolute maximum point near midgap without suffering a significant drop in the ratio (τ_{HL}/τ_{LL}). The leakage current can then be significantly reduced without compromising the forward voltage drop. A similar behavior for the variation of the (τ_{HL}/τ_{LL}) ratio with recombination center position is observed for the case of the low doping level at low temperatures, but the ratio does not exhibit as broad a maxima as in the case of the higher doping concentration. In fact, at higher temperatures, the (τ_{HL}/τ_{LL}) ratio exhibits a sharp peak at a value less than $(1 + \zeta)$.

An optimum recombination level position that will provide a large (τ_{HL}/τ_{LL}) ratio while minimizing the leakage current can be chosen by selecting a location when the (τ_{HL}/τ_{LL}) is 10% below its maximum value for all cases where the broad maxima are exhibited. Alternatively, at low doping levels and higher temperatures where the ratio (τ_{HL}/τ_{LL}) remains below 90 percent of the maximum achievable value of $(1 + \zeta)$, the optimum location will occur when the (τ_{HL}/τ_{LL})

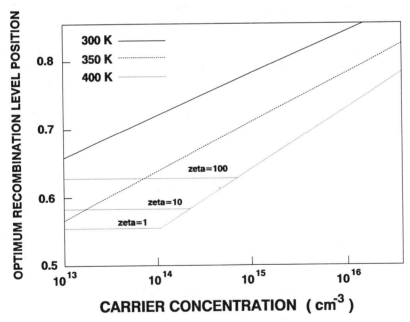

Fig. 2.32 Optimum recombination level position for bipolar devices operating under high-level injection conditions.

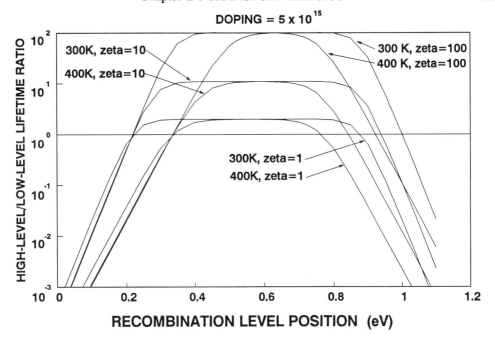
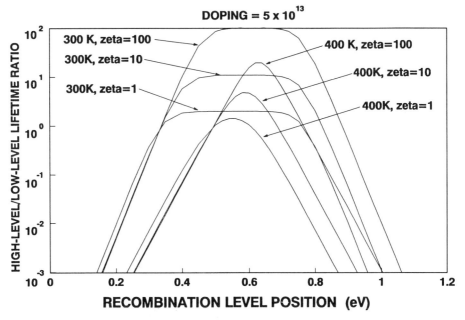

Fig. 2.31 Variation of the high-level to low-level lifetime ratio with recombination level position for doping concentration of 5×10^{15} and 5×10^{13} per cm^3.

is at its peak value. The optimum recombination center location obtained in this manner is shown in Fig. 2.32 as a function of the donor concentration. It should be noted that although two optimum recombination level locations can be defined on either side of the maximum in Fig. 2.31, only the position above the intrinsic level must be used because it produces the desired increase in lifetime with injection level as shown in Fig. 2.23. The optimum recombination level location obtained from Fig. 2.32 can be seen to be closer to midgap when the doping level is reduced. Thus, it becomes more difficult to achieve a low leakage current in devices designed for operation at higher voltages.

In addition to the above optimization criteria, other constraints can be defined for the deep level impurity [32]. The introduction of a deep level impurity causes compensation effects which results in changes in the background resistivity. When the deep level impurity concentration approaches the background donor concentration, electrons are transferred from the donor into the deep level instead of the conduction band. The decrease in net free electron concentration in the conduction band causes an increase in the resistivity. Changes in resistivity have serious effects on both breakdown voltage and current conduction. The compensation can be minimized by achieving the lowest possible deep level concentration for any desired lifetime. The lifetime is related to the deep level concentration via equations (2.63) and (2.64). From these expressions, it can be concluded that the least compensation will be achieved when the capture cross- sections for electrons and holes are large.

Another technological constraint is that the same deep level impurity should be useful for

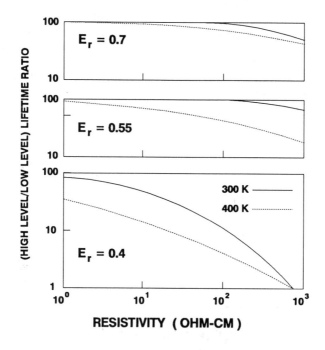

Fig. 2.33 Variation of the high-level to low-level lifetime ratio with resistivity.

a broad range of resistivities. This allows a single lifetime control process to be used for a variety of devices. To use the same lifetime control process over a broad range of resistivities, the ratio (τ_{HL}/τ_{LL}) should remain high over a wide range of doping levels. In Fig. 2.33, the variation of the (τ_{HL}/τ_{LL}) ratio with background resistivity is shown for three recombination level positions. It can be seen that the desired characteristics are exhibited when the deep level is located at 0.7 eV above the valency band. This technologically desirable feature is consistent with the optimum recombination center location obtained by using Fig. 2.32 over most of the range of doping levels.

2.3.4 Lifetime Control

Two fundamental processing methods have been developed to control lifetime in power devices. The first method involves the thermal diffusion of an impurity that exhibits deep levels in the energy gap of silicon. The second method is based upon the creation of lattice damage in the form of vacancies and interstitial atoms by bombardment of the silicon wafers with high energy particles. Historically, the introduction of recombination centers in silicon by the diffusion of impurities was adapted to power device processing well before the use of the high energy radiation technique. A very large number of impurities have been found to exhibit deep levels in silicon. A catalog of these impurities has been compiled [31, 33]. Although many of these impurities could be used in the fabrication of power devices, only gold and platinum have been extensively used to control the characteristics of commercial devices. For this reason, the properties of gold and platinum in silicon will be treated here.

The diffusion of gold and platinum into silicon occurs much more rapidly than that of dopant atoms such as phosphorus or boron. Consequently, the diffusion of these impurities is always performed after the fabrication of the device structure but prior to metallization. In order to obtain a homogeneous distribution of the recombination centers throughout the wafer, it is necessary to perform the diffusion at temperatures between 800°C and 900°C. The diffusion temperature determines the solid solubility of the impurity atoms in silicon and can be used to control the impurity density. A higher gold or platinum diffusion temperature should be used to obtain a lower lifetime. It must be kept in mind that, because these diffusion temperatures exceed the eutectic point between silicon and the aluminum metallization used for power devices, the lifetime control must precede device metallization. Consequently, the devices cannot be tested immediately prior to or after impurity diffusion. Further, small changes in the diffusion temperature are observed to cause a large variation in the device characteristics. These problems can be overcome by radiation processes using high energy particles.

The bombardment of silicon with high energy particles has been demonstrated to create lattice damage. This lattice damage consists of silicon atoms displaced from their lattice sites into interstitial positions leaving behind a vacancy. It has been found that the vacancy is highly mobile in silicon even at low temperatures. Consequently, the defects that remain after irradiation are composed of complexes of the vacancy with impurity atoms such as phosphorus or oxygen and of two adjacent vacancy sites in the lattice (called the *divacancy*). In heavily phosphorus doped silicon, the phosphorus-vacancy defect becomes predominant. However, in the high resistivity material used to fabricate power devices, the divacancy defect has been shown to be the dominant one in both P- and N-type silicon [34, 35].

For controlling lifetime in power devices, both electron irradiation and gamma irradiation have been shown to be effective processing techniques [36, 37]. The advantages of using these techniques to control the lifetime in power devices are as follows: (a) the irradiation can be performed at room temperature after complete device fabrication and initial testing of the characteristics; (b) the lifetime is controlled by the radiation dose which can be accurately metered by monitoring the electron current during irradiation; (c) a much tighter distribution in the device characteristics can be achieved because of the improved control over the recombination center density; (d) the irradiation process allows trimming the device characteristics to the desired value by using several irradiation steps because the radiation is performed after complete device fabrication, thus allowing device testing between consecutive irradiations; (e) the irradiation damage can be annealed out by heating the devices above 400°C, thus allowing the recovery of devices that may have had an overdose during irradiation; and (f) the irradiation process is a cleaner and simpler process as compared with impurity diffusion and avoids any possible contamination between the processing of devices requiring high lifetime and those requiring low lifetime. These advantages have made the electron irradiation process very attractive for manufacturing power devices.

The concentration of deep levels created by electron irradiation is proportional to the total flux of electrons that bombard the silicon wafer. The change in the lifetime following the electron irradiation is also dependent upon the initial lifetime before performing the electron irradiation. This relationship can be expressed as:

$$\frac{1}{\tau_f} = \frac{1}{\tau_i} + K\phi \qquad (2.76)$$

where K is a constant that depends upon the energy of the electron beam. For the case of 3 MeV electrons, K has a value of about 10^{-8} cm^2/sec if the fluence ϕ is expressed in electrons per cm^2. By monitoring the electron beam current, it is possible to precisely control the electron beam dose. Thus, if the initial lifetime is much larger than the desired lifetime after the electron irradiation, it is possible to obtain good control over the final lifetime. In addition, the final lifetime becomes uniform across even a device with a large area, as well as between devices. For these reasons, it is general practice to try and obtain the highest possible lifetime in the silicon after fabrication of the devices. This is achieved by performing processing under clean conditions and by using gettering techniques to prevent accumulation of deep level impurities in the silicon during fabrication.

However, the application of gold diffusion is still in use today because of two reasons. Firstly, the nature of the recombination centers introduced by gold diffusion are different from those produced by electron irradiation, and superior conduction characteristics are observed in gold doped devices. Secondly, the annealing of electron irradiation induced defects at a few hundred degrees above room temperature has been of concern with regard to the long term stability of the power devices. Annealing studies of the defect levels produced by electron irradiation in silicon have, however, demonstrated that, although the divacancy does anneal out at 300°C, the lifetime does not change appreciably due to the creation of a new recombination center during the annealing [34]. Thus, electron irradiation does not result in long term instability under typical device operating conditions.

Since gold/platinum diffusion and electron/gamma irradiation are being used today to control the characteristics of power devices, the recombination centers introduced by these techniques in silicon are provided in Fig. 2.34. It has been found that gold diffusion into silicon

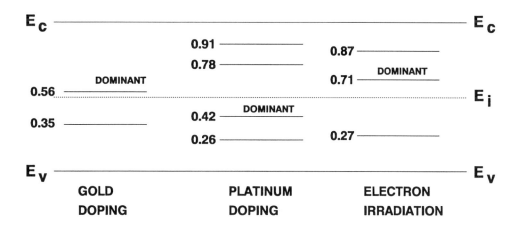

Fig. 2.34 Recombination levels in silicon produced by gold doping, platinum doping, and electron irradiation.

introduces an acceptor and a donor deep level in the energy gap [38]. The center controlling the minority carrier lifetime in N-type silicon is the one at 0.54 eV below the conduction band. In the case of platinum diffusion into silicon, four deep levels have been found [39]. Of these levels, the one lying at 0.42 eV above the valency band edge has been found to control the minority carrier lifetime in N-type silicon. In the case of electron irradiation, three deep levels have been observed. The divacancy induced deep level at 0.40 eV below the conduction band has been shown to control the minority carrier lifetime in n-type silicon [34]. In order to calculate the lifetime variation with resistivity and temperature, it is necessary to know the capture cross-section for holes and electrons at these dominant recombination centers. These data are provided in Table 2.1 together with the capture cross-section ratios (ζ).

Using the data provided in Fig. 2.34 and Table 2.1, the ratio (τ_{HL}/τ_{LL}) can be calculated for each of the above lifetime control techniques by assuming that only the dominant recombination center is present. This is a reasonable approximation for these multiple deep level centers because in all three cases the level closest to midgap is found to control the recombination and generation rates.

The calculated variation of the ratio (τ_{HL}/τ_{LL}) with resistivity is shown in Fig. 2.35 for several temperatures. Note that the ratio (τ_{HL}/τ_{LL}) is the lowest for electron irradiation and the highest for gold doping. From this calculation, it can be concluded that gold doping will provide

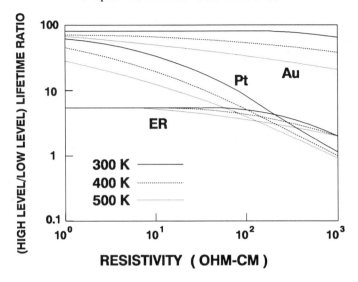

Fig. 2.35 Ratio of high-level to low-level lifetime as a function of silicon resistivity for gold doping, platinum doping, and electron irradiation.

Table 2.1 Dominant levels for gold doping, platinum doping, and electron irradiation induced deep levels in silicon.

Impurity	Energy Level Position (eV)	Capture Cross Section		Ratio ζ
		Holes (cm^2)	Electrons (cm^2)	
Gold	0.56	6.08×10^{-15}	7.21×10^{-17}	69.7
Platinum	0.42	2.70×10^{-12}	3.2×10^{-14}	69.8
Electron Irradiation	0.71	8.66×10^{-16}	1.61×10^{-16}	4.42

a superior trade-off between current conduction and switching speed when compared with electron irradiation. Experimental measurements performed using power rectifiers confirm the results of these calculations [36].

To evaluate the impact of these lifetime control methods, it is useful to calculate the ratio of the space-charge generation lifetime to the high level lifetime. The calculated ratio (τ_{SC}/τ_{HL}) for each of the three cases is shown as a function of temperature in Fig. 2.36. The calculated ratio (τ_{SC}/τ_{HL}) for electron irradiation and for platinum doping are many orders of magnitude larger than for gold doping especially at lower temperatures. For equal forward conduction

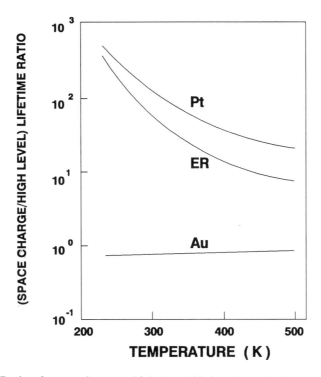

Fig. 2.36 Ratio of space-charge to high-level lifetime for gold doping, platinum doping, and electron irradiation.

characteristics as controlled by the high level lifetime, the leakage current of electron irradiated and platinum doped devices will therefore be much lower than for gold doped devices. Experimental measurements performed on rectifiers have confirmed the results of these calculations [36]. Based upon the theoretical calculations and the experimental data, it can be concluded that the advantages of electron irradiation (i.e., lower leakage currents, superior uniformity, and processing simplicity) make it a preferred technique for lifetime control.

2.3.5 Auger Recombination

The Auger recombination process occurs by the transfer of the energy and momentum released by the recombination of an electron-hole pair to a third particle that can be either an electron or a hole. This process becomes significant in heavily doped P- and N-type silicon such as the diffused end regions of power devices. It is also an important effect in determining recombination rates in the lightly doped base regions of power devices operating at high injection levels during forward conduction because of the simultaneous presence of a high concentration of holes and electrons injected into this region. Theoretical investigations of Auger recombination [40, 41] in indirect gap semiconductors indicate that the process can occur with or without phonon assistance, and may occur directly from band to band or via traps.

Measurements of Auger recombination have been performed by observing the decay of excess minority carriers at low injection levels in heavily doped N- and P-type silicon, and by observing the decay of excess free carriers at high injection levels in high resistivity silicon [42, 43]. The measured Auger coefficients for these cases are not equal due to the difference in the structure of the conduction and the valency bands. In the case of heavily doped N-type silicon, the Auger process transfers energy and momentum to an electron in the conduction band and the Auger lifetime is given by:

$$\tau_A^N = \frac{1}{2.8 \times 10^{-31} \, n^2} \qquad (2.77)$$

where n is the majority carrier (electron) concentration. In the case of P-type silicon, the Auger process occurs by the transfer of energy and momentum to a hole in the valency band and the measured Auger lifetime is given by

$$\tau_A^P = \frac{1}{1 \times 10^{-31} \, p^2} \qquad (2.78)$$

where p is the majority carrier (hole) concentration. In the case of Auger recombination occurring at high injection levels where the density of electrons and holes is simultaneously large, the Auger lifetime, as measured in heavily excited silicon by laser radiation, is given by:

$$\tau_A^{\Delta n} = \frac{1}{3.4 \times 10^{-31} \, \Delta n^2} \qquad (2.79)$$

where Δn is the excess carrier concentration. The variation of the Auger lifetime with carrier concentration is shown in Fig. 2.37 for all three cases. These plots are extended to doping levels of up to 2×10^{20} for the heavily doped cases since such high dopant concentrations are prevalent in the diffused regions of power devices. The extremely low Auger lifetimes at the high doping concentrations have a strong influence upon the injection efficiency and recombination in the end regions of power devices operating with minority carrier injection. The Auger lifetime for the case of high injected carrier densities in lightly doped regions is even lower than in the case of heavily doped regions because the Auger processes requiring two electrons plus a hole and two holes plus an electron can occur simultaneously. At surge current conditions, the injected carrier concentration near the junctions of power devices can exceed 10^{18} per cc. Since these devices are usually processed to achieve a high Shockley-Read-Hall lifetime (usually in excess of 10 microseconds), it can be seen from Fig. 2.37 that at high injection levels the Auger recombination process can play a significant role in affecting current conduction and switching speed. It should be noted that the Auger recombination process occurs simultaneously with the Shockley-Read-Hall recombination process. The effective lifetime is then given by

$$\frac{1}{\tau_{eff}} = \frac{1}{\tau_{SRH}} + \frac{1}{\tau_A} \qquad (2.80)$$

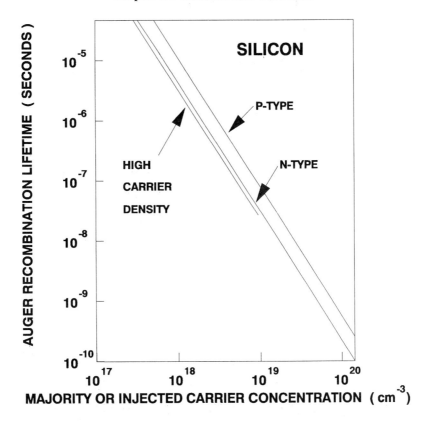

Fig. 2.37 Auger recombination lifetime as a function of injected carrier density.

where τ_{SRH} is the Shockley-Read-Hall recombination lifetime discussed in Section 2.3.1.

REFERENCES

1. B.J. Baliga, "Semiconductors for High Voltage Vertical Channel, Field Effect Transistors," J. Applied Physics, Vol. 53, pp. 1759-1764 (1982).

2. C. Canali, C. Jacoboni, F. Nava, G. Ottaviani, and A.A. Quaranta, "Electron Drift Velocity in Silicon," Phys. Rev., Vol. B12, pp. 2265-2284 (1975).

3. G. Ottaviani, L. Reggiani, C. Canali, F. Nava, and A.A. Quaranta, "Hole Drift Velocity in Silicon," Phys. Rev., Vol. B12, pp. 3318-3329 (1975).

4. F.J. Morin and J.P. Maita, "Electrical Properties of Silicon Containing Arsenic and Boron," Phys. Rev., Vol. 96, pp. 28-35 (1954).

5. C. Jacoboni, C. Canali, G. Ottaviani, and A.A. Quaranta, "A review of some charge transport properties of silicon," Solid State Electronics, Vol. 20, pp. 77-89 (1977).

6. C. Canali, G. Majni, R. Minder, and G. Ottaviani, "Electron and hole drift velocity measurements in silicon," IEEE Trans. Electron Devices, Vol. ED-22, pp. 1045-1047 (1975).

7. N.J. Murphy, F. Berz, and I. Flinn, "Carrier mobility in silicon MOST's," Solid State Electronics, Vol. 12, pp. 775-786 (1969).

8. A.A. Guzev, G.L. Kurishev, and S.P. Sinitsa, "Scattering mechanisms in inversion channels of MIS structures on silicon," Phys. Status Solidi, Vol. A14, pp. 41-50 (1972).

9. S.C. Sun and J.D. Plummer, "Electron mobility in inversion and accumulation layers on thermally oxidized silicon surfaces," IEEE Trans. Electron Devices, Vol. ED-27, pp. 1497-1508 (1980).

10. Y.C. Cheng and E.A. Sullivan, "Relative importance of phonon scattering to carrier mobility in silicon surface layer at room temperature," J. Appl. Phys., Vol. 44, pp. 3619-3625 (1973).

11. O. Leistiko, A.S. Grove, and C.T. Sah, "Electron and hole mobilities in inversion layers on thermally oxidized silicon surfaces," IEEE Trans. Electron Devices, Vol. ED-12, pp. 248-254 (1965).

12. T. Sato, Y. Takeishi, H. Hara, and Y. Okamoto, "Mobility anisotropy of electrons in nversion layers on oxidized silicon surfaces," Physical Review, Vol. B-4, pp. 1950-1960 (1971).

13. A. Ohwada, H. Maeda, and T. Tanaka, "Effect of the crystal orientation upon electron mobility at the Si/SiO$_2$ interface," Japan. J. Applied Physics, Vol. 8, pp. 629-630 (1969).

14. T. Sato, Y. Takeishi, H. Tango, H. Ohnuma, and Y. Okamoto, "Drift-velocity saturation of holes in silicon inversion layers," J. Phys. Soc. Japan, Vol. 31, p. 1846 (1971).

15. F.F. Fang and A.B. Fowler, "Hot electron effects and saturation velocities in silicon inversion layers," J. Applied Physics, Vol. 41, pp. 1825-1831 (1970).

16. G.G. McFarlane, J.P. McLean, J.E. Quarrington, and V. Roberts, "Fine structure in the absorption edge spectrum of ilicon," Phys. Rev., Vol. 111, pp. 1245-1254 (1958).

17. H.P.D. Lanyon and R.A. Tuft, "Bandgap narrowing in heavily doped silicon," IEEE Int. Electron Devices Meeting Digest, Abs. 13.3, pp. 316-319 (1978).

18. J.W. Slotboom and H.C. DeGraf, "Measurements of bandgap narrowing in silicon bipolar transistors," Solid State Electronics, Vol. 19, pp. 857-862 (1976).

19. A.W. Wieder, "Arsenic emitter effects," IEEE Int. Electron Devices Meeting Digest, Abs. 18.7, pp. 460-462 (1978).

20. R. Mertens and R.J. Van Overstraeten, "Measurement of the minority carrier transport parameters in heavily doped silicon," IEEE Int. Electron Devices Meeting Digest, Abs. 13.4, pp. 320-323 (1978).

21. G.E. Possin, M.S. Adler, and B.J. Baliga, "Measurements of the pn product in heavily doping epitaxial emitters," IEEE Trans. Electron Devices, Vol. ED-31, pp. 3-17 (1984).

22. F. Mousty, P. Ostoja, and L. Passari, "Relationship between resistivity and phosphorus concentration in silicon," J. Applied Physics, Vol. 45, pp. 4576-4580 (1974).

23. D.M. Caughey and R.F. Thomas, "Carrier mobilities in silicon empirically related to doping and field," Proc. IEEE, Vol. 55, pp. 2192-2193 (1967).

24. M. Tanenbaum and A.D. Mills, "Preparation of uniform resistivity N-type silicon by nuclear transmutation," J. Electrochemical Society, Vol. 108, pp. 171-176 (1961).

25. H.M. Janus and O. Malmos, "Application of thermal neutron irradiation for large scale production of homogeneous phosphorus doping of floatzone silicon," IEEE Trans. Electron Devices, Vol. ED-23, pp. 797-802 (1976).

26. "Neutron transmutation doping in semiconductors," Edited by J.M. Meese, Plenum Press, New York (1979).

27. R.N. Hall, "Electron-hole recombination in germanium," Physical Review, Vol. 87, p. 387 (1952).

28. W. Shockley and W.T. Read, "Statistics of the recombination of holes and electrons," Physical Review, Vol. 87, pp. 835-842 (1952).

29. C.T. Sah, R.N. Noyce, and W. Shockley, "Carrier generation and recombination in P-N junctions and P-N junction characteristics," Proc. I.R.E., Vol. 45, pp. 1228-1243 (1957).

30. B.J. Baliga, "Recombination level selection criteria for lifetime reduction in integrated circuits," Solid State Electronics, Vol. 21, pp. 1033-1038 (1978).

31. B.J. Baliga and S. Krishna, "Optimization of recombination levels and their capture cross-section in power rectifiers and thyristors," Solid State Electronics, Vol. 20, pp. 225-232 (1977).

32. B.J. Baliga, "Technological constraints upon the properties of deep levels used for lifetime control in the fabrication of power rectifiers and thyristors," Solid State Electronics, Vol. 20, pp. 1029-1032 (1977).

33. A.G. Milnes, "Deep impurities in semiconductors," Wiley, New York (1973).

34. A.O. Evwaraye and B.J. Baliga, "The dominant recombination centers in electron-irradiated semiconductor devices," J. Electrochemical Society, Vol. 124, pp. 913-916 (1977).

35. B.J. Baliga and A.O. Evwaraye, "Correlation of lifetime with recombination centers in electron irradiated P-type silicon," J. Electrochemical Society, Vol. 130, pp. 1916-1918 (1983).

36. B.J. Baliga and E. Sun, "Comparison of gold, platinum, and electron irradiation for controlling lifetime in power rectifiers," IEEE Trans. Electron Devices, Vol. ED-24, pp. 685-688 (1977).

37. R.O. Carlson, Y.S. Sun, and H.B. Assalit, "Lifetime control in silicon power devices by electron or gamma irradiation," IEEE Trans. Electron Devices, Vol. ED-24, pp. 1103-1108 (1977).

38. J.M. Fairfield and B.V. Gokhale, "Gold as a recombination center in silicon," Solid State Electronics, Vol. 8, pp. 685-691 (1965).

39. K.P. Lisiak and A.G. Milnes, "Platinum as a lifetime control deep impurity in silicon," J. Applied Physics, Vol. 46, pp. 5229-5235 (1975).

40. L. Huldt, "Band-to-band Auger recombination in indirect gap semiconductors," Phys. Status Solidi, Vol. A8, pp. 173-187 (1971).

41. A. Haug, "Carrier density dependence of Auger recombination," Solid State Electronics, Vol. 21, pp. 1281-1284 (1978).

42. J. Dziewior and W. Schmid, "Auger coefficients for highly doped and highly excited silicon," Applied Physics Letters, Vol. 31, pp. 346-348 (1977).

43. K.G. Svantesson and N.G. Nilson, "Measurement of Auger recombination in silicon by laser excitation," Solid State Electronics, Vol. 21, pp. 1603-1608 (1978).

44. P.U. Calzolari and S. Graffi, "A theoretical investigation of the generation current in silicon PN junctions under reverse bias," Solid State Electronics, Vol. 15, pp. 1003-1011 (1972).

Chapter 2 : TRANSPORT PHYSICS

PROBLEMS

2.1 Calculate the values for the electron and hole mobilities at 27, 100, and 200 °C for a doping concentration of 1 x 10^{15} per cm^3.

2.2 Determine the ratio of the electron to hole mobility for a doping concentration of 1 x 10^{15} per cm^3 at 27, 100, and 200 °C.

2.3 Calculate the minority carrier density at 300 °K for homogeneously doped silicon with a doping concentration of 1 x 10^{15}, 1 x 10^{17}, 1 x 10^{18}, 1 x 10^{19}, and 1 x 10^{20} per cm^3 with and without the influence of band gap narrowing. Determine the ratio of the minority carrier density obtained with and without including the band gap narrowing effect.

2.4 Compare the resistivity variation for a neutron transmutation doped silicon wafer of 6 cm in diameter with and without rotation of the ingot during the irradiation process.

2.5 Determine the radiation time required to obtain a resistivity of 50 ohm-cm in silicon for a nuclear reactor with a thermal neutron flux density of 1 x 10^{14} per (cm^2 second) by using the neutron transmutation doping process.

2.6 An N-type silicon region has a doping concentration of 1 x 10^{15} per cm^3 and is doped with gold at a concentration of 1 x 10^{13} per cm^3. Calculate the minority carrier low-level lifetime, high-level lifetime, and space-charge-generation lifetime at 300 °K. Assume that only a single dominant deep level exists. What is the ratio of the high-level to low-level lifetime, and the space-charge-generation to low-level lifetime ?

2.7 An N-type silicon region has a doping concentration of 1 x 10^{15} per cm^3 and is electron irradiated to produce a deep level concentration of 1 x 10^{13} per cm^3. Calculate the minority carrier low-level lifetime, high-level lifetime, and space-charge-generation lifetime at 300 °K. Assume that only a single dominant deep level exists. What is the ratio of the high-level to low-level lifetime, and the space-charge-generation to low-level lifetime ?

2.8 Calculate the minority carrier lifetime for an N-type silicon region with a doping concentration of 1 x 10^{19} per cm^3 containing 1 x 10^{13} atoms of gold per cm^3.

Chapter 3

BREAKDOWN VOLTAGE

A unique distinguishing feature of all semiconductor power devices is their high voltage blocking capability. Depending upon the application, the breakdown voltage can range from 25 volts for applications such as power supplies to over 6000 volts for applications in power transmission and distribution. The ability to support high voltages is determined by the onset of avalanche breakdown, which occurs when the electric field within the device structure becomes large. In power devices, large electric fields can occur both within the interior regions of the device where current transport takes place and at the edges of the devices. Proper design of devices requires careful attention to field distributions both at the interior and at the edges to ensure high voltage blocking capability. Since the forward voltage drop during current conduction is larger for devices with higher breakdown voltage capability, it is important to obtain a device breakdown voltage as close as possible to the intrinsic capability of the semiconductor material for optimum device performance.

In power devices, the voltage is supported across a depletion layer formed across either a P-N junction, a metal-semiconductor (Schottky barrier) interface, or a metal oxide semiconductor (MOS) interface. The electric field that exists across the depletion layer is responsible for sweeping out any holes or electrons (mobile carriers) that enter this region by the process of either space charge generation or by diffusion from the neighboring quasi-neutral regions. When voltage is increased, the electric field in the depletion region increases and the mobile carriers are accelerated to higher velocities. In the case of silicon, the mobile carriers attain a saturated drift velocity of about 1×10^7 cm/sec when the electric field becomes larger than 1×10^5 V/cm. At even higher electric fields, the mobile carriers have sufficient kinetic energy that their collisions with the atoms in the lattice can excite electrons from the valency band to the conduction band. This process for the generation of electron-hole pairs is called *impact ionization*. Since the electron-hole pairs created in the depletion region by impact ionization undergo acceleration by the high electric field, they participate in the creation of further pairs of electrons and holes. Consequently, impact ionization is a multiplicative phenomenon that produces a cascade of mobile carriers which are transported through the depletion layer. This leads to an increase in the current flow through the depletion region. The device is considered to undergo *avalanche breakdown* when the rate of impact ionization approaches infinity because the device cannot support any further increase in the applied voltage. Consequently, avalanche breakdown is a fundamental limitation to the maximum operating voltage for power devices.

This chapter firt discusses the physics of avalanche breakdown and then analyzes its impact upon the maximum voltage that can be supported by semi-infinite abrupt and graded P-N junctions. This breakdown voltage is a measure of the highest voltage that can be supported by device structures. In practical devices, the junction must be terminated at the edges because of their finite size. In general, the breakdown voltage at the termination has been found to be well below that within the interior of power devices. Consequently, a large variety of junction termination designs have been proposed in order to raise the breakdown voltage to a value close to that for the semi-infinite junction. These edge terminations are discussed in detail in the chapter because proper design of a power device requires great care to prevent poor breakdown at the edges.

3.1 AVALANCHE BREAKDOWN

Avalanche breakdown occurs when the impact ionization process attains an infinite rate. The impact ionization process results in the creation of electron-hole pairs within the depletion region, producing current flow. In order to characterize this process, it is useful to define an ionization coefficient.

3.1.1 Ionization Coefficients

The *impact ionization coefficient for holes* (α_p) is defined as the number of electron-hole pairs created by a hole traversing 1 cm through the depletion layer along the direction of the electric field. In a similar manner, the *impact ionization coefficient for electrons* (α_n) is defined as the number of electron-hole pairs created by an electron traversing 1 cm through the depletion layer along the direction of the electric field. Based upon extensive measurements of the ionization coefficients in silicon, their variation with electric field has been found to be given by the following relationships:

$$\alpha_n = a_n \cdot \exp(-b_n/E) \qquad (3.1)$$

and

$$\alpha_p = a_p \cdot \exp(-b_p/E) \qquad (3.2)$$

where $a_n = 7 \times 10^5$ per cm, $b_n = 1.23 \times 10^6$ V/cm for electrons, and $a_p = 1.6 \times 10^6$ per cm, $b_p = 2 \times 10^6$ V/cm for holes, for electric fields ranging between 1.75×10^5 to 6×10^5 V/cm. It is important to note that the ionization coefficients increase very rapidly with increasing electric field. Plots of α_n and α_p as a function of the electric field are given in Fig. 3.1. Due to this very strong dependence of the ionization coefficients upon the electric field, the breakdown voltage of devices can be severely reduced by the presence of a high localized electric field within the

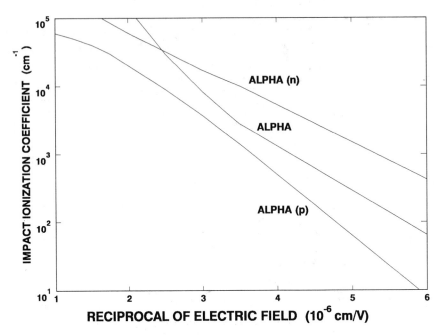

Fig. 3.1. Impact ionization coefficients for silicon.

structure.

An approximation for the impact ionization coefficient that is useful for deriving analytical solutions for the breakdown voltage of devices is :

$$\alpha = 1.8 \times 10^{-35} \ E^7 \qquad (3.3)$$

The dependence of this approximation for the impact ionization coefficient upon the electric field is compared with the more accurate coefficients given by Eq. (3.1) and (3.2) in Fig. 3.1. Although the approximation should not be used for numerical analysis, it is very useful for understanding the physics of breakdown in device structures.

3.1.2 Multiplication coefficient

In order to compute the avalanche breakdown voltage, it is necessary to determine the condition under which the impact ionization achieves an infinite rate. In order to determine this condition, consider an N+/P junction with a positive bias applied to the N+ region. This reverse bias across the junction creates a depletion layer extending primarily in the P region if the doping concentration in the N+ region is far greater than that in the P region. In order to perform the avalanche breakdown analysis, assume that an electron-hole pair is generated at a distance x from the junction as illustrated in Fig. 3.2. Avalanche breakdown occurs when the

Chapter 3 : BREAKDOWN VOLTAGE

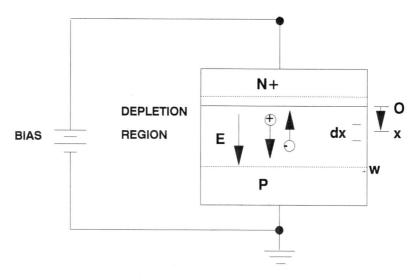

Fig. 3.2. Reverse biased P-N junction.

impact ionization achieves an infinite rate. Due to the strong electric field in the depletion region, the hole will be swept towards the P region and the electron will be swept toward the N+ region. At high electric fields pertinent to the breakdown analysis, these carriers will be accelerated by the electric field until they gain sufficient energy to create hole-electron pairs during collisions with the lattice atoms. Using the definitions for the ionization coefficients, the hole will create ($\alpha_p.dx$) electron-hole pairs when traversing a distance dx. Similarly, the electron will create ($\alpha_n.dx$) electron-hole pairs when traversing a distance dx. The total number of electron-hole pairs created in the depletion region due to a single electron-hole pair initially generated at a distance x from the junction is then given by :

$$M(x) = 1 + \int_0^x \alpha_n M(x)\, dx + \int_x^W \alpha_p M(x)\, dx \qquad (3.4)$$

where W is the depletion layer width. A solution for this differential equation is :

$$M(x) = M(0) \cdot \exp\left[\int_0^x (\alpha_n - \alpha_p)\, dx\right] \qquad (3.5)$$

where M(0) is the total number of electron-hole pairs at the edge of the depletion region. Substituting this expression into Eq. (3.4) for the case of x = 0 and solving for M(0) gives :

$$M(0) = \left\{1 - \int_0^W \alpha_p \exp\left[\int_0^x (\alpha_n - \alpha_p)\, dx\right] dx\right\}^{-1} \qquad (3.6)$$

Substituting this expression in Eq. (3.5) gives :

$$M(x) = \frac{\exp[\int_0^x (\alpha_n - \alpha_p) \, dx]}{1 - \int_0^W \alpha_p \exp[\int_0^x (\alpha_n - \alpha_p) \, dx] \, dx} \qquad (3.7)$$

If the electric field distribution along the impact ionization path is known, this equation can be used to determine the total number of electron-hole pairs created as a result of the generation of a single electron-hole pair at any distance x from the junction. The term M is known as the *multiplication coefficient*. The avalanche breakdown condition occurs when the total number of electron-hole pairs generated within the depletion region approaches infinity, which also corresponds to M tending to infinity. From Eq. (3.7), it can be concluded that this condition occurs when :

$$\int_0^W \alpha_p \exp[\int_0^x (\alpha_n - \alpha_p) \, dx] \, dx = 1 \qquad (3.8)$$

The expression on the left-hand side of Eq. (3.8) is known as the *ionization integral*. In the case of the approximation given in Eq. (3.3) for the ionization coefficient, the avalanche breakdown condition corresponds to :

$$\int_0^W \alpha \, dx = 1 \qquad (3.9)$$

The analysis of the breakdown voltage of devices can be performed by evaluation of this ionization integral in either closed form or by using numerical techniques. However, it is important to emphasize that this solution corresponds to the case of the multiplication coefficient approaching infinity. In many power device structures, the current flowing through the depletion region is amplified by gain of internal transistors. In these cases, it is necessary to solve for the multiplication coefficient instead of using the ionization integral.

The multiplication coefficient can be calculated by using :

$$M_n = \frac{1}{[1 - (V/V_A)^4]} \qquad (3.10)$$

for the case of a N$^+$/P diode, and

$$M_p = \frac{1}{[1 - (V/V_A)^6]} \qquad (3.11)$$

for the case of a P$^+$/N diode. In these equations, V_A is the avalanche breakdown voltage and V is the applied reverse bias supported by the junction. These expressions for the multiplication coefficient are useful for analysis of the blocking voltage capability of devices containing internal open base transistors. Some examples of these devices are thyristors and IGBTs.

Chapter 3 : BREAKDOWN VOLTAGE

3.2 ABRUPT JUNCTION DIODE

The most basic P-N junction structure that can be analyzed is the parallel-plane (semi-infinite), abrupt junction diode in which the doping concentration on one side of the junction is very large when compared with the other side. Junctions fabricated with shallow, high concentration diffusions into homogeneous, lightly-doped substrates approach the behavior of an abrupt junction diode because the depletion region extends primarily on only one (the lightly-doped) side of the junction.

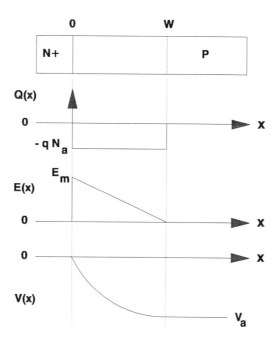

Fig. 3.3. Electric field and potential distribution for an abrupt P-N junction.

Consider the case of a parallel-plane, abrupt, N+/P junction illustrated in Fig. 3.3 with a reverse bias of V_a applied to the P region. Since the doping concentration in the N+ region is very large, the depletion region can be assumed to extend only in the P-region. In order to solve for the potential distribution, the Poisson's equation must be solved. The Poisson's equation in the P-region is given by :

$$\frac{d^2V}{dx^2} = -\frac{dE}{dx} = -\frac{Q(x)}{\epsilon_s} = \frac{qN_A}{\epsilon_s} \qquad (3.12)$$

where Q(x) is the charge in the depletion region due to the presence of ionized acceptors, ϵ_s is the dielectric constant of the semiconductor, q is the electron charge, and N_A is the acceptor concentration on the homogeneously doped P-region. An integration of Eq. (3.12) with the boundary condition that the electric field must go to zero at the edge (W) of the depletion region, provides a solution for the electric field distribution :

$$E(x) = \frac{q N_A}{\epsilon_s} (W - x) \qquad (3.13)$$

For the case of a homogeneously doped P-region, as illustrated in Fig. 3.3, the electric field decreases linearly from its maximum value at the junction. An integration of the electric field through the depletion region with the boundary condition that the potential is zero in the N+ region provides the potential distribution :

$$V(x) = \frac{q N_A}{\epsilon_s} (Wx - \frac{x^2}{2}) \qquad (3.14)$$

As shown in Fig. 3.3, the potential in the depletion region varies quadratically from 0 in the N+ region to V_a at the edge of the depletion region. The thickness of the depletion region (W) can be obtained by using the boundary condition $V = V_a$ at $x = W$:

$$W = \sqrt{\frac{2 \epsilon_s V_a}{q N_A}} \qquad (3.15)$$

From this expression, it can be seen that the depletion region width increases with increasing applied bias. Further, it is worth pointing out that the depletion region width is larger for junctions with lower doping concentration on the lightly doped side. Consequently, reducing the doping concentration allows the diode to support higher voltages.

As noted earlier, the maximum electric field occurs at $x = 0$ for the abrupt junction. By using Eq. (3.13) with Eq. (3.15), the maximum electric field can be related to the applied bias:

$$E_m = \sqrt{\frac{2 q N_A V_a}{\epsilon_s}} \qquad (3.16)$$

From this equation, it can be seen that the maximum electric field in the depletion region increases with increasing applied bias. When this field approaches values at which the ionization coefficients become large, the junction approaches the breakdown condition. From this equation, it can also be seen that, for any applied bias, the electric field is smaller for junctions with lower doping concentrations on the lightly doped side. Thus, the breakdown voltage can be increased by reducing the doping.

A closed form solution for the breakdown voltage of the abrupt junction diode can be obtained by using the electric field distribution defined by Eq. (3.13) in the ionization integral given by Eq. (3.9) with Eq. (3.3) to relate the ionization coefficient to the electric field :

Chapter 3 : BREAKDOWN VOLTAGE

$$\int_0^W 1.8 \times 10^{-35} \left[\frac{q\, N_A}{\epsilon_s} (W - x) \right]^7 dx = 1 \qquad (3.17)$$

The solution of this equation gives the critical depletion region width ($W_{c,PP}$) at the point of breakdown for the abrupt parallel plane junction:

$$W_{c,PP} = 2.67 \times 10^{10}\, N_A^{-7/8} \qquad (3.18)$$

The avalanche breakdown voltage for the abrupt, parallel-plane junction can be obtained by setting $x = W$ in Eq. (3.14) and then using $W = W_{c,PP}$:

$$BV_{PP} = 5.34 \times 10^{13}\, N_A^{-3/4} \qquad (3.19)$$

The breakdown voltage and the depletion region width obtained by using these expressions are plotted as a function of the doping concentration in Fig. 3.4.

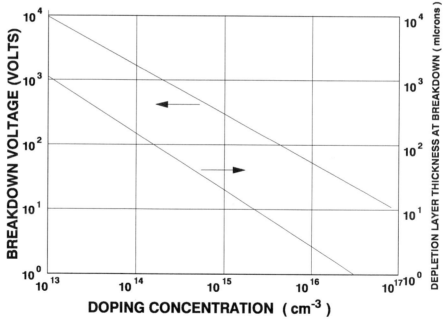

Fig. 3.4. Avalanche breakdown voltage and depletion region thickness at breakdown for an abrupt junction diode.

These equations and the graph in Fig. 3.4 can be used to determine the doping concentration and width of the lightly doped region for obtaining any desired breakdown voltage. As an example, for a breakdown voltage of 600 volts, it is necessary to use a doping concentration for the lightly doped region of 4×10^{14} per cm³, and its width should be 45 microns. From Fig. 3.4, it can be noted that high voltage power devices require drift regions with relatively low doping concentrations and large thicknesses.

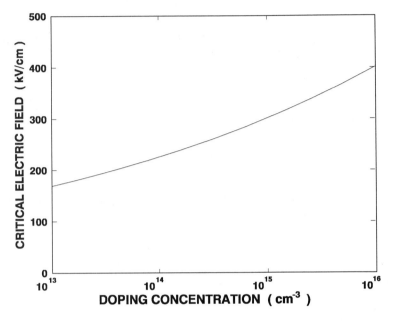

Fig. 3.5. Critical electric field for breakdown of abrupt parallel-plane junctions.

The maximum electric field under the breakdown condition is referred to as the *critical electric field*. By substituting the breakdown voltage Eq. (3.19) into Eq. (3.16), it can be shown that :

$$E_{c,PP} = 4010 \, N_A^{1/8} \qquad (3.20)$$

The critical electric field is a useful parameter for identifying the onset of avalanche breakdown within device structures. Whenever the local electric field approaches this value, avalanche breakdown can be expected. It is important to note that this provides only an indication of breakdown, and the exact breakdown voltage must be determined by performing the ionization integral through the depletion region. A plot of the critical electric field for silicon is shown in Fig. 3.5 as a function of the doping concentration. It can be seen that the critical electric field for breakdown is a weak function of the doping concentration and lies in the range of 1 to 4 x 10^5 V/cm.

The expressions derived in this section for the abrupt junction diode represent the upper bounds for breakdown within device structures. Consequently, they are useful normalization parameters for analysis of device structures and edge terminations to determine how close they come to the ideal value.

3.3 PUNCH-THROUGH DIODE

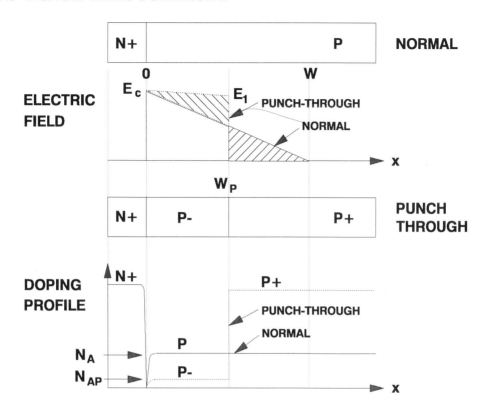

Fig. 3.6. Comparison of punch-through diode structure with abrupt junction diode.

In the case of the abrupt junction diode treated in section 3.2, it was assumed that the lightly doped region is thicker than the maximum depletion region thickness at breakdown. In some power devices, such as P-i-N rectifiers, it is preferable to use a punch-through structure to support the voltage. The punch-through structure is compared with the abrupt junction diode structure in Fig. 3.6. In general, the punch-through structure has a lower doping concentration on the lightly doped side with a high concentration contact region, and the thickness of the lightly doped side is smaller than that for the abrupt junction case. This alters the electric field distribution as illustrated in Fig. 3.6. In the case of the punch-through structure, the electric field varies more gradually with distance within the lightly doped region due to its lower doping concentration. The net result is a rectangular electric field profile for the punch-through case as compared with a triangular electric field profile for the abrupt junction case.

An expression for the breakdown voltage for the punch-through diode can be derived by using the critical electric field to define the onset of breakdown. As indicated in Fig. 3.6, the electric field has a maximum value at the N+/P junction and decreases linearly within the lightly doped region to a value of E_1 at the P-/P+ interface. The voltage supported by the depletion region, which has a thickness equal to the thickness (W_P) of the lightly doped (P-) region, is given by :

$$V_{PT} = \frac{1}{2} (E_c + E_1) W_P \qquad (3.21)$$

By using Eq. (3.13) for the variation in electric field with distance :

$$E_1 = E_c - \frac{q N_{AP}}{e_s} W_P \qquad (3.22)$$

Substituting this expression in Eq. (3.21), the breakdown voltage for the punch-through diode is obtained :

$$V_{PT} = E_c W_P - \frac{q N_{AP} W_P^2}{2 e_s} \qquad (3.23)$$

The breakdown voltages calculated for punch-through diodes with various lightly doped region thicknesses are shown in Fig. 3.7 as a function of the doping concentration in the lightly

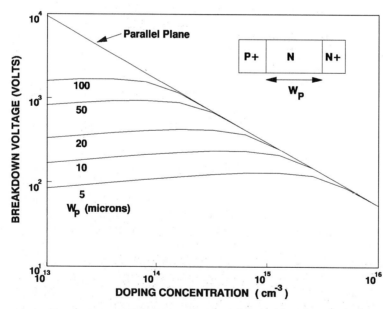

Fig. 3.7. Avalanche breakdown voltage for punch-through diodes.

doped region. Note that, whenever the doping concentration and the thickness become large, the breakdown voltage becomes equal to that for the abrupt junction case. Further, the breakdown

Chapter 3 : BREAKDOWN VOLTAGE

voltage of the punch-through diode is a weak function of the doping concentration in the lightly doped region if its thickness is small.

The most important feature of the punch-through structure is that its thickness W_P is smaller than that for the abrupt junction case for equal breakdown voltages. As an example, consider the case of a diode to be designed with a breakdown voltage of 600 volts. The abrupt junction design would require a lightly doped region with a doping concentration of 4×10^{14} per cm^3 and thickness of 45 microns, while the punch-through diode would require a lightly doped region with a thickness of only 25 microns for a doping concentration of 5×10^{13} per cm^3. A reduced thickness of the lightly doped region is preferable for bipolar power devices, such as P-i-N rectifiers operating under high level injection conditions, as discussed later.

3.4 LINEARLY GRADED DIODE

In the case of the abrupt junction diode, the doping concentration on the heavily doped side of the junction is assumed to be far greater than that on lightly doped side of the junction. Junctions fabricated by ion implantation and diffusion techniques with high surface concentration and shallow depth can be approximated by the abrupt junction analysis. However, in many devices, a significant fraction of the reverse bias voltage applied across the P-N junction is supported on the diffused side because the difference in concentration between the two sides of the junction is not very large. The breakdown voltage of a linearly graded junction is of interest for the analysis of diffused junctions with shallow concentration gradients.

In the case of a linearly graded junction, the doping concentration is assumed to vary linearly with distance from the N to the P side as illustrated in Fig. 3.8. The charge in the depletion region then increases linearly with distance on both sides of the junction:

$$Q(x) = -q G x \tag{3.24}$$

where G is called the *grade constant*. Note that the charge is negative on the P-side of the junction because of the presence of ionized acceptors and positive on the N-side of the junction because of the presence of ionized donors. In the case of the linearly graded junction, Poisson's equation on the P-side is given by:

$$\frac{d^2V}{dx^2} = -\frac{dE}{dx} = -\frac{Q(x)}{\varepsilon_s} = \frac{q G x}{\varepsilon_s} \tag{3.25}$$

An integration of Eq. (3.25) with the boundary condition that the electric field must go to zero at the edge (W) of the depletion region, provides a solution for the electric field distribution:

$$E(x) = \frac{q G}{2 \varepsilon_s} (W^2 - x^2) \tag{3.26}$$

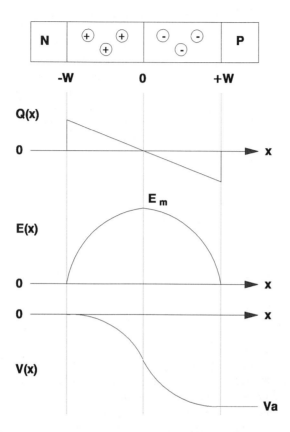

Fig. 3.8. Electric field and potential distribution for linearly graded junction.

As illustrated in Fig. 3.8, the electric field decreases parabolically with distance from its maximum value at the junction. An integration of the electric field through the depletion region with the boundary condition that the potential is zero at edge (-W) of the depletion region on the N-side of the junctions provides the potential distribution :

$$V(x) = \frac{q\,G}{e_s} \left(\frac{x^3}{6} - \frac{W^2\,x}{2} - \frac{W^3}{3} \right) \qquad (3.27)$$

As shown in Fig. 3.8, the potential in the depletion region is a cubic function of distance and varies from 0 at the edge of the depletion region on the N-side to V_a at the edge of the depletion region on the P-side. The thickness of the depletion region (W) can be obtained by using the boundary condition $V = V_a$ at $x = W$:

Chapter 3 : BREAKDOWN VOLTAGE

$$W = \left(\frac{3 \, \epsilon_s \, V_a}{q \, G} \right)^{1/3} \qquad (3.28)$$

Note that the depletion region extends equally on both sides of the junction to a thickness W. The extension of the depletion region on the diffused side of junctions must be properly accounted for in designing contacts and to avoid reach-through breakdown.

A closed form solution for the breakdown voltage of the linearly graded junction diode can be obtained by using the electric field distribution defined by Eq. (3.26) in the ionization integral given by Eq. (3.9) with Eq. (3.3) to relate the ionization coefficient to the electric field:

$$\int_{-W}^{W} 1.8 \times 10^{-35} \left[\frac{q \, G}{2 \, \epsilon_s} (W^2 - x^2) \right]^7 dx = 1 \qquad (3.29)$$

The solution of this equation gives the critical depletion region width ($W_{c,L}$) at the point of breakdown for the linearly graded, parallel plane junction:

$$W_{c,L} = 9.1 \times 10^5 \, G^{-7/15} \qquad (3.30)$$

The avalanche breakdown voltage for the linearly graded, parallel-plane junction can be obtained by setting x = W in Eq. (3.27) and then using W = $W_{c,L}$:

$$BV_{LPP} = 9.2 \times 10^9 \, G^{-2/5} \qquad (3.31)$$

A graded diffusion profile across the P-N junction enhances the breakdown voltage because of the extension of the depletion region into the diffused side. This behavior can be understood from the above analysis of the linearly graded junction.

3.5 DIFFUSED JUNCTION DIODE

Power semiconductor devices are fabricated by diffusion of dopants introduced into the semiconductor surface by either gas phase transport or by ion-implantation. When the dopant is diffused under constant surface concentration conditions, a complementary error function doping profile is observed while a Gaussian distribution profile is observed for the case of ion-implanted junctions with constant dopant charge. Both dopant profiles have an exponential variation with depth. This results in a net impurity density that varies gradually with distance at the metallurgical junction in a manner similar to that for linearly graded junctions. However, deeper in the semiconductor, the compensation of the background impurity by the diffused dopant becomes negligible and the dopant concentration is essentially uniform with depth. Under reverse bias conditions, the depletion extends in the linearly graded region at low bias values as

80   Chapter 3 : BREAKDOWN VOLTAGE

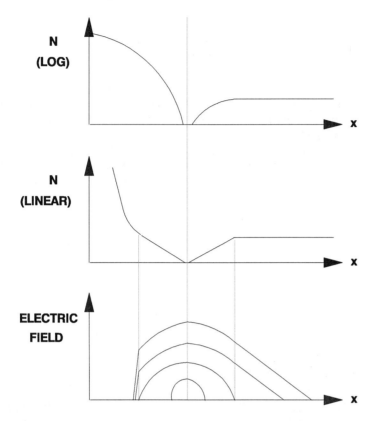

Fig. 3.9. Doping and electric profile for a diffused junction.

illustrated in Fig. 3.9. However, as the applied bias is increased, the depletion region extends into the uniformly doped region on the lightly doped side. It can be seen that although most of the applied voltage is supported on the lightly doped side, the breakdown voltage of a diffused junction will be larger than that for the abrupt junction because an additional voltage is supported on the diffused side.

The breakdown voltage of diffused, parallel-plane junctions can be computed by using numerical analysis. In order to obtain a quantitative evaluation of the benefits of grading the diffusion profile, the breakdown voltages for the cases of a high surface concentration, shallow diffusion are compared in Fig. 3.10 with those for a low surface concentration, deep diffusion. For the low surface concentration, deep diffusion, the grading of the junction becomes significant at higher background concentrations. Consequently, its breakdown voltage is larger than that for the high surface concentration, shallow junction. As an example, an increase in breakdown voltage by about 30 percent is observed for a background doping concentration of 1×10^{16} per cm^3. This increase in the breakdown voltage is of significance for power MOSFETs designed to operate at low blocking voltages.

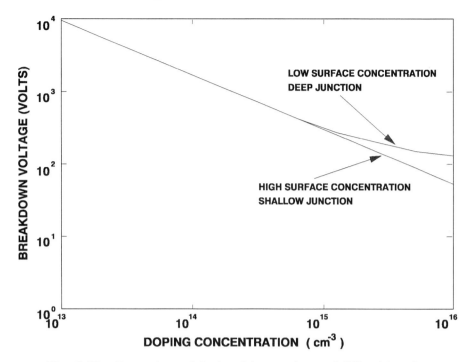

Fig. 3.10. Comparison of the breakdown voltage of diffused junctions.

3.6 EDGE TERMINATIONS

The analysis of a parallel-plane (semi-infinite) junction provides an upper bound for the breakdown voltage in power devices. The breakdown voltage of practical devices can be reduced by the occurrence of high electric fields either within the interior portion of the device structure or at the edges of the device. The design of the interior cells is discussed in later sections for each device structure. In this section, methods for terminating the device are described with an analysis of the breakdown voltage that can be obtained for each edge termination. It is worth emphasizing that proper design of the edge termination is crucial not only for obtaining a high breakdown voltage but also for optimization of the drift region in order to reduce the on-state voltage drop and the switching times.

Although many specific edge termination designs have been discussed in the literature, they can be classified into two basic types, namely, the *planar terminations* based upon masked diffusion processes and *beveled terminations* based upon the selective removal of material from the vicinity of the junction. In both cases, a floating field plate has been found to be useful in enhancing the breakdown voltage. The design of these terminations is usually accomplished by two-dimensional numerical analysis. However, in order to provide physical insight into the mechanism responsible for limiting the breakdown voltage, analytical expressions are derived

82   Chapter 3 : BREAKDOWN VOLTAGE

in this section based upon simplifying assumptions.

3.6.1 Planar Junction Terminations

Planar diffusion technology is widely used for the fabrication of power devices and integrated circuits. It is based upon the selective introduction of dopants into the semiconductor surface either through the vapor phase using an oxide as the mask or by ion implantation of the dopant using photoresist as a mask. The junction is then formed by the diffusion of the dopant into the semiconductor. With this method, a parallel-plane junction is formed within the diffusion window. However, the dopant diffuses laterally at the edges of the diffusion window. The lateral diffusion has been found to extend to about 85 percent of the vertical depth. For purposes of analytical solution, it can be assumed that the lateral diffusion is equal to the vertical diffusion.

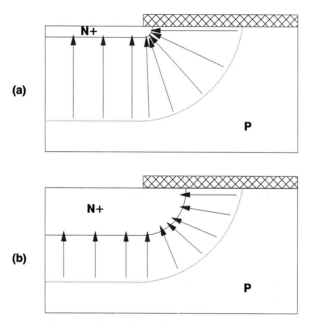

Fig. 3.11 Electric field crowding for (a) shallow and (b) deep planar diffused junctions.

When a planar diffusion is performed with a rectangular diffusion window, a cylindrical junction is formed at the straight edges of the diffusion window and a spherical junction is formed at each of the four corners. A cross-section through the junction is illustrated in Fig. 3.11. When a reverse bias is applied across the junction, the depletion region contour follows the edge of the junction as indicated by the dashed lines. In the interior portion of the junction, the electric field lines are uniform and the breakdown voltage is equal to that for the parallel-plane (semi-infinite) junction treated in previous sections. However, the electric field lines become crowded at the edge as shown in the figure. Note that the field crowding is worse

Chapter 3 : BREAKDOWN VOLTAGE

for the case of junction depths that are small *in comparison with* the depletion layer thickness. The electric field crowding is synonymous with a local increase in the electric field which results in a reduction of the breakdown voltage to a value below that for the parallel-plane junction. The increase in the electric field is greater for the case of the spherical junction when compared with the cylindrical junction as discussed below. Consequently, it is essential to avoid the formation of sharp corners in diffusion windows when designing power devices.

3.6.1.1 Cylindrical Junction: A cross-section of the cylindrical junction formed at the straight edges of a diffusion window used to fabricate planar junctions is illustrated in Fig. 3.12 for the case of a very high doping concentration on the diffused side of the junction and a uniform doping concentration on the lightly doped side of the junction. The cylindrical junction has a

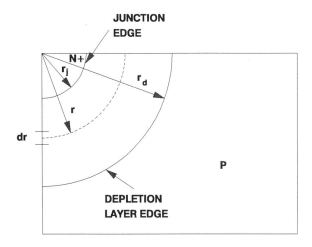

Fig. 3.12. Cross-section of a cylindrical junction.

depth r_j and the depletion region extends into the lightly doped side to a radius r_d. In order to solve for the breakdown voltage, it is necessary to use Poisson's equation in cylindrical coordinates :

$$\frac{1}{r}\frac{d}{dr}\left(r\frac{dV}{dr}\right) = -\frac{1}{r}\frac{d}{dr}(r\,E) = -\frac{Q(r)}{\epsilon_s} = \frac{qN_A}{\epsilon_s} \qquad (3.32)$$

where the potential distribution V(r) and the electric field distribution E(r) are defined along a radius vector r extending into the depletion region. Integration of this equation with the boundary condition that the electric field is zero at the depletion region boundary (r_d) gives :

$$E(r) = \frac{q\,N_A}{2\,\epsilon_s}\left(\frac{r_d^2 - r^2}{r}\right) \qquad (3.33)$$

The maximum electric field occurs at the metallurgical junction (r_j) :

$$E_m(r_j) = \frac{q N_A}{2 e_s}\left(\frac{r_d^2 - r_j^2}{r_j}\right) \qquad (3.34)$$

The maximum electric field for the cylindrical junction can be compared with that for the parallel-plane junction by considering the case of a junction depth much smaller than the depletion region thickness :

$$E_{m,CYL}(r_j) \approx \frac{q N_A}{2 e_s}\frac{r_d^2}{r_j} \qquad (3.35)$$

while for the parallel-plane portion with the same depletion region thickness :

$$E_{m,PP} = \frac{q N_A}{e_s} r_d \qquad (3.36)$$

By taking the ratio of these expressions :

$$\frac{E_{m,CYL}}{E_{m,PP}} \approx \frac{r_d}{2 r_j} \qquad (3.37)$$

This equation indicates that, for shallow junctions with small radii of curvature, the maximum electric field at the cylindrical edge is much larger than for the parallel plane portion. As an example, if the junction depth is 1 micron and the depletion region thickness is 50 microns, the maximum electric field at the cylindrical edge is 25 times larger than that for the parallel-plane junction. The high electric field at the cylindrical junction will result in avalanche breakdown being confined to the edges of the window because the impact ionization coefficient is a very strong function of the electric field.

The potential distribution at the cylindrical junction can be obtained by integration of the electric field distribution :

$$V(r) = \frac{q N_A}{2 e_s}\left[\left(\frac{r_j^2 - r^2}{2}\right) + r_d^2 \ln\left(\frac{r}{r_j}\right)\right] \qquad (3.38)$$

By using the boundary condition that the potential is zero on the heavily doped side and V_a on the lightly doped side of the junction, the depletion region width (r_d) can be obtained. In order to solve for the breakdown voltage, the ionization integral must be evaluated with the electric field distribution given by Eq. (3.33). A closed form analytical solution for the ionization integral can be derived by making the assumption that the electric field varies inversely with distance :

$$E(r) \approx \frac{q N_A}{2 e_s}\frac{r_d^2}{r} = \frac{K}{r} \qquad (3.39)$$

Chapter 3 : BREAKDOWN VOLTAGE

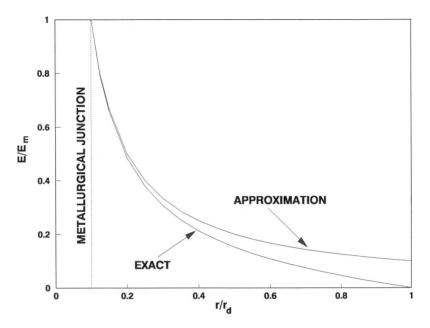

Fig. 3.13. Comparison of the hyperbolic approximation of electric field distribution with actual distribution for the cylindrical junction.

The hyperbolic electric field distribution given by Eq. (3.39) is compared with the actual field distribution given by Eq. (3.33) in Fig. 3.13 for the case of $r_j = 0.1\, r_d$. Note that the two field distributions are similar in the high electric field region at the metallurgical junction. However, the hyperbolic distribution implies that the electric field extends to infinity. Consequently, the ionization integral must also be evaluated from r_j to infinity when using this approximation. Using the hyperbolic field distribution with the ionization coefficient given by Eq. (3.3), a solution for the breakdown condition of the cylindrical junction is obtained :

$$K = \left(\frac{6\, r_j^6}{1.85 \times 10^{-35}} \right)^{1/7} \qquad (3.40)$$

Substituting this expression into Eq. (3.35) for the maximum electric field, provides the critical electric field at breakdown for the cylindrical junction:

$$E_{c,CYL} = \left(\frac{3.25 \times 10^{35}}{r_j} \right)^{1/7} \qquad (3.41)$$

A comparison between the maximum electric field at breakdown for the cylindrical junction with that for the parallel-plane case is obtained by taking the ratio of Eq. (3.41) and (3.20), with

Eq. (3.18) to eliminate N_A:

$$\frac{E_{c,CYL}}{E_{c,PP}} = \left(\frac{3}{4}\frac{W_{c,PP}}{r_j}\right)^{1/7} \tag{3.42}$$

From this expression, it can be concluded that the maximum electric field at breakdown for the cylindrical junction is larger than that for the parallel-plane junction with the same doping level on the lightly doped side. However, the breakdown voltage for the cylindrical junction is lower than that for the parallel-plane case. In order to obtain the breakdown voltage, the depletion layer thickness (r_d) at breakdown for the cylindrical junction must first be obtained by using Eq. (3.41) and Eq. (3.34). Using this value for r_d in Eq. (3.38), the breakdown voltage for the cylindrical junction is obtained.

In order to simplify analysis, it has been found that normalization of the breakdown voltage for the cylindrical junction with that for the parallel-plane case allows the derivation of an expression that is valid for all doping concentrations on the lightly doped side. This expression can be derived by elimination of the doping concentration (N_A) from the equations and expressing the breakdown voltage in terms of the ratio of the radius of curvature (r_j) and the depletion region thickness at breakdown for the parallel-plane case (W_c):

$$\frac{BV_{CYL}}{BV_{PP}} = \frac{1}{2}\left[\left(\frac{r_j}{W_c}\right)^2 + 2\left(\frac{r_j}{W_c}\right)^{6/7}\right] \ln\left[1 + 2\left(\frac{W_c}{r_j}\right)^{8/7}\right] - \left(\frac{r_j}{W_c}\right)^{6/7} \tag{3.43}$$

Due to the assumptions made during the derivation of this expression, it is valid only when the junction depth or radius of curvature is smaller than the depletion region thickness at breakdown for the parallel-plane case. For these cases, which are the most often encountered in practical devices, this expression has been found to be in good agreement with the results obtained by two-dimensional numerical analysis.

A plot of the normalized breakdown voltage for cylindrical junctions is provided in Fig. 3.14 as a function of the normalized radius of curvature. It can be seen that the breakdown voltage increases as the normalized radius of curvature increases. This behavior is consistent with the electric field crowding discussed earlier with reference to Fig. 3.11. From this behavior, it can be concluded that shallower junctions can be used for devices with high background doping levels (low breakdown voltage designs) than for devices being designed for high breakdown voltages. As an example, if the junction depth is 3 microns and the background doping level is 1×10^{15} per cm^3 (depletion region thickness at breakdown of 20 microns), the normalized breakdown voltage is about 0.4. In contrast, if the background doping level is 4×10^{14} per cm^3 (depletion region thickness at breakdown of 60 microns), the normalized breakdown voltage is reduced to only 0.25. In general, it is difficult to obtain breakdown voltages above 50 percent of the parallel-plane case with a cylindrical edge termination. It is necessary to incorporate floating field rings and guard rings to address this problem as discussed later.

3.6.1.2 Spherical Junction: The cross-section of the spherical junction formed at the sharp corners of a diffusion window used to fabricate planar junctions is identical to that for the

Chapter 3 : BREAKDOWN VOLTAGE   87

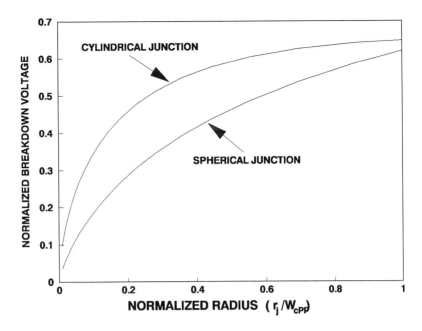

Fig. 3.14. Normalized breakdown voltage for cylindrical and spherical junctions.

cylindrical junction as illustrated in Fig. 3.12 for the case of a very high doping concentration on the diffused side of the junction and a uniform doping concentration on the lightly doped side of the junction. However, the electric field crowding can be expected to be worse than that for the cylindrical junction because the electric field lines approach a point from three dimensions in the spherical junction while they approach a line from two dimensions in the case of the cylindrical junction.

Consider a spherical junction with a depth r_j and the depletion region extending into the lightly doped side to a radius r_d. In order to solve for the breakdown voltage, it is necessary to use Poisson's equation in spherical coordinates :

$$\frac{1}{r^2}\frac{d}{dr}\left(r^2 \frac{dV}{dr}\right) = -\frac{1}{r^2}\frac{d}{dr}(r^2 E) = -\frac{Q(r)}{e_s} = \frac{qN_A}{e_s} \qquad (3.44)$$

Integration of this equation with the boundary condition that the electric field is zero at the depletion region boundary (r_d) gives :

$$E(r) = \frac{q N_A}{3 e_s}\left(\frac{r_d^3 - r^3}{r^2}\right) \qquad (3.45)$$

The maximum electric field occurs at the metallurgical junction (r_j) :

$$E_m(r_j) = \frac{q N_A}{3 \epsilon_s} \left(\frac{r_d^3 - r_j^3}{r_j^2} \right) \qquad (3.46)$$

For the same radius of curvature and background doping concentration, the maximum electric field for the spherical junction is even larger than that for the cylindrical junction. This can be demonstrated by considering the case of a junction depth much smaller than the depletion region thickness:

$$E_{m,SP}(r_j) \approx \frac{q N_A}{3 \epsilon_s} \frac{r_d^3}{r_j^2} \qquad (3.47)$$

By taking the ratio of Eq. (3.47) and (3.35):

$$\frac{E_{m,SP}}{E_{m,CYL}} \approx \frac{2}{3} \frac{r_d}{r_j} \qquad (3.48)$$

This equation indicates that, for shallow junctions with small radii of curvature, the maximum electric field at the spherical junctions formed at the corners is much larger than that at the cylindrical junction formed at the straight edges. As an example, if the junction depth is 1 micron and the depletion region thickness is 50 microns, the maximum electric field at the spherical corner is 33 times larger than that for the cylindrical edge. The high electric field at the spherical corners will result in avalanche breakdown being confined to the sharp corners of the diffusion window because the impact ionization coefficient is a very strong function of the electric field.

The potential distribution at the spherical junction can be obtained by integration of the electric field distribution:

$$V(r) = \frac{q N_A}{3 \epsilon_s} \left[\left(\frac{r_j^2 - r^2}{2} \right) + r_d^3 \left(\frac{1}{r_j} - \frac{1}{r} \right) \right] \qquad (3.49)$$

By using the boundary condition that the potential is zero on the heavily doped side and V_a on the lightly doped side of the junction, the depletion region width (r_d) can be obtained. In order to solve for the breakdown voltage, the ionization integral must be evaluated with the electric field distribution given by Eq. (3.45). A closed form analytical solution for the ionization integral can be derived by making the assumption that the electric field varies inversely as the square of the distance:

$$E(r) \approx \frac{q N_A}{3 \epsilon_s} \frac{r_d^3}{r^2} = \frac{K_s}{r^2} \qquad (3.50)$$

The electric field distribution given by Eq. (3.50) is compared with the actual field distribution

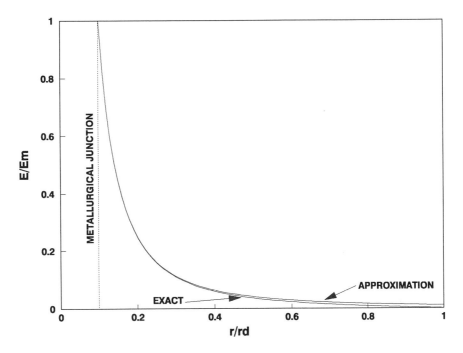

Fig. 3.15. Comparison of the approximation of electric field distribution with actual distribution for the spherical junction.

given by Eq. (3.45) in Fig. 3.15 for the case of $r_j = 0.1\, r_d$. Note that the two field distributions are similar in the high electric field region at the metallurgical junction. However, the distribution given by Eq. (3.50) implies that the electric field extends to infinity. Consequently, the ionization integral must also be evaluated from r_j to infinity when using this approximation. Using this field distribution with the ionization coefficient given by Eq. (3.3), a solution for the depletion region thickness (radius r_d) at the breakdown condition of the spherical junction is obtained. Substituting this expression into Eq. (3.46) for the maximum electric field, provides the critical electric field at breakdown for the spherical junction:

$$E_{c,SP} = \left(\frac{7 \times 10^{35}}{r_j}\right)^{1/7} \quad (3.51)$$

A comparison between the maximum electric field at breakdown for the spherical junction with that for the parallel-plane case is obtained by taking the ratio of Eq. (3.51) and (3.20), with Eq. (3.18) to eliminate N_A:

$$\frac{E_{c,SP}}{E_{c,PP}} = \left(\frac{13\, W_{c,PP}}{8\, r_j}\right)^{1/7} \quad (3.52)$$

From this expression, it can be concluded that the maximum electric field at breakdown for the spherical junction is larger than that for the parallel-plane junction with the same doping level on the lightly doped side. However, the breakdown voltage for the spherical junction is lower than that for the parallel-plane case. In order to obtain the breakdown voltage, the depletion layer thickness (r_d) at breakdown for the spherical junction must first be obtained by using Eq. (3.51) and Eq. (3.46). Using this value for r_d in Eq. (3.49), the breakdown voltage for the spherical junction is obtained.

In order to simplify analysis, it has been found that normalization of the breakdown voltage for the spherical junction with that for the parallel-plane case allows the derivation of a single expression that is valid for all doping concentrations on the lightly doped side. This expression can be derived by elimination of the doping concentration (N_A) from the equations and expressing the breakdown voltage in terms of the ratio of the radius of curvature (r_j) and the depletion region thickness at breakdown for the parallel-plane case (W_c):

$$\frac{BV_{SP}}{BV_{PP}} = \left(\frac{r_j}{W_c}\right)^2 + 2.14\left(\frac{r_j}{W_c}\right)^{6/7} - \left[\left(\frac{r_j}{W_c}\right)^3 + 3\left(\frac{r_j}{W_c}\right)^{13/7}\right]^{2/3} \quad (3.53)$$

Due to the assumptions made during the derivation of this expression, it is valid only when the junction depth or radius of curvature is smaller than the depletion region thickness at breakdown for the parallel-plane case. For these cases, which are the most often encountered in practical devices, this expression has been found to be in good agreement with the results obtained by two-dimensional numerical analysis.

A plot of the normalized breakdown voltage for spherical junctions as a function of the normalized radius of curvature is included in Fig. 3.14 for comparison with cylindrical junctions. The reduction in the breakdown voltage with smaller radius of curvature is consistent with the electric field crowding discussed earlier in reference to Fig. 3.11. From this behavior, it can be concluded that shallower junctions can be used for devices with high background doping levels (low breakdown voltage designs) than for devices being designed for high breakdown voltages. As an example, if the junction depth is 3 microns and the background doping level is 1×10^{15} per cm^3 (depletion region thickness at breakdown of 20 microns), the normalized breakdown voltage is about 0.2. In contrast, if the background doping level is 4×10^{14} per cm^3 (depletion region thickness at breakdown of 60 microns), the normalized breakdown voltage is reduced to only 0.1. In general, it is difficult to obtain breakdown voltages above 25 percent of the parallel-plane case with a spherical edge termination. This breakdown voltage is approximately half that at the cylindrical junction formed at the straight edges of the diffusion window. It is necessary to round the corners of all diffusion windows at the edges of power devices to address this problem. It has been shown that a radius of two times the maximum depletion thickness at breakdown for the parallel-plane junction is sufficient to increase the breakdown voltage at the corners so that they approach the breakdown voltage at the straight edges. Since the depletion region thickness is relatively small (usually less than 100 microns), the loss in area at the corners is not a major issue during device design.

Chapter 3 : BREAKDOWN VOLTAGE

3.6.2 Floating Field Ring Terminations

It has been shown in the previous section that power devices made using shallow

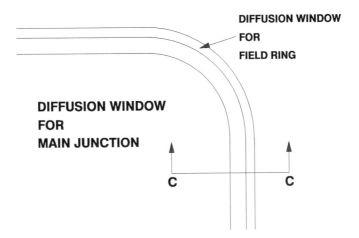

Fig. 3.16. Top view of diffusion window used to fabricate a junction with single floating field ring.

junctions with planar edge termination would suffer from a much lower breakdown voltage than the intrinsic capability of the lightly doped region. The low breakdown voltage has been found to be associated with electric field crowding at the cylindrical junction formed at the straight edges. An elegant approach to reducing the electric field at the edges is by the placement of floating field rings. The term *floating field ring* refers to the absence of any electrical contact to these rings. Their potential is instead determined by the bias applied to the main junction. When the floating field ring is placed within a depletion region thickness from the main junction, its potential will lie at an intermediate value between the applied bias (V_a) and zero. The objective is to extend the depletion boundary along the surface and reduce the electric field at the cylindrical junction.

In general, the depth of the floating field ring can be greater or smaller than the depth of the main junction. However, floating field rings are almost invariably fabricated simultaneously with the main junction by simply including a diffusion window surrounding the main junction as illustrated in Fig. 3.16. This allows the fabrication of the floating field ring with no additional processing steps. Note that a uniform spacing is maintained between the main junction and the floating field ring at all locations around the main junction. A cross-section through the edge of the device is illustrated in Fig. 3.17 together with electric field lines. Since some of the electric field lines are terminated by the floating field ring, the electric field crowding observed in Fig. 3.11 for the cylindrical junction can be alleviated.

The potential of the floating field ring can be related to the bias applied to the main junction by making the assumption that the floating field ring does not perturb the depletion extension from the main junction. Under this assumption, the variation in the potential with

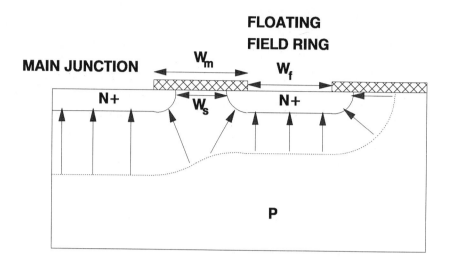

Fig. 3.17. Cross-section of edge termination with a single floating field ring.

distance is given by Eq. (3.14). For the case of a floating field ring spacing of W_s and a depletion region thickness of W_d, the floating field ring potential is given by:

$$V_{ffr} = \frac{q N_A}{\epsilon_s} \left(W_d W_s - \frac{W_s^2}{2} \right) \qquad (3.54)$$

Combining this relationship with Eq. (3.15) between W_d and the applied bias V_a:

$$V_{ffr} = \sqrt{\frac{2 q N_A}{\epsilon_s} W_s^2 V_a + \frac{q N_A}{2 \epsilon_s} W_s^2} \qquad (3.55)$$

It can be seen from this expression that the potential of the floating field ring varies as the square root of the applied bias to the main junction.

Although the floating field ring can be intuitively expected to increase the breakdown voltage of the cylindrical junction, its effectiveness depends upon its exact spacing from the main junction. If the floating field ring is very far away from the main junction, the depletion layer will not extend to it and the breakdown voltage will be the same as that for the cylindrical junction without the field ring. If the floating field ring is placed very close to the main junction, its potential will be essentially equal to that of the main junction. Consequently, the breakdown voltage will be limited by a high electric field at the floating field ring junction. This argument indicates that the breakdown voltage will exhibit a maximum value corresponding to an optimum floating field ring position as illustrated in Fig. 3.18. The enhancement in breakdown voltage and the optimum floating field ring position can be analytically calculated as discussed below.

An analytical solution for the breakdown voltage of a cylindrical junction with an

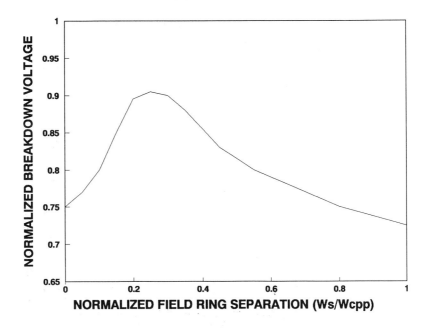

Fig. 3.18. Effect of floating field ring position upon the breakdown voltage.

optimally placed single floating field ring termination has been derived under the following assumptions : (1) The electric field at the edge of the floating field ring is the same as that for a cylindrical junction; (2) Under the breakdown condition, the electric field at the edge of the floating field ring and the main junction simultaneously become equal to the critical breakdown field for the cylindrical junction; and (3) The electric field at the main junction is determined by the voltage difference between the main junction and the floating field ring potential. Based upon the first assumption, the potential of the floating field ring at breakdown (V_F) must be equal to that for the cylindrical junction at breakdown :

$$(V_F / BV_{PP}) = (BV_{CY} / BV_{PP}) \qquad (3.56)$$

The ratio on the right hand side can be calculated by using Eq. (3.43). The voltage difference between the main junction and the floating field ring can be related by using Eq. (3.38) :

$$(V_M - V_F) = \frac{q\,N_A}{2\,\epsilon_s}\left[\left(\frac{W_s^2 - r_d^2}{2}\right) + r_d^2\,\ln\left(\frac{r_d}{r_j}\right)\right] \qquad (3.57)$$

Under the breakdown condition, the main junction potential (V_M) is equal to the breakdown voltage with an optimally spaced floating field (BV_{FFR}). By making use of Eq. (3.35) and (3.42), and normalizing to the breakdown voltage of the parallel plane junction given by :

$$BV_{PP} = \frac{1}{2} E_{cPP} W_c = \frac{q N_A}{2 e_s} W_c^2 \qquad (3.58)$$

it can be shown that:

$$\left[\frac{BV_{FFR} - V_F}{BV_{PP}}\right] = \frac{1}{2}\left(\frac{r_j}{W_c}\right)^2 - 0.96\left(\frac{r_j}{W_c}\right)^{6/7}$$

$$+ 1.92\left(\frac{r_j}{W_c}\right)^{6/7} \ln\left[1.386\left(\frac{W_c}{r_j}\right)^{4/7}\right] \qquad (3.59)$$

Since the floating field ring potential at breakdown is given by Eq. (3.56), the breakdown voltage of the optimum floating field ring termination can be calculated by using this equation. Due to the assumptions made during the derivation of this equation, it is valid only for small values of the ratio (r_j/W_c), which is applicable for most junctions encountered in power devices. A comparison of the breakdown voltage for the optimum floating field ring termination is made with the cylindrical junction in Fig. 3.19. It is important to note that the breakdown voltage with

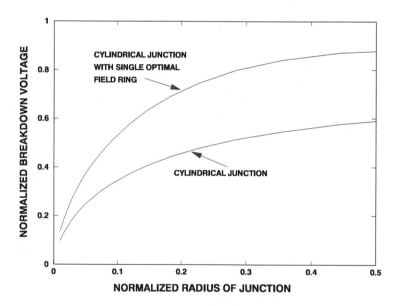

Fig. 3.19. Comparison of the breakdown voltage of an optimum single floating field ring design with that for a cylindrical junction.

the single optimum floating field ring is about twice that for the cylindrical junction. As an example, if the junction depth is 3 microns and the background doping level is 1×10^{15} per cm^3

(depletion region thickness at breakdown of 20 microns), the normalized breakdown voltage is about 0.75 as compared with 0.4 for the cylindrical junction. Thus, the inclusion of a floating field ring during device edge termination design can be used to enhance the breakdown voltage significantly without any additional processing steps during device fabrication.

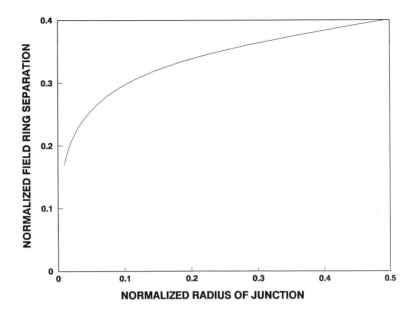

Fig. 3.20. Optimum floating field ring spacing for a single floating field ring design.

In order to obtain the improvement in breakdown voltage, it is necessary to determine the precise location of the floating field ring. As discussed earlier in conjunction with Fig. 3.18, a non-optimum spacing between the main junction and the floating field ring may not result in any increase in the breakdown voltage. An analytical solution for the optimum spacing between the main junction and the floating field ring can be derived by using Eq. (3.57) to obtain W_s for the case when V_F is equal to BV_{CY}. Using Eq. (3.58) to eliminate N_A:

$$W_s^2 - 2\sqrt{\frac{BV_{FFR}}{BV_{PP}}} W_c W_s + \left(\frac{BV_{CY}}{BV_{PP}}\right) W_c^2 = 0 \tag{3.60}$$

This quadratic equation has an elegant solution expressed in terms of the normalized breakdown voltages of the cylindrical and floating field ring terminations. In order to obtain this solution, the optimum spacing is normalized to the depletion thickness at breakdown for the parallel plane junction:

$$\frac{W_s}{W_c} = \sqrt{\frac{BV_{FFR}}{BV_{PP}} - \sqrt{\left(\frac{BV_{FFR}}{BV_{PP}}\right) - \left(\frac{BV_{CY}}{BV_{PP}}\right)}} \qquad (3.61)$$

The calculation of the optimum floating field ring position is most conveniently performed by using the appropriate value for the ratio (r_j/W_c). For this reason, a plot of the normalized optimum floating field ring spacing is provided in Fig. 3.20 as a function of (r_j/W_c). It is worth noting that the optimum normalized spacing is approximately 0.25. As an example, if the junction depth is 3 microns and the background doping level is 1×10^{15} per cm^3 (depletion region thickness at breakdown of 20 microns), the optimum spacing is 5 microns. This spacing can be easily provided with modern power device design rules. It is important to note that the design of the spacing between the diffusion windows in the mask must take into account the lateral diffusion from both the main junction and the floating field ring. Consequently, for the above example, the spacing of windows on the mask should be about 10 microns if the lateral diffusion is assumed to be 85% of the vertical depth and no photolithographic or etching biases are included.

The optimal spacing plotted in Fig. 3.20 was calculated under the assumption that there is no charge at the surface of the semiconductor over the lightly doped region. The presence of surface charge has a strong influence upon the depletion layer spreading at the surface because this charge complements the charge due to the ionized acceptors inside the depletion layer. In order to have a significant influence upon the depletion region, the surface charge must be comparable in magnitude to the charge within the depletion layer underlying the surface. This corresponds to a charge of about 1×10^{12} ions per cm^2. The alteration of the depletion layer shape for the case of positive, zero, and negative surface charge is illustrated in Fig. 3.21 for a planar junction. A positive surface charge causes the depletion layer at the surface of the lightly doped side to extend further, while negative charge will tend to retard the spreading of the depletion layer. The opposite effect will apply to a junction with a lightly doped N-type region. For the cylindrical junction illustrated in Fig. 3.21, the presence of positive surface charge will cause a decrease in the electric field crowding and result in raising the breakdown voltage while negative surface charge will have the opposite effect. Thus, positive charge is beneficial for the simple planar junction with a lightly doped P-region.

In the case of junctions with floating field rings, surface charge of either polarity can lower the breakdown voltage because it affects the reach-through voltage to the floating field ring. In the case of a P-type lightly doped region, the reach-through voltage will become smaller in the presence of a positive charge. This indicates that the optimum spacing should be increased from the values calculated by Eq. (3.61). Since the breakdown voltage distribution is sharply peaked around the optimum spacing as shown in Fig. 3.18, the presence of surface charge can alter the breakdown voltage by a factor of up to 2 times. If the surface charge is precisely known, it is possible to analyze for the optimal floating field ring location including the effect of this charge and then design the mask spacing W_m. In practice, the charge at the surface of thermally oxidized silicon is positive and can vary from 10^{10} to 10^{12} per cm^2. Even with a well controlled, clean fabrication sequence for a complex device such as a power MOSFET, the surface charge will vary from wafer to wafer and even across a wafer by 1×10^{11} per cm^2. This

Chapter 3 : BREAKDOWN VOLTAGE

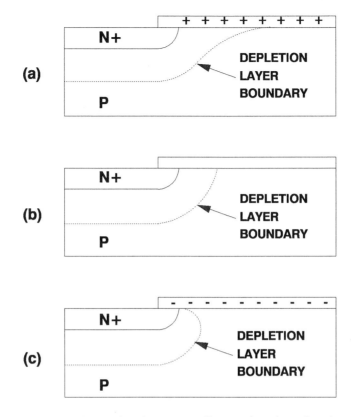

Fig. 3.21. Alteration of depletion layer spreading at the edge of a planar junction due to: (a) positive charge, (b) zero charge, and (c) negative charge.

represents an inherent practical limitation to designing the breakdown of devices using single floating field rings.

In addition to the optimum spacing of floating field ring, another design parameter that is important is the width W_f of the window through which the floating field ring is diffused into the semiconductor. When the width of the window W_f is very small, the floating field ring can become ineffective in reducing the depletion layer curvature even when it is at the optimum location. The effect of the size of the diffusion window upon the depletion region is illustrated in Fig. 3.22 (a) and (b). In order to be fully effective in raising the breakdown voltage, it is necessary to make the width (W_f) of the floating field ring comparable to the depletion layer width at breakdown (W_c). Making the floating field ring width much larger than the depletion layer width is not recommended because no improvement in the breakdown voltage occurs and it results in wasting space at the edge of chip. The floating field ring concept, with its extensions using multiple rings or field plates, represents the most widespread device termination technique used in modern power devices. Its optimal design is particularly important for achieving the high performance ratings that have been reported for power MOSFETs.

98   Chapter 3 : BREAKDOWN VOLTAGE

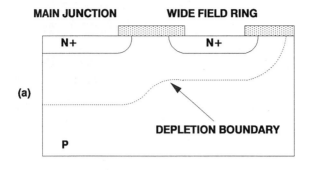

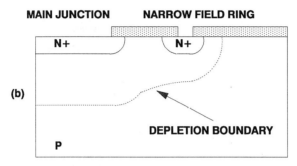

Fig. 3.22. Change in depletion layer spreading with width of floating field ring.

3.6.3 Multiple Floating Field Ring Terminations

A single field ring reduces depletion layer curvature, reduces the electric field crowding, and enhances the breakdown voltage. An even greater improvement in the breakdown voltage can be obtained by using multiple floating field rings working in conjunction with each other. As in the case of the single floating field ring, it is desirable to fabricate the multiple floating field rings simultaneously with the main junction by designing the mask with multiple windows surrounding the main junction.

Two basic philosophies have been used for the design of the multiple floating field ring termination. In the first case, illustrated in Fig. 3.23, the spacings between individual floating field rings are varied together with their widths. In this approach, both the floating field ring spacing and the width are reduced with increasing distance from the main junction because the depletion layer thickness below them gets progressively smaller. A gradual extension of the depletion layer occurs away from the main junction. The space occupied by the edge termination at the device periphery is reduced by decreasing the spacing and width of the floating field rings. An underlying assumption in this approach is that the surface charge is precisely known, allowing the design of the field ring spacings using this charge. In an optimal design, these field rings will all share the applied voltage in such a manner that avalanche breakdown occurs

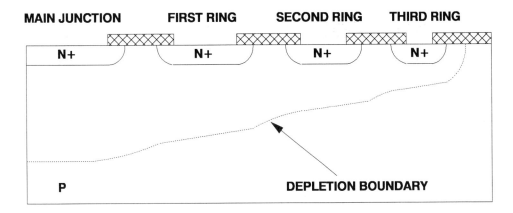

Fig. 3.23. Multiple field ring termination design with gradually decreasing field ring width and spacing.

simultaneously at the outer edges of all the floating field rings. In the case of a junction with a P-type lightly doped side, if the surface charge in the actual device is more positive than that assumed during device design, the inner field rings can become ineffective transferring all the voltage to the outer field rings resulting in premature breakdown at the outer field ring.

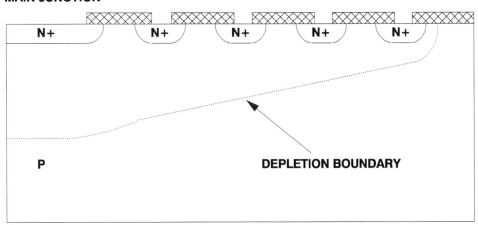

Fig. 3.24. A multiple field ring termination with equal field ring width and spacing.

In the second design approach, illustrated in Fig. 3.24, all the floating field rings are made narrow and equally spaced. The smaller width of these floating field rings and their closer spacing allows more of them to be accommodated within a given edge area. This design also produces a much finer gradation in the depletion layer at the edge of the device. The design of this termination is easier because it is essentially based upon using the minimum design rules when laying out the windows and spacings that control the field ring spacing and width. This termination is less sensitive to surface charge variations because a larger number of rings reduces the impact of surface charge variations when compared to the first approach with optimally spaced multiple field rings.

The multiple floating field ring design allows increase of the breakdown voltage to values arbitrarily close to the parallel plane case by using a very large number of rings. The main limitation to applying this approach is the space taken up at the edge of the chip. In practical devices, it is usual to use three floating field rings. However, designs with up to ten floating field rings have been reported when achieving close to the parallel plane junction breakdown is important.

3.6.4 Field Plates

In a previous section, it has been shown that the electric field at the surface of a planar diffused junction is greater than for the parallel plane junction due to depletion layer curvature effects. Floating field rings are one method for reducing this curvature. Another method for reducing the depletion layer curvature is by altering the surface potential. The most straight forward method for achieving this is by placing a metal field plate at the edge of the planar junction as illustrated in Fig. 3.25. Since the shape of the depletion region depends upon the surface potential, the depletion layer spreading at the surface is sensitive to the potential applied to the field plate.

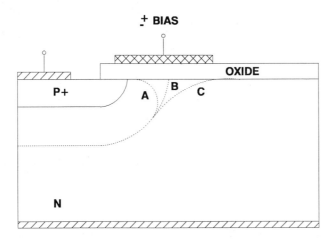

Fig. 3.25. Influence of bias on a field plate upon depletion layer spreading at the edge of a planar junction.

When a positive bias is applied to the metal field plate with respect to the N-type substrate, it will attract electrons to the surface and cause the depletion layer to shrink as illustrated by case A. If a negative bias is applied to the field plate, it will drive away electrons from the surface causing the depletion layer to expand as illustrated in case C. The latter phenomenon can be expected to increase the breakdown voltage. This has been experimentally observed [18]. It has been found that the breakdown voltage of the diode with field plate is related to the field plate potential (V_{FP}) by:

$$V_D = m V_{FP} + C \tag{3.62}$$

where m is approximately equal to one and C is a constant. The value for m is closer to unity for small oxide thicknesses. With sufficient bias on the field plate, the breakdown voltage of the planar diode can be made to approach that of the parallel plane junction.

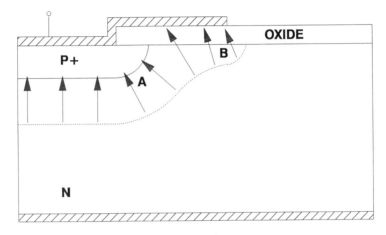

Fig. 3.26. Planar junction with field plate.

In actual power devices, it is impractical to provide a separate bias to control the potential on the field plate. Instead, the field plate is created by extending the junction metallization over the oxide as shown in Fig. 3.26. Since the field plate has the same bias applied to it as the main junction, a depletion region forms below it. This extends the depletion region along the surface to the edge of the field plate, which results in a reduction of the electric field crowding at the main junction (point A). However, a high electric field can occur at the edge of the field plate at point B.

Two dimensional numerical simulations of the potential distribution for a planar junction with a field plate tied to its highly doped side have been performed for a variety of substrate doping concentrations and oxide thicknesses [19]. The analysis shows a high electric field arising at the edge of the field plate if the oxide thickness is small. A simple physical understanding of this behavior can be obtained by considering the edge of the metal field plate as being similar to a cylindrical junction with the oxide under the field plate akin to the highly doped side of the junction. This analogy requires taking into account the difference in the dielectric constants for the dielectric layer and the semiconductor. Based upon Gauss's law, at the interface, the electric field in the semiconductor is related to the electric field in the oxide by the ratio of their permittivities. For the case of thermally grown oxide, the relative permittivity is 3.85 compared with 11.7 for silicon. This results in a decrease in the electric field, which is equivalent to a junction with depth of:

$$x_j = \frac{e_{si}}{e_{ox}} t_{ox} \approx 3\, t_{ox} \qquad (3.63)$$

Thus, an oxide thickness of 1 micron under the field plate can be considered to be equivalent to a junction depth of 3 microns. This junction depth can be used in conjunction with the normalized breakdown voltage Eq. (3.43) to calculate the breakdown voltage at the edge of the field plate. This analysis is valid only when the edges of the termination have a large radius. A metal field plate with a sharp corner would behave like a spherical junction with a radius of curvature given by Eq. (3.63). Since this breakdown voltage is much lower than for the cylindrical junction, it is important to avoid any sharp corners on the metal field plate during device design.

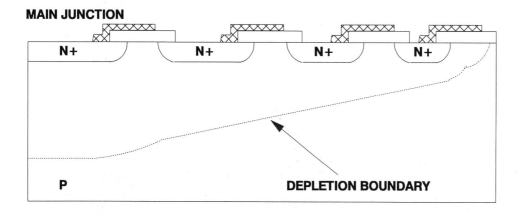

Fig. 3.27. Edge termination design combining floating field rings with field plates.

A common practice during device design is to use the field plate in conjunction with floating field rings, as illustrated in Fig. 3.27, to achieve high breakdown voltage. A field plate is connected to each floating field plate and extends beyond its edge. The field plates not only reduce the electric field crowding at each of the floating field rings but prevent charges on the top surface of the device from altering the surface potential in the semiconductor. This increases the stability of the breakdown voltage.

3.6.5 Bevel Edge Terminations

The successful application of solid state technology to power control required the development of thyristors capable of operating at high voltages. It was apparent during the early stages of thyristor development that it is necessary to design edge terminations which promote bulk breakdown. These high voltage devices were fabricated from an entire wafer to provide high current handling capability, which resulted in the junctions extending to the edges of the wafer. Further, since gallium and aluminum diffusions were used to obtain highly graded, deep junctions, the planar termination techniques discussed in earlier sections were not considered feasible. It was instead necessary to control the shape of the edge of the wafer and provide adequate surface passivation to prevent premature surface breakdown.

The simplest approach to shaping the edge of the wafer is to make a cut perpendicular to the wafer surface. This can be done either by sawing or by scribing and breaking the wafer. This method creates considerable surface damage at the edges of the device which leads to a high leakage current. Further, it was discovered that making the cut at the edge of the wafer at an angle can result in an increase in the breakdown voltage. The formation of a controlled flat edge at an angle to the wafer surface is referred to as a *bevel*.

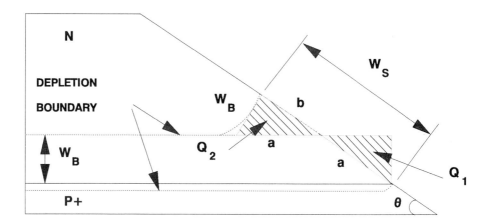

Fig. 3.28. Depletion boundaries for the positive bevel edge termination.

Consider the case of a junction illustrated in Fig. 3.28 whose area at the edge decreases when proceeding from the heavily doped side to the lightly doped side. This edge contour is known as a *positive bevel junction*. In order to maintain charge balance on the opposite sides of the junction, the depletion layer on the lightly doped side of the junction is forced to expand near the surface as indicated in Fig. 3.28. This expansion of the depletion layer causes a reduction in the electric field at the surface. Since the depletion layer width along the surface is much larger than in the bulk, it can be concluded that the electric field along the surface will be much smaller than in the bulk. This is a good design for the termination of junctions because it ensures bulk breakdown prior to surface breakdown if the surface electric field is sufficiently low. It should be noted that even if the surface electric field is lower than in the bulk, surface breakdown may precede bulk breakdown because the ionization coefficients at the surface are generally larger than in the bulk for the same electric field strength due to the presence of defects at the surface.

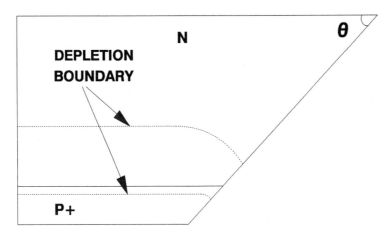

Fig. 3.29. Depletion layer boundary for a negative bevel edge termination.

If the edge of the junction is cut in the opposite direction so that the area of the junction increases when proceeding from the highly doped side towards the lightly doped side, the termination is called a *negative bevel junction*. The depletion layer shape for this case is illustrated in Fig. 3.29. The establishment of charge balance on the opposite sides of the junction causes the depletion layer at the surface of the lightly doped side to decrease while the depletion layer on the heavily doped side expands. If the diffused side of the junction is heavily doped, the depletion layer shrinkage on the lightly doped side will have the dominant influence. Since the reverse bias across the junction is being supported across a narrower depletion layer at the surface than in the bulk, the electric fields at the surface can be expected to be higher than in the bulk. Consequently, surface breakdown will precede bulk breakdown in the negative bevel junction and its breakdown voltage will be lower than for the parallel plane case. This leads to the conclusion that negative bevel contours are undesirable and should be avoided during device processing.

However, when the junction is highly graded and a very shallow negative bevel angle is created, this argument is no longer found to be valid because the surface depletion width can be increased as illustrated in Fig. 3.30. Here, the gradual change in doping level on the diffused (P^+) side of the junction and the large amount of material removed from the diffused side forces the depletion layer to expand considerably along the surface on the diffused side of the junction. It should be noted that the depletion layer on the lightly doped side gets pinned to the metallurgical junction edge at the surface under these circumstances and the expansion of the depletion layer on the diffused side of the junction lowers the surface electric field. Since the shallow angle for the negative bevel results in the consumption of a large peripheral surface area

Chapter 3 : BREAKDOWN VOLTAGE

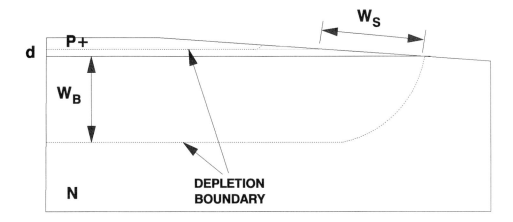

Fig. 3.30 Depletion layer boundary for a shallow negative bevel edge contour.

in devices, negative bevel edge contours are used primarily in devices (such as power thyristors) containing two back-to-back junctions. In these cases, a positive bevel is used for the reverse blocking junction and the negative bevel for the forward blocking junction.

Bevel junction terminations are widely used for the fabrication of high voltage, large area rectifiers and thyristors. They are infrequently used for small area devices because the treatment of the edges of individual devices by the lapping and polishing technology used to prepare bevel edge terminations is only economical when the pellet size is large.

3.6.5.1 Positive Bevel Junction. The positive bevel junction is one in which the junction area decreases when proceeding from the more highly doped side to the lightly doped side. Two-dimensional numerical analysis of the positive bevel junction has confirmed the surface field reduction described with the aid of Fig. 3.28. From this analysis, it has been found that the maximum electric field along the surface is always lower than the maximum electric field in the bulk for all positive bevel angles. The point at which the maximum electric field occurs moves away from the metallurgical junction into the lightly doped side as the positive bevel angle is reduced. When the positive bevel angle becomes small, the depletion layer on the diffused side of the junctions gets pinned at the metallurgical junction. In this case, the electric field drops to zero where the metallurgical junction hits the surface.

The maximum electric field at the surface of the positive bevel junction decreases with decreasing bevel angle. The variation in the electric field from the surface towards the bulk has also been examined. It has been found that the electric field decreases monotonically when going from the bulk towards the surface. Thus, the positive bevel junction exhibits the ideal characteristics for achieving bulk breakdown. It is worth noting that this is the only device termination technique that has so far been developed that allows achieving bulk breakdown or, in other words, the avalanche breakdown voltage of the parallel plane junction.

Although the above analysis indicates that bulk breakdown will occur for a positive

106 Chapter 3 : BREAKDOWN VOLTAGE

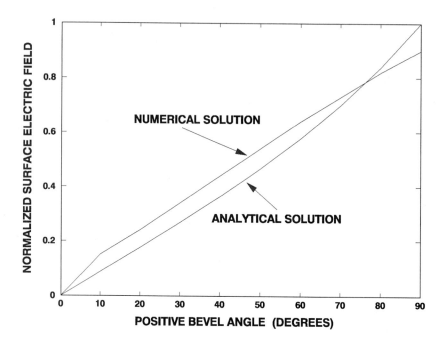

Fig. 3.31 Reduction in the surface electric field with decrease in positive bevel angle.

junction if the impact ionization coefficients at the surface are assumed to be equal to those in the bulk, in practice it is important to reduce the maximum surface electric field to at least 50 percent lower than the maximum electric field in the bulk due to the presence of surface defects which enhance impact ionization. The maximum surface electric field of a variety of positive bevel junctions with different background doping levels and diffusion profiles has been analyzed, and it has been found that all of these cases can be represented by a single curve by using a normalization scheme similar to that adopted for the planar junctions (i.e., by normalizing to the electric field in the parallel-plane portion of the junction). This curve is provided in Fig. 3.31. From this curve, it can be concluded that positive bevel angles ranging from 30 to 60° are adequate for ensuring bulk breakdown. Shallower positive bevel angles are not recommended due to the wastage of space at the edge of the chip.

An improved understanding of the reduction in the surface electric field can be obtained by examining the depletion region extension along the bevel surface as shown in Fig. 3.28 based upon charge balance considerations. Since the charge within the depletion region on both sides of the junction must be equal, the removal of the charge Q_1 from the semiconductor region on the lightly doped side by the bevel must be compensated for by the extension of the depletion region further into the lightly doped side. This extension can be analyzed by equating the charge Q_2 in the cross-hatched region to the charge Q_1. The depletion region extension along the surface of the positive bevel is then given by:

Chapter 3 : BREAKDOWN VOLTAGE 107

$$W_S = a + b = \left(\frac{W_B}{\sin \theta}\right)(1 + \cos \theta) \qquad (3.64)$$

Since the same voltage is supported along the surface as in the bulk, the maximum electric field at the bevel surface is related to the maximum bulk electric field by :

$$\frac{E_{mPB}}{E_{mB}} = \frac{2 \sin \theta}{(1 + \cos \theta)} \qquad (3.65)$$

The calculated ratio of the surface to bulk electric field obtained by using this analytical formula is also plotted in Fig. 3.31 for comparison with the numerically calculated values. The above model can be seen to exhibit a reasonable agreement with the simulations.

The reduced surface electric field of positive bevel junctions makes them relatively stable and easy to passivate. The fabrication of the positive bevel angle in high voltage, large area devices is usually performed by grit blasting using an abrasive powder emanating from a nozzle placed at the appropriate angle to the wafer edge while the wafer is being rotated. The crystal damage caused by the grit blasting is subsequently removed using a chemical etch just prior to coating the wafer edge with the passivant. In the case of a large number of small area devices fabricated on a single wafer, a positive bevel can be achieved by using a V-shaped diamond saw blade to cut through the wafer from the lightly doped side. Again, a chemical etch is essential after the saw cut to remove the residual damage. Such positive bevelled pellets are a problem to handle during packaging due to their sharp edges. Further, the passivation of these pellets is difficult since it must be done after the die mount down process, which can contaminate the surface. For these reasons, bevel edge terminations are not usually used for the fabrication of small devices.

3.6.5.2 Negative Bevel Junction. In a negative bevel junction, the area of the junction decreases when proceeding from the lightly doped side towards the highly doped side. The variation of maximum electric field as a function of the negative bevel angle over the full range from 0 to 90° is illustrated in Fig. 3.32 for a specific case. Note that the maximum surface electric field is higher than in the bulk for negative bevel angles between 30 and 80 degrees. This occurs due to the reduction in depletion layer width on the lightly doped side as illustrated in Fig. 3.29. It is only when the negative bevel angle is made very small that the maximum surface electric field becomes less than in the bulk.

It was pointed out earlier with the aid of Fig. 3.30 that the surface electric field will be reduced below the parallel plane case for a negative bevel junction if the angle is small and the diffusion is highly graded. Despite the reduction of the maximum surface electric field to less than the bulk electric field, the avalanche breakdown voltages of negative bevel junctions do not equal the breakdown voltage of the parallel plane junction. The reason for this is the occurrence of an increase in the electric field beneath the surface. The electric field distribution proceeding into the bulk from the surface of the negative bevel junction is shown for several specific cases in Fig. 3.33. In all of these cases, the electric field at the surface (E_{mS}) is less than the maximum bulk electric field (E_{mB}). However, the maximum electric field below the surface

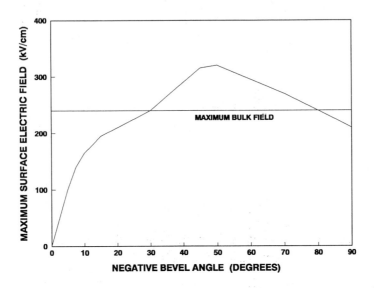

Fig. 3.32. Variation of maximum surface electric field with negative bevel angle.

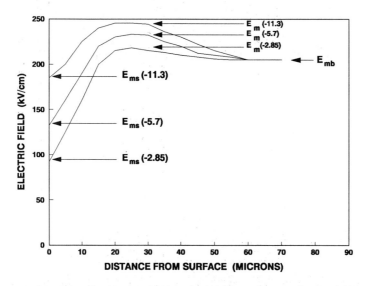

Fig. 3.33. Electric field distribution below the surface of several negative bevel junctions.

(E_m) is always greater than the maximum bulk electric field (E_{mB}). This implies that the breakdown voltage of negative bevel junctions will always be lower than for the parallel plane junction.

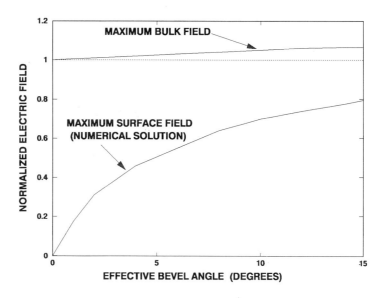

Fig. 3.34. Maximum surface and bulk electric fields for negative bevel junctions.

Numerical simulations of the breakdown voltage of a variety of diffused junctions with a broad range of surface concentrations and junction depths have been performed as a function of the background doping and the negative bevel angle [14]. It has been found that the breakdown voltage of all of these cases can be represented by a single curve by using a normalization scheme similar to that used for planar diffused junctions. In this case, a bevel parameter (ϕ) called the effective bevel angle, must be defined as :

$$\phi = (0.04) \cdot \theta \cdot \left(\frac{W}{d}\right)^2 \tag{3.66}$$

where θ is the actual physical bevel angle in degrees and W, d are the depletion layer widths on the lightly and highly doped sides of the junction, respectively. To achieve a breakdown voltage approaching the parallel plane case it is necessary to obtain a small effective negative bevel angle in the range of 2 to 6° in order to reduce the surface electric field as shown in Fig. 3.34. Junctions with large W/d ratios will require smaller absolute bevel angles to achieve a given reduction in surface field. As an example, if the W/d ratio is 5, an absolute bevel angle of 5 degrees is required to reduce the surface electric field to 50 percent of the bulk value. If the W/d ratio is 10, an absolute bevel angle of 1 degree will be required to achieve the same reduction in surface electric field. This indicates that highly graded diffused junctions are needed to increase d relative to W for obtaining high breakdown voltage in negative bevel junctions.

The breakdown voltage of negative bevel junctions can be presented in a normalized form by using the effective bevel angle as shown in Fig. 3.35. In order to obtain a breakdown voltage

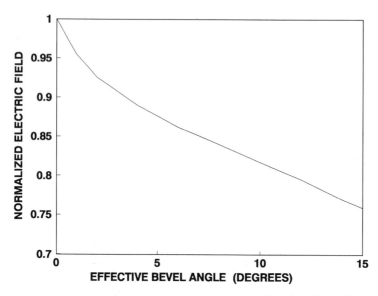

Fig. 3.35. Normalized breakdown voltage for negative bevel junctions.

within 90 percent of the parallel plane case, it is necessary to use effective bevel angles below 3 degrees. For a high voltage (3,000 volt) thyristor, an actual physical negative bevel angle of 2 to 4° is typically used with a diffusion gradient that achieves a ratio (W/d) of 4 to 5. These small negative bevel angles are formed using a lapping process followed by chemical etching to remove the surface damage. This technique results in consuming a large amount of space on the device edges due to the shallow bevel angles required to reduce the surface electric field. Consequently, the application of negative bevels is generally confined to large area, high current devices with symmetric blocking characteristics.

3.6.6 Etch Contour Terminations

The concept of using chemical etches to remove surface damage and improve the breakdown voltage or reduce surface leakage current has been known since the early days of high voltage device development. A moat etch such as that illustrated in Fig. 3.36 was typically used to terminate devices. This moat was used to separate adjacent devices. The etching was done at wafer level using masking material such as photoresist or even aluminum metallization. The scientific application of etching to reduce surface electric field by the precise removal of silicon from selective areas of the high voltage P-N junction near the surface was more recently developed [15,16].

The chemical etches used for forming these etch contours are mixtures of nitric, hydrofluoric and acetic acid. The composition and temperature of the acid must be carefully controlled in order to regulate the etch rate. In some etch contours, the depth must be controlled

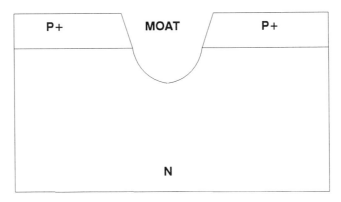

Fig. 3.36. Moat etch termination.

to within 0.1 microns. Such precision is often difficult to achieve with good uniformity across an entire wafer by using chemical etching. With the advent of dry etch technologies, such as plasma and reactive ion etching, much greater precision of etch depth and uniformity are achievable. Further, these dry etching processes are less sensitive to doping concentration in the silicon than chemical etches. However, the resulting edge termination has a non-planar topology which is difficult to passivate and the chip becomes mechanically fragile. For these reasons, it is preferable to use planar edge terminations and a detailed discussion of etch terminations is, therefore, not provided here.

3.6.7 Surface Implantation Termination

In the previous sections on bevel and etch contours, the breakdown voltage was enhanced by removal of material (and hence charge) from either the heavily doped or lightly doped side of the junction. A similar result can be achieved by the complementary process of adding precisely controlled charge to the junction along the surface at its edges. This approach has been called junction termination extension [17]. This termination relies upon the ability to introduce charge at the surface near the junction with an accuracy of better than one percent by using ion implantation. The control over the charge introduced by ion implantation is significantly superior to that achievable by the etching techniques and can be performed with much better uniformity.

The application of the junction termination extension technique to the planar junction is illustrated in Fig. 3.37. The charge required to obtain high breakdown voltages must be accurately controlled. If the dose of the implanted charge is too low, it will have little influence on the electric field distribution and the maximum in the electric field will occur at point A as in the case of the termination without the junction termination extension. If the dose of the implanted charge is too high, the junction will have been simply extended to point B. Since the depth of the ion implanted region is small, the extension region has a very small radius of curvature. The termination will then breakdown due to the high electric field at point B as determined by planar junction breakdown (see Section 3.6.1). To obtain a reduction in the

112   Chapter 3 : BREAKDOWN VOLTAGE

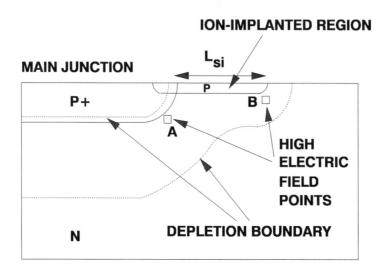

Fig. 3.37. Planar junction with ion-implanted edge termination.

electric field and an increase in the breakdown voltage, it is essential to control the implanted charge so that the implanted region gets completely depleted under reverse bias. For the case of a homogeneously doped junction, the charge within the depletion region on either side of the junction is related to the maximum electric field at the metallurgical junction by :

$$Q = \int_0^W q \, N_D \, dx = q \, N_D \, W = e_S \, E_m \qquad (3.67)$$

At the breakdown condition, the maximum electric field becomes equal to the critical electric field (E_c) given by Eq. (3.20). Using a value of 2×10^5 V/cm for E_c, the depletion region charge is found to be about 1.3×10^{12} per cm^3. When the implanted charge in the P-region is equal to this value, it will become completely depleted under the reverse bias conditions. Consequently, the voltage will be supported along the surface along the entire length (L_{si}) of the ion-implanted region. If this length is designed to be much longer than the depletion layer thickness in the bulk, the surface electric field can be reduced to below the bulk electric field, resulting in a breakdown voltage close to that for the parallel plane junction.

The breakdown voltage with a single implanted zone at the edge of a planar junction has been found to be strongly dependent upon the precise dose of dopants introduced into the edge region. Although the ion-implantation dose can be precisely controlled, the net dopant dose in the semiconductor after device processing can vary due to segregation of the dopant in the surface oxide. In addition, this termination has been found to be sensitive to surface charge. In order to address these problems, multiple zones of ion-implanted regions have been tried [18]. By using three zones with charge reduction by a factor of 2 when proceeding from the interior (main junction) to the exterior, breakdown voltages of over 90 % of the parallel plane value can be obtained. Although such a three zone ion-implanted region can be formed by using three masking and ion-implantation steps, a more elegant and cost effective method has been developed based upon designing a single mask with variable implant window size when proceeding from the interior (main junction) to the exterior. This results in a reduced effective charge within the semiconductor with increasing distance from the main junction in spite of the fact that a single implantation is performed. The spacing between the implantation windows is chosen so that the implanted regions diffuse together during device fabrication.

3.7 OPEN BASE TRANSISTOR BREAKDOWN

The previous sections have considered the breakdown voltage of P-N junctions with various junction terminations. In these cases, the maximum achievable breakdown voltage is that of a parallel plane P-N junction. However, many power devices contain back-to-back P-N junctions with floating middle regions. The maximum voltage blocking capability of these devices is limited by the open-base transistor breakdown of the parasitic transistor formed in these structures [20]. The influence on the breakdown voltage of the resistivity and the spacing between the P-N junctions in these devices is discussed below.

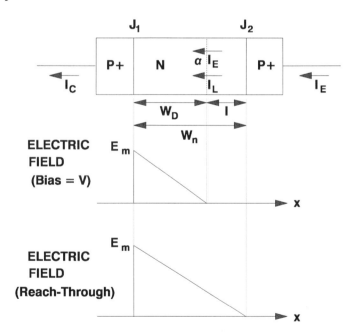

Fig. 3.38. Open-base transistor structure and its electric field distribution.

The open-base transistor structure is shown in Fig. 3.38 with one of the P-N junctions

(J_1) reverse biased. This junction (J_1) then acts as the collector and the other junction (J_2) acts as the emitter. The applied reverse bias creates a depletion region that extends from the reverse biased collector junction (J_1) towards the forward biased emitter junction (J_2), and the maximum electric field occurs at junction J_1. When the width of the N-base layer is large compared to the depletion width (W_D), the breakdown voltage of the collector junction is limited by avalanche breakdown, which occurs when the maximum electric field at junction J_1 becomes equal to the critical electric field for breakdown. As discussed earlier, avalanche breakdown occurs when the multiplication factor of the junction becomes infinitely large. This represents one limit for the breakdown voltage with changes in the resistivity of the N-base region and is shown in Fig. 3.39 as the avalanche breakdown limit. The avalanche breakdown voltage decreases with increasing N-base doping level.

When the emitter junction (J_2) is brought closer to the collector junction (J_1), open-base transistor breakdown occurs due to the injection of electrons from the emitter junction when the collector depletion layer approaches the emitter junction. If the minority carrier diffusion length in the N-base region is extremely small, the open-base transistor breakdown occurs when the depletion layer touches junction J_2 as illustrated at the bottom of Fig. 3.38. This condition is referred to as *reach-through*. The maximum electric field at the collector junction (J_1) can be much less than the critical electric field for breakdown under these conditions. Consequently, the open-base transistor breakdown voltage is smaller than the avalanche breakdown voltage for any particular doping concentration for the N-base region. The reach-through voltage is given by the expression:

$$BV_{RT} = \frac{q \, N_D}{2 \, \epsilon_s} W_n^2 \qquad (3.68)$$

The reach-through voltage increases linearly with increasing N-base doping concentration and as the square of the thickness of the N-base region (W_n). This limit to the open-base transistor breakdown voltage is shown in Fig. 3.39 for several values of the N-base thickness.

In actual devices with finite diffusion length for minority carriers in the N-base region, it is necessary to take into account the gain of the P-N-P transistor for determining the open-base transistor breakdown voltage. The leakage current due to space charge generation and diffusion currents is amplified by the gain of the transistor leading to a change in the condition for the breakdown. For this analysis, consider the collector and emitter currents indicated in the open-base transistor structure shown in Fig. 3.38. Using Kirchhoff's law:

$$I_C = \alpha_{PNP} I_E + I_L = I_E \qquad (3.69)$$

from which

$$I_E = I_C = \frac{I_L}{(1 - \alpha_{PNP})} \qquad (3.70)$$

From this equation, it can be concluded that the emitter and collector currents will approach a large value when the current gain approaches unity. Consequently, the condition for breakdown

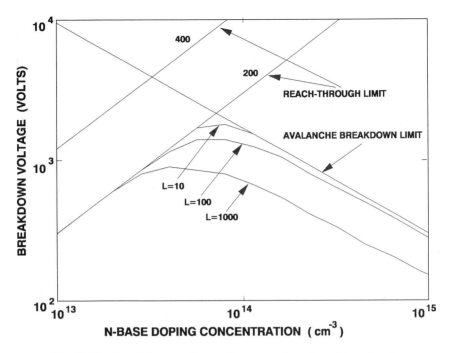

Fig. 3.39. Breakdown voltage of an open-base transistor structure.

of the open-base transistor is given by :

$$\alpha_{PNP} = \gamma_E \cdot \alpha_T \cdot M = 1 \qquad (3.71)$$

where γ_E is the emitter injection efficiency, α_T is the base transport factor, and M is the avalanche multiplication coefficient. Due to the low doping concentration of the N-base region in power devices designed to support high voltages, the injection efficiency can be assumed to be equal to unity. The base transport factor for a bipolar transistor can be related to the ratio of the undepleted base width (l) to the minority carrier diffusion length (L_p) by :

$$\alpha_T = \frac{1}{\cosh\ (l/L_p)} \qquad (3.72)$$

In this equation, the undepleted region width is dependent upon the applied bias:

$$l = W_n - \sqrt{\frac{2\ \epsilon_s\ V}{q\ N_D}} \qquad (3.73)$$

The multiplication coefficient can be related to the applied bias (V) and the avalanche breakdown voltage (BV_{PP}) by :

$$M = \frac{1}{[1 - (V/BV_{PP})^n]} \tag{3.74}$$

Using these equations, the breakdown voltage for the open-base transistor can be solved for any combination of N-base doping concentration and thickness. The solutions are dependent upon the diffusion length of the minority carriers in the N-base region. An example of the variation of the breakdown voltage as a function of the N-base doping concentration for the case of a width of 200 microns is provided in Fig. 3.39 for various diffusion lengths. It is worth pointing out that the maximum breakdown voltage is always smaller than the avalanche breakdown voltage, and that a smaller diffusion length is desirable for obtaining a higher breakdown voltage.

The presence of back-to-back P-N junctions imposes new design criteria for the termination of high voltage devices. The terminations discussed in earlier sections are oriented to lowering the maximum surface field for a single junction. To achieve a similar surface field reduction at two back-to-back junctions, these terminations must be combined as discussed below.

3.7.1 Negative/Positive Bevel Combination

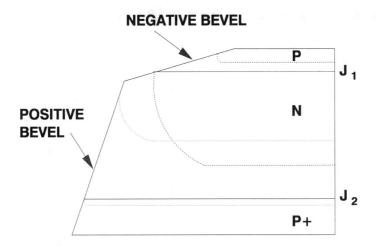

Fig. 3.40 Combination of positive and negative bevel edge termination for an open-base transistor structure.

The termination of back-to-back P-N junctions using a combination of negative and positive bevel contours is illustrated in Fig. 3.40. The depletion layer positions on both sides of junction J_1 under reverse blocking conditions for the upper junction are shown by the dashed lines. Note the extension of the depletion layer on the diffused (P^+) side of the junction near

Chapter 3 : BREAKDOWN VOLTAGE   117

the surface due to the negative bevel and its compression on the lightly doped (N) side. The design of the breakdown voltage at the termination in this case can be done as described before for negative bevel contours.

When junction J2 is reverse biased, the depletion layer takes the shape indicated by the dotted lines in Fig. 3.40. In this case, the depletion layer expands on the lightly doped (N) side of the junction near the surface due to the positive bevel. The design of this termination is not as straightforward as in the case of the simple positive bevel angle due to the possibility of premature breakdown at the surface if the depletion layer were to reach-through to the forward biased junction J1. The tendency for reach-through breakdown at the surface becomes greater as the positive bevel angle is reduced because the depletion layer extends further on the lightly doped sided as the positive bevel angle gets smaller. Very small positive bevel angles should not be used for this reason despite the advantage of reduced surface fields. In practice, the positive bevel angles used for the case of back-to-back P-N junctions range from 45° to 60°.

Due to the large area consumed at the edge by the shallow negative bevel required to achieve high breakdown voltages, the negative/positive bevel combination is only used for large area devices. This termination also relies upon highly graded diffusions which are applicable to very high voltage devices with breakdown voltages of over 2000 volts.

3.7.2 Double Positive Bevel Contours

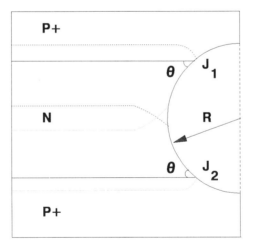

Fig. 3.41. Double positive bevel edge termination for an open-base transistor structure.

Since the positive bevel contour produces the most effective surface field reduction among all device termination techniques and offers the opportunity to achieve the breakdown voltage of the parallel plane junction, the combination of positive bevels at both the back-to-back junctions is certainly attractive. A technique to achieve this type of edge termination is illustrated in Fig. 3.41. A circular groove is formed at the edge of the wafer usually by using lapping with slurry applied to a wire of radius R that is pressed against the edge of the wafer as it is rotated. Alternately, a groove can be formed at the edge of the wafer by grit blasting using a nozzle directed orthogonally to the wafer edge while it is being rotated. In both cases, a positive bevel angle (θ) is formed locally at each of the junctions. When either junction is reverse biased, the depletion layer expands on the lightly doped side of the junction as illustrated by the dashed lines for junction J_1 and dotted lines for junction J_2.

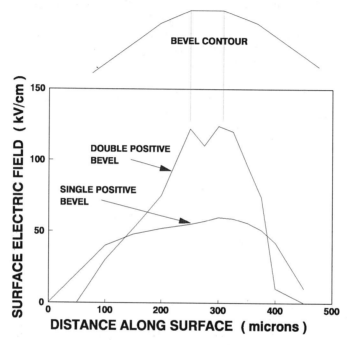

Fig. 3.42 Comparison of surface electric fields for double positive bevel with single positive bevel.

The design of this termination is complicated by the interaction between the upper and lower terminations. In order to reduce the total thickness of the N-base region, the depletion layers extending from junctions J_1 and J_2 can be designed to overlap as illustrated by the dashed and dotted lines. However, this increases the effective positive bevel angle for both junctions. A comparison of the surface electric fields for the double positive bevel and the single positive bevel has been obtained by numerical analysis [21] as illustrated in Fig. 3.42. The actual surface contour modeled is also shown in this figure because it was not the idealized circular shape shown in Fig. 3.41. Important points to note are the significantly higher surface electric fields observed for the double positive bevel case and the presence of field maxima located at regions of discontinuity in the contour geometry. It can be seen that such discontinuities are highly undesirable and should either be prevented from forming during the contour formation or be subsequently removed by using anisotropic etching techniques.

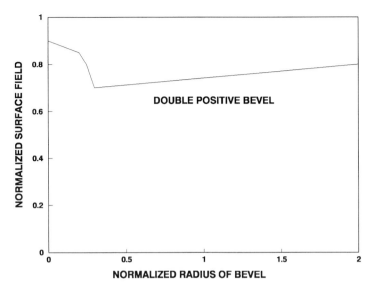

Fig. 3.43. Normalized maximum surface electric field for double positive bevel junction.

Investigation of a variety of surface contour geometries by numerical analysis indicates that it is very difficult to reduce the maximum surface electric field to below 60 percent of the bulk for the double positive bevel. For example, the impact of changing the radius of curvature (R) of the circular double positive bevel (Fig. 3.41) is provided in Fig. 3.43 [12]. In this figure, the surface electric field has been normalized to the peak electric field at the parallel-plane portion, and the radius of curvature normalized to the depletion layer thickness on the lightly doped side of the parallel-plane portion. The maximum surface electric field achievable by using the double positive bevel contour is significantly higher than the typical maximum surface field achievable by using shallow negative bevel contours. The higher surface electric fields obtained with the double positive bevel contour can create greater difficulty in device passivation. However, the space consumed at the edge of the wafer is much smaller than for the negative bevel angle and it is worth pointing out that the double positive bevel still offers the possibility of achieving bulk breakdown at the parallel plane portion of the junction as in the case of the simple positive bevel because no field maximum occurs underneath the surface.

3.8 HIGH VOLTAGE SURFACE PASSIVATION

The passivation of the surface of discrete power devices plays an important role in their fabrication because it determines the ability of the device to withstand high surface electric fields. In addition, the surface charge induced by the passivant can strongly influence the electric fields in the bulk. Methods of passivation extend from the application of rubber-like coatings on the bevelled edges of the device to the chemical vapor deposition of films that are impervious to moisture.

The most commonly used passivation process for high voltage, large area devices with bevel contoured surfaces is by means of rubberized coatings or organic polymers [22]. These techniques consist of the initial removal of surface damage resulting from the mechanical surface bevelling techniques by using chemical etching immediately followed by the application of an organic material such as polyamide in a solution of dimethylacetamide. The passivant is then cured by heating in a nitrogen ambient up to temperatures in the range of 250°C. This technique of device passivation has been applied for the fabrication of devices with breakdown voltages exceeding 5,000 volts to obtain stable device characteristics. To ensure long term stability it is still necessary to enclose these devices in a hermetically sealed package. This problem can be overcome by using coatings such as silicon nitride that are impervious to the migration of ions and moisture.

Silicon dioxide and silicon nitride films applied to surfaces of devices have been observed to provide stable operation of high voltage devices with both positive bevelled [22] and negative bevelled [23] surfaces. It has been found that the cleaning procedure for the surface of the device prior to the growth of the silicon dioxide layer by thermal oxidation is a critical factor in obtaining sharp breakdown characteristics. The best results are obtained by using the following steps: (a) 30 second dip in 40% hydrofluoric acid, (b) water rinse, (c) perchlorate etch for 10 seconds to remove surface oxides, (d) water rinse and dry by spinning, and (e) immediate loading in furnace for oxidation. The oxide growth temperature has been found to strongly influence the device leakage current, with the best results obtained by growth at 800°C followed by slow cooling of the furnace (2°C/min) from 800° to 720°C. In addition, the use of oxidation tubes and boats made of pure silicon to hold the wafers is recommended for obtaining low leakage currents. Further, the use of wet oxidation is acceptable only if preceded by a dry oxidation cycle. Since these silicon dioxide films are not impervious to moisture or ion migration, they are generally covered by a film of silicon nitride obtained by using chemical vapor deposition techniques. The successful application of this process has been reported for devices with breakdown voltages up to 3,000 volts [22,23].

The silicon dioxide-silicon nitride sandwich has also been used for the passivation of planar junctions. However, superior results can be obtained in these junctions by using semi-insulating polycrystalline silicon films (SIPOS). Polycrystalline silicon films with high resistivities can be obtained by the addition of oxygen or nitrogen during the deposition [24]. The resulting films have a resistivity of about 10^{10} ohm-cm compared with 10^5 ohm-cm for undoped polysilicon and 10^{14} ohm-cm for silicon dioxide. As discussed earlier, the breakdown voltage of planar junctions is lower than that of parallel plane junctions due to higher electric fields in the curved portion of the junction. These fields can be decreased by the application of the semi-insulating SIPOS layer on the junction surface to result in higher breakdown voltages. In this method of passivation, the negative potential of the P-diffused region of the junction is transmitted to the surface region of the N-type substrate by the ohmic current flow in the SIPOS film. This results in an extension of the surface depletion layer which lowers the surface electric field. The improvement in the breakdown voltage of the junction depends upon the conductivity of the SIPOS film, with larger voltages being achieved at lower oxygen doping levels. These films, however, have a lower resistivity which results in an additional junction leakage current. An optimum oxygen content of 15 to 35 percent is found to produce the best device characteristics. The oxygen doped SIPOS films have low charge densities and can be applied for the passivation of both P- and N-type substrates, but are not impervious to moisture or ionic contamination. Consequently, it is necessary to coat them with a film of nitrogen-doped SIPOS. A layer of silicon dioxide is also added on top of the nitrogen-doped SIPOS layer to increase the breakdown strength of the composite dielectric film. Power bipolar transistors with collector breakdown voltage of up to 10,000 volts have been fabricated with the SIPOS passivation technique.

Another technique that has been used to passivate high voltage junctions in power devices is by the electrophoretic deposition of zinc-boro-silicate glasses containing additives such as Ta_2O_5 or Sb_2O_3 [25]. This technique is capable of providing thick glass coatings in selective areas with expansion coefficients matched to that of silicon. The process consists of the preparation of a suspension of glass particles in a medium such as acetone or isopropyl alcohol. The glass particles are then deposited on the silicon substrates by passage of current through the suspension with the silicon as the cathode. After the deposition of the glass, it is first densified to remove the solvent and then fired at temperatures between 800°C and 900°C in an oxygen containing ambient. The addition of Ta_2O_5 to the glass has been found to improve junction stability and the addition of Sb_2O_3 to the glass has been found to improve the dielectric breakdown strength. Successful application of this technique has been demonstrated with junction breakdown voltages of over 1,000 volts.

3.9 COMPARISON OF TERMINATIONS

A large variety of high voltage device termination techniques have been described above. These techniques are compared in Table 3.1 on the basis of the typical breakdown voltage achievable using each technique and the surface electric field reduction [26]. It can be seen that the positive bevel contour offers the best results. However, the selection of a specific technique for any particular application is dependent on several other important factors.

Firstly, it depends upon the size of the device. In the case of small devices, bevelling techniques are not practical. The most commonly used field termination method for small dies is the use of planar junctions with field rings. This allows achieving up to 80 percent of the ideal breakdown voltage of the parallel plane junction. This is adequate for most bipolar devices. In the case of power MOSFET and JFET devices, where their on-resistance is strongly influenced by the breakdown voltage, the junction termination extension offers the best promise, albeit at higher leakage currents. Alternately, the positive etch contour can be used to achieve close to 90 percent of the ideal breakdown voltage.

Secondly, the planar junction with field ring and the planar junction with junction termination extension can be used to simultaneously process a large number of small devices on each wafer at the same time. In contrast, the bevelling methods require handling individual devices and are not cost effective for small devices. For large area devices, fabricated for instance from individual silicon wafers, the bevelling techniques are the most promising due to the nearly ideal breakdown voltages achievable with this technique. The bevelling technique ensures a large surface electric field reduction which makes surface passivation less critical. Today most high current rectifiers and thyristors are fabricated by using combinations of positive and negative bevel contours. Devices with breakdown voltages of up to 6,500 volts and current handling capability of thousands of amperes are commercially available.

In summary, the choice of the edge termination technique should be made based upon the die size, upon the sensitivity of the other electrical characteristics of the device upon the breakdown voltage, and upon the ease of surface passivation of the edges. Until recently, the termination of high voltage devices was considered an art rather than a science. The development of two-dimensional numerical solution techniques now allows the design of edge terminations with a high degree of confidence.

Technique	Breakdown Voltage (%)	Peak Surface Electric Field (%)	Typical Device Size	Device Types
Planar junction	50	80	Small (<100 mils)	BJT, MOSFET
Planar junction with field ring	80	80	Medium (< 1 inch)	BJT, MOSFET, SCR, GTO, IGBT, MCT
Planar junction with field plate	60	80	Medium (< 1 inch)	BJT, MOSFET, IGBT, MCT
Positive bevel	100	50	Large (> 1 inch)	Rectifier, SCR, GTO, MCT
Negative bevel	90	60	Large (> 1 inch)	SCR, GTO, MCT
Double positive bevel	100	80	Large (> 1 inch)	SCR, GTO, MCT
Surface ion implanted edge	95	80	All	BJT, MOSFET, SCR, GTO, IGBT, MCT

Table 3.1 Comparison of high voltage edge termination techniques.

REFERENCES

1. R. Van Overstraeten and H. DeMan, "Measurements of the ionization rates in diffused silicon P-N junctions," Solid State Electronics, Vol. 13, pp. 583-608 (1970).

2. N.R. Howard, "Avalanche multiplication in silicon junctions," J. Electron. Control., Vol. 13, pp. 537-544 (1962).

3. W. Fulop, "Calculation of avalanche breakdown of silicon P-N junctions," Solid State Electronics, Vol. 10, pp. 39-43 (1967).

4. R.A. Kokosa and R.L. Davies, "Avalanche breakdown of diffused silicon P-N junctions," IEEE Trans. Electron Devices, Vol. ED-13, pp. 874-881 (1966).

5. M.S. Adler and V.A.K. Temple, "Semiconductor avalanche breakdown design manual," GE Technology Marketing Operation, Schenectady, NY (1979).

6. B.J. Baliga and S.K. Ghandhi, "Analytical solutions for the breakdown voltage of abrupt cylindrical and spherical junctions," Solid State Electronics, Vol. 19, pp. 739-744 (1976).

7. S.M. Sze and G. Gibbons, "Effect of junction curvature on breakdown voltage in semiconductors," Solid State Electronics, Vol. 9, pp. 831-845 (1966).

8. V.A.K. Temple and M.S. Adler, "Calculation of the diffusion curvature related avalanche breakdown in high voltage planar P-N junctions," IEEE-Trans. Electron Devices, Vol. ED-22, pp. 910-916 (1975).

9. Y.C. Kao and E.D. Wolley, "High voltage planar P-N junctions," Proc. IEEE, Vol. 55, pp. 1409-1414 (1967).

10. M.S. Adler, V.A.K. Temple, A.P. Ferro, and R.C. Rustay, "Theory and breakdown voltage for planar devices with a single field limiting ring," IEEE Trans. Electron Devices, Vol. ED-24, pp. 107-113 (1977).

11. R.L. Davies and F.E. Gentry, "Control of electric field at the surface of P-N junctions," IEEE Trans. Electron Devices, Vol. ED-11, pp. 313-323 (1964).

12. M.S. Adler and V.A.K. Temple, "Maximum surface and bulk electric fields at breakdown for planar and beveled devices," IEEE Trans. Electron Devices, Vol. ED-25, pp. 1266-1270 (1978).

13. J. Cornu, "Field distribution near the surface of beveled P-N junctions of high voltage devices," IEEE Trans. Electron Devices, Vol. ED-20, pp. 347-352 (1973).

14. M.S. Adler and V.A.K. Temple, "A general method for predicting the avalanche breakdown voltage of negative bevelled devices," IEEE Trans. Electron Devices, Vol. ED-23, pp. 956-960 (1976).

15. V.A.K. Temple, B.J. Baliga, and M.S. Adler, "The planar junction etch for high voltage and low surface fields in planar devices," IEEE Trans. Electron Devices, Vol. ED-24, pp. 1304-1310 (1977).

16. V.A.K. Temple and M.S. Adler, "The theory and application of a simple etch contour for near ideal breakdown voltage in plane and planar P-N junctions," IEEE Trans. Electron Devices, Vol. ED-23, pp. 950-955 (1976).

17. V.A.K. Temple, "Junction termination extension, a new technique for increasing avalanche breakdown voltage and controlling surface electric fields in P-N junctions," IEEE Int. Electron Devices Meeting Digest, Abs. 20.4, pp. 423-426 (1977).

18. A.S. Grove, O. Leistiko, and W.W. Hooper, "Effect of surface fields on the breakdown voltage of planar silicon P-N junctions," IEEE Trans. Electron Devices, Vol. ED-14, pp. 157-162 (1967).

19. F. Conti and M. Conti, "Surface breakdown in silicon planar diodes equipped with field plate," Solid State Electronics, Vol. 15, pp. 93-105 (1972).

20. A. Herlet, "The maximum blocking capability of silicon thyristors," Solid State Electronics, Vol. 8, pp. 655-671 (1965).

21. J. Cornu, S. Schweitzer and O. Kuhn, "Double positive bevel: a better edge contour for high voltage devices," IEEE Trans. Electron Devices, Vol. ED-21, pp. 181-184 (1974).

22. R.R. Verderber, G.A. Gruber, J.W. Ostrowski, J.E. Johnson, K.S. Tarneja, D.M. Gillott, and B.J. Coverston, "SiO_2/Si_3N_4 passivation of high power rectifiers," IEEE Trans. Electron Devices, Vol. ED-17, pp. 797-799 (1970).

23. R.E. Blaha and W.R. Fahrner, "Passivation of high breakdown voltage P-N-P structures by thermal oxidation," J. Electrochemical Society, Vol. 123, pp. 515-518 (1976).

24. T. Matsushita, T. Aoki, T. Ohtsu, H. Yamoto, H. Hayashi, M. Okayama, and Y. Kawana, "Highly reliable high voltage transistors by use of the SIPOS process," IEEE Trans. Electron Devices, Vol. ED-23, pp. 826-830 (1976).

25. K. Miwa, M. Kanno, S. Kawashima, S. Kawamura, and T. Shibuya, "Glass passivation of silicon devices by electrophoresis," Denki Kagaku, Vol. 40, pp. 478-484 (1972).

26. B.J. Baliga, "High voltage device termination techniques - A comparative review," IEE Proc., Vol. 129, pp. 173-179 (1982).

27. S.K. Ghandhi, "Semiconductor power devices," Wiley, New York (1977).

28. R. Stengl and U. Gosele, "Variation of lateral doping - a new concept to avoid high voltage breakdown of planar junctions," IEEE Int. Electron Devices Meeting Digest, Abs. 6.4, pp. 154-157 (1985).

29. B.J. Baliga, "Closed form analytical solutions for the breakdown voltage of planar junctions terminated with a single floating field ring," Solid State Electronics, Vol. 33, pp. 485-488 (1990).

PROBLEMS

3.1 What is the background doping concentration required to obtain a breakdown voltage of 60, 200, 600, 1200 and 2000 volts for an abrupt parallel-plane junction? Determine the corresponding depletion layer thickness at breakdown.

3.2 A punch-through diode has an N- drift region with a doping concentration of 5×10^{13} per cm^3 and thickness of 20 microns. Determine its breakdown voltage. What is the doping concentration and thickness of the drift region to obtain the same breakdown voltage for the non-punch-through case?

3.3 A P^+/N junction with cylindrical edge termination is produced by planar diffusion into an N-type region with doping concentration of 1×10^{14} per cm^3. Determine the junction depth required to obtain a breakdown voltage of 800 volts.

3.4 A P^+/N junction with cylindrical edge termination is produced by planar diffusion into an N-type region with doping concentration of 1×10^{14} per cm^3. Determine the junction depth required to obtain a breakdown voltage of 800 volts if a single optimally spaced field ring is used around the junction.

3.5 Determine the positive bevel angle required to reduce the surface electric field to less than 50 percent of the bulk value. Calculate the percentage area consumed at the edge of a device 5 cm in diameter for a wafer thickness of 40 mils (0.1 cm).

3.6 A negative bevel angle is used to obtain a breakdown voltage of 90 percent of the parallel-plane value for a diffused junction with depletion width on the lightly doped side being 5 times that on the diffused side. Determine the negative bevel angle. Calculate the percentage area consumed at the edge of a device 5 cm in diameter if the junction has a depth of 100 microns, assuming that the negative bevel extends just to the edge of the junction.

Chapter 3 : BREAKDOWN VOLTAGE

3.7 Consider an open-base P$^+$NP$^+$ transistor with an N-base doping concentration of 1×10^{14} per cm^3 and thickness of 100 microns. Determine the breakdown voltage of the transistor if the base region contains 1×10^{13} gold atoms per cm^3.

3.8 A planar diffused edge termination for a power device is fabricated with an N-type drift region with a doping concentration of 1×10^{14} per cm^3 and a P$^+$ diffusion with a depth of 5 microns. Assume that the lateral diffusion is equal to the junction depth. Determine the breakdown voltage obtained when the diffusion is performed using a diffusion mask with a square shaped window having an edge of 2000 microns. What is the ratio of the active device area (defined as the area within the diffusion window) to the total die area (area including the edges of the device) if a space of 100 microns is left at the edge beyond the depletion boundary?

3.9 A planar diffused edge termination for a power device is fabricated with an N-type drift region with a doping concentration of 1×10^{14} per cm^3 and a P$^+$ diffusion with a depth of 5 microns. Assume that the lateral diffusion is equal to the junction depth. Determine the breakdown voltage obtained when the diffusion is performed using a diffusion mask with a square shaped window having a side of 2000 microns with corners rounded with a radius of twice the depletion thickness at breakdown. What is the ratio of the active device area (defined as the area within the diffusion window) to the total die area (area including the edges of the device) if a space of 100 microns is left at the edge beyond the depletion boundary?

3.10 A planar diffused edge termination for a power device is fabricated with an N-type drift region with a doping concentration of 1×10^{14} per cm^3 and a P$^+$ diffusion with a depth of 5 microns. Assume that the lateral diffusion is equal to the junction depth. Determine the breakdown voltage obtained when the diffusion is performed using a diffusion mask with a square shaped window having a side of 2000 microns with corners rounded with a radius of twice the depletion thickness at breakdown and inclusion of an optimally spaced floating field ring. What is the ratio of the active device area (defined as the area within the diffusion window) to the total die area (area including the edges of the device) if a space of 100 microns is left at the edge beyond the depletion boundary?

Chapter 4

POWER RECTIFIERS

In power electronic applications, there has been a continuous trend toward higher operating frequency especially in motor control and switch mode power supplies. In motor control circuits, operation at frequencies above the acoustic range is attractive for consumer applications. In power supplies, operation at higher frequencies is attractive because of the reduction in size and power losses in the passive components (inductors and capacitors) which leads to a more efficient, compact system design. To accomplish higher frequency operation in power circuits, it is essential to use power rectifiers with improved switching performance.

In the past, only P-i-N rectifiers were available for use in power circuits. The performance of these diodes has been continually improving due to optimization of device structure and the lifetime control process that is used to adjust the switching speed. The maximum operating frequency of these bipolar devices is ultimately limited by the reverse recovery process. During reverse recovery, a large reverse current flows through the diodes, which produces an undesirable stress upon the power transistors operating in the circuits.

The on-state voltage drop of a P-i-N rectifier is determined by the energy band gap of the semiconductor. For silicon devices, the on-state voltage drop of the P-i-N rectifier cannot be reduced below about 0.8 volts even if the device is not required to block high reverse voltage. In the case of the output rectifiers used in power supplies for computers and telecommunications, this voltage drop is a large fraction of the output voltage (typically 5 volts). This results in a loss in the power supply efficiency by 20 percent. As the output voltage of the power supply is reduced to suit VLSI needs, this problem becomes aggravated. The development of Schottky barrier rectifiers with low on-state voltage drops has been important for these applications. In a Schottky barrier rectifier, the minimum on-state voltage drop is determined by the barrier height of the metal-semiconductor junction. This can be varied by proper choice of the work function of the metal. Schottky barrier rectifiers are of importance for applications that require diodes with low on-state voltage drop and fast switching speed.

During the 1980s, another rectifier structure, based upon combining the physics of the P-i-N and Schottky barrier rectifiers, was proposed. These diodes have been demonstrated to exhibit much better reverse recovery behavior while maintaining an on-state voltage drop superior to that for either the P-i-N or Schottky barrier rectifiers. The use of these diodes in high frequency circuits that require large blocking voltages is expected to result in significant

Chapter 4 : POWER RECTIFIERS   129

reduction in power losses.
    In this chapter, the physics of operation of the Schottky barrier rectifier will first be discussed with emphasis on the characteristics of devices suitable for power switching applications. This is followed by a discussion of the physics of operation of the P-i-N rectifiers. The third portion of the chapter discusses the physics of operation of the merged P-i-N/Schottky (MPS) rectifier.

4.1 SCHOTTKY BARRIER RECTIFIERS

The non-linear current transport across a metal semiconductor contact has been known for a very long time. The potential barrier responsible for this behavior was ascribed to presence of a stable space charge layer by Schottky in 1938. The fundamental principles that describe current transport in the metal-semiconductor contacts have been treated in detail in several places. In this section, the application of the metal-semiconductor barrier to achieve high power rectification will be given emphasis

4.1.1 Metal-Semiconductor Contact

In general, the position of the Fermi level in a metal and a semiconductor will have different energy when they are isolated from each other. The band structure for the metal and

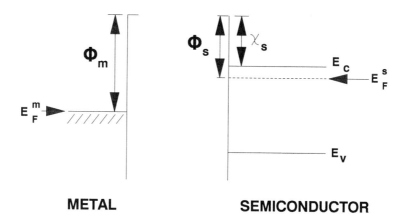

Fig. 4.1 Energy band diagrams for a metal and a semiconductor in isolation.

the semiconductor, when they are electrically isolated from each other, is shown in Fig. 4.1. In this illustration, the Fermi level of the N-type semiconductor region lies above that for the metal. The work function of the metal (ϕ_m) is defined as the energy required to raise an electron from the Fermi level position in the metal to a state at rest outside the surface of the metal, i.e. to the vacuum level. Similarly, the work function of the semiconductor (ϕ_s) is defined as the energy required to raise an electron from the Fermi level position in the semiconductor to a state at rest outside the surface of the semiconductor, i.e. to the vacuum level. Since there are no electrons located at the Fermi level position in the semiconductor, it is useful to define an electron affinity of the semiconductor (χ_s) as the energy required to raise an electron from the bottom of the conduction band to a state at rest outside the surface of the semiconductor, i.e. to the vacuum level. The semiconductor work function and electron affinity are related by :

$$\phi_s = \chi_s + (E_c - E_F^s) \qquad (4.1)$$

The potential difference between the Fermi levels in the semiconductor and the metal is called the *contact potential* (V_c) given by :

$$q V_c = (E_F^s - E_F^m) = \phi_m - \phi_s = \phi_m - (\chi_s + E_c - E_F^s) \qquad (4.2)$$

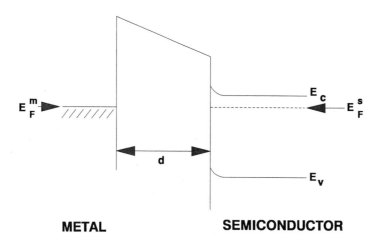

Fig. 4.2. Energy band diagram for metal/semiconductor contact after electrical connection.

When the metal and the semiconductor are electrically connected, electrons are transferred from the semiconductor to the metal to establish thermal equilibrium. The transfer of electrons creates a negative charge in the metal and a depletion layer at the semiconductor surface. The band structure under these conditions is illustrated in Fig. 4.2 for case of a separation d between the metal and the semiconductor. The band structure when the metal and the semiconductor are brought into intimate contact, i.e. when d is reduced to zero, is illustrated in Fig. 4.3. The depletion region formed in the semiconductor now supports the contact potential difference. This voltage is also referred to as *the built-in potential* (V_{bi}) of the metal-semiconductor contact. Thus, the Schottky barrier height (ϕ_{bn}) is given by:

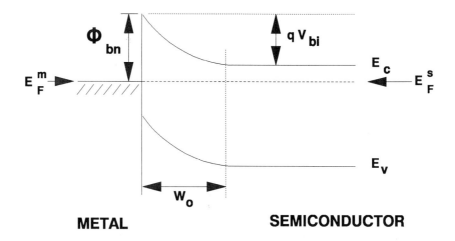

Fig. 4.3. Band diagram for metal/semiconductor contact after intimate contact.

$$q\,\phi_{bn} = q\,V_{bi} + (E_c - E_F^S) = (\phi_m - \phi_s) + (E_c - E_F^S) \quad (4.3)$$

The zero bias depletion region thickness is given by :

$$W_0 = \sqrt{\frac{2\,\epsilon_s\,V_{bi}}{q\,N_D}} \quad (4.4)$$

4.1.2 Forward Conduction

When a positive bias is applied to the metal with respect to the N-type semiconductor, the band diagram changes as shown in Fig. 4.4. Current transport across the metal/semiconductor contact can then occur via four basic processes : (a) by the transport of electrons from the semiconductor over the potential barrier into the metal, called *thermionic emission current*, (b) by quantum mechanical tunnelling of carriers through the barrier, called *tunnelling current*, (c) by *recombination current* flow in the space charge region, and (d) by the injection of holes from the metal into the semiconductor, called *minority carrier current*. In the case of Schottky power rectifiers, due to the relatively low doping levels in the semiconductor required to support high reverse blocking voltages, the potential barrier is not sharp enough to produce significant current flow via the tunnelling process. The space charge recombination current is similar to that observed in a P-N junction diode and is observed only at very low current densities. The current transport due to minority carrier injection is negligible unless the barrier height is made large. A large barrier height is usually not encountered in the case in Schottky barrier power rectifiers because it would result in a high forward voltage drop.

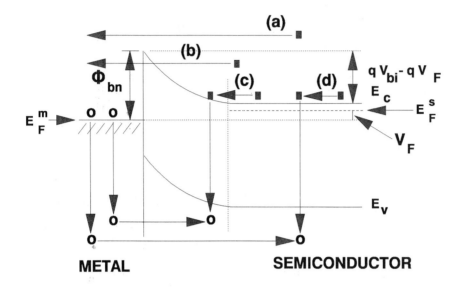

Fig. 4.4. Band diagram for metal-semiconductor contact under forward bias conditions.

Consequently, the current flow via the thermionic emission process is the dominant current transport mechanism in Schottky barrier power rectifiers.

In the case of a high mobility semiconductor, such as silicon and gallium arsenide at relatively low doping levels, the thermionic emission theory can be used to describe current flow across the Schottky barrier interface:

$$J = A T^2 e^{-(q\phi_{bn}/kT)} [e^{(qV/kT)} - 1] \qquad (4.5)$$

where A is the effective Richardson constant, T is the absolute temperature, q is the electron charge, k is Boltzmann's constant, ϕ_{bn} is the barrier height between the metal and N-type semiconductor, and V is the applied voltage. For N-type silicon, an effective Richardson constant of 110 Amperes/cm²/°K² can be used. For N-type gallium arsenide, an effective Richardson constant of 140 Amperes/cm²/°K² is more appropriate. When a forward bias is applied, the first term in the square brackets becomes dominant and the current flow across the Schottky barrier under forward conduction is given by:

$$J = A T^2 e^{-(q\phi_{bn}/kT)} e^{(qV_{FB}/kT)} \qquad (4.6)$$

where V_{FB} is the voltage drop across the Schottky barrier. In the case of Schottky barrier power rectifiers, a thick lightly doped drift layer must be used for the semiconductor region to support the reverse blocking voltage as illustrated in Fig. 4.5. The diode current flows through the drift

Chapter 4 : POWER RECTIFIERS

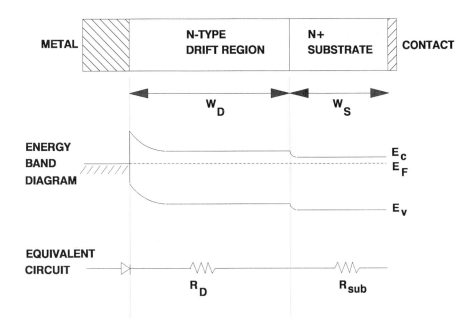

Fig. 4.5. Band diagram for a Schottky power rectifier illustrating the series resistance in the semiconductor.

layer producing a resistive voltage drop. It should be noted that there is no modulation of the drift region resistance in these devices due to the negligible minority carrier injection. Since the drift layer thickness is relatively small (typically less than 10 microns), the Schottky barrier power rectifier is made by growing a thin epitaxial layer upon a thick, highly doped (N$^+$) substrate. The resistance of the substrate can be significant in low voltage rectifiers for which the drift layer resistance becomes small. The substrate resistance component is indicated as R_{sub} in Fig. 4.5. Using Eq. (4.6) and including the resistive voltage drop, it can be shown that the total forward voltage drop in the Schottky rectifier will be given by:

$$V_F = \frac{kT}{q} \ln\left(\frac{J_F}{J_S}\right) + R_s J_F \qquad (4.7)$$

where R_S is the total series resistance per cm^2, which is referred to as the *specific resistance* of the rectifier. It includes contributions from the drift region, the substrate, and any contact resistance.

In Eq. (4.7), the term J_s refers to the saturation current of the Schottky barrier, which is related to the Schottky barrier height by :

$$J_S = A\ T^2\ e^{-(q\phi_{bn}/kT)} \qquad (4.8)$$

134 Chapter 4 : POWER RECTIFIERS

It should be noted that the saturation current is a strong function of the barrier height and temperature. Its value determines the forward conduction and reverse leakage current (as discussed later) characteristics of the Schottky barrier diode. The calculated saturation current

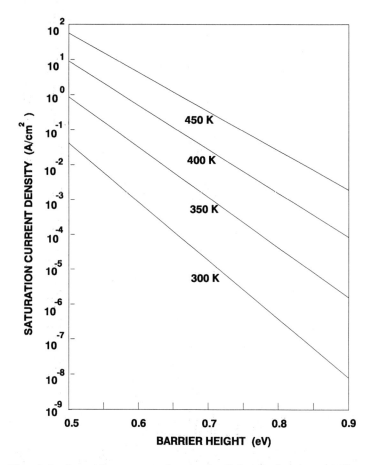

Fig. 4.6. Saturation current density for Schottky barrier rectifiers.

densities at four temperatures are provided in Fig. 4.6 as a function of the Schottky barrier height (ϕ_{bn}) for the range of 0.5 to 0.9 volts which is of interest to power rectifiers. In performing this calculation, a Richardson constant of 120 Amperes/cm^2/°K^2 was assumed.

In order to calculate the forward conduction characteristics of a Schottky barrier rectifier, the specific resistance of the diode must first be evaluated. For a N$^+$ substrate with a typical resistivity of 0.01 ohm-cm and thickness of 500 microns (20 mils), the contribution from the substrate (R_{sub}) can be calculated to be 5 x 10^{-4} ohm cm^2. Although this may appear at first sight to be negligible, it can contribute to an increase in forward voltage drop by 50 millivolts at a typical forward conduction current density of 100 amperes per cm², which is not insignificant for rectifiers designed to operate at forward voltage drops of less than 500 millivolts. The contribution from the substrate can be made much smaller by increasing its doping level and reducing its thickness, but this requires the acquisition of specially prepared wafers with arsenic doping.

The contribution to the series resistance of the diode from the drift region (R_D) is dependent upon the reverse blocking voltage. To achieve higher reverse blocking capability, it is essential to increase the resistivity and thickness of the drift region as discussed in Chapter 3. For the ideal case, where the edge effects are assumed not to influence the breakdown voltage, the specific resistance of the drift region is given by:

$$R_{D,sp} = \frac{W_{C,PP}}{q \mu_n N_D} = 5.93 \times 10^{-9} (BV_{PP})^{2.5} \qquad (4.9)$$

for an N-type drift region. For the case of a typical Schottky power rectifier with a reverse breakdown voltage of 50 volts, the drift region specific resistance is calculated to be 1.05×10^{-4} ohm cm². The voltage drop contributed by the series resistance of the drift region is only 10 millivolts for these relatively low breakdown voltage devices.

However, from Eq. (4.9), it is obvious that the series resistance contribution from the

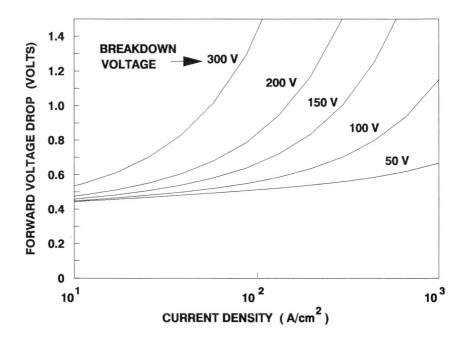

Fig. 4.7   Impact of breakdown voltage on forward conduction characteristics of silicon Schottky barrier rectifiers.

drift region will increase very rapidly as the reverse breakdown voltage increases, resulting in a large increase in the forward voltage drop. This is a serious limitation to increasing the blocking voltage capability of silicon Schottky barrier rectifiers. In order to assess the impact of the breakdown voltage upon the forward conduction characteristics of a silicon Schottky barrier rectifier, consider a typical case with a barrier height of 0.8 volts. The calculated forward conduction characteristics at room temperature are provided in Fig. 4.7 for several breakdown voltages. In performing these calculations, it has been assumed that the voltage drop contributions from the substrate and contact can be made negligible. These contributions can be factored into the characteristics by adding them to the voltage drop. From Fig. 4.7, it can be seen that the forward voltage drop of a Schottky power rectifier designed with reverse blocking capability of 50 volts is between 0.5 and 0.6 volts at a typical forward conduction current density of 100 amperes per cm^2. This low forward voltage drop compared to a P-i-N rectifier (whose forward drop is typically 0.9 volts), makes the Schottky rectifier attractive as an output rectifier in switch mode power supplies. As the reverse blocking capability is increased to 200 volts, the forward voltage drop of the Schottky rectifier approaches that of the P-i-N rectifier. In addition, as discussed later, the reverse blocking characteristics become soft, making the Schottky rectifier generally unacceptable for use in high voltage power circuits.

The forward voltage drop of the Schottky barrier rectifier is a function of temperature. As temperature increases, the saturation current density (J_s) increases and the voltage drop across

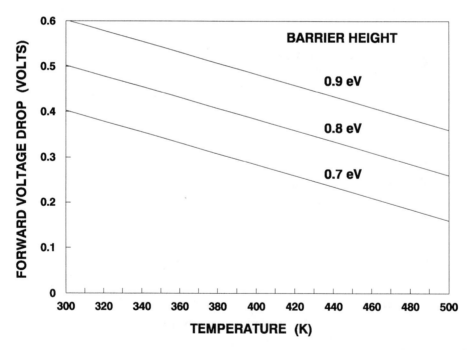

Fig. 4.8. Change in forward voltage drop of a low breakdown voltage Schottky barrier rectifier with temperature.

Chapter 4 : POWER RECTIFIERS

the metal-semiconductor barrier decreases. Concurrently, the series resistance of the drift region (R_D) increases because of a decrease in the mobility. In the case of low reverse blocking rectifiers (e.g., 50 volts), the contribution from the series resistance is small. The calculated variation of the forward voltage drop of a Schottky rectifier between 300°K and 500°K is provided in Fig. 4.8 for such low breakdown voltage rectifiers. By using Eq. (4.7) and (4.8), it can be shown that :

$$V_F = \phi_{bn} + \frac{kT}{q} \ln\left(\frac{J_F}{A T^2}\right) \qquad (4.10)$$

if the series resistance is negligible. Since the logarithmic term is nearly constant and negative, the forward voltage drop decreases linearly with increasing temperature.

4.1.3 Reverse Blocking

When a reverse voltage is applied to the Schottky rectifier, the voltage is supported

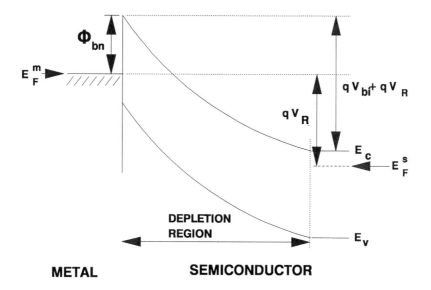

Fig. 4.9. Metal-semiconductor contact with reverse bias.

across a depletion layer which extends into the semiconductor as illustrated in Fig. 4.9. Since no voltage can be supported in the metal, the reverse blocking voltage of the Schottky barrier rectifier is equal to that of an abrupt P-N junction rectifier discussed in Chapter 3. The doping concentration (N_D) and width ($W_{c,PP}$) of the drift region must be scaled with the breakdown voltage (BV_{PP}) as follows:

$$N_D = 2 \times 10^{18} (BV_{PP})^{-4/3} \tag{4.11}$$

and

$$W_{c,PP} = 2.58 \times 10^{-6} (BV_{PP})^{7/6} \tag{4.12}$$

In general, the actual breakdown voltage of the Schottky rectifier is typically about one-third that for the abrupt, parallel-plane junction due to high electric fields at the edges. The doping level and width of the drift region must be adjusted to compensate for this unless special care is taken to suppress high electric fields at the edges.

The depletion region thickness at a reverse bias (V_R) below the breakdown voltage of the diode is given by:

$$W(V_R) = \sqrt{\frac{2 \, \epsilon_s}{q \, N_D} (V_R + V_{bi})} \tag{4.13}$$

because all the applied bias is supported in the semiconductor. This depletion width can be used to calculate the capacitance per unit area of the Schottky barrier rectifier:

$$C(V_R) = \frac{\epsilon_s}{W(V_R)} \tag{4.14}$$

The leakage current of Schottky rectifiers consists of the space charge generation current from the depletion region, the diffusion current from the semiconductor neutral region, and the thermionic current across the barrier. The first two components are the same as those observed in P-i-N rectifiers. The third component is substantially greater than that for P-i-N rectifiers. This has been an important limitation to the high temperature performance of Schottky rectifiers. When a reverse bias voltage is applied to the Schottky barrier, Eq. (4.5) can be rewritten as:

$$J_R = A T^2 e^{-(q\phi_{bn}/kT)} \left[e^{-(qV_R/kT)} - 1 \right] \tag{4.15}$$

Since the applied reverse bias voltage (V_R) is much greater than the thermal energy (kT/q), the exponential term in the square brackets becomes negligible. Consequently,

$$J_R = - A T^2 e^{-(q\phi_{bn}/kT)} = - J_s \tag{4.16}$$

As discussed earlier with the aid of Fig. 4.6, the saturation current density increases with reduction in the Schottky barrier height and with increase in temperature.

In Eq. (4.16), the barrier height (ϕ_{bn}) is given as a constant. Under the application of the reverse bias voltage, it has been found that there is a reduction in the Schottky barrier height due to image force lowering. In order to analyze this phenomenon, consider the band diagram shown in Fig. 4.10. When an electron in the semiconductor approaches the metal at a distance x, a

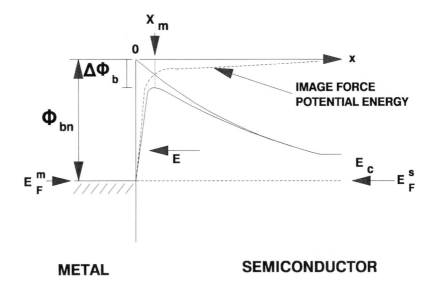

Fig. 4.10. Schottky barrier lowering due to the image force.

positive mirror image charge of the same magnitude occurs in the metal at a distance -x. This results in an electrostatic force on the electron given by :

$$F(x) = \frac{q^2}{4 \pi \varepsilon_s (2x)^2} \tag{4.17}$$

The attractive force between these particles creates a negative potential energy. This is the energy required to take the electron from position x to infinity. The corresponding image force potential (V_I) is given by :

$$q V_I = - \frac{q}{16 \pi \varepsilon_s} \int_x^\infty \frac{dx}{x^2} = \frac{q}{16 \pi \varepsilon_s x} \tag{4.18}$$

The image force potential adds to the potential due to the Schottky barrier. At the distance x_m from the metal-semiconductor interface, the potential goes through a maximum. At this point the image force potential becomes equal to the potential drop across the depletion region due to the electric field (E). Thus :

$$\frac{q}{16 \pi \varepsilon_s x_m} = E_m x_m \tag{4.19}$$

if the assumption is made that the electric field in the vicinity of 0 to x_m is approximately equal to the maximum electric field E_m. The reduction in the barrier height due to image force lowering, indicated as $\Delta\phi_b$ in Fig. 4.10, is given by :

$$\Delta\phi_b = V_I + E_m x_m \qquad (4.20)$$

Using Eq. (4.19) :

$$\Delta\phi_b = \sqrt{\frac{q E_m}{4 \pi \epsilon_s}} \qquad (4.21)$$

where the maximum electric field is related to the applied reverse bias by :

$$E_m = \sqrt{\frac{2 q N_D}{\epsilon_s}(V_R + V_{bi})} \qquad (4.22)$$

As an example, for the case of a drift region doping concentration (N_D) of 1×10^{16} per cm^3, the reduction in barrier height is about 65 millivolts. Although this appears to be a small change, it is an important phenomenon which contributes to an increase in the reverse leakage current with increasing reverse bias voltage because the leakage current is an exponential function of the barrier height. The reverse leakage current including the Schottky barrier lowering effect is given by :

$$J_R(V_R) = - A T^2 e^{-q(\phi_{bn} - \Delta\phi_b)/kT} \qquad (4.23)$$

In addition, the leakage current of the Schottky rectifier contains the space charge generation component (which increases with reverse bias voltage) and the diffusion component, as in the case of the P-N junction rectifiers. These components to the leakage current represent only a small fraction of the total leakage current and can usually be neglected unless the barrier height is very large and the lifetime in the drift region is very low. The calculated leakage current of a Schottky rectifier with and without the Schottky barrier lowering effect are compared in Fig. 4.11. Note the considerable increase in the reverse leakage current due to the Schottky barrier lowering at high reverse voltages. At high reverse voltages, the actual leakage current of Schottky rectifiers is even worse than that predicted by the Schottky barrier lowering.

In order to account for this larger reverse leakage current at high reverse bias voltages, it is necessary to take into account the pre-avalanche multiplication of carriers at the high electric fields when the applied reverse bias approaches close to the breakdown voltage. The impact ionization process can be treated as a purely electron initiated process using the electron current injected across the metal-semiconductor interface by thermionic emission. The total electrons that reach the edge of the depletion region will be larger than those injected at the metal-semiconductor interface by a factor M_n, the electron multiplication factor. An analytical expression for the electron multiplication factor can be obtained by using a power series approximation for the impact ionization coefficients α_n and α_p similar to that given in Eq. (3.3) with the following values :

Chapter 4 : POWER RECTIFIERS

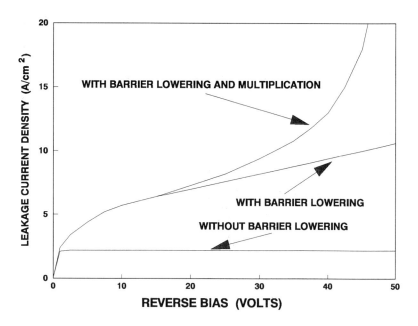

Fig. 4.11 Reverse leakage characteristics of a Schottky barrier rectifier.

$$\alpha_n = 6.6 \times 10^{-24} E^{4.93} \qquad (4.24)$$

and

$$\alpha_p = 2.3 \times 10^{-24} E^{4.93} = 0.344 \alpha_n \qquad (4.25)$$

where it is assumed that the ratio of the ionization coefficients for electrons and holes remains independent of the electric field. The multiplication coefficient is given by:

$$M_n = \{ 1 - 1.52 [1 - \exp(-\frac{4.33 \times 10^{-24} E_m^{4.93} W}{1 + n})] \}^{-1} \qquad (4.26)$$

The calculated reverse leakage current including the pre-avalanche multiplication effect is also shown in Fig. 4.11. This curve is in good agreement with the measured data on Schottky barrier power rectifiers.

The maximum operating temperature of a Schottky barrier power rectifier is determined by the rapid increase in the leakage current with temperature. In order to stress the importance of this phenomenon, a plot of the calculated leakage current density (J_s) as a function of temperature is provided in Fig. 4.12 using a Richardson constant for silicon of 110 amperes/cm²/°K². For any given barrier height, the leakage current exhibits a sharp increase

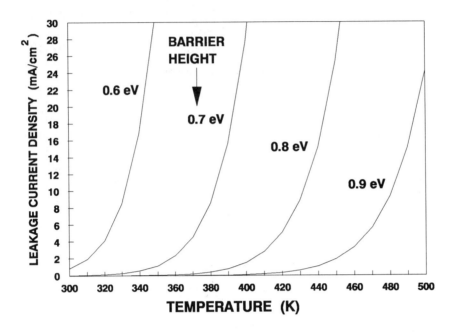

Fig. 4.12 Increase in the leakage current of Schottky barrier rectifiers at higher temperatures.

beyond a certain temperature. When the leakage current becomes sufficiently large, the power dissipation in the reverse blocking mode can exceed that due to the on-state (conduction) losses and the switching losses. This can result in thermal runaway because the power dissipation increases the temperature and establishes a positive feedback mechanism for increasing the leakage current. It is worth pointing out that if the Schottky barrier height is reduced to reduce the on-state power dissipation, the leakage current shows a sharp increase at lower temperatures. This reduces the maximum operating temperature of the Schottky rectifier as discussed below.

4.1.4 Trade-off Curves

From the previous sections, it can be concluded that the optimization of the characteristics of the Schottky power rectifier requires a trade-off between forward voltage drop and reverse leakage current. As the Schottky barrier height (ϕ_{bn}) is reduced, the forward voltage drop decreases but the leakage current increases and the maximum operating temperature decreases. Consequently, low barrier heights should be used for Schottky barrier power rectifiers intended for high current operation with large duty cycles, where the power losses during forward conduction are dominant. Larger barrier heights are necessary for Schottky barrier power rectifiers intended for use in applications with higher reverse bias stress and higher

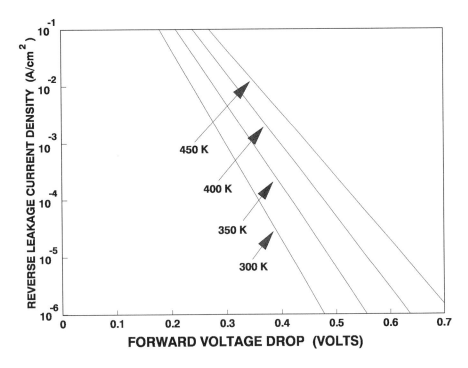

Fig. 4.13 Fundamental trade-off curves between forward voltage drop and reverse leakage current for Schottky barrier rectifiers.

ambient temperatures. A trade-off curve that can be useful during device design is the relationship between the forward voltage drop and the reverse leakage current. This relationship can be derived using Eqs. (4.10) and (4.16) to eliminate ϕ_{bn}:

$$J_R = J_F \, e^{-(qV_F/kT)} \qquad (4.27)$$

In deriving this general relationship, the increase in leakage current due to Schottky barrier lowering and pre-avalanche multiplication has been neglected. In addition, the series resistance of the drift region has been assumed to have a negligible contribution to the forward voltage drop. As discussed earlier, this is true only for devices with low breakdown voltages. The calculated trade-off curves for Schottky rectifiers at several ambient temperatures are provided in Fig. 4.13 by using the barrier height as a parametric variable. It is worth pointing out that the trade-off curves are not dependent upon the semiconductor material used for device fabrication (except for small changes in the Richardson constant A). Since the trade-off curves between forward voltage drop and reverse leakage current are determined solely by the Schottky barrier height, no improvement can be expected by replacing silicon with other semiconductors, such as gallium arsenide or silicon carbide, for these low breakdown voltage rectifiers.

4.1.5. Power Dissipation

The ultimate limiting factor that determines the choice of the Schottky barrier height is the power dissipation in the rectifier. The power dissipation during forward conduction depends upon the forward conduction current, forward voltage drop and duty cycle. The power dissipation during reverse blocking depends upon the leakage current, the reverse bias voltage, and the duty cycle. In choosing the Schottky barrier height, it is important to calculate the total power dissipation as a function of temperature by using the following equation:

$$P_D = J_F V_F \frac{t_{on}}{T} + J_L V_R \frac{(T - t_{on})}{T} \qquad (4.28)$$

where t_{on} is the time period during which the diode is in its on-state; T is the total period; J_F and J_R are the forward (on-state) and reverse leakage current densities, respectively; and V_F and V_R are the forward (on-state) and reverse bias voltages, respectively. The switching losses have been neglected in this equation.

The calculated power dissipation as a function of temperature for the case of four

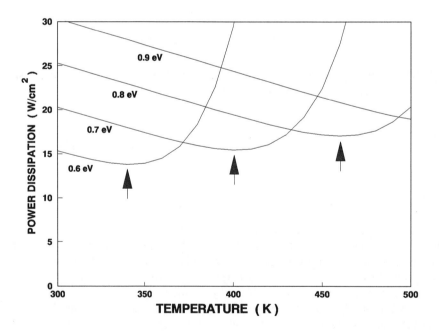

Fig. 4.14. Power dissipation as a function of temperature for Schottky barrier power rectifiers.

Schottky barrier heights is provided in Fig. 4.14. These curves were calculated using a forward current density of 100 amperes per cm², a reverse blocking voltage of 20 volts and a duty cycle of 0.5. It can be seen that the total power dissipation can be reduced by lowering the Schottky barrier height but this is accompanied by a reduction in the maximum operating temperature as indicated by the arrows. Although the arrows are indicated at the minimum power dissipation temperature, the actual maximum operating temperature is slightly higher as determined by the thermal resistance of the heat sink.

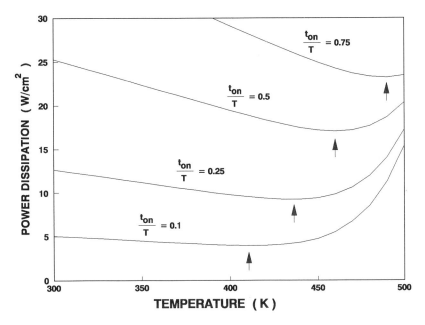

Fig. 4.15. Effect of duty cycle on power dissipation and maximum operating temperature of Schottky barrier power rectifiers.

The effect of changing the duty cycle (t_{on}/T) is shown in Fig. 4.15. In performing these calculations, a fixed Schottky barrier height of 0.8 eV was used with other parameters remaining the same as used for calculating the curves in Fig. 4.14. An important point to note from these curves is that although the total power dissipation decreases as the duty cycle (t_{ON}/T) is reduced, the maximum operating temperature also comes down. Consequently, to maintain a high operating temperature, it becomes necessary to raise the barrier height and incur an increase in total power dissipation. Calculations performed for a variety of switching power supply cases indicate that the lowest Schottky barrier height for these rectifiers should be in excess of 0.7 eV.

4.1.6. Switching Behavior

Forward current transport in the Schottky rectifier occurs primarily via majority carriers with negligible minority carrier injection because of the low values of ϕ_{bn}. As a result, these devices exhibit extremely fast reverse recovery speed. Further, there is no forward overvoltage transient which is experienced with P-i-N rectifiers as discussed later. However, a problem that has been encountered during the application of these very fast switching diodes is severe ringing in the current and voltage waveforms during the turn-off transient. This behavior is not unique to the Schottky rectifiers. It is even observed when using very fast recovery P-i-N rectifiers. The ringing occurs due to the existence of a series resonant circuit formed by the capacitance of the rectifier and any series inductance in the circuit. The ringing phenomenon can be reduced by the connection of an R-C snubber in parallel with the rectifier.

4.1.7. Device Technology

The fabrication of Schottky barrier power rectifiers has been accomplished by using a variety of metals by different manufacturers. The Schottky barrier height (ϕ_{bn}) is dependent upon the metal via its work function. The commonly used metals for Schottky power rectifiers are

Table 4.1. Work functions and Schottky barrier heights of metals used to make silicon Schottky power rectifiers.

Metal	Al	Cr	Mo	Pt	W
Work Function (eV)	4.28	4.50	4.60	5.65	4.55
Barrier Height (eV)	0.60-0.80	0.61	0.68	0.90	0.67

listed in Table 4.1. In many cases, the Schottky barrier height can be obtained using these work functions and the electron affinity for silicon (3.92 eV). The measured barrier height for these metals is also given in Table 4.1. These values have been found to follow the relationship

$$\phi_{bn} = (0.27 \phi_m - 0.55) \tag{4.29}$$

Table 4.2. Schottky barrier heights for silicides formed on N-type silicon.

Silicide	$CrSi_2$	$MoSi_2$	$PtSi_2$	WSi_2
Barrier Height (eV)	0.57	0.55	0.78	0.65

When the metal reacts with the underlying silicon, a metal silicide forms which has different barrier height than the metal. The barrier heights measured for some silicides on N-type silicon are provided in Table 4.2.

The trade off between forward voltage drop and reverse leakage current for Schottky power rectifiers is generally performed by changing the metal used to form the barrier. An alternative processing approach is the use of a very shallow ion implant at the surface of the semiconductor. The introduction of a thin layer, with thickness less than the electron mean free path (typically, less than 100 °A), at the surface of the drift layer with a carefully controlled dose can be used to change the effective barrier height between the metal and semiconductor. For an N-type drift layer, an N-type layer at the surface will lower the barrier height while a P-type layer at the surface will raise the barrier height. This processing approach is attractive because it allows the selection of a metal based upon the metallurgical properties of the interface which will produce the most reliable operation, while allowing the tailoring of the barrier height by controlling the ion implant dose.

Since the optimization of the Schottky barrier can be best achieved by starting with a larger Schottky barrier height and reducing it, consider the case of an N-type semiconductor with a thin N^+ layer at the surface. The resulting electric field and energy band profiles are shown

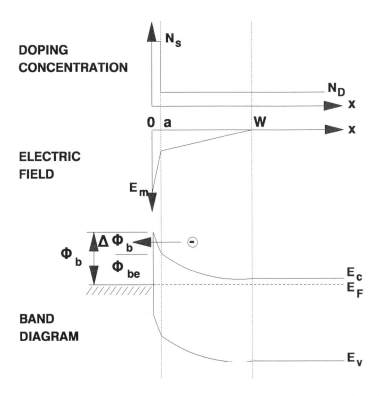

Fig. 4.16. Metal-semiconductor contact with thin, highly doped surface layer.

in Fig. 4.16. The electric field profile can be described by the equations :

$$E(x) = -E_m + \frac{q N_s x}{\epsilon_s} \qquad (4.30)$$

from x = 0 to x = a, and

$$E(x) = -\frac{q N_D}{\epsilon_s}(W - x) \qquad (4.31)$$

from x = a to x = W, where E_m is the maximum electric field at the metal-semiconductor interface given by :

$$E_m = \frac{q}{\epsilon_s}[N_s a + N_D(W - a)] \qquad (4.32)$$

The Schottky barrier lowering ($\Delta\phi_b$) due to the presence of the surface layer can be obtained by substituting this maximum electric field expression into Eq. (4.21). For the case where the implanted charge ($N_s.a$) is much greater than the charge in the zero bias depletion layer formed in the lightly doped drift region, it can be shown that :

$$\Delta\phi_b \simeq \frac{q}{\epsilon_s}\sqrt{\frac{a N_s}{4 \pi}} \qquad (4.33)$$

The implantation of a shallow charge at the surface with doses ranging from 10^{12} to 10^{13} cm^{-2} can be used to decrease the barrier height by 0.05 to 0.2 eV.

It is important to maintain the implanted charge close to the surface. Antimony implantation at energies of 5 to 10 keV is very effective for accomplishing barrier height tailoring in N-type silicon because its large mass allows implantation close to the surface and its low diffusion coefficient prevents redistribution during subsequent implant activation or other high temperature processing steps. It has been observed that any residual ion implant damage can also produce a reduction in Schottky barrier height indicating that care must be taken to ensure complete annealing of the damage if reproducible results are to be achieved.

4.1.8. Edge Terminations

The method of termination of the Schottky barrier rectifier has been found to be of importance to obtaining acceptable reverse blocking characteristics. A severe electric field crowding occurs at the edge of the metal in the absence of a edge termination structure leading to very high leakage and soft breakdown well below the parallel plane breakdown voltage. For this reason, special edge termination techniques have been developed compatible with the Schottky rectifier fabrication process. The more commonly used approaches are illustrated in Fig. 4.17. The metal field plate structure, illustrated in Fig. 4.17(a), is based upon extending

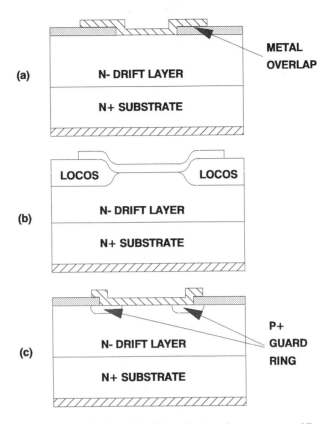

Fig. 4.17. Edge terminations for Schottky barrier power rectifiers.

the Schottky barrier metal over an oxide layer at the edges. The design of this structure is discussed in Chapter 3. A tapered gate oxide achieved by using a high etch rate phosphosilicate glass has been found to provide improved breakdown characteristics. An alternate approach uses the LOCOS (local-oxidation-of-silicon) process to create a tapered oxide at the edges as shown in Fig. 4.17(b). The most commonly used approach is to place a P-type diffused guard ring at the edges of the Schottky barrier metal. The guard ring greatly reduces the electric field crowding at the metal edges and makes the breakdown voltage that for the cylindrical junction as discussed in Chapter 3. It should be noted that a P-N junction is now formed in parallel with the Schottky rectifier. Since the forward voltage drop of the Schottky rectifier is in the range of 0.5 to 0.6 volts, the P-N junction does not inject any significant amount of minority carriers which would adversely impact the high speed switching capability of the Schottky rectifier.

The low forward voltage drop and high switching speed of the Schottky barrier rectifier have made it attractive for high frequency inverters. Until recently, the reverse breakdown voltage of these rectifiers has been limited to less than 50 volts. Devices with breakdown voltages approaching 100 volts are now emerging. The relatively low breakdown voltage of the Schottky rectifier has limited its application mainly as an output rectifier for high frequency switching power supplies. For this application, devices with current ratings of over 50 amperes are available.

4.1.9 High Voltage Schottky Barrier Rectifiers

In the previous sections of this chapter, only silicon device structures were considered. It was pointed out that the silicon Schottky rectifier suffers from a high reverse leakage current and soft breakdown, which has limited its maximum reverse blocking voltage to less than 100 volts. In addition, the high series resistance of the drift layer at higher voltages limits the maximum forward conduction current density. These two drawbacks of the silicon Schottky rectifier can be addressed by fabricating the devices by using other semiconductor materials, such as gallium arsenide and silicon carbide.

As discussed earlier in the chapter, the specific resistance of the drift region is dependent upon the depletion layer width at breakdown, the mobility, and the doping level in the drift region:

$$R_{D,sp} = \frac{W_{c,PP}}{q\,\mu_n\,N_D} \tag{4.34}$$

For the case of the parallel plane junction, the depletion layer width at breakdown is given by

$$W_{c,PP} = 2\,\frac{BV_{PP}}{E_c} \tag{4.35}$$

and the doping concentration in the drift region is given by:

$$N_D = \frac{\epsilon_s\,E_c^2}{2\,q\,BV_{PP}} \tag{4.36}$$

From these expressions, the specific on-resistance can be related to the critical electric field for breakdown:

$$R_{on,sp} = \frac{4\,BV_{PP}^2}{\epsilon_s\,\mu\,E_c^3} \tag{4.37}$$

The demoninator of this expression is referred to as *Baliga's figure of merit*. It can be seen that the specific on-resistance decreases with increasing mobility and critical electric field for breakdown. There are many semiconductors that offer improved specific on-resistances due to either a higher mobility or higher breakdown field strength as compared with silicon.

The mobility for electrons in gallium arsenide at low electric fields is larger than in silicon by a factor of 5.6 times. In addition, because of the larger energy band gap, the critical electric field for breakdown in gallium arsenide is larger than in silicon. These two effects result in a reduction in the specific on-resistance by a factor of 13 times. In addition, the pinning of the Fermi level at the gallium arsenide surface makes the fabrication of stable Schottky barriers relatively easy, resulting in the achievement of breakdown voltages close to the ideal parallel plane case without resorting to special edge terminations.

The calculated forward conduction characteristics of gallium arsenide Schottky barrier

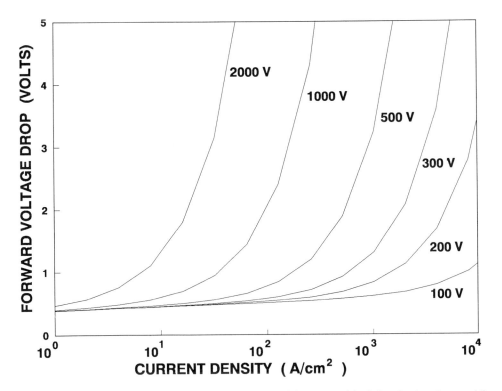

Fig. 4.18 Forward conduction characteristics of gallium arsenide Schottky barrier rectifiers obtained by using a barrier height of 0.8 eV.

rectifiers are provided in Fig. 4.18 for a variety of reverse blocking voltages. In performing these calculations, a Schottky barrier height of 0.8 eV was used. It can be seen that the gallium arsenide Schottky barrier rectifiers are expected to have a lower forward drop than silicon P-i-N rectifiers for breakdown voltages of up to about 500 volts at a typical operating current density of 100 to 200 amperes per cm^2. In this voltage range, the gallium arsenide Schottky barrier rectifiers offer a clear advantage over P-i-N rectifiers due to their inherently higher switching speed because of the absence of the reverse recovery current.

The fabrication of gallium arsenide Schottky barrier power rectifiers has been accomplished by using a relatively simple fabrication process. After the growth of the N-type epitaxial layer upon an N^+ substrate, the ohmic contact is formed on the back of the wafer. This is followed by the evaporation of the Schottky barrier metal onto the surface of the drift layer through a shadow mask. To obtain diodes with breakdown voltages within 90 percent of the ideal parallel plane case, it is important to etch the gallium arsenide surface with Caro's solution (a mixture of sulfuric acid, hydrogen peroxide and water), followed by a dip in warm water and rinse in hydroflouric acid, just prior to metal deposition. In addition, the best yield of high voltage diodes has been obtained by using aluminum as the Schottky metal with the evaporation conducted while the gallium arsenide wafers are heated to 200 °C. Gallium arsenide Schottky barrier rectifiers with breakdown voltages up to 200 volts have been fabricated by using this process. These diodes exhibit excellent high speed switching characteristics.

An even more promising material for power devices is silicon carbide because of its much greater critical electric field for breakdown. There are many poly-types in SiC of which the 3C-SiC and 6H-SiC are the most promising. The critical electric field for breakdown in 3C-SiC is about 1×10^6 V/cm while that for 6H-SiC is about twice as large. In comparison, the critical breakdown field for silicon is an order of magnitude smaller. The electron mobility for 3C-SiC is comparable to that in silicon while that for 6H-SiC is smaller by a factor of two. Using these values, it can be shown that the specific on-resistance for both poly-types is nearly equal and about 200 times smaller than for silicon.

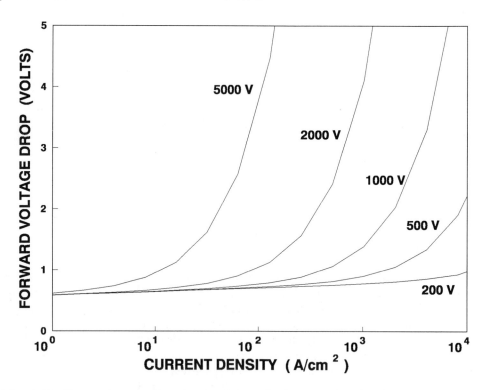

Fig. 4.19 Forward conduction characteristics of silicon carbide Schottky barrier rectifiers obtained by using a barrier height of 1.2 eV.

The calculated on-state characteristics of silicon carbide Schottky barrier rectifiers are shown in Fig. 4.19 for breakdown voltages ranging up to 5000 volts. From these plots, it can be concluded that the on-state voltage drop of the silicon carbide Schottky barrier rectifier is superior to that for even the silicon P-i-N rectifier for blocking voltages of up to 3000 volts. Schottky barrier rectifiers with breakdown voltages of up to 500 volts made by using platinum and titanium have been successfully fabricated with forward voltage drops of 1 volt at an on-state current density of 200 amperes per cm^2. These rectifiers have been found to have excellent reverse recovery and reverse bias leakage characteristics even at high operating temperatures. These rectifiers are likely to replace silicon P-i-N rectifiers in high voltage power electronic circuits.

4.2 P-i-N RECTIFIERS

The P-i-N rectifier was one of the very first semiconductor devices developed for power circuit applications. In this device, the i-region is flooded with minority carriers during forward conduction. Due to this, the resistance of the i-region becomes very small during current flow allowing these diodes to carry a high current density during forward conduction. For this reason, the development of P-i-N rectifiers with very high breakdown voltages ranging up to 5000 volts has been possible. However, the injection of a high concentration of minority carriers into the i-region also creates problems during switching of the P-i-N rectifier.

Two important drawbacks of the P-i-N rectifier are worth pointing out. First, when a P-i-N rectifier is turned on with a high di/dt, its forward voltage drop has been found to initially exceed its voltage drop during current conduction at the same current level under steady state conditions. This phenomenon is called *forward voltage overshoot* during turn-on of the rectifier. The forward voltage overshoot in a P-i-N rectifier arises from the existence of the highly resistive i-region. Under steady state current conduction, the i-region resistance is drastically reduced by the injected minority carriers. However, during turn-on under high di/dt conditions, the current rises at a faster rate than the diffusion of the minority carriers injected from the junction. A high voltage drop develops across the i-region for a short time until the minority carriers can swamp out the i-region resistance. A typical example of the diode current and voltage waveforms observed during the forward recovery process is provided in Fig. 4.20. Note that the voltage overshoot can be an order of magnitude larger than the forward voltage drop (V_F). Under very high di/dt's, voltage overshoots in excess of 30 volts have been observed.

The magnitude of the voltage overshoot is dependent upon the resistivity and thickness of the i-region. In general, the i-region should be designed to minimize its resistance within the constraints of achieving the necessary reverse blocking capability. A high forward overshoot in the rectifier can be a serious problem in power circuits because this voltage may appear across the emitter-base junction of a bipolar transistor used as the active element and exceed its breakdown voltage.

The second and more serious drawback of the P-i-N rectifier is its poor reverse recovery characteristic. Reverse recovery is the process whereby the rectifier is switched from its on-state

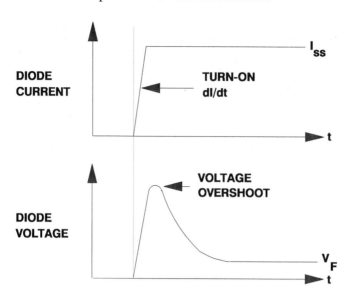

Fig. 4.20 Forward recovery waveforms for a P-i-N rectifier.

characteristic. Reverse recovery is the process whereby the rectifier is switched from its on-state to its reverse blocking state. To undergo this transition, the minority carrier charge stored in the i-region during forward conduction must be removed. The removal of the stored charge occurs via two phenomena, namely, the flow of a large reverse current followed by recombination. The reverse recovery waveforms for a P-i-N rectifier are schematically illustrated in Fig. 4.21. Note the existence of a large reverse current pulse during the turn-off process. The peak reverse current I_{RP} is typically equal to the forward current I_F. This current flows through the transistors used in the circuit adding to power dissipation and degrading their reliability. An additional concern is the large voltage overshoot represented by the peak (V_{RP}). This voltage overshoot is caused by the reverse recovery di/dt current flow through inductances in the circuit.

When the switching frequency of a power circuit increases, the turn-off dI/dt must be increased. It has been found that this causes an increase in both the peak reverse recovery current (I_{RP}) and the ensuing reverse recovery dI/dt. If the reverse recovery dI/dt is large, an increase in the breakdown voltage of all the circuit components becomes essential. Raising the breakdown voltage capability causes an increase in the forward voltage drop of power transistors which degrades system efficiency. Consequently, much of the recent work on P-i-N rectifiers has been focused upon improving the reverse recovery characteristics.

A trade-off between the switching speed and the forward voltage drop is essential during power rectifier design. This trade-off is dependent upon a number of factors such as the N-base width, the recombination center position in the energy gap, the distribution of the deep level impurities and the doping profile in the i-region as discussed in Chapter 2.

Chapter 4 : POWER RECTIFIERS

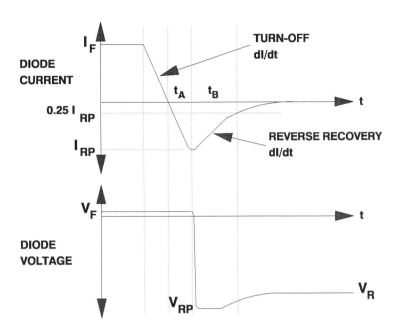

Fig. 4. 21 Reverse recovery waveforms for a P-i-N rectifier.

4.2.1 Forward Conduction

In this section, the physics of current flow in the on-state of a P-i-N rectifier is discussed with emphasis on the characteristics at high current densities. Since the current transport across a P-N junction at low current densities has been discussed in many textbooks on semiconductor devices, it will not be discussed in detail here. As in the case of the basic P-N junction, the injection of minority carriers plays a major role in determining the forward conduction characteristic of the P-i-N rectifier. In this case, the nature of the current-voltage characteristic depends strongly upon the current level.

4.2.1.1 Recombination Current: At very low injection levels, the current flow is dominated by recombination occurring within the space charge layer of the P-N junction due to the presence of deep levels. If the applied bias is V_a, the minority carrier concentration on the P-side of the junction is given by :

$$n_p = \frac{n_i^2}{N_A} e^{\frac{q V_a}{k T}} \qquad (4.38)$$

Consequently, the pn product on the P region is given by :

$$pn = n_i^2 e^{\frac{qV_a}{kT}} \quad (4.39)$$

A similar analysis indicates that the pn product on the other side of the junction is also given by this expression, which allows the assumption that it must be approximately constant throughout the depletion region. Based upon the Shockley-Read-Hall theory for recombination via a deep level impurity, the recombination rate is given by:

$$U = \frac{np - n_i^2}{\tau_{sc}(n + p + 2n_i)} \quad (4.40)$$

Substituting Eq. (4.39) into Eq. (4.40) and making the assumption that p = n, it can be shown that:

$$U = \frac{n_i}{2\tau_{sc}}(e^{\frac{qV_a}{2kT}} - 1) \quad (4.41)$$

Each of the carriers generated within the depletion region will be swept out creating a current flow given by:

$$J_F = \frac{qn_i W_D}{2\tau_{sc}}(e^{\frac{qV_a}{2kT}} - 1) \quad (4.42)$$

where W_D is the depletion layer width. This form of current dependence upon the forward bias voltage is observed at very low current levels in P-i-N rectifiers.

4.2.1.2 Low Level Injection Current Flow. As the forward bias on the P-N junction is increased, the current flow is determined by recombination of minority carriers injected into the neutral regions on either side of the P-N junction. Consider the case where the minority carrier density injected into the neutral regions has a concentration well below that of the majority carrier density. This is referred to as *low level injection*. Under these conditions, the effect of perturbation of the majority carrier concentration can be neglected. A cross-section of the P-N junction and the minority carrier density is illustrated in Fig. 4.22. Using the *law of the junction* derived under Boltzmann quasi-equilibrium assumptions:

$$p_N(0) = p_{oN} e^{\frac{qV_a}{kT}} \quad (4.43)$$

where $p_N(0)$ is the electron concentration at the edge of the depletion boundary on the N-region. Since the injected carriers diffuse away from the junction with a characteristic length referred to as the *diffusion length* (L_p), the minority carrier distribution in the N-region is given by:

Chapter 4 : POWER RECTIFIERS

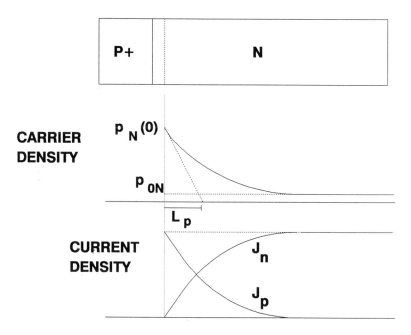

Fig. 4.22 P-N junction under low level injection conditions.

$$p_N = p_N(0)\ e^{-\frac{x}{L_p}} \qquad (4.44)$$

For a uniformly doped N-region, the current flow at low injection levels occurs exclusively by diffusion. The electron current due to the diffusion of these carriers is then given by:

$$J_p = q\ D_p \left(\frac{dp_N}{dx}\right)_{x=0} = \frac{q\ D_p\ p_{0N}}{L_p}\ (e^{\frac{q V_a}{kT}} - 1) \qquad (4.45)$$

It is worth pointing out that the minority carrier diffusion current component decreases with distance x away from the junction, while the majority carrier drift current increases with x. This majority carrier drift current is supported by a small electric field in the N-region. Note that the rate of change in current with forward bias changes from that for recombination current in the depletion region.

In deriving the above expression, it was assumed that the width of the N-region is very large as compared with the diffusion length. If the width of the N-region is comparable to the diffusion length and an ohmic contact is assumed at its end, then the minority carrier density must reduce to zero at this boundary. This distorts the exponential carrier distribution shown in Fig. 4.22, resulting in a modification of the diffusion current to:

$$J_p = \frac{q \, D_p \, p_{0N}}{L_p \tanh(W/L_p)} (e^{\frac{q V_a}{kT}} - 1) \qquad (4.46)$$

where W is the width of the N-region. A similar expression can be derived for the diffusion current arising from the injected carriers on the P-side of the junction.

4.2.1.3 High Level Injection Current Flow. Consider the cross-section of a P-i-N rectifier, as shown in Fig. 4.23, with an N-type drift (i) region. In order to support large reverse

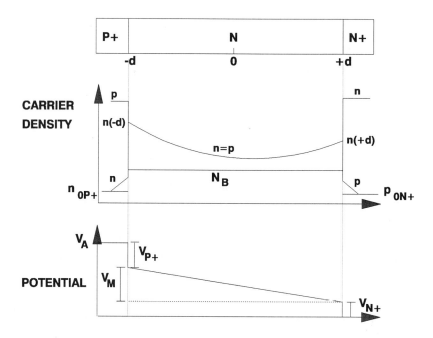

Fig. 4.23 Carrier and potential distribution profiles in a P-i-N rectifier under high level injection conditions.

blocking voltages, it is necessary to use a low doping concentration (N_B) as well as a large thickness (2d) for the drift region as discussed for the punch-through breakdown case in Chapter 3. During on-state current flow, as the current density increases, the injected carrier density also increases and ultimately exceeds the relatively low background doping of the N-base region. This condition is called *high level injection*. When the injected hole density becomes much greater than the background doping, charge neutrality in the N-base region requires that the concentrations of holes and electrons become equal:

Chapter 4 : POWER RECTIFIERS 159

$$n(x) = p(x) \quad (4.47)$$

These concentrations can become far greater than the background doping level resulting in a large decrease in the resistance of the i-region. This phenomenon, called *conductivity modulation*, is an extremely important effect that allows transport of a high current density through the P-i-N rectifier with low on-state voltage drop.

Under steady state conditions, the current flow in the P-i-N rectifier can be accounted for by the recombination of holes and electrons in the N-base region and the anode/cathode end regions. First, consider the case where the recombination in the end regions is negligible, i.e., the end regions have unity injection efficiency. The current density is then determined by recombination in the N-base region:

$$J = \int_{-d}^{+d} q R \, dx \quad (4.48)$$

where R is the recombination rate given by:

$$R = \frac{n(x)}{\tau_{HL}} \quad (4.49)$$

If the high level lifetime (τ_{HL}) is assumed to be independent of the carrier density :

$$J = \frac{2 q n_a d}{\tau_{HL}} \quad (4.50)$$

where n_a is the average carrier density and d is half the N-base width. From this equation, it can be seen that the free carrier density in the drift region increases in proportion to the current density. Since this increase in carrier density will result in a proportional increase in the conductivity of the drift region, it can be concluded that the voltage drop in the drift region will be independent of the current density. This is important for maintaining a low on-state voltage drop even at high operating current densities in P-i-N rectifiers.

In order to solve for the actual free carrier distribution in the drift region, it is necessary to use the continuity equation :

$$\frac{dn}{dt} = 0 = -\frac{n}{\tau_{HL}} + D_a \frac{d^2n}{dx^2} \quad (4.51)$$

where D_a is the ambipolar diffusion coefficient. Defining an ambipolar diffusion length as :

$$L_a = \sqrt{D_a \tau_{HL}} \quad (4.52)$$

the preceding equation can be written as :

$$\frac{d^2n}{dx^2} - \frac{n}{L_a^2} = 0 \tag{4.53}$$

The boundary conditions needed to solve this equation are obtained by the current transport occurring at the P⁺ and N⁺ ends of the diode. At the N+ boundary, the current transport is due to electrons and the hole current goes to zero. Thus:

$$J_p(+d) = q \mu_p p(+d) E(+d) - q D_p \left(\frac{dp}{dx}\right)_{x=+d} = 0 \tag{4.54}$$

Using the high level approximation $n(x) = p(x)$, the electric field at $x = +d$ can be related to the carrier density:

$$E(+d) = \frac{kT}{q\, n(+d)} \left(\frac{dn}{dx}\right)_{x=+d} \tag{4.55}$$

The current density at $x = +d$ due to electron transport is given by:

$$J = J_n(+d) = q \mu_n n(+d) E(+d) + q D_n \left(\frac{dn}{dx}\right)_{x=+d} \tag{4.56}$$

Substituting Eq. (4.55) in Eq. (4.56), it can be shown that

$$J = 2 q D_n \left(\frac{dn}{dx}\right)_{x=+d} \tag{4.57}$$

Analysis of the boundary condition at the opposite end of the device gives:

$$J = J_p(-d) = -2 q D_p \left(\frac{dn}{dx}\right)_{x=-d} \tag{4.58}$$

The solution of Eq. (4.53) with these boundary conditions is given by:

$$n = p = \frac{\tau_{HL} J}{2 q L_a} \left[\frac{\cosh(x/L_a)}{\sinh(d/L_a)} - \frac{\sinh(x/L_a)}{2\cosh(d/L_a)}\right] \tag{4.59}$$

This catenary carrier distribution is illustrated in Fig. 4.23. The hole and electron concentrations are the highest at the P⁺/N(-d) and N/N⁺(+d) junctions, with the minimum closer to the cathode side due to the difference in the mobility of electrons and holes. The extent of the drop in the carrier concentration away from the junctions is dependent upon the ambipolar diffusion length. At medium current densities, this diffusion length is controlled by the high level lifetime.

To determine the voltage drop across the rectifier, it is necessary to first obtain the electric field distribution. The current flow in the base region is related to the electric field by:

$$J = J_p + J_n \tag{4.60}$$

with

$$J_p = q\, \mu_p \left(p\, E - \frac{kT}{q}\, \frac{dp}{dx} \right) \tag{4.61}$$

and

$$J_n = q\, \mu_n \left(n\, E + \frac{kT}{q}\, \frac{dn}{dx} \right) \tag{4.62}$$

From these equations and the charge neutrality condition defined by Eq. (4.47), it can be shown that:

$$E = \frac{J}{q\,(\mu_n + \mu_p)\, n} - \frac{kT}{2qn}\, \frac{dn}{dx} \tag{4.63}$$

In this equation, the first term on the right hand side accounts for the ohmic drop due to current flow and the second term accounts for the asymmetric concentration gradient produced by the unequal electron and hole mobilities in silicon. The voltage drop across the N-base region (V_m) can be obtained by integrating the electric field distribution using the carrier distribution given by Eq. (4.59). The following approximations can be used to calculate the voltage drop in the middle region:

$$V_m = \frac{3\,kT}{q} \left(\frac{d}{L_a} \right)^2 \qquad \text{for} \quad d \leq L_a \tag{4.64}$$

and

$$V = \frac{3\,\pi\,kT}{8\,q}\, e^{d/L_a} \qquad \text{for} \quad d \geq L_a \tag{4.65}$$

It is important to note that the voltage drop in the middle region is independent of the current density because the free carrier concentration increases in proportion to the current density. A normalized plot of the voltage drop across the middle (N-base) region is provided as a function of the normalized base width in Fig. 4.24. Note the very rapid increase in the ohmic voltage drop with increasing ratio (d/L_a). As the half-base width (d) becomes longer than the diffusion length, the voltage drop across the middle region causes an appreciable increase in the forward voltage drop due to a reduction in the conductivity modulation in the central portion of the N-base region. In devices that must operate at higher switching frequencies, the lifetime must be reduced to such an extent that the middle region voltage drop becomes an important contributor to the forward drop.

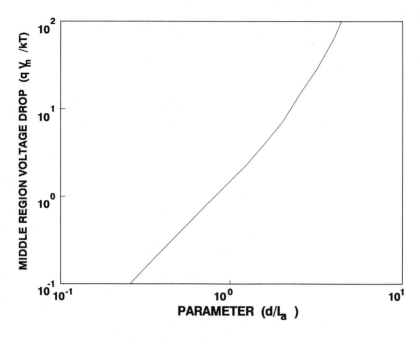

Fig. 4.24 Voltage drop in middle (i) region of P-i-N rectifier.

The voltage drops across the anode (P^+-N) and cathode (N^+-N) junctions also contribute to the forward voltage drop as illustrated in the potential profile shown in the lower portion of Fig. 4.23. The voltage drop across these junctions can be related to the injected minority carrier density at the junction:

$$p(-d) = p_o \, e^{\frac{q V_{P+}}{kT}} \qquad (4.66)$$

where p_o is the minority carrier density in the N-base region at thermal equilibrium and V_{P+} is the voltage drop across the anode junction. From this expression, the voltage drop across the anode junction is given by:

$$V_{P+} = \frac{kT}{q} \ln\left[\frac{p(-d) \, N_D}{n_i^2}\right] \qquad (4.67)$$

Similarly, at the cathode junction

$$n(+d) = n_o \, e^{\frac{q V_{N+}}{kT}} \qquad (4.68)$$

resulting in

Chapter 4 : POWER RECTIFIERS 163

$$V_{N+} = \frac{kT}{q} \ln\left[\frac{n(+d)}{N_D}\right] \qquad (4.69)$$

The total voltage drop across the end regions is obtained from Eqs. (4.67) and (4.69):

$$V_{P+} + V_{N+} = \frac{kT}{q} \ln\left[\frac{n(+d)\, n(-d)}{n_i^2}\right] \qquad (4.70)$$

In deriving this equation, the charge neutrality condition has been used. Combining with Eq. (4.59), the current density of a forward biased diode at high level injection in the absence of recombination in the end regions is given by:

$$J = \frac{2\,q\,D_a\,n_i}{d}\, F\left(\frac{d}{L_a}\right)\, e^{\frac{qV_a}{2kT}} \qquad (4.71)$$

where

$$F\left(\frac{d}{L_a}\right) = \frac{(d/L_a)\, \tanh(d/L_a)}{\sqrt{1 - 0.25\, \tanh^4(d/L_a)}}\, e^{-\frac{qV_M}{2kT}} \qquad (4.72)$$

A plot of the function $F(d/L_a)$ is provided in Fig. 4.25. From Eq. (4.71), it can be concluded that a low forward drop occurs when the function F is large. From Fig. 4.25, it can be seen that the highest value for F occurs at (d/L_a) values close to unity.

For a fixed value of d, as required to obtain the desired forward and reverse blocking capability, the forward drop goes through a minimum as the diffusion length (L_a) increases. The variation in the forward drop with increasing (d/L_a) ratio is shown in Fig. 4.26 for a typical operating current density of 280 amperes per cm^2. The lowest forward drop occurs when the (d/L_a) ratio becomes equal to unity. The minimum in the forward drop arises because the voltage drop across the middle region decreases with increasing diffusion length. Higher diffusion lengths imply larger high level lifetime, which increases the middle region conductivity modulation [see Eq. (4.50)] and reduces the ohmic drop across it. However, the resulting higher injected free carrier concentration at the anode and cathode junctions must be provided by a higher voltage drop across these junctions as described by Eq. (4.70). At small values of (d/L_a), the junction drops are predominant and the forward drop decreases with increasing (d/L_a) ratio. At large values of (d/L_a) the middle region ohmic drop becomes predominant and the forward drop increases with increasing (d/L_a) ratio. The minimum occurs at (d/L_a) of unity. It should be noted that this discussion applies only when the end region and Auger recombinations are neglected.

From the above discussion of the forward conduction characteristics as a function of injection level, it can be concluded that (a) at extremely low current densities, the current should vary as $\exp(qV_A/2kT)$ due to space-charge generation current flow, (b) at low current densities where low level injection prevails, the current should vary as $\exp(qV_A/kT)$ due to diffusion

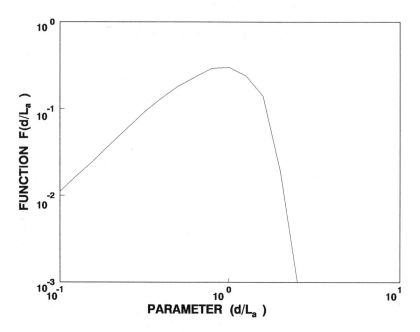

Fig. 4.25 Variation of the function $F(d/L_a)$ with d/L_a.

limited current flow, and (c) at medium current densities when high level injection prevails, the current should vary as $\exp(qV_A/2kT)$. At very high current levels, the output characteristic is found to deviate from the exponential behavior discussed above. This occurs because of the onset of significant recombination in the end regions as well as a reduction in the diffusion length due to carrier-carrier scattering.

At high current densities, the injection of minority carriers into the anode and cathode regions becomes significant. Since the minority carrier lifetime decreases rapidly with increasing doping level, recombination in the end regions adds additional components to the forward current:

$$J = J_{P+} + J_M + J_{N+} \qquad (4.73)$$

The current density due to injected carriers in the end regions can be derived from low level injection theory because minority carrier density is far smaller than the high doping levels of the end regions. Thus:

$$J_{P+} = \frac{q\, D_{nP+}\, n_{0P}}{L_{nP+}\, \tanh(W_P/L_{nP+})}\, e^{\frac{q V_{P+}}{kT}} = J_{P+S}\, e^{\frac{q V_{P+}}{kT}} \qquad (4.74)$$

and

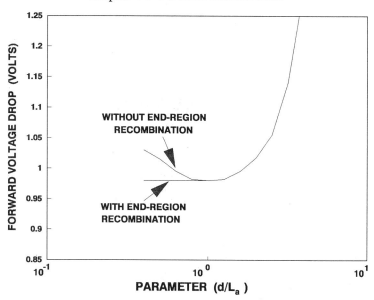

Fig. 4.26 Calculated variation in the forward voltage drop of a P-i-N rectifier with increase in (d/L$_a$).

$$J_{N+} = \frac{q\, D_{pN+}\, p_{0N}}{L_{pN+}\, \tanh(W_N/L_{pN+})}\, e^{\frac{q\, V_{N+}}{k\, T}} = J_{N+S}\, e^{\frac{q\, V_{N+}}{k\, T}} \quad (4.75)$$

where W_P and W_N are the thickness of the anode and cathode regions. The parameters J_{P+S} and J_{N+S} are called the *saturation current densities* of heavily doped P$^+$ and N$^+$ regions. The saturation current densities are a measure of the quality of the end regions and depend upon the method of preparation. Typical values for the saturation current densities are in the range of 1×10^{-13} to 4×10^{-13} amperes per cm^2.

Using quasi-equilibrium at the P$^+$/N and N$^+$/P junctions, the injected carrier concentrations on either side of the junction are related by:

$$\frac{p_{P+}(-d)}{p(-d)} = \frac{n(-d)}{n_{P+}(-d)} \quad (4.76)$$

where p_{P+} and n_{P+} are hole and electron concentration on the anode side (P$^+$) of the junction. Due to the low injection level on the heavily doped side:

$$p_{P+}(-d) = p_{0P+} \quad (4.77)$$

The injected electron concentration in the P$^+$ anode region is related to the voltage across the anode junction:

$$n_{P+}(-d) = n_{0P+} \, e^{\frac{q V_{P+}}{kT}} \tag{4.78}$$

Using these two expressions in Eq. (4.76):

$$n(-d)\, p(-d) = p_{0P+}\, n_{0P+}\, e^{\frac{q V_{P+}}{kT}} = n_{ieP+}^2 \, e^{\frac{q V_{P+}}{kT}} \tag{4.79}$$

where n_{ieP+} is the effective intrinsic concentration in the P^+ anode including the effect of band gap narrowing discussed in Chapter 2. Using the charge neutrality condition:

$$e^{\frac{q V_{P+}}{kT}} = \left[\frac{n(-d)}{n_{ieP+}}\right]^2 \tag{4.80}$$

Substituting this expression into Eq. (4.74) gives:

$$J_{P+} = J_{P+S} \left[\frac{n(-d)}{n_{ieP+}}\right]^2 \tag{4.81}$$

A corresponding derivation at the cathode end of the rectifier yields:

$$J_{N+} = J_{N+S} \left[\frac{n(+d)}{n_{ieN+}}\right]^2 \tag{4.82}$$

where n_{ieN+} is the intrinsic carrier concentration on the cathode side including heavy doping effects. In deriving these equations, it has been assumed that the anode and cathode junctions are abrupt and that these end regions are uniformly doped. An important conclusion that can be made from the last two equations is that, if the recombination in the middle region is negligible and end region recombination becomes dominant, the injected carrier density in the N-base region will no longer increase linearly with current density but will increase as the square root of the current density. The forward drop will then increase more rapidly with increasing current density than $\exp(qV_a/kT)$.

At high current densities, the recombination in the end regions is not the only phenomenon responsible for the deviation in the output characteristics from an exponential behavior. Two additional phenomena that impact the current conduction characteristics are carrier-carrier scattering and Auger recombination. These phenomena were discussed in Chapter 2. Carrier-carrier scattering occurs in the middle region at high current densities due to the simultaneous presence of a high concentration of both holes and electrons. The greater probability of mutual Coulombic scattering causes a reduction in the mobility and diffusion length for both carriers. The reduction in diffusion length with increasing current density produces a decrease in the conductivity modulation in the central portion of the middle region, which in turn results in a higher ohmic voltage drop. The effect of including carrier-carrier scattering in the calculation of the forward conduction characteristics is shown in Fig. 4.27 by

Chapter 4 : POWER RECTIFIERS

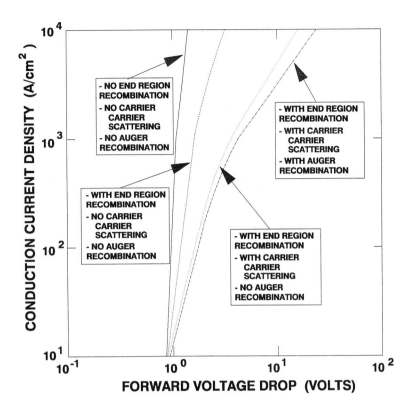

Fig. 4.27 Effect of end region recombination, carrier-carrier scattering, and Auger recombination on the forward conduction characteristics of a P-i-N rectifier.

the dashed line. At typical operating current densities of 200 to 300 Amperes per cm², the forward drop is increased by more than half a volt by carrier-carrier scattering. The effect of carrier-carrier scattering is even greater under current surge conditions during which the current density exceeds 1,000 Amperes per cm².

At high current densities, the carrier density in the N-base region becomes sufficiently large that Auger recombination begins to affect the carrier statistics. In addition to the Shockley-Read-Hall recombination described by Eq. (4.49), the rate of recombination must include the additional Auger recombination process:

$$R = \frac{n(x)}{\tau_{HL}} + C_A \, [n(x)]^2 \qquad (4.83)$$

where C_A is the Auger recombination coefficient. The inclusion of the Auger recombination term modifies the carrier distribution. Consequently, Eq. (4.51) must be rewritten in the form:

$$D \frac{d^2 n(x)}{dx^2} = R = \frac{n(x)}{\tau_{HL}} + C_A [n(x)]^2 \qquad (4.84)$$

A solution for this differential equation has been obtained:

$$n(x) = n_0 \left[cn\left(\frac{x - x_0}{L} \mid m\right) \right]^{-1} \qquad (4.85)$$

where cn(u/m) is a Jacobian elliptic function of argument u. In this expression, n_0 is the value of x at which $dn/dx = 0$ and $n_0 = n(x_0)$. The parameter m is given by:

$$m = \frac{(C_A \tau_{HL} n_0^2 / 2) + 1}{C_A \tau_{HL} n_0^2 + 1} \qquad (4.86)$$

and

$$L = \sqrt{D_a \tau_{HL} (2m - 1)} \qquad (4.87)$$

is the modified ambipolar diffusion length.

The effect of including the Auger recombination upon the output conduction characteristics is shown in Fig. 4.27 by the heavy dashed line. To perform these calculations, an Auger coefficient of 2×10^{-31} cm^6 per sec was used. The impact of Auger recombination is small at operating current densities of 200 to 300 amperes per cm^2 and becomes significant only at very high current densities.

Auger recombination also influences current transport in the end regions. These regions are doped to very high concentrations, generally exceeding 10^{19} atoms per cm^3. Due to the high majority carrier density, Auger recombination alters the minority carrier lifetime. The resulting decrease in the minority carrier diffusion lengths (L_{nP+}, L_{pN+}) increases the recombination current flow (J_{P+}, J_{N+}) into the end regions. The effect of Auger recombination in the end regions is included in the output characteristics shown in Fig. 4.27 by factoring it into the end region saturation current (J_s).

It was previously shown with the aid of Fig. 4.26 that, for a fixed base width, the forward drop goes through a minimum when the diffusion length (L_a) is varied. This minimum was found to occur at (d/L_a) of unity when no end region recombination or carrier-carrier scattering is included in the analysis. When these phenomena are included, the forward voltage drop no longer increases rapidly with increasing diffusion length at low (d/L_a) values. Instead, the forward voltage drop becomes nearly independent of the lifetime in the N-base region. This is illustrated in Fig. 4.26. Note that in addition to the flattening out of the curves at higher base region lifetimes when the end region recombination is included, there is also an increase in the minimum achievable forward voltage drop when the saturated end region current density exceeds 10^{-14} amperes per cm^2.

The processing of the end regions of the P-i-N rectifier must be tailored to obtain the optimum depth and concentration profile that will minimize the saturation current densities in order to obtain the lowest achievable forward voltage drop. Although a high doping of the anode and cathode regions is to be recommended in order to obtain a high injection efficiency, the onset of band gap narrowing and Auger recombination in these regions with increasing doping levels leads to a decrease in the saturation current densities J_{Ps} and J_{Ns}. Calculations of the variation of the saturation current density with increasing doping of the end regions for a step junction indicate that an optimum doping level for the end region exists at which the saturated current density reaches its minimum value. However, the measured saturated current densities in devices fabricated using various processing techniques (including alloying and diffusion) indicate only a weak dependence of the saturated current density upon the doping concentration in the end region as well as the process used to fabricate these regions.

4.2.2 Reverse Blocking

The maximum voltage that can be supported by the P-i-N rectifier depends upon the doping and thickness of the i-region. Since the on-state voltage drop of the P-i-N rectifier can be reduced by decreasing the i-region thickness, it is advantageous to use a punch through design. The analysis of this breakdown has been discussed in Chapter 3 in Section 3.3 on punch-through diodes. In this section, the leakage current under reverse blocking conditions will be discussed. There are two basic mechanisms responsible for current flow under reverse bias conditions. The first is associated with the generation of hole-electron pairs within the depletion region, referred to as *space-charge-generation*, and the second is associated with the generation of hole-electron pairs in the neutral regions that then diffuse to the junction, referred to as *diffusion leakage current*.

4.2.2.1 Space-Charge-Generation Leakage Current. Any hole-electron pairs generated within the depletion region of the reverse biased P-i-N rectifier by thermal energy will be swept out due to the strong electric field. The generation of hole-electron pairs is described by the generation rate:

$$U = \frac{n_i}{\tau_{sc}} \quad (4.88)$$

where τ_{sc} is the space charge generation lifetime defined in Chapter 2 in Section 2.3.2.

Consider a reverse biased P/N junction, as illustrated in Fig. 4.28, with the depletion region extending primarily on the N-side. Each generation event results in the flow of an electron from the N-side to the P-side, with the complementary flow of a hole in the opposite direction to complete the circuit. If the generation process is assumed to be uniform over the depletion region, the leakage current due to space generation will be given by:

$$J_{sc} = \frac{q W n_i}{\tau_{sc}} \quad (4.89)$$

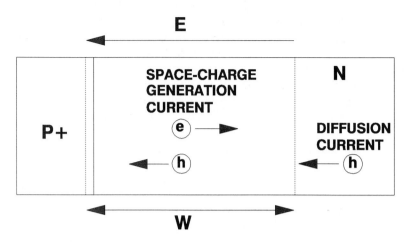

Fig. 4.28 Leakage current flow in a reverse biased P-N junction.

where W is the depletion layer thickness. As discussed earlier, the depletion region thickness increases with increasing reverse bias:

$$W = \sqrt{\frac{2 \, \epsilon_s \, V_a}{q \, N_D}} \qquad (4.90)$$

Consequently, the leakage current due to space charge generation will increase with increasing reverse bias until the depletion layer no longer expands due to punch-through.

The space charge generation current is a strong function of temperature because the generation of hole-electron pairs is enhanced by the additional thermal energy at higher temperatures. This effect is inherent in Eq. (4.89) via the temperature dependence of n_i upon temperature as discussed previously in Chapter 2. However, the space charge generation current is dominant at lower power device operating temperatures (typically at below 100 degree Centigrade) because the diffusion current component is an even stronger function of temperature.

4.2.2.2 Diffusion Leakage Current: The generation of electron-hole pairs in the neutral regions also produces a leakage current flow in the reverse biased P-N junction. The minority carriers generated in the neutral region located in proximity with the depletion boundaries diffuse to the depletion region and are then swept to the opposite side by the electric field in the depletion region. The diffusion current can be calculated from the equation for P-N junction current flow under low level injection conditions described by Eq. (4.45) by setting V_a as a negative value. Since the exponential term in the equation then becomes extremely small, the diffusion leakage current is given by:

Chapter 4 : POWER RECTIFIERS

$$J_{DL} = \frac{q\, D_p\, p_{0N}}{L_p} = \frac{q\, D_p\, n_i^2}{L_p\, N_D} \tag{4.91}$$

where L_p is the diffusion length on the N-side of the junction. A similar expression can be written for the diffusion leakage current due to minority carriers generated in the neutral P-side of the junction.

The diffusion leakage current is not a function of the applied reverse bias except for very small values. Further, it can be seen that the diffusion component of the leakage current increases more rapidly with temperature than the space charge generation current because it varies as the square of the intrinsic concentration (n_i). As a consequence, the diffusion leakage current becomes dominant in power devices at temperatures above about 100 °C.

4.2.2.3 Total Leakage Current. The total reverse leakage current for a P-i-N rectifier with N-type drift region is then given by:

$$J_L = J_{DP} + J_{SC} + J_{DN} \tag{4.92}$$

where J_{DP} is the diffusion current component from the P+ region, J_{SC} is the space charge generation current from the depletion region, and J_{DN} is the diffusion current component from the N or N+ region. From the previous sections, the total leakage current is given by:

$$J_L = \frac{q\, D_n\, n_i^2}{L_n\, N_A} + \frac{q\, W\, n_i}{\tau_{sc}} + \frac{q\, D_p\, n_i^2}{L_p\, N_D} \tag{4.93}$$

This expression can be used to calculate the leakage current for the reverse biased P-i-N rectifier as a function of bias and temperature. A low leakage current is obtained when the space charge and low level lifetime (which determines the diffusion length) are large. This condition is not satisfied in fast recovery rectifiers because of the use of lifetime control to reduce switching time.

4.2.3 Reverse Recovery

A major limitation to the performance of P-i-N rectifiers at high frequencies is the losses that occur during switching from the on-state to the off-state. This process of switching from the on-state to the off-state is called *reverse recovery*. As illustrated in Fig. 4.21, a large reverse transient current occurs in P-i-N rectifiers during reverse recovery. Since the voltage across the rectifier is also large during the second portion of the recovery following the peak in the reverse current, a large power dissipation occurs in the rectifier. In addition, the peak reverse current adds to the average current flowing through the transistors that are controlling the current flow in the circuit. This not only produces an increase in the power dissipation in the transistors but also creates a high internal stress that can cause second breakdown induced failure. A simplified analysis of the reverse recovery of the P-i-N rectifier is presented here for understanding the underlying physics.

Consider the linearized turn-off current waveform shown in Fig. 4.29 with a constant

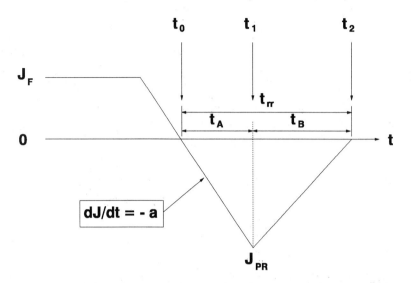

Fig. 4.29 Linearized reverse recovery waveform for a P-i-N rectifier.

di/dt during turn-off from an initial current density J_F. Using charge control analysis:

$$\frac{dQ(t)}{dt} = J(t) - \frac{Q(t)}{\tau_{HL}} \quad (4.94)$$

where Q(t) is the stored charge per unit area in the P-i-N rectifier due to the on-state current flow. Since the current density decreases linearly with time:

$$Q(t) = \int_0^{t_1} J(t) \, dt = \int_0^{t_1} (J_F - at) \, dt \quad (4.95)$$

The solution for this integral is:

$$Q(t) = J_F t - \frac{a t^2}{2} \quad (4.96)$$

In order to relate this charge to the change in the charge distribution within the P-i-N diode structure, consider the carrier density profile at the P+/N junction illustrated in Fig. 4.30 at various time intervals. In this figure, the carrier density in the on-state is the same as that derived previously for high level injection conditions. It is assumed that during the ramp recovery, the carrier density decreases at the P+/N junction while remaining relatively constant within the N-region at a distance b from the junction. The on-state current density is then given

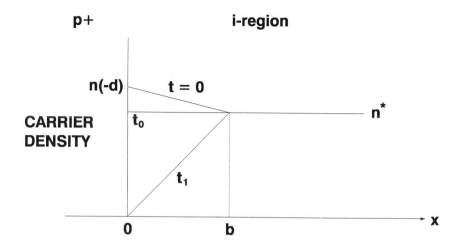

Fig. 4.30 Change in carrier distribution within the P-i-N rectifier during reverse recovery.

by:

$$J_F = 2 q D_n \frac{dn}{dx}\bigg|_{x=-d} \quad (4.97)$$

with

$$\frac{dn}{dx}\bigg|_{x=-d} = \frac{n(-d) - n^*}{b} \quad (4.98)$$

Using these expressions, the distance b can be related to the on-state current density:

$$b = \frac{2 q D_n [n(-d) - n^*]}{J_F} \quad (4.99)$$

The peak reverse recovery current occurs at time t_1 when the P-i-N rectifier begins to support the reverse voltage. At this time, the carrier density at the P+/N junction becomes equal to zero as illustrated in Fig. 4.30. Under the assumption that the carrier density varies linearly over the distance b from 0 to n^*, the peak reverse recovery current density is given by:

$$J_{PR} = 2 q D_n \frac{n^*}{b} \quad (4.100)$$

Making the assumption that the on-state current density is determined by recombination in the middle region:

$$n^* = \frac{J_F \tau_{HL}}{2 q d} \tag{4.101}$$

Substituting for n^* in Eq. (4.100):

$$J_{PR} = \frac{\tau_{HL} D_n}{b d} J_F \tag{4.102}$$

From this equation, it can be concluded that the peak reverse recovery current can be reduced by decreasing the high level lifetime in the middle region.

The reverse recovery time t_{rr} can be determined by relating the charge removed during the reverse recovery process to the initial stored charge within the middle region. Since the charge removed during the reverse recovery process is the area under the reverse recovery current waveform and the initial stored charge is given by the product of the average carrier concentration n^* in the middle region and its thickness (2d):

$$\frac{1}{2} J_{PR} t_{rr} = Q_s = q n^* 2d = J_F \tau_{HL} \tag{4.103}$$

By using Eq. (4.102) and Eq. (4.103), it can be shown that:

$$t_{rr} = 2 \tau_{HL} \frac{J_F}{J_{PR}} = \frac{2 b d}{D_n} \tag{4.104}$$

From this expression, it can be concluded that a smaller reverse recovery time can be obtained by reducing the high level lifetime and by increasing the peak reverse current relative to the on-state current. The latter is achieved by increasing the di/dt (a in Fig. 4.29) during turn-off.

The time t_1 taken for the current to reach its peak reverse value can be solved for by using Eq. (4.96) and the charge removed from the middle region during the time interval from t_0 to the time t_1. The charge removed during this time interval can be obtained from Fig. 4.30 by using:

$$\frac{1}{2} J_{PR} t_A = Q_R(t_A) = \frac{1}{2} q b n^* \tag{4.105}$$

Substituting for n^* in this equation and solving for t_A:

$$t_A = \frac{b \tau_{HL}}{2 d} \frac{J_F}{J_{PR}} \tag{4.106}$$

The second portion of the reverse recovery occurs over a duration t_B which can be obtained by:

$$t_B = t_{rr} - t_A = \left(2 - \frac{b}{2 d}\right) \tau_{HL} \frac{J_F}{J_{PR}} \tag{4.107}$$

Chapter 4 : POWER RECTIFIERS

In general, it is desirable to have a larger ratio of t_A to t_B in order to reduce the di/dt during the period t_B. Using the above equations, it can be shown that:

$$\frac{t_B}{t_A} = \left(\frac{4\,d}{b} - 1\right) \qquad (4.108)$$

Thus, a smaller value of b will favor a softer reverse recovery waveform.

4.2.4 Lifetime Control

From the above discussion, it can be concluded that a faster reverse recovery can be obtained in a P-i-N rectifier when the lifetime in the middle region is reduced. The most popular approach to controlling the switching speed of rectifiers has been by the introduction of recombination centers in the i-region. In on-state analysis, it was shown that an approximate relationship between the injected carrier density in the i-region and the forward conduction current density is:

$$n^* = \frac{\tau_{HL}}{2\,q\,d}\,J_F \qquad (4.109)$$

From this equation, it can be seen that, as the high level lifetime (τ_{HL}) decreases, the average carrier density in the i-region will decrease proportionately, if the operating current density (J_F) is maintained constant. The net charge stored in the i-region

$$Q_s = 2\,q\,d\,n^* = \tau_{HL}\,J_F \qquad (4.110)$$

will also decrease as the lifetime is reduced.

The time taken for the reverse recovery process is directly dependent upon the stored charge. If the reverse recovery current waveform is approximated as a triangle of base ($t_A + t_B$) and height I_{RP}, the reverse recovery charge is given by:

$$Q_R = \frac{1}{2} I_{RP}\,(t_A + t_B) \qquad (4.111)$$

Equating the reverse recovery charge to the stored charge, a relationship given by Eq. (4.104) between the reverse recovery time and the lifetime is obtained. From this equation, the reverse recovery time can be expected to be linearly proportional to the high level lifetime and inversely proportional to the peak reverse current.

A reduction in the lifetime can be accomplished by the introduction of recombination centers into the i-region. Among the many possible approaches, the diffusion of gold and platinum, and the use of high energy electron irradiation have been most carefully studied. A discussion of the relative merits of these process techniques was provided in Chapter 2. A comparison of the trade-off curve between forward voltage drop and reverse recovery time for

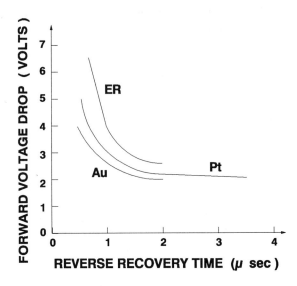

Fig. 4.31 Trade-off curves for three lifetime control methods.

these three cases is shown in Fig. 4.31. The reasons for the differences between these methods of lifetime control was analyzed in Chapter 2. Note that the trade-off curve is the best for gold doping and the worst for electron irradiation performed at room temperature using 1.5 MeV electron energy. By using higher electron energies and performing the irradiation at elevated temperatures (typically 300 °C), it has been found that the electron irradiation trade-off curve can be made to approach that for gold doping and superior to that for platinum doping.

The introduction of recombination centers into the i-region also leads to an undesirable increase in leakage current due to enhanced space-charge-generation at high operating temperatures. The magnitude of the increase in leakage current is dependent upon the position of the recombination level in the energy gap. The leakage current for gold and platinum doped rectifiers have been compared to electron irradiated devices. The electron irradiated devices exhibit an order of magnitude lower leakage current than the gold doped devices but have a higher leakage current than for platinum doped devices. In addition, it has been reported that electron irradiated rectifiers exhibit a 'snappy' recovery, i.e., their reverse recovery di/dt is large compared with the 'soft' recovery exhibited by platinum doped devices. This factor must be traded-off with the significant processing convenience offered by electron irradiation.

An interesting approach to improving the trade-off curve between forward drop and reverse recovery speed is to use an inhomogeneous distribution of recombination centers. Computer simulation of various lifetime profiles indicates that the preferred location for the positioning of the recombination centers is in the middle of the N-base region and away from the P-N junction. This can be intuitively deduced by examining the carrier distribution within the N-base during reverse recovery as shown in Fig. 4.30. From the time decay of the carriers, it is obvious that the carriers in the proximity of the junction will be first removed leaving a high concentration of carriers in the central portion of the N-base region during reverse recovery. By locating a high density of recombination centers at the peak of the carrier distribution, the reverse recovery process can be accelerated with less penalty upon forward voltage drop. Computer modeling indicates that a two-fold improvement in switching speed can be obtained over a uniform recombination center distribution for the same forward drop. Unfortunately, it is difficult to achieve such a narrow distribution of recombination centers because the deep level impurities have a very high diffusion coefficient in silicon and electron irradiation produces a uniform distribution of recombination centers. A promising approach is the use of high energy proton implantation. The use of protons with energies in excess of 1 MeV allows placing the implantation damage at 100 microns below the surface. This has been found to result in a better trade off curve between the on-state voltage drop and the reverse recovery time. However, such implants require a very high energy implantation source which may not be commercially viable.

The above discussion has been directed towards improving the reverse recovery characteristics. The introduction of recombination centers also impacts the forward recovery performance. As discussed earlier, the forward voltage overshoot can be minimized by lowering the resistivity of the i-region. Due to the higher mobility for electrons, an N-type drift region is preferable. The presence of a high concentration of recombination centers in the band gap leads to a compensation effect which raises the N-base resistance. For a given lifetime reduction, the compensation effect is dependent upon the capture cross-section of the recombination level as discussed in Chapter 2. In this regard, gold doping produces a larger change in N-base resistivity than platinum for achieving the same reverse recovery speed.

4.2.5 Doping Profile

The analysis of the reverse recovery of a P-i-N rectifier indicates that a P^+-π-N^+ diode (π refers to a lightly doped P-type i-region) will exhibit a faster recovery than a P^+-ν-N^+ diode (ν refers to lightly doped N-type i-region). The difference between these cases arises from the faster removal of charge in the P^+-π-N^+ diode. Although this may at first appear to favor the use of a π-type i-region for power rectifiers, the opposite is generally done for two reasons. Firstly, the reverse recovery of the π-base device is unacceptably abrupt leading to a 'snappy' recovery. The ν-base diodes exhibit the desired 'soft' recovery characteristic which is favored despite the longer reverse recovery time. The second reason for preference of a ν-base is that it can be passivated much more easily than a π-base structure to achieve high, stable breakdown characteristics.

Apart from the semiconductor type used for the i-region, the tailoring of its doping profile has been used to improve the reverse recovery characteristics. An important technique for improving the reverse recovery speed is to use very abrupt profiles for the P^+ anode and N^+ cathode regions. In the past, power rectifiers were made by using high resistivity bulk, N-type material and diffusing the P^+ and N^+ regions from opposite sides of the wafer. This process produces highly graded diffusion profiles and a relatively wide N-base region. These graded profiles are desirable for increasing the breakdown voltages in the bulk and edges as discussed in Chapter 3. However, during forward conduction, there is a significant amount of minority carrier injection into the end regions. The existence of the stored charge in the diffused region slows down the reverse recovery process. This effect is particularly important for low voltage rectifiers with narrow N-base regions.

With the improvement of silicon epitaxial growth technology, the development of power rectifiers with abrupt profiles is feasible. The doping profile of a rectifier fabricated using an N-type epitaxial layer grown on an N^+ substrate has an abrupt interface with high doping concentration in the end regions. This results in a much smaller injected carrier density in the end regions. Epitaxial diodes, with breakdown voltages of 400 volts, have been found to exhibit a reduction in stored charge by over a factor 2 and a decrease in reverse recovery time by 50 percent. The epitaxial process is particularly attractive for low breakdown voltage diodes which have narrow N-base regions.

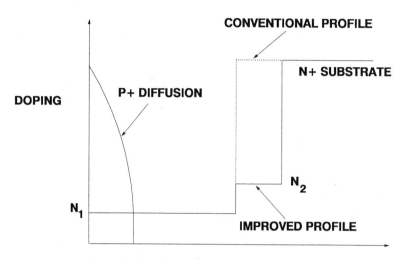

Fig. 4.32 Two layer drift region for soft recovery rectifiers.

A further improvement in the reverse recovery characteristics has been achieved by the use of an N-base consisting of two portions. A comparison between the conventional doping profile and the improved two region profile is provided in Fig. 4.32. The improved N-base design consists of the conventional lightly doped region (N_1) designed to support the reverse blocking voltage plus a more heavily doped region (N_2). The doping of the second region (N_2) has to be high enough to limit depletion layer spreading but low enough to still allow conductivity modulation. A typical doping level for this region is in the mid-10^{14} per cm^3 range. During reverse recovery, charge stored in the second region (N_2) is not swept out rapidly resulting in the desired soft recovery. Devices made with this profile by using epitaxial growth have t_B larger than t_A, which is a criterion used to identify a soft recovery rectifier.

4.2.6 Ideal Ohmic Contact

In the past, the contact to the drift region of the P^+-N rectifier has been made by using a high concentration (N$^+$) layer, which is formed by either diffusion in the case of the double diffused rectifiers or by using a highly doped substrate in the case of epitaxial rectifiers. The N-N$^+$ interface produces an 'ohmic' contact for majority carriers, i.e., electrons in this case. The N-N$^+$ contact allows the transport of electrons across the interface but creates an electric field due to the concentration gradient:

$$E = \frac{kT}{qN_D} \frac{dN_D}{dx} \qquad (4.112)$$

The presence of the electric field reflects any minority carriers (holes in this case) that approach

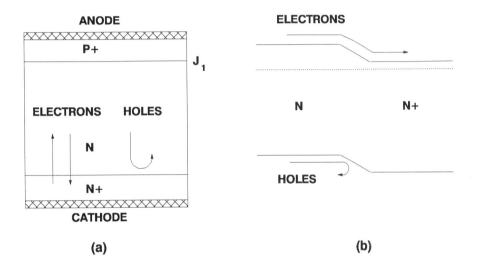

Fig. 4.33 Conventional N/N$^+$ ohmic contact : (a) Structure, (b) Band Diagram.

the interface as illustrated in the band diagram shown in Fig. 4.33. The minority carrier reflecting property of a highly doped contact layer has been utilized to improve the performance of solar cells. In the case of power rectifiers, the reflection of minority carriers can be detrimental to the achievement of fast turn-off because minority carriers become trapped in the N-base region.

To obtain an 'ideal' ohmic contact that can simultaneously allow the transport of holes and electrons across its interface, a contact consisting of a mosaic of P$^+$ and N$^+$ regions has been proposed. In this contact, the P$^+$ regions of the contact provide an ohmic path for holes (minority carriers here) while the N$^+$ regions provide the conventional ohmic path for electrons (majority carriers here). The 'ideal' ohmic contact formed with the mosaic of P$^+$ and N$^+$ regions and its composite band diagram are shown in Fig. 4.34. It should be noted that the ideal ohmic contact differs from the well known shorted emitter structure used for improving the performance of thyristors. In the case of the shorted emitter structure, holes are injected from

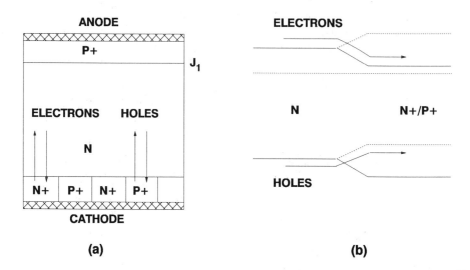

Fig. 4.34 Ideal ohmic contact with mosaic of P^+ and N^+ regions : (a) Structure, (b) Band diagram.

the P^+ region into the N layer during forward conduction. To achieve this, the transverse ohmic voltage drop across the P^+ regions must be sufficient to forward bias the P^+-N junction. Since the injection of minority carriers in the shorted emitter structure is essential for obtaining conductivity modulation of the drift layer of the thyristor, the width W_P must be made large. In contrast, for the ideal ohmic contact, the minority carrier injection used to modulate the resistance of the N-base region occurs via the forward biased junction (J_1) of the rectifier. The ideal ohmic contact is designed with small widths (W_P and W_N) for the P^+ and N^+ regions to suppress any transverse voltage drop across the P^+ region since no injection from the P^+ region is necessary.

A modification of the P^+/N^+ mosaic contact which achieves the same purpose is the substitution of the P^+ regions with a Schottky contact. This ideal ohmic contact is illustrated in Fig. 4.35. It is worth pointing out that the Schottky contact does not have to support any reverse voltage, which alleviates the stringent process requirements generally needed to suppress the soft, high leakage, breakdown characteristics of Schottky diodes.

In the case of rectifiers fabricated using the double diffusion process, the ideal contact can be achieved by masking portions of the N^+ contact diffusion and performing a P^+ diffusion in these portions. As discussed in previous sections, the epitaxial process offers significant improvement in rectifier characteristics especially for low breakdown voltage diodes. To create the ideal ohmic contact with an epitaxial process, the conventional process with N-type epitaxial layer grown on an N^+ substrate must be altered to a N-type epitaxial layer grown on a P^+ substrate. This makes the surface of the N-type drift layer accessible to subsequent diffusion of the P^+ and N^+ regions to form the mosaic contact on the top surface. The drawback of the process is the need to form grooves around the edge of each device extending to the interface between the N-type epitaxial layer and the P^+ substrate to create individual diodes.

Chapter 4 : POWER RECTIFIERS

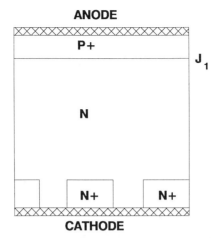

Fig. 4.35 Ideal ohmic contact with Schottky interface.

The ideal ohmic contact concept has been experimentally tested by the fabrication of epitaxial diodes for which a P-type epitaxial layer was grown on an N^+ substrate. Ideal ohmic contacts with the P^+/N^+ mosaic structure and the N^+-Schottky structure were both evaluated. The contact geometry used consisted of long thin stripes with a P^+ region width of 10 microns and a 30 micron width for the N^+ or Schottky regions. No lifetime control was used. The characteristics of these diodes were compared with those of diodes simultaneously fabricated with the conventional P^+ contact to the P-type drift region. The reverse recovery time for the diodes with ideal ohmic contact was found to be 60 nanoseconds compared with about 500 nanoseconds for the diodes with conventional contacts. A peak reverse recovery current (I_{RP}) equal to the forward conduction current (I_F) was used during these measurements. In addition to the improved reverse recovery characteristic, the forward voltage drop of the diodes with the ideal ohmic contact was found to be 0.1 volts lower than for diodes with conventional ohmic contacts which were processed with gold doping to achieve fast reverse recovery. Further, due to the absence of gold doping, diodes with ideal ohmic contacts exhibit a lower leakage current at elevated temperatures when compared with fast recovery diodes. These features allow their operation at higher ambient temperatures.

4.2.7 Maximum Operating Temperature

The maximum operating temperature of a rectifier is limited by thermal runaway. The temperature rise is determined by the power dissipation in the diode. This is comprised of three components - the power dissipation due to current conduction in the on-state, the power dissipation during reverse blocking, and the power dissipation during switching. As the switching speed improves, the steady state power losses become dominant. The steady state power dissipation in the on-state is determined by the forward voltage drop. As temperature increases, the forward drop of a P-i-N rectifier decreases leading to a reduction in the power dissipation. Concurrently, the leakage current grows exponentially with increasing temperature. The leakage current arises from two sources - space charge generation in the depletion layer and the generation of carriers in the neutral base region within a diffusion length from the depletion layer edge as discussed earlier. At high temperatures, the diffusion current term becomes predominant and the leakage current grows exponentially with temperature with an activation energy equal to the energy gap (E_g).

The power dissipation due to the on-state voltage drop and the reverse blocking leakage current determines the total power dissipation:

$$P_D = I_F V_F \frac{t_{on}}{T} + I_L V_R \frac{(T - t_{on})}{T} \qquad (4.113)$$

where t_{on} is the time the diode is in the on-state and T is the total period; I_F, V_F are the forward conduction current and forward voltage drop; and I_L, V_R are reverse leakage current and applied voltage. At low temperatures, the leakage current (I_L) is small and the first term dominates. The power dissipation in the rectifier decreases with temperature in this temperature region due to a decrease in forward voltage drop. At high temperatures, the leakage current becomes large and the second term grows until it exceeds the first term in Eq. (4.113). In this temperature range, the power dissipation in the diode increases with increasing temperature. This creates a thermal runaway situation. The maximum operating temperature of the rectifier is determined by the point at which the curve relating power dissipation and temperature goes through a minimum.

4.3 JBS RECTIFIERS

With the trend towards lower operating voltages for VLSI circuits, there is an increasing demand to reduce the forward voltage drop in rectifiers used for switch mode power supplies. The forward voltage drop of a Schottky rectifier can be reduced by decreasing the Schottky barrier height. Unfortunately, a low barrier height results in a severe increase in leakage current and reduction in maximum operating temperature. Further, Schottky power rectifiers fabricated with barrier heights of less than 0.7 eV have been found to exhibit extremely soft breakdown characteristics which makes them prone to failure.

The *junction barrier controlled Schottky (JBS) rectifier* is a Schottky rectifier structure with a P-N junction grid integrated into its drift region. This device structure has also been called a *pinch rectifier*. A cross-section of the device structure is provided in Fig. 4.36. The junction grid is designed so that its depletion layers do not pinch-off under zero and forward bias conditions. When designed in this manner, the device contains multiple conductive channels under the Schottky barrier through which current can flow during forward biased operation. Under a positive applied bias to the N$^+$ substrate, the P-N junctions and the Schottky barrier become reverse biased. The depletion layers formed at the P-N junctions spread into the

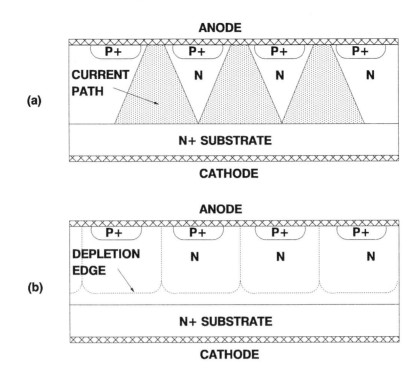

Fig. 4.36 JBS rectifier structure under (a) forward conduction mode; (b) reverse blocking mode.

channel. In the JBS rectifier, the junction grid is designed so that the depletion layers will intersect under the Schottky barrier when the reverse bias exceeds a few volts. After depletion layer pinch-off, a potential barrier is formed in the channel. Once the potential barrier is formed, further increase in applied voltage is supported by it with the depletion layer extending toward the N^+ substrate, and the potential barrier shields the Schottky barrier from the applied voltage. This shielding prevents the Schottky barrier lowering phenomenon and eliminates the large increase in leakage current observed for conventional Schottky rectifiers. Once the pinch-off condition is established, the leakage current remains constant except for the small increasing contribution from the space-charge generation components. This allows the JBS rectifier to be operated right up to the avalanche breakdown point without the onset of the thermal runaway experienced in Schottky rectifiers due to their very soft breakdown characteristics.

Due to the suppressed leakage current, the Schottky barrier height used in the JBS rectifier can be significantly less than for the conventional Schottky rectifiers. This has allowed a reduction in the forward voltage drop while maintaining an acceptable reverse blocking characteristic. For the same leakage current, the JBS rectifier has been found to provide a forward voltage drop of 0.25 volts compared with 0.5 volts for the Schottky rectifier.

4.3.1 Forward Conduction Characteristics

Analysis of the forward conduction characteristics of the JBS rectifier can be performed along the same manner as used for the Schottky rectifier by allowing for the increase in the series resistance of the drift region due to current constriction in the channel structure. Consider the case of a P-N junction grid with stripe geometry and cross-sectional dimensions defined in

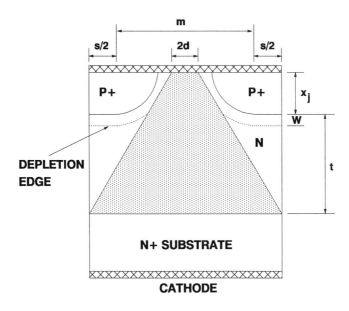

Fig. 4.37 Current path in JBS rectifier used for on-state analysis.

Fig. 4.37. In this structure, the junction grid is formed by planar diffusion through a diffusion window of width 's' with a masked region of width 'm'. The lateral diffusion of the junction is assumed to be 85 percent of the vertical depth (x_j).

The relationship between the forward voltage drop and current density for a Schottky barrier must be modified to allow for the area taken up by the P$^+$ regions in the JBS rectifier structure. Within the JBS rectifier structure, the current density across the Schottky barrier (J_{FS}) is given by:

$$J_{FS} = \frac{(m + s)}{2d} J_{FC} \qquad (4.114)$$

where J_{FC} is the cell current density obtained by dividing the total JBS rectifier current by the cell area. Using this expression in Eq. (4.10):

$$V_{FS} = \phi_B + \frac{kT}{q} \ln\left(\frac{(m + s)}{2d} \frac{J_{FC}}{AT^2} \right) \qquad (4.115)$$

Chapter 4 : POWER RECTIFIERS

In this equation, the dimension (2d), which defines the actual area of the Schottky barrier, is related to the structural dimensions and the junction depletion width:

$$2d = m - 2W - 1.7 x_j \tag{4.116}$$

where the junction depletion width is given by:

$$W = \sqrt{\frac{2\,\epsilon_s}{q\,N_D}(V_{bi} - V_F)} \tag{4.117}$$

In addition to the voltage drop across the Schottky barrier, the voltage drop across the drift region must also be accounted for. Based upon current spreading from a cross-section at the top with a width of 2d to a cross-section at the bottom with a width of (m + s), the drift region resistance is found to be given by:

$$R_D = \rho\,\frac{(x_j + t)(m + s)}{(m + s - 2d)}\ln\!\left(\frac{m+s}{2d}\right) \tag{4.118}$$

where ρ and t are the resistivity and depletion layer width required to obtain the desired breakdown voltage. Combining the voltage drop across the Schottky barrier and the drift region, the forward voltage drop of the JBS rectifier is obtained:

$$V_{FS} = \phi_B + \frac{kT}{q}\ln\!\left(\frac{(m+s)}{2d}\frac{J_{FC}}{AT^2}\right) + \rho\,\frac{(x_j+t)(m+s)}{(m+s-2d)}\ln\!\left(\frac{(m+s)}{2d}\right)J_{FC} \tag{4.119}$$

The forward conduction characteristics can be calculated by using this equation. For an exact analysis, an iterative procedure is required because of the dependence of 'd' upon the forward voltage drop. In the case of JBS rectifiers intended for operation at very low forward voltage drops, the depletion layer width 'W' can be assumed to be constant allowing a closed form analytical solution of the forward conduction characteristics.

4.3.2 Reverse Blocking Characteristics

As in the case of the Schottky rectifier, the reverse leakage current of the JBS rectifier consists of two components. The first component arises from the injection of carriers across the Schottky barrier. From Eq. (4.16), after accounting for the Schottky barrier lowering described by Eq. (4.21), and by including the effect of area taken up by the P$^+$ diffusion, it can be shown that:

$$J_L = \left(\frac{2d}{m+s}\right) A\,T^2 \exp\!\left[-\left(\frac{q\,\phi_B}{kT}\right)\right]\exp\!\left(\frac{q}{kT}\sqrt{\frac{q\,E}{4\pi\,\epsilon_s}}\right) \tag{4.120}$$

where E is the electric field at the Schottky barrier as given by Eq. (4.22). An important feature of the JBS rectifier is that the electric field at the Schottky barrier remains approximately constant (independent of the reverse bias) once the channel potential barrier forms. Its value then corresponds to the voltage at which the channel pinch-off occurs:

$$E = \sqrt{\frac{2 q N_D}{e_s} (V_P + V_{bi})} \qquad (4.121)$$

where V_P is the channel pinch-off voltage given by:

$$V_P = \frac{q N_D}{8 e_s} (m - 1.7 x_j)^2 - V_{bi} \qquad (4.122)$$

The second component of the leakage current arises from the space charge generation and diffusion currents. This component is:

$$J_{LD} = q \sqrt{\frac{D}{\tau} \frac{n_i^2}{N_D}} + \frac{q n_i W}{\tau} \qquad (4.123)$$

where W is the depletion layer width given by:

$$W = \sqrt{\frac{2 e_s}{q N_D} (V_R + V_{bi})} \qquad (4.124)$$

This leakage current component is a small fraction of the total leakage current but must be included in the analysis of the reverse blocking characteristics to account for the slight increase in leakage current with reverse bias voltage beyond the pinch-off point. An excellent agreement between the calculated and measured device leakage characteristics have been observed.

4.3.3 Device Characteristics

From the on-state current flow pattern illustrated in Fig. 4.37, it can be concluded that the lowest on-state voltage drop can be obtained by making the width (s) of the junction diffusion window as small as possible. This minimizes the dead space below the junction where the current does not flow. The best JBS rectifier characteristics can therefore be expected when submicron lithography is used to pattern the diffusion windows for the P^+ diffusions. Experimental results have been obtained on JBS rectifiers capable of supporting 30 volts in the reverse direction with 0.5 micron diffusion windows. It has been shown that a much better trade-off curve between the on-state voltage drop and the reverse leakage current can be obtained in this case when compared with diffusion windows of 2 microns in size.

4.4 MPS RECTIFIERS

In the case of high voltage P-i-N rectifiers, the current flow during the reverse recovery transient has been found to be a significant source of power losses in power electronic circuits. The reverse recovery current flow produces high power dissipation not only in the rectifier but also in the transistor controlling the recovery transient because the peak reverse recovery current adds to the steady state current flowing through the transistor. The reverse recovery current can be reduced by the use of lifetime control methods. However, this produces an increase in the on-state voltage drop, resulting in a trade-off between on-state losses and switching losses.

The *merged P-i-N/Schottky (MPS) rectifier* is an alternate approach to reducing the switching losses in high voltage power rectifiers without increasing the on-state voltage drop.

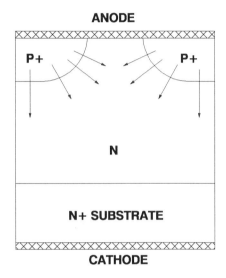

Fig. 4.38 Merged P-i-N/Schottky rectifier structure.

The device structure, shown in Fig. 4.38, is similar in appearance to that of the JBS rectifier. However, the operating physics of the two are quite different. In the JBS rectifier, the P-N junction is used exclusively to reduce the leakage current by preventing Schottky barrier lowering during reverse blocking. This feature is also made use of in the MPS rectifier to obtain a high breakdown voltage in spite of the presence of the Schottky region. However, in the MPS rectifier, the P-N junction becomes forward biased in the on-state, unlike in the case of the JBS rectifier, because the drift region has a very high resistance due to its design for supporting high voltages during reverse blocking. The forward bias on the P-N junction produces the injection of holes into the N- drift region. This results in conductivity modulation of the drift region in a manner similar to the P-i-N rectifier, which drastically reduces its resistance to current flow. This allows large current flow via the Schottky region with a low series resistance. The injection level required to reduce the resistance in series with the Schottky region is not as large as that observed in the P-i-N rectifier. Consequently, the stored charge in the MPS rectifier is found to be much smaller than that in the P-i-N rectifier.

4.4.1 Forward Conduction Characteristics

The on-state characteristics of the MPS rectifier have been determined by two-dimensional numerical simulation and compared with those for the P-i-N rectifier and Schottky rectifier with the same high voltage drift region parameters designed to support 900 volts. These

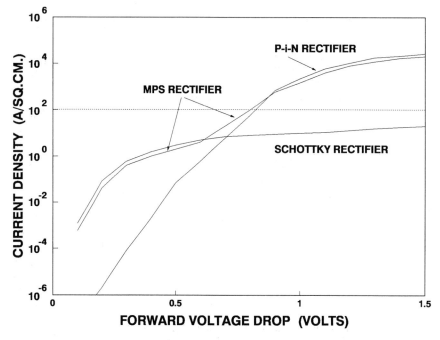

Fig. 4.39 Comparison of the on-state characteristics of the MPS rectifier with that of the PiN and Schottky rectifiers.

characteristics are compared in Fig. 4.39. Although the current flow in the case of the Schottky rectifier begins to occur at relatively low forward bias voltages across the barrier, the current becomes limited by the high series resistance of the unmodulated drift region. This results in a low on-state current density with high forward voltage drop for the Schottky rectifier. In contrast, for the P-i-N rectifier, very little current flow occurs at forward bias voltages below 0.5 volts. However, at larger forward bias voltages, the P-N junction begins to inject holes into the drift region and reduces its series resistance. This results in a low on-state voltage drop in the range of 1 volt at high on-state current densities. Under typical on-state conditions, it has been found that the injected carrier concentration is about 1×10^{17} per cm^3. This results in a

Chapter 4 : POWER RECTIFIERS 189

high stored charge in the drift region in the on-state.

In the case of the MPS rectifier, current flow begins to occur at low forward bias voltages as in the case of the Schottky rectifiers. At on-state biases below 0.5 volts, the current flow occurs primarily across the Schottky region. Since the area of the Schottky region is about half the total cell area, the on-state current density in the MPS rectifier is about half that for the Schottky rectifier at these low forward bias values. However, when the forward bias is increased to above 0.6 volts, the P-N junction in the MPS rectifier begins to inject holes into the drift region. This produces conductivity modulation of the drift region resulting in a low resistance in series with the Schottky region. As a consequence, the on-state current density in the MPS rectifier becomes even larger than that for the P-i-N rectifier for forward bias voltages between 0.6 and 0.9 volts. For a typical on-state current density of 100 amperes/cm^2 as indicated by the dashed line, the on-state voltage drop of the MPS rectifier is even lower than that for the P-i-N rectifier. The on-state losses for the MPS rectifier are, therefore, even lower than those for the P-i-N rectifier. Although it is interesting to note that the on-state voltage drop of the MPS rectifier is the lowest among all the high breakdown voltage rectifiers, the difference between the MPS rectifier and the P-i-N rectifier is less than 0.1 volts.

When the forward bias voltage is increased beyond 0.9 volt, it can be seen from Fig. 4.39 that the curves for the MPS and P-i-N rectifiers cross each other. This takes place because at very high current densities typical of surge conditions, the current flow occurs primarily via the P-N junction within the MPS rectifier structure. Since the area of the P-N junction in the MPS rectifier is about half that for the P-i-N rectifier, the current density flowing within the MPS rectifier also becomes half that for the P-i-N rectifier under these conditions. Due to the relatively small difference in the on-state voltage drop, the surge current handling capability of the MPS rectifier is essentially equal to that for the P-i-N rectifier.

4.4.2 Stored Charge

As discussed in the previous section, the on-state voltage drop of the MPS rectifier is slightly lower than that of the P-i-N rectifier. It is important to demonstrate that under these conditions, the stored charge in the MPS rectifier is substantially smaller than that in the P-i-N rectifier. This can be done by performing an analysis of the carrier distribution in the MPS rectifier in the on-state in a manner analogous to that done earlier for the P-i-N rectifier. As in the case of the P-i-N rectifier, the continuity equation for electrons in the drift region is given by:

$$\frac{d^2n}{dx^2} - \frac{n}{L_a^2} = 0 \qquad (4.125)$$

However, the boundary conditions for the MPS rectifier in the Schottky region are different because it acts as an interface with infinite recombination rate making the carrier concentration go to zero. Based upon this, the boundary conditions along the drift region for the Schottky portion of the MPS rectifier can be written as:

$$n(-d) = 0 \tag{4.126}$$

and

$$\left(\frac{dn}{dx}\right)_{x=+d} = \frac{J_F}{2qD_n} \tag{4.127}$$

where J_F is the on-state current density, and d is half the thickness of the N- drift region. The solution for the carrier distribution is:

$$n(x) = \frac{J_F L_a}{2qD_n} \frac{\sinh[(x+d)/L_a]}{\cosh(2d/L_a)} \tag{4.128}$$

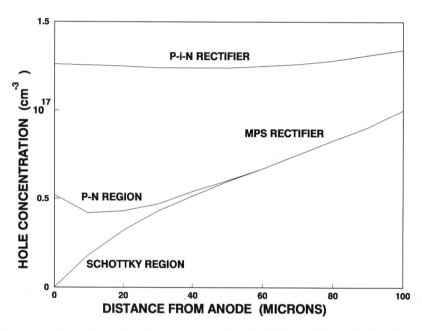

Fig. 4.40 Comparison of carrier distribution profiles in MPS rectifier with P-i-N rectifier.

This carrier profile is compared with that within a P-i-N rectifier in Fig. 4.40. In this figure, the carrier profile is shown for the MPS rectifier along a line through the Schottky region and through the P-N junction region. The above equation describes the profile through the Schottky region. It can be seen that the carrier concentration at the P-N junction interface is also smaller than that for the P-i-N rectifier. From these carrier profiles, it can be concluded that the stored charge in the MPS rectifier is substantially smaller than that in the P-i-N rectifier. An estimate for the stored charge in the MPS rectifier can be obtained by integration of the carrier profile

Chapter 4 : POWER RECTIFIERS

given by Eq. (4.128):

$$Q_s = q \int_{-d}^{+d} p(x)\ dx = \frac{J_F L_a^2}{2 D_n} \left[1 - \frac{1}{\cosh(2d/L_a)} \right] \qquad (4.129)$$

As an example, consider a rectifier with a N- drift region thickness of 100 microns corresponding to a breakdown voltage of about 1000 volts. If the lifetime in the drift region is 1 microsecond, the stored charge in the Schottky region predicted by Eq. (4.129) is approximately one fifth of that for the P-i-N rectifier operating at the same on-state current density. In actual MPS rectifiers, the stored charge will be slightly larger due to the higher carrier concentration at the P-N junction as shown in Fig. 4.40.

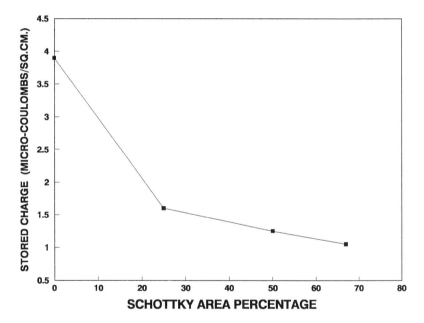

Fig. 4.41 Measured stored charge in MPS rectifiers.

MPS rectifiers have been fabricated with N- drift layers of 96 microns in thickness to verify the operating physics. The on-state characteristics of these devices were found to be in good agreement with those predicted by the numerical analysis (shown in Fig. 4.39). One of the design parameters for the MPS rectifier is the relative area of the Schottky and P-N junction regions. It has been found that the stored charge can be reduced by increasing the area of the Schottky region relative to the P-N junction region as shown in Fig. 4.41. It can be seen that, for a MPS rectifier with a Schottky area of 50 percent of the total cell area, the stored charge extracted from reverse recovery measurements is one fourth of that measured for a P-i-N rectifier fabricated simultaneously with the MPS rectifiers. This is in good agreement with the theoretical predictions based upon numerical simulations and the analytical model.

4.4.3 Reverse Recovery

The reverse recovery behavior of the MPS rectifiers has been found to be significantly superior to those for P-i-N rectifiers because of the reduced stored charge. First, the peak reverse recovery current has been found to be reduced by a factor of four. This not only reduces the power loss within the rectifier during reverse recovery but also reduces power losses and stresses in the transistor controlling the reverse recovery of the rectifier in power circuits. Another important parameter of interest for applications is the reverse recovery di/dt (see Fig. 4.21). A smaller reverse recovery di/dt is important for preventing high voltage spikes from developing in power circuits. The MPS rectifiers have been found to exhibit lower reverse di/dt values than P-i-N rectifiers for two reasons. The first reason is that the peak reverse recovery current is smaller. The second reason is the different carrier distribution within the drift region. The low carrier density at the blocking junction in the MPS rectifier results in the device supporting the reverse applied voltage earlier than in the case of the P-i-N rectifier. This results in a larger proportion of the stored charge remnant in the drift region after the junction becomes reverse biased. Since the peak in the reverse recovery current occurs at this time, there is a larger proportion of the stored charge removed by recombination after the peak in the reverse recovery waveform, producing a smaller di/dt or a *soft recovery* behavior.

4.4.4 Reverse Blocking

In the case of Schottky rectifiers, it has been found that the reverse blocking characteristics are poor due to the Schottky barrier lowering effect. This problem is not observed in MPS rectifiers if the space between the P-N junctions within the cell is made sufficiently small so as to obtain a potential barrier under the Schottky region. The relatively low doping concentration of the drift region required to obtain high breakdown voltages favors this barrier formation without resorting to small spacings between the P-N junctions. It has been found that a spacing of between 5 to 15 microns can be used to obtain breakdown voltages essentially equal to that of the P-i-N rectifier. However, the high temperature leakage current for the MPS rectifier is larger than that for the P-i-N rectifier due to the Schottky regions. As discussed in section 4.1, the leakage current across the Schottky barrier can be reduced by increasing the barrier height. Although a low barrier height is needed in conventional Schottky barrier rectifiers to reduce the on-state voltage drop, this is not essential for the MPS rectifier. It is possible to use a barrier height of 0.8 volts for the Schottky region in the MPS rectifier to suppress the high temperature leakage current without significantly altering the on-state voltage drop or the stored charge.

4.5 TRENDS

As VLSI technology moves towards smaller device dimensions in order to achieve a higher packing density, there will be an increasing need to lower the power supply voltage. At present, most circuits are operated from a 5 volt power supply. In the future, the power supply voltage is expected to drop to about 3 volts. The forward voltage drop of the diodes used as output rectifiers in switching power supplies then becomes increasingly important. The JBS rectifiers discussed in this chapter are expected to make rectifiers with lower forward voltage drops available, resulting in major reduction in the power losses within the power supply. This will aid the power supply designer in achieving higher system efficiency, together with reduction in size and weight of the power supplies.

In addition, the availability of high speed, high voltage (100 to 600 volt) rectifiers that match the performance of the MOS-gated power devices discussed in later chapters will aid the trend towards increasing circuit operating frequency. The slow evolutionary progress in improving the performance of the P-i-N rectifiers in the past has been replaced by the introduction of several new device concepts that can be expected to have a revolutionary impact on power systems in the future. In the near term, the MPS rectifiers based upon silicon technology hold promise. For the long term, the Schottky rectifiers made using silicon carbide are promising candidates for high voltage circuit applications.

REFERENCES

1. S.K. Ghandhi, "Semiconductor Power Devices", Wiley, New York (1977).

2. B.J. Baliga and E. Sun, "Comparison of gold, platinum and electron irradiation for controlling lifetime in power rectifiers," IEEE Trans. Electron Devices, Vol. ED-24, pp. 685-688 (1977).

3. R.O. Carlson, Y.S. Sun, and H.B. Assalit, "Lifetime control in silicon power devices by electron or gamma irradiation," IEEE Trans. Electron Devices, Vol. ED-24, pp. 1103-1108 (1977).

4. E.D. Wolley and S.F. Bevacqua, "High speed, soft recovery, epitaxial diodes for power inverter circuits," IEEE Industrial Application Society Meeting Digest, pp. 797-800 (1981).

5. D. Silber, D.W. Novak, W. Wondrak, B. Thomas, and H. Berg, "Improved dynamic properties of GTO-thyristors and diodes by proton implantation," IEEE Int. Electron Devices Meeting Digest, Abs. 6.6, pp. 162-165 (1985).

6. H. Benda and E. Spenke, "Reverse recovery processes in silicon power rectifiers," Proc. IEEE, Vol. 55, pp. 1331-1354 (1967).

7. R.J. Grover, "Epi and Schottky diodes," in 'Semiconductor Devices for Power Conditioning,' pp. 331-356, Edited by R. Sittig and P. Roggwitter, Plenum Press, New York (1982).

8. J.R. Hauser and P.M. Dunbar, "Minority carrier reflecting properties of semiconductor high-low junctions," Solid State Electronics, Vol. 18, pp. 715-716 (1975).

9. Y. Amemiya, T. Sugeta, and Y. Mizushima, "Novel low-loss and high speed diode utilizing an ideal ohmic contact," IEEE Trans. Electron Devices, Vol. ED-29, pp. 236-243 (1982).

10. E.H. Rhoderick, 'Metal-Semiconductor Contacts', Clarendon Press, Oxford (1978).

11. S.M. Sze, 'Physics of Semiconductor Devices', Chapter 5, Wiley, New York (1981).

12. C.R. Crowell and S.M. Sze, "Current transport in metal-semiconductor barriers," Solid State Electronics, Vol. 9, pp. 1035-1048 (1966).

13. J.M. Andrews and M.P. Lepselter, "Reverse current-voltage characteristics of metal-silicide Schottky diodes," Solid State Electronics, Vol. 13, pp. 1011-1023 (1970).

14. C.R. Crowell, "The Richardson constant for thermionic emission in Schottky barrier diodes," Solid State Electronics, Vol. 8, pp. 395-399 (1965).

15. D.J. Page, "Theoretical performance of the Schottky barrier power rectifier," Solid State Electronics, Vol. 15, pp. 505-515 (1972).

16. B. Bixby, B. Hikin, and V. Rodov, "Application considerations for very high speed fast recovery power diodes," IEEE Industrial Applications Society Meeting Digest, pp. 1023-1027 (1977).

17. B.R. Pelly, "Power semiconductor devices - A status review," IEEE Industrial Semiconductor Power Conversion Conference Digest, pp. 1-19 (1982).

18. J.M. Shannon, "Reducing the effective height of a Schottky barrier using low energy ion implantation," Appl. Phys. Letters, Vol. 24, pp. 369-371 (1974).

19. S. Ashok and B.J. Baliga, "Effect of antimony ion implantation on Al-silicon Schottky diode characteristics," J. Appl. Phys., Vol. 56, pp. 1237-1239 (1984).

20. Y.I. Choi, "Enhancement of breakdown voltages of Schottky diodes with a tapered window," IEEE Trans. Electron Devices, Vol. ED-28, pp. 601-602 (1981).

21. N.G. Anantha and K.G. Ashar, "Planar mesa Schottky barrier diode" IBM J. Research and Development, Vol. 15, pp. 442-445 (1971).

22. M.P. Lepselter and S.M. Sze, "Silicon Schottky barrier diode with near-ideal I-V characteristics," Bell Syst. Tech. J., Vol. 47, pp. 195-208 (1968).

23. R. Severns, "The power MOSFET as a rectifier," Power Conversion International, pp. 49-50, March-April (1980).

24. R.P. Love, P.V. Gray, and M.S. Adler, "A large area power MOSFET designed for low conduction losses," IEEE Int. Electron Devices Meeting, Abs. 17.4, pp. 418-421 (1981).

25. D. Ueda, H. Takagi, and G. Kano, "A new vertical power MOSFET structure with extremely reduced on-resistance," IEEE Trans. Electron Devices, Vol. ED-32, pp. 2-6 (1985).

26. B.J. Baliga, "The pinch rectifier: a low forward drop, high speed power diode," IEEE Electron Device Letters, Vol. EDL-5, pp. 194-196 (1984).

27. B.J. Baliga, "Analysis of junction barrier controlled Schottky rectifier characteristics," Solid State Electronics, Vol. 28, pp. 1089-1093 (1985).

28. B.M. Wilamowski, "Schottky diodes with high breakdown voltages," Solid State Electronics, Vol. 26, pp. 491-493 (1983).

29. Y. Shimuzu, M. Naito, S. Murakami, and Y. Terasawa, "High speed, low loss, PN diode having a channel structure," IEEE Trans. Electron Devices, Vol. ED-31, pp. 1314-1319 (1984).

30. B.J. Baliga, A.R. Sears, M.M. Barnicle, P.M. Campbell, W. Garwacki, and J.P. Walden, "Gallium arsenide Schottky power rectifiers," IEEE Trans. Electron Devices, Vol. ED-32, pp. 1130-1134 (1985).

31. B.J. Baliga, A.R. Sears, M.M. Barnicle, P.M. Campbell, and W. Garwacki, "Gallium arsenide Schottky power rectifiers," IEEE Trans. Electron Devices, Vol. ED-32, pp. 229-232 (1983).

32. V.A.K. Temple and F.W. Holroyd, "Optimizing carrier lifetime profiles for improving trade-off between turn-off time and forward drop," IEEE Trans. Electron Devices, Vol. ED-30, pp. 782-790 (1983).

33. B.J. Baliga, "New Power Rectifier Concepts", Proc. Third Int. Workshop on the Physics of Semiconductor Devices, New Delhi (1985).

34. M. Mehrotra and B. J. Baliga, "Very low forward drop JBS rectifiers fabricated using submicron technology," IEEE Trans. Electron Devices, Vol. ED-30, pp. 2131-2132 (1993).

PROBLEMS

4.1 A Schottky barrier power rectifier, with a breakdown voltage of 250 volts, is fabricated using a metal contact with a barrier height of 0.8 eV. Calculate its forward voltage drop at 25 °C for an on-state current density of 200 amperes per cm^2. What are the contributions from the series resistance and the Schottky barrier? (For solving this problem, assume that ideal breakdown voltage is achieved and neglect the resistances of the substrate and its contact.)

4.2 A Schottky barrier power rectifier, with a breakdown voltage of 250 volts, is fabricated using a metal contact with a barrier height of 0.8 eV. Calculate its forward voltage drop at 200 °C for an on-state current density of 200 amperes per cm^2. What are the contributions from the series resistance and the Schottky barrier? (For solving this problem, assume that ideal breakdown voltage is achieved and neglect the resistances of the substrate and its contact.)

4.3 A Schottky barrier power rectifier, with a breakdown voltage of 50 volts, is fabricated using a metal contact with a barrier height of 0.75 eV. Calculate its leakage current density at 125 °C for reverse bias voltages of 1 volt and 40 volts. (For solving this problem, assume that ideal breakdown voltage is achieved.)

4.4 A Schottky barrier power rectifier, with a breakdown voltage of 250 volts, is fabricated using a metal contact with a barrier height of 0.8 eV. Compare its leakage current density at 125 °C for a reverse bias of 200 volts with that for a P-N junction rectifier with the same breakdown voltage if the space charge generation lifetime in the drift region is 1 microsecond.

4.5 A Schottky barrier power rectifier, with a breakdown voltage of 100 volts, is fabricated using a metal contact with a barrier height of 0.7 eV. The device is operated at an on-state current density of 100 amperes per cm^2 in a power circuit with a duty cycle of 50 percent. Determine the on-state power dissipation, off-state power dissipation, and total power dissipation per cm^2 within the rectifier at 27, 125, and 150 °C.

Chapter 4 : POWER RECTIFIERS   197

4.6 A P^+NN^+ diode is fabricated with an N-region thickness of 100 microns. The diode has been optimized to obtain the lowest possible on-state voltage drop when operated at an on-state current density of 200 amperes per cm^2. Calculate the reverse recovery time and the peak reverse recovery current when switching off the rectifier.

4.7 Compare the on-state voltage drop of a P-i-N rectifier with a breakdown voltage of 1000 volts when it is fabricated with (a) a punch-through drift region design with drift layer doping concentration of 1×10^{13} per cm^3, and (b) a non-punch-through design with optimum doping and width for the drift region. For the solution, use an on-state current density of 200 amperes per cm^3 and a high level lifetime of 1 microsecond in the drift region.

4.8 A P-i-N rectifier is designed with an i-region width of 300 microns. Calculate the trade-off curve between the on-state voltage drop and the reverse recovery time by using lifetime values of 20, 10, 5, 2, 1.5, 1.0, 0.5, and 0.1 microseconds. Determine the peak reverse recovery current density for each of these cases. In solving this problem, make the following assumptions: (a) the on-state current density is 300 amperes per cm^2, (b) recombination occurs only in the mid-region, (c) Auger recombination and carrier-carrier scattering can be neglected, and (d) recombination during the reverse recovery process can be neglected.

Chapter 5

BIPOLAR TRANSISTORS

Bipolar power transistors have been commercially available for more than 30 years. Although the operating physics for bipolar power transistors is essentially the same as that for signal transistors, their characteristics differ due to the need for supporting a high collector voltage in the forward blocking mode. The high voltage capability of the bipolar power transistor is obtained by the incorporation of a high resistivity, thick drift region into the collector structure. In addition, the base region must be carefully designed to prevent reach-through breakdown. These differences have a strong influence upon the current gain of the device.

Another distinguishing feature of bipolar power transistors is that they operate at relatively high current densities. This produces high level injection in not only the collector but also the base region. The high level injection produces severe degradation of the current gain. In general, power bipolar transistors have a low current gain at typical operating current levels. This is a problem for applications because of the need to provide bulky and expensive control circuits using many discrete components. This problem was partially solved by the development of Darlington bipolar transistors. However, even with the Darlington configuration, the current gain is insufficient for control by using integrated circuits. For this reason, the bipolar power transistor has been recently displaced by the insulated gate bipolar transistor (IGBT) for high voltage applications. Despite the diminishing use of power bipolar transistors in applications, it is important to have a complete understanding of their operating principles because the basic structure is incorporated within the IGBT.

In this chapter, the physics of operation of the bipolar power transistor is discussed. After introducing the basic structure, the current gain at low and high current levels is discussed. This discussion is followed by the analysis of the output characteristics, including the quasi-saturation region. The next portion of this chapter discusses the switching behavior and the forward/reverse biased safe-operating-area. These parameters define the current-voltage locus within the output characteristics that the device can tolerate without destructive failure. The last portion of chapter discusses the Darlington configuration proposed to improve the current gain of bipolar power transistors.

Chapter 5 : BIPOLAR TRANSISTORS

5.1 DEVICE OPERATION

The bipolar power transistor is fundamentally a current controlled three terminal switch. It is usually operated in the common emitter configuration with the collector current controlled by the magnitude and polarity of the input base current. For ease of control, it is desirable to operate the device with small base currents to control large collector currents. This allows operation of the device with a high power gain, which is of course also dependent upon the ratio of the collector and gate operating voltages. In some power circuits, it becomes necessary to operate the bipolar transistor in the common base configuration as well. This operation results in a current gain of less than unity.

The N-P-N and P-N-P bipolar transistors are shown in Fig. 5.1 in the common emitter

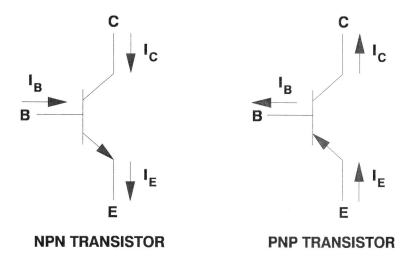

Fig. 5.1 Bipolar transistors in common emitter configuration.

configurations. The emitter terminal serves as the reference. In this figure, the directions for the emitter, base, and collector current flow are indicated. In this configuration, the input variable is the base current and the output variable is the collector current. Based upon Kirchoff's law:

$$I_E = I_C + I_B \tag{5.1}$$

where I_E, I_B, and I_C are the emitter, base and collector currents, respectively. The ratio of the collector current to the base current is called the *common emitter current gain or β (beta)* of the bipolar transistor, given by:

$$\beta = \frac{I_C}{I_B} \tag{5.2}$$

Combining these two equations:

$$I_E = \left(1 + \frac{I_C}{I_B}\right) I_B = I_B (1 + \beta) \qquad (5.3)$$

From these relationships, it can be concluded that a large beta is desirable in order to operate the device with a high power gain. The device physical parameters that lead to a higher current gain are discussed in a later section of this chapter.

The common base configurations for the N-P-N and P-N-P bipolar transistors are shown

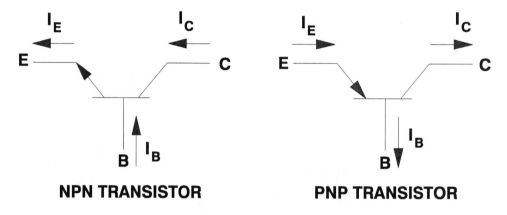

Fig. 5.2 Common base configuration for bipolar transistors.

in Fig. 5.2 together with the directions for the emitter, base, and collector currents. In this case, the base terminal serves as the reference. In the common base configuration, the input variable is the emitter current and the output variable is the collector current. Based upon Kirchhoff's law:

$$I_E = I_C + I_B \qquad (5.4)$$

The ratio of the collector current to the emitter current, called the *common base current gain* or α *(alpha)* of the bipolar transistor, is given by:

$$\alpha = \frac{I_C}{I_E} \qquad (5.5)$$

The common emitter and common base current gains are related by:

Chapter 5 : BIPOLAR TRANSISTORS

$$\alpha = \frac{I_C}{(I_B + I_C)} = \frac{\beta I_B}{(I_B + \beta I_B)} = \frac{\beta}{(1 + \beta)} \quad (5.6)$$

In a similar manner, it can be shown that:

$$\beta = \frac{\alpha}{(1 - \alpha)} \quad (5.7)$$

The device parameters that govern the current gain will be discussed in subsequent sections of this chapter.

5.2 CURRENT TRANSPORT

In both the common-base and common-emitter configurations, the collector current flow is controlled by the application of a voltage between the base and emitter terminals so as to forward bias the emitter-base junction. For the case of the NPN transistor, the forward bias results in the injection of electrons from the N$^+$ emitter into the P-base region, and the injection of holes from the P-base region into the emitter region. The emitter current is comprised of both of these components. However, the portion of the emitter current due to hole injection into the N+ emitter region does not contribute to the collector current. The electrons injected into the P-base region diffuse through it and are collected at the reverse biased base-collector junction. Due to the finite minority carrier lifetime in the P-base region, some of the injected electrons are lost due to recombination.

The current transport between the emitter and collector for an NPN transistor in the common base configuration is shown in Fig. 5.3 with depletion boundaries and the internal current components indicated. The emitter-base junction is assumed to be under forward bias, while the collector-base junction is assumed to be under reverse bias. The current I_{nE} is the emitter current component due to electron injection at the emitter-base junction. The current I_{nC} is the electron current at the collector-base junction. Using these current components, the common base current gain can be written as:

$$\alpha = \frac{\delta I_C}{\delta I_E} = \frac{\delta I_{nE}}{\delta I_E} \frac{\delta I_{nC}}{\delta I_{nE}} \frac{\delta I_C}{\delta I_{nC}} \quad (5.8)$$

From this expression, it can be seen that the change in electron current flow between the emitter and collector can be partitioned into three portions. The first term, referred to as the *emitter injection efficiency or gamma*, given by:

$$\gamma_E = \frac{\delta I_{nE}}{\delta I_E} \quad (5.9)$$

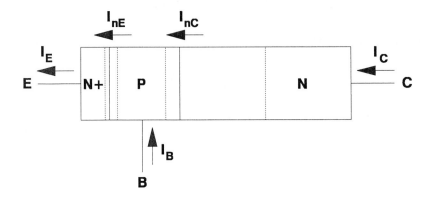

Fig. 5.3 Internal current components in a NPN Bipolar Transistor.

is the fraction of the total emitter current that comprises the electron current. As the name implies, it is a measure of the ability of the emitter to inject electrons into the base region in an efficient manner. Since the emitter current due to hole injection into the emitter region does not contribute to the collector current, the injection efficiency term is always less than unity.

The second term in Eq. (5.8), referred to as the *base transport factor*, given by:

$$\alpha_T = \frac{\delta I_{nC}}{\delta I_{nE}} \qquad (5.10)$$

is a measure of the ability for electrons that are injected into the base from the emitter to reach the collector-base junction. In the presence of a finite recombination in the base region, some electrons will be lost during transport through the base region. Consequently, the base transport factor is always less than unity.

The third term in Eq. (5.8), referred to as the *collector efficiency*, given by:

$$\gamma_C = \frac{\delta I_C}{\delta I_{nC}} \qquad (5.11)$$

is a measure of the ability for electrons to transport through the collector region. In the case of a reverse biased collector-base junction, a strong electric field is established within a depletion region at this junction. The electrons transported through the base region are swept out by this electric field into the collector region. At collector biases well below the avalanche breakdown voltage of the collector-base junction, this process occurs without loss of electrons. Consequently, the collector efficiency can be assumed to be equal to unity. However, at large collector biases, the electric field within the depletion region approaches the critical breakdown electric field. This results in the initiation of the avalanche multiplication process. In the

Chapter 5 : BIPOLAR TRANSISTORS

presence of avalanche multiplication, the collector current can exceed the electron current arriving at the collector-base junction. Under these conditions, the collector efficiency becomes greater than unity, and equal to the multiplication factor (M).

In the case of a power bipolar transistor, it is necessary to consider both the emitter injection efficiency and the base transport factor in analyzing the current gain. This is because, unlike signal transistors, the power transistors must be designed with relatively large base thickness to prevent reach-through breakdown. In addition, the operation of the power bipolar transistor at high current levels leads to high level injection in the base region, which has a strong impact on the current gain. These effects are described in the next section.

5.2.1 Emitter Injection Efficiency

In order to analyze the emitter injection efficiency, consider an NPN transistor with a narrow base region. As shown in Fig. 5.4, the emitter-base junction is forward biased while

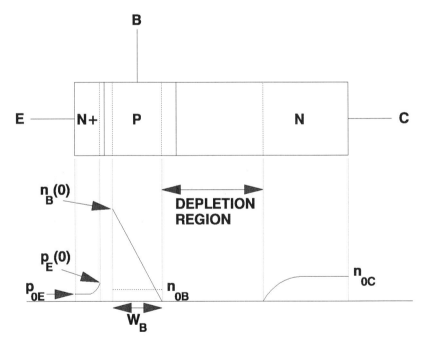

Fig. 5.4 Minority carrier distribution in a narrow base NPN transistor.

the collector-base junction is reverse biased. The applied forward bias across the emitter-base junction (V_{BE}) produces injection of electrons into the base region and holes into the emitter region. Based upon the law of the junction, the injected concentration can be related to the equilibrium concentrations of the minority carriers (n_{0B} and p_{0E}) on either side of the junction:

$$n_B(0) = n_{0B} \exp\left(\frac{q V_{BE}}{kT}\right) \quad (5.12)$$

$$p_E(0) = p_{0E} \exp\left(\frac{q V_{BE}}{kT}\right) \quad (5.13)$$

Due to the high doping concentration in the emitter, the diffusion length for the holes (L_p) is small. The continuity equation for holes in the emitter is given by:

$$\frac{d^2 p}{dx^2} - \frac{p}{L_p^2} = 0 \quad (5.14)$$

If the thickness of the N$^+$ emitter is much larger than the diffusion length, the hole concentration decays exponentially with distance from the junction to its equilibrium value as shown in Fig. 5.4. The solution for the hole distribution is given by:

$$p(x) = p_E(0) \exp\left(-\frac{x}{L_p}\right) \quad (5.15)$$

with x increasing away from the junction edge. The hole current flowing at the emitter-base junction is then given by:

$$J_p(0) = q D_{pE} \left[\frac{dp}{dx}\right]_{x=0} \quad (5.16)$$

Using Eqs. (5.15) and (5.16),

$$J_p(0) = \frac{q D_{pE}}{L_{pE}} p_{0E} \exp\left(\frac{q V_{BE}}{kT}\right) \quad (5.17)$$

Now, consider the electron distribution in the base region. The continuity equation for electrons in the base is given by:

$$\frac{d^2 n}{dx^2} - \frac{n}{L_n^2} = 0 \quad (5.18)$$

where L_n is the diffusion length for electrons in the base region. In the case of the narrow base transistor, the diffusion length is assumed to be much longer than the thickness (W_B) of the base region. Further, the reverse bias at the collector-base junction forces the electron concentration to go to zero at W_B. Using this boundary condition, the electron concentration is found to decrease linearly from $n_B(0)$ to 0 across the base region as shown in Fig. 5.4. Thus:

Chapter 5 : BIPOLAR TRANSISTORS

$$n(x) = n_B(0)\left(1 - \frac{x}{W_B}\right) \qquad (5.19)$$

The electron current flowing at the emitter-base junction is then given by:

$$J_n(0) = q\, D_{nB} \left[\frac{dn}{dx}\right]_{x=0} \qquad (5.20)$$

Using Eqs. (5.19) and (5.20),

$$J_n(0) = \frac{q\, D_{nB}}{W_B}\, n_{0B}\, \exp\!\left(\frac{q\, V_{BE}}{k\, T}\right) \qquad (5.21)$$

It is worth pointing out that these currents have been derived under the assumption that no recombination occurs in the base region. Consequently, all the base current supplies the recombination in the emitter region. The common emitter current gain, as determined by the emitter efficiency, is then given by:

$$\beta_E = \frac{I_C}{I_B} = \frac{J_n(0)}{J_p(0)} \qquad (5.22)$$

Using Eqs. (5.17) and (5.21),

$$\beta_E = \frac{D_{nB}\, n_{0B}\, L_{pE}}{D_{pE}\, p_{0E}\, W_B} \qquad (5.23)$$

It is important to note that, with the exception of the physical base width, all the other parameters are a strong function of the doping in the emitter and base regions. The dopant concentration dependence of the mobility (discussed in Chapter 2) must be taken into account when determining the diffusion coefficients. In the case of the minority carrier diffusion length in the emitter, in addition to the reduced diffusion coefficient, the effect of Auger recombination becomes very important in determination of the lifetime. Further, the equilibrium values of the minority carrier concentrations in the emitter and base regions must be determined after including the band gap narrowing phenomenon. If the intrinsic concentrations in the emitter and base regions are denoted as n_{ieE} and n_{ieB}, respectively, the injection efficiency limited current gain can be related to the doping concentrations in the emitter (N_{DE}) and base (N_{AB}) regions:

$$\beta_E = \left(\frac{D_{nB}}{D_{pE}}\right)\left(\frac{L_{pE}}{W_B}\right)\left(\frac{N_{DE}}{N_{AB}}\right)\left(\frac{n_{ieB}^2}{n_{ieE}^2}\right) \qquad (5.24)$$

It is desirable to design the power bipolar transistor to obtain a high current gain in order to reduce the base drive current. From the above equation, it can be concluded that this is

achievable by using a very low base doping concentration and a very high emitter doping concentration. However, this approach must be tempered by the following considerations:
(a) An increase in the emitter doping is accompanied by a reduction in the diffusion length due to Auger recombination and by an increase in the emitter intrinsic carrier concentration due to band gap narrowing. These effects counteract the increased doping concentration.
(b) A reduction in the base doping concentration leads to a low reach-through breakdown voltage and a poor output conductance due to depletion of the base region. Although these effects can be prevented by increasing the base width, this will also result in reduction of the current gain.
(c) A low base doping concentration results in high level injection in the base region at relatively low current densities. This produces a reduction in current gain as discussed in the next section.
(d) A low base doping concentration results in a high base sheet resistance which degrades current distribution under the emitter in the on-state and the storage time during turn-off.

The equation [Eq. (5.24)] for the emitter injection efficiency limited current gain was derived under the assumptions of a thick emitter region and no recombination in the base region. If the diffusion length for holes in the emitter is not much smaller than the emitter width (W_E), the holes injected into the emitter region can diffuse through it and reach the emitter (ohmic) contact. The effect of this upon the hole current can be taken into account by replacing the term ($D_{pE} \cdot n_{0E}/L_{pE}$) by:

$$\int_0^{W_E} D_{pE}(x) \frac{n_{ieE}^2(x)}{N_{DE}(x)} \exp\left(-\frac{x}{L_{pE}}\right) dx \qquad (5.25)$$

An optimum doping level of about 1×10^{19} per cm^{-3} for the emitter has been found to result in the highest gain for the bipolar transistor. The diffusion profile for the N$^+$ emitter region is usually empirically optimized to obtain the highest gain.

5.2.2 Emitter Efficiency including High Level Injection in Base

The analysis in the previous section indicates that a low base doping level is desirable to obtain a high emitter injection efficiency. However, if the base doping level is low, the injected electron concentration in the base at higher current densities can become comparable to and even exceed the base doping level. This is referred to as *high level injection in the base*. Under high level injection conditions in the base, the majority carrier concentration increases in order to satisfy charge neutrality. The enhancement of the majority carrier concentration in the base produces an increase in the injection of holes into the N$^+$ emitter region. This results in a reduction in the injection efficiency and current gain of the transistor, which is referred to as the *Rittner effect*.

Consider the emitter-base junction shown in Fig. 5.5 operating under high level injection conditions in the base region. *High level injection* is defined to take place when the injected minority carrier concentration exceeds the doping concentration. Since the doping concentration in the emitter region is very high, it will be assumed that low level injection conditions prevail in the emitter. In the base region, the majority carrier concentration is shown to exceed the

Chapter 5 : BIPOLAR TRANSISTORS

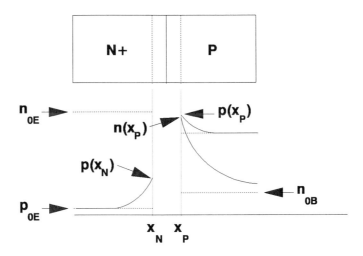

Fig. 5.5 Emitter-base junction of an N-P-N transistor operating under high level injection in the base region.

doping level (p_{0B}). Using the Boltzmann quasi-equilibrium boundary condition for the P-N junction :

$$\frac{p(x_N)}{p(x_P)} = \frac{n(x_P)}{n(x_N)} = \exp\left(-\frac{q\,\Delta\Psi}{kT}\right) \tag{5.26}$$

where $\Delta\Psi$ is the potential barrier across the junction under the forward bias conditions. Note that under low level injection conditions :

$$n(x_N) = n_{0E} \tag{5.27}$$

and

$$p(x_P) = p_{0B} \tag{5.28}$$

Eq. (5.27) is still valid under high level injection conditions in the base because of the high doping concentration in the emitter, but Eq. (5.28) no longer holds true due to high level injection in the base region. Consequently,

$$p(x_N)\cdot n_{0E} = p(x_P)\cdot n(x_P) \tag{5.29}$$

In order to satisfy the charge neutrality condition in the base region :

$$p(x_P) = p_{0B} - n_{0B} + n(x_P) \tag{5.30}$$

Since the equilibrium electron concentration in the base (n_{0B}) is small compared with the equilibrium majority carrier concentration in the base (p_{0B}):

$$p(x_P) = p_{0B} + n(x_P) \qquad (5.31)$$

Combining Eqs.(5.29) and (5.31) gives:

$$p(x_N) = [p_{0B} + n(x_P)] \frac{n(x_P)}{n_{0E}} \qquad (5.32)$$

According to the law of the junction:

$$n(x_P) = n_{0B} \exp\left(\frac{q V_{BE}}{k T}\right) \qquad (5.33)$$

where V_{BE} is the forward bias applied to the base-emitter junction. Combining this equation with Eq. (5.32) gives:

$$p(x_N) = \frac{p_{0B}}{n_{0E}}\left[1 + \frac{n(x_P)}{p_{0B}}\right] n_{0B} \exp\left(\frac{q V_{BE}}{k T}\right) \qquad (5.34)$$

The minority, majority, and intrinsic concentrations in the emitter and base regions are related by:

$$p_{0B} \cdot n_{0B} = n_{iB}^2 \qquad (5.35)$$

and

$$p_{0E} \cdot n_{0E} = n_{iE}^2 \qquad (5.36)$$

Using these relationships in Eq. (5.34):

$$p(x_N) = p_{0E}\left(\frac{n_{iB}}{n_{iE}}\right)^2 \left[1 + \frac{n(x_P)}{p_{0B}}\right] \exp\left(\frac{q V_{BE}}{k T}\right) \qquad (5.37)$$

In this expression, the term within the square brackets represents the increase in the injection of holes into the emitter region due to high level injection in the base region. If the electron injection into the base [$n(x_P)$] is small compared with the base doping concentration (p_{0B}), this terms becomes equal to unity corresponding to the case for low level injection in the base region. The increase in hole injection into the emitter by the high level injection into the base degrades the injection efficiency.

An expression for the injection efficiency under high level injection conditions in the base can be derived by assuming that the injected carriers in the emitter decay exponentially with distance away from the junction with a diffusion length L_{pE}. Then:

$$p(x) = p(x_N) \exp\left(-\frac{x}{L_p}\right) \qquad (5.38)$$

with $p(x_N)$ given by Eq. (5.37). The base current that supplies recombination current for the emitter is then given by:

$$I_B = \frac{A\, q\, D_{pE}}{L_{pE}} p(x_N) \exp\left(\frac{q\, V_{BE}}{q\, T}\right) \qquad (5.39)$$

If recombination in the base is neglected, the injected carrier concentration decreases linearly from $n(x_P)$ to zero across the base region as shown in Fig. 5.4. Consequently, the collector current is given by:

$$I_C = A\, q\, D_{nB} \frac{n(x_P)}{W_B} \qquad (5.40)$$

Using this equation:

$$n(x_P) = \frac{I_C\, W_B}{A\, q\, D_{nB}} = \frac{J_C\, W_B}{q\, D_{nB}} \qquad (5.41)$$

Combining these relationships, it can be shown that:

$$I_B = \frac{A\, q\, D_{pE}}{L_{pE}} \left(\frac{n_{iB}}{n_{iE}}\right)^2 p_{0E} \left[1 + \frac{J_C\, W_B}{q\, D_{nB}\, p_{0B}}\right] \exp\left(\frac{q\, V_{BE}}{k\, T}\right) \qquad (5.42)$$

From Eq. (5.40), the collector current is given by:

$$I_C = \frac{A\, q\, D_{nB}}{W_B} n_{0B} \exp\left(\frac{q\, V_{BE}}{k\, T}\right) \qquad (5.43)$$

The common emitter current gain determined by the emitter injection efficiency is then obtained by taking the ratio of the collector to the base current:

$$\beta = \frac{L_{pE}}{W_B} \frac{D_{nB}}{D_{pE}} \frac{n_{0E}}{p_{0B}} \frac{1}{[1 + (J_C\, W_B)/(q\, D_{nB}\, p_{0B})]} \qquad (5.44)$$

This equation indicates that, at high current densities, the current gain will decrease inversely with increasing current density. A schematic illustration of this behavior of the current gain is provided in Fig. 5.6, where the current gain and current density have been plotted on a logarithmic scale. A fall-off in the current gain with increasing collector current is highly

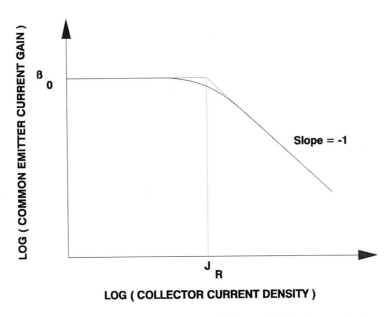

Fig. 5.6 Variation of current gain due to high level injection in the base.

disadvantageous because the base drive current for the transistor increases as the square of the collector current.

The current density at which the current gain decreases by a factor of 2 due to high level injection in the base has been referred to as the *Rittner current density* (J_R). Based upon this definition:

$$\beta = \frac{\beta_0}{[1 + (J_C/J_R)]} \qquad (5.45)$$

An expression for the Rittner current density can be obtained from Eq. (5.44):

$$J_R = \frac{q\, D_{nB}\, p_{0B}}{W_B} \qquad (5.46)$$

It is advantageous to obtain a high value for the Rittner current density because this moves the fall-off in current gain to higher current densities. It can be seen that the Rittner current density can be increased by increasing the base doping concentration (p_{0B}) and decreasing the base width (W_B). However, an increase in the base doping concentration decreases the current gain at low current levels, resulting in no improvement in the overall gain. Any decrease in the base width must be done after taking into account its impact on the reach-through breakdown voltage.

5.2.3 Base Transport Factor

The base transport factor is a measure of the ability of the injected carriers to get from the emitter/base junction to the base/collector junction. Since the transport of the carriers occurs by diffusion through the base region, the base transport factor (α_T) is given by :

$$\alpha_T = \frac{I_{nC}}{I_{nE}} = \frac{[\delta n/\delta x]_{x=W_B}}{[\delta n/\delta x]_{x=0}} \qquad (5.47)$$

If recombination in the base region is negligible :

$$\left[\frac{\delta n}{\delta x}\right]_{x=W_B} = \left[\frac{\delta n}{\delta x}\right]_{x=0} \qquad (5.48)$$

and the base transport factor becomes equal to unity. In this case the minority carrier concentration decreases linearly from the emitter to the collector junction.

When the recombination in the base region cannot be neglected, the carrier concentration

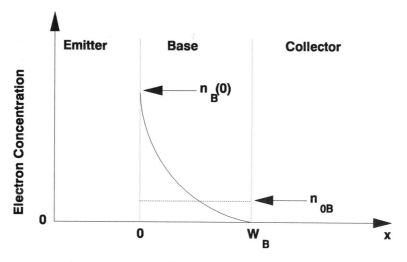

Fig. 5.7 Distribution of injected carriers in the base region.

in the base region no longer decreases linearly as shown in Fig. 5.7. In order to analyze the carrier distribution, consider the continuity equation for electrons in the steady state:

$$\frac{d^2n}{dx^2} - \frac{n}{L_n^2} = 0 \qquad (5.49)$$

The solution for this equation has the form :

$$n = A\, e^{-x/L_n} + B\, e^{x/L_n} \tag{5.50}$$

where the constants A and B are determined by the boundary conditions:

$$n = n_B(0) = n_{0B}\, e^{qV_{BE}/kT} \tag{5.51}$$

at $x = 0$, and

$$n = 0 \tag{5.52}$$

at $x = W_B$. The solution for the carrier distribution in the base is then obtained as:

$$n = n_{0B}\, \frac{\sinh[(W_B - x)/L_n]}{\sinh(W_B/L_n)}\, \exp\left(\frac{q\, V_{BE}}{k\, T}\right) \tag{5.53}$$

By using this carrier distribution to obtain $\delta n/\delta x$ at the emitter and collector ends of the base region, the base transport factor can be obtained:

$$\alpha_T = \frac{1}{\cosh(W_B/L_n)} \tag{5.54}$$

If the diffusion length (L_n) is much larger than the base width (W_B), the base transport factor can be calculated using the following approximation:

$$\alpha_T \approx 1 - \frac{W_B^2}{2\, L_n^2} \tag{5.55}$$

This expression indicates that a higher current can be obtained by reducing the base width and maintaining a large minority carrier diffusion length in the base.

5.2.4 Collector Bias Effects

The above analysis of the base transport factor was performed under the assumption that the base/collector junction is reverse biased. The current gain determined by the base transport factor is then dependent upon the *undepleted base width (W_B)*. When the collector bias is increased, the voltage is supported across a larger depletion region thickness. In the case of a bipolar power transistor, a lightly doped drift region is used in the collector region to support high collector voltages. In spite of the lightly doped collector drift region, a portion of the collector bias is also supported across a depletion region in the base region. The thickness of the depletion region in the base increases with increasing collector bias. This results in a reduction in the thickness of the undepleted base width. As a consequence, the base transport

Chapter 5 : BIPOLAR TRANSISTORS

factor increases with increasing collector bias. This produces a decrease in the output conductance of the bipolar power transistor.

A typical shape for the output characteristics of a bipolar transistor at large collector

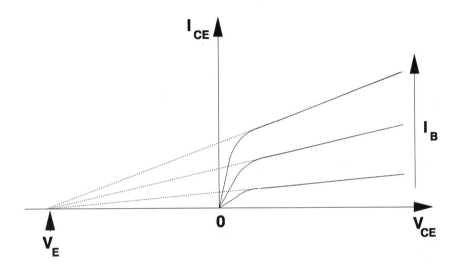

Fig. 5.8 Output Characteristics of a bipolar transistor.

biases is illustrated in Fig. 5.8. As an approximation, the collector current is shown to increase linearly with increasing collector voltage. An extrapolation of all the output characteristics to zero collector current is indicated to lead to a common intercept on the x-axis. The voltage at which this intercept occurs is referred to as the *Early Voltage* (V_E). In actual devices, the output characteristics are not linear particularly at higher collector biases. This behavior can be analyzed by considering depletion of the base region at higher collector voltages.

Consider the N-P-N bipolar power transistor illustrated in Fig. 5.9 at two collector bias voltages V_1 and V_2, with the bias V_1 larger than the bias V_2. The thickness of the depletion region in the base is related to the base doping concentration N_{AB} and the voltage supported in the base region (V_B) by:

$$W_D = \sqrt{\frac{2 \epsilon V_B}{q N_{AB}}} \tag{5.56}$$

Due to the relatively low doping concentration in the N- drift region, most of the collector bias is supported across the drift region. The voltage supported across the base region can be determined by taking into account the difference in the doping concentrations in the base and drift regions :

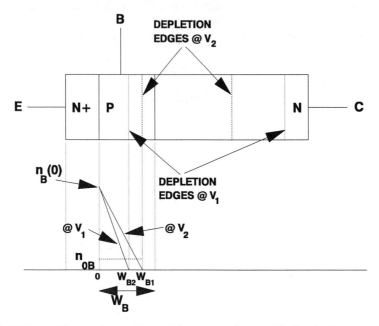

Fig. 5.9 Effect of increasing collector bias on carrier profile in the base region.

$$V_B = V_{CB}\left(\frac{N_{Dv}}{N_{AB} + N_{Dv}}\right) \quad (5.57)$$

where N_{Dv} is the doping concentration in the N- drift region. It should be noted that either a lower doping concentration in the collector or a higher doping concentration in the base region will result in less voltage being supported across the base region. The undepleted base thickness is then given by :

$$W = W_B - W_D = W_B - \sqrt{\frac{2\,\epsilon\,V_B}{q\,N_{AB}}} \quad (5.58)$$

Assuming that recombination in the base region is negligible, the collector current is given by:

$$J_c = q\,D_n\,\frac{n_B(0)}{W} \quad (5.59)$$

Combining Eq. (5.58) and (5.59):

Chapter 5 : BIPOLAR TRANSISTORS

$$J_c(V_{CB}) = q\, D_n \frac{n_B(0)}{W_B} \frac{1}{[1 - \sqrt{(2\,\epsilon\, V_B)/(q\, N_{AB}\, W_B^2)}\,]} \quad (5.60)$$

At low collector voltages, the depletion width in the base region is small. Under these conditions, the collector current is given by:

$$J_{c0} = q\, D_n \frac{n_B(0)}{W_B} \quad (5.61)$$

For a good transistor design, the quantity inside the square root is usually small. By using a series expansion for the square root and neglecting higher order terms, it can then be shown that:

$$J_c = J_{c0} \left[1 + \frac{\epsilon\, V_B}{q\, N_{AB}\, W_B^2} \right] \quad (5.62)$$

Using Eq. (5.57):

$$J_c = J_{c0} \left[1 + \frac{\epsilon}{q\, N_{AB}\, W_B^2} \frac{V_{CB}}{1 + (N_{AB}/N_{Dv})} \right] \quad (5.63)$$

This equation indicates that the collector current will increase linearly with increasing collector voltage. Using this equation, the Early voltage, at which the collector current extrapolates to zero, can be determined:

$$V_E = \frac{q\, N_{AB}\, W_B^2}{\epsilon} \left[1 + \frac{N_{AB}}{N_{Dv}} \right] \quad (5.64)$$

The Early voltage predicted by this equation is independent of the collector current at low biases, which implies that all the output characteristics should extrapolate to a common intercept as illustrated in Fig. 5.8. A large value for the Early voltage is desirable to obtain good output characteristics. From Eq. (5.64), it can be concluded that this is achievable by using a base region with large width and doping concentration. Unfortunately, this is in conflict with obtaining a high current gain.

5.2.5 Voltage Saturation Region

In the previous section, the collector bias was assumed to be sufficiently large to maintain the collector junction under reverse bias. This does not hold true when the collector voltage is reduced to values approaching the base drive voltage (about 1 volt). When the collector base

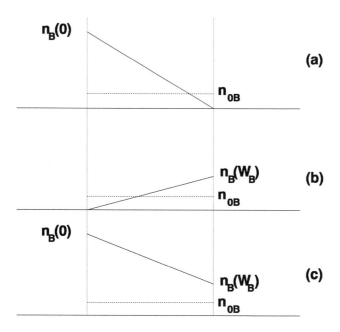

Fig. 5.10 Minority carrier profiles in the base region for (a) forward active region, (b) reverse active region, and (c) voltage saturation region.

junction becomes forward biased, the minority carrier profile in the base region becomes altered from that shown in Fig. 5.4. Since both the emitter-base junction and the collector-base junction are forward biased, the minority carrier profile can be obtained by using the superposition principle. This assumes reciprocity between the emitter-base and collector-base junctions. The superposition principle is valid only if the system is linear. Although the bipolar transistor is highly non-linear in terms of relating the terminal currents and voltages, it is fortunately linear in terms of relating the injected minority carrier concentration in the base to the terminal currents. Based upon these considerations, the minority carrier profile in the base region can be obtained by using the minority carrier profiles for operation of the transistor in the forward active region and the reverse active region. The *forward active region* is the case of a forward biased emitter-base junction and a reverse biased collector-base junction as discussed in the previous section. The *reverse active region* is the case of a forward biased collector-base junction and a reverse biased emitter-base junction. The minority carrier profiles for these cases are compared in Fig. 5.10. Superposition of these carrier profiles provides the carrier profile in the voltage saturation region. It can be seen that the slope of the carrier profile is reduced when the collector-base junction becomes forward biased. The collector current is then also reduced:

$$J_c = \frac{n_B(0) - n_B(W_B)}{W_B} \qquad (5.65)$$

Chapter 5 : BIPOLAR TRANSISTORS

From the above discussion, the on-state voltage drop across the bipolar transistor in the voltage saturation region is less than the junction potential (typically 0.7 volts). This is an advantage from the point of view of reducing the power dissipation in the on-state. It is important to note that this voltage drop is the internal voltage drop from the emitter-base junction to the base-collector junction. It does not include the voltage drop across the N- drift region in the power bipolar transistor. The doping profile of a power bipolar transistor is illustrated in Fig. 5.11 for the case of a device fabricated using emitter and base diffusion into

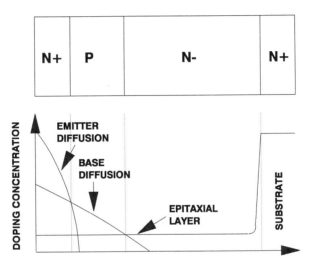

Fig. 5.11 Doping profile for an epitaxial power bipolar transistor.

a high resistivity epitaxial layer grown on a highly doped substrate. The thickness of the epitaxial layer is chosen to support large collector voltages.

The specific resistance of the collector drift region can be related to its resistivity (ρ_D) and thickness (W_D):

$$R_{D,s} = \rho_D W_D = \frac{W_D}{q \mu_n N_D} \tag{5.66}$$

Due to the relatively high resistivity of the N- drift region (typically 50 ohm-cm) and its large thickness (typically 50 microns), the specific resistance of the drift region is large (typically 0.25 ohm-cm^2). This would result in a high voltage drop across the drift region at typical on-state current density in the collector (50 amperes per cm^2). Fortunately, in the case of the power bipolar transistor, this resistance is reduced by the injection of minority carriers from the base region into the drift region. Due to the low doping concentration in the drift region, high level injection occurs, resulting in conductivity modulation of the drift region. The enhancement in the conductivity of the drift region by the injected carriers results in a reduction in the on-state voltage drop. When the collector bias is sufficiently small, the base-collector junction operates with a sufficient forward bias to produce high level injection of carriers throughout the drift region. This regime of operation is called the *voltage saturation region*. As the collector bias is increased at any fixed base drive current, the forward bias across the base-collector junction is reduced. This results in a decrease in the injection of carriers into the drift region. Under these conditions, only a portion of the drift region near the base-collector junction operates under high level injection while the rest of the drift region is not modulated. This regime of operation is called the *voltage quasi-saturation region*.

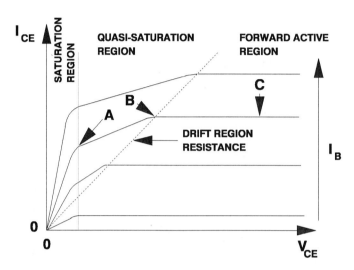

Fig. 5.12 Output characteristics of a power bipolar transistor showing the saturation and quasi-saturation regions of operation.

The different operating regions are illustrated in Fig. 5.12 at relatively small collector voltages close to the on-state voltage drop. Note that the current gain of the transistor is decreasing when proceeding from the forward active region through the quasi-saturation region into the saturation region. This is because of the additional recombination current that must be supplied by the base electrode for sustaining carriers that are injected into the collector drift region. Consider the bias points shown in Fig. 5.12 for a fixed base drive current. From 0 to A, the device is in the voltage saturation region with the conductivity of the entire drift region modulated by the injected carriers. From A to B, the device is in the quasi-saturation region. At collector voltages above bias point A, the conductivity of a part of the drift region is no longer modulated by the injected carriers, resulting in an additional voltage drop across the device. At collector voltages above bias point B, the device enters the forward active region. The bias point B is the point of transition from forward biased to reverse biased operation for the base-collector junction.

The analysis of the saturation and quasi-saturation regions can be performed by determination of the minority carrier distribution profile in the drift region. Consider the cross-section of the power bipolar transistor shown in Fig. 5.13 with the minority carrier distribution

Chapter 5 : BIPOLAR TRANSISTORS

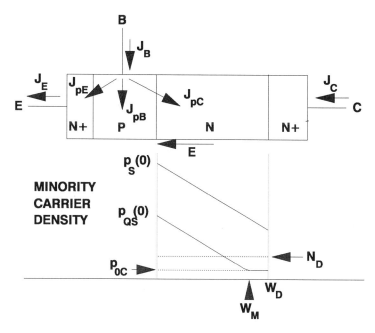

Fig. 5.13 Minority carrier distribution in the collector drift region under saturation and quasi-saturation operation.

in the drift region indicated for the case of the saturation and quasi-saturation regions. The base-collector junction is assumed to be forward biased resulting in the injection of minority carriers (holes) into the drift region. In this figure, the minority carrier concentration at the base-collector junction is defined as p(0) for both cases. In the drift region, charge neutrality demands:

$$n = p + N_D \qquad (5.67)$$

Due to the low doping concentration in the drift region to obtain high blocking voltage capability, high level injection conditions can be assumed leading to:

$$\frac{dn}{dx} = \frac{dp}{dx} \qquad (5.68)$$

The continuity equations in the drift region can be written as:

$$J_{pC} = q \mu_p p E - q D_p \frac{dp}{dx} \qquad (5.69)$$

$$J_{nC} = q \mu_n n E + q D_n \frac{dn}{dx} \tag{5.70}$$

Under saturation and quasi-saturation conditions, the base current supplies recombination in the emitter, base and collector regions as indicated in Fig. 5.13. The recombination currents supplied to the emitter and base regions have been previously discussed. The recombination current supplied to the collector drift region is represented by J_{pC}. If the current gain is assumed to be large, then this current can be assumed to be small (i.e., approximately equal to zero). Then, from Eq. (5.69):

$$E = \frac{kT}{q} \frac{1}{p} \frac{dp}{dx} \tag{5.71}$$

Further, since the hole current is assumed to be small, the electron current (J_{nC}) can be assumed to be equal to the total collector current (J_C). By using Eq. (5.71) in Eq. (5.70) with the relationships in Eqs. (5.67) and (5.68), it can be shown that:

$$J_C = q \mu_n (p + N_D) \left(\frac{kT}{q} \frac{1}{p} \frac{dp}{dx} \right) + q D_n \frac{dp}{dx} \tag{5.72}$$

Making use of the Einstein relationship between the mobility and diffusion coefficient:

$$J_C = 2 q D_n \left(1 + \frac{N_D}{2p} \right) \frac{dp}{dx} \tag{5.73}$$

The solution of this equation, with the boundary condition that the hole concentration at the base-collector junction is p(0), gives the minority carrier distribution in the drift region:

$$p(x) = p(0) - \frac{J_C x}{2 q D_n} + \frac{N_D}{2} \ln\left[\frac{p(0)}{p(x)} \right] \tag{5.74}$$

Due to the low doping concentration in the drift region, the last term in Eq. (5.74) can be neglected. Thus:

$$p(x) = p(0) - \frac{J_C x}{2 q D_n} \tag{5.75}$$

This equation indicates a linear decrease in the minority carrier concentration from p(0) at the base-collector junction as illustrated in Fig. 5.13 for both the saturation and quasi-saturation cases. In the saturation case, the injected minority carrier concentration remains much larger than the majority carrier concentration (N_D) throughout the drift region. In the quasi-saturation case, the minority carrier concentration remains much higher than the majority carrier concentration over only a portion (W_M) of the drift region. This distance (W_M) can be determined by assuming that the minority carrier concentration becomes equal to the equilibrium value at this position:

$$p(W_M) = \frac{n_i^2}{N_D} \quad (5.76)$$

Substituting this into Eq. (5.75) gives:

$$W_M = \frac{2\,q\,D_n\,[p(0) - N_D]}{J_C} = \frac{2\,q\,D_n\,p(0)}{J_C} \quad (5.77)$$

The voltage drop in the collector drift region can be obtained by integration of the electric field distribution as determined by Eq. (5.71):

$$V_M = -\int_0^{W_M} E\,dx = -\frac{kT}{q}\int_{p(0)}^{p(W_M)} \frac{dp(x)}{p(x)} \quad (5.78)$$

Using Eq. (5.75):

$$V_M = \frac{kT}{q}\ln\left[\frac{p(0)}{N_D}\right] \quad (5.79)$$

The magnitude of this voltage drop in the modulated region is relatively small (typically between 100 and 200 millivolts). Since the modulated region extends throughout the drift region under saturation conditions, the on-state voltage drop across the power bipolar transistor is also small under voltage saturation conditions.

At the onset of quasi-saturation (point A in Fig. 5.12), the thickness of the modulated region becomes equal to the total drift region thickness (W_D). The collector current density at which quasi-saturation begins to occur can then be obtained by using Eq. (5.77) with W_M equal to W_D:

$$J_{QS} = \frac{2\,q\,D_n\,p(0)}{W_D} \quad (5.80)$$

In the quasi-saturation region, an ohmic voltage drop occuring in the unmodulated portion of the drift region must be added to the voltage drop discussed above for the modulated portion. The voltage drop in the unmodulated portion of the drift region is given by:

$$V_U = \frac{J_C\,(W_D - W_M)}{q\,\mu_n\,N_D} \quad (5.81)$$

The stored charge per unit area in the collector drift region due to high level injection is obtained by integration of the minority carrier profile:

$$Q_{sD} = \frac{q\,p(0)\,W_M}{2} \qquad (5.82)$$

and the base current supplied to support recombination in the collector drift region is given by:

$$J_{pC} = \frac{q\,p(0)\,W_M}{2\,\tau_{HL}} \qquad (5.83)$$

This current is responsible for the reduced current gain observed in the saturation and quasi-saturation regions.

The saturation and quasi-saturation regions are important because the bipolar power transistor is operated in these regions during the on-state. From the point of view of reducing the power dissipation in the on-state, it is desirable to operate the transistor in the saturation region. However, as shown above, the stored charge in the collector drift region becomes much larger when the transistor is driven into the saturation region. During turn-off, this stored charge must be removed before the transistor can begin to support voltage. A larger stored charge in the collector drift region results in a longer storage phase during turn-off for a fixed reverse base drive. This is undesirable when the transistor must be operated at higher frequencies.

5.2.6 Base Widening at High Current Densities

Another phenomenon of importance to power bipolar transistors is an increase in the effective base width through which the carriers must diffuse when the device is operated at high collector current densities. This phenomenon occurs in the forward active region of operation where the base-collector junction is reverse biased. The high collector current density in the drift region is supported by the drift of the majority carriers under the influence of a large electric field created by the large collector bias. Due to the high electric fields prevalent in the drift region, the majority carriers can be assumed to be moving at their saturated drift velocity (v_s) which is approximately 10^7 cm/sec. When the collector current density is large, there is a correspondingly large majority carrier density in the drift region given by:

$$n = \frac{J_C}{q\,v_s} \qquad (5.84)$$

The presence of these majority carriers (electrons) alters the electric field distribution because their charge (negative) subtracts from the charge (positive) due to the donors. The Poisson's equation for the collector depletion region is then:

$$\frac{dE(x)}{dx} = -\frac{q}{e_s}[N_D - n] \qquad (5.85)$$

Substituting for n from Eq. (5.84) and performing an integration:

$$E(x) = E(0) - \frac{q}{e_s}\left[N_D - \frac{J_c}{q\,v_s}\right]x \qquad (5.86)$$

It can be seen that the electric field in the drift region varies linearly with distance as in the case of a reverse biased junction but its rate of variation is dependent upon the current density. At low current densities, the electric field distribution is the same as that for the case of static blocking conditions with the peak of the electric field located at the base-collector junction (i.e.

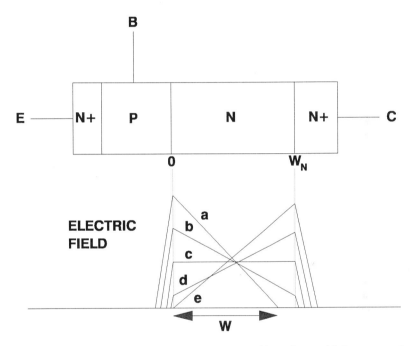

Fig. 5.14 Electric field distribution in the collector drift region at high current densities.

at x = 0). This is indicated by the curve labeled 'a' in Fig. 5.14. As the current density increases, the contribution from the majority carriers in the drift region becomes significant and, for a fixed applied collector bias, the slope of the electric field profile becomes smaller as shown by curve 'b'. It should be noted that, for the case of the epitaxial power bipolar transistor with a highly doped substrate, the depletion region is truncated by the high doping concentration in the N+ substrate. As the current density increases, a value is reached at which the slope of the electric field profile becomes zero as shown by curve 'c'. This current density is given by:

$$J_{c0} = q\,v_s\,N_D \qquad (5.87)$$

At this current density, the number of electrons per cm³ moving through the collector drift region becomes exactly equal to the doping concentration in the drift region. At even higher current densities, the electron density in the drift region exceeds the doping concentration. This produces a change in the polarity of the net charge from positive to negative, which reverses the slope of the electric field profile as shown in curve 'd'. The peak of the electric field no longer occurs at the base-collector junction but now occurs at the interface between the lightly doped drift region and the highly doped substrate. At a sufficiently large current density, the electric field at the base-collector junction can become equal to zero. This condition is shown in Fig. 5.14 as curve 'e'. The current density at which this takes place is called the *Kirk current density*. Solution of Poisson's equation for this case gives:

$$J_K = q \, v_s \, N_D + \frac{2 \, \epsilon_s \, V_{CB}}{W_N^2} \tag{5.88}$$

This current density is important because, at even higher values, the electric field profile shifts

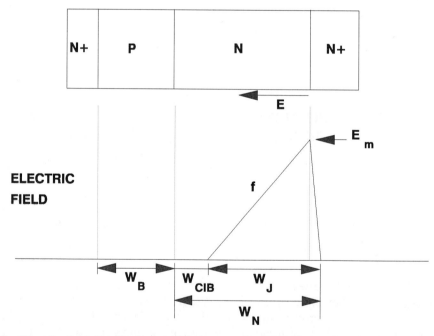

Fig. 5.15 Electric field distribution at high current densities illustrating the formation of a current induced base region within the drift region.

to that shown in Fig. 5.15 as curve 'f'. Note that, in this case, there is an undepleted portion of the drift region on the right hand side of the base-collector junction. This undepleted portion of the drift region is referred to as *the current induced base region* because it is formed by the presence of a high collector current density flowing through the drift region and because the injected electrons from the emitter-base junction must diffuse not only through the physical base width (W_B) but must traverse the undepleted portion of the drift layer (W_{CIB}) before they can be swept out by the high electric field within the rest of the drift region of width W_J. This increase in the effective base width leads to a reduction in the current gain at higher current densities.

As indicated by Eq. (5.88), the formation of the current induced base is related to the device drift region parameters. For the case of bipolar power transistors with lower breakdown voltages, the drift region doping (N_D) is relatively large and its width (W_N) is small. This leads to the first term in Eq. (5.88) becoming dominant and the Kirk current density becomes independent of the collector bias (V_{CB}). On the other hand, for the case of bipolar transistors with high breakdown voltages, the drift region doping (N_D) is relatively small and its width (W_N) is large. This leads to the second term in Eq. (5.88) becoming dominant and the Kirk current density becomes strongly dependent on the collector bias (V_{CB}). This is illustrated in Fig. 5.16

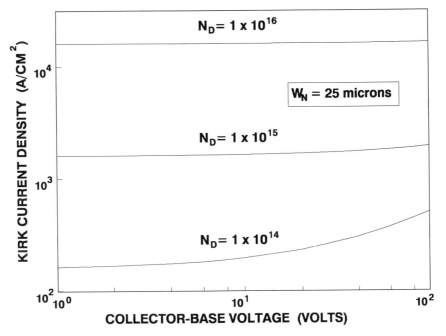

Fig. 5.16 Dependence of the Kirk current density on collector-base bias.

for three cases of drift region doping.

The width (W_{CIB}) of the current induced base region can be determined by using the electric field distribution illustrated in Fig. 5.15. This electric field distribution occurs at high current densities. Consequently, the majority carrier density in the drift region given by Eq. (5.84) is much larger than the doping concentration in the drift region. Under these conditions, the electric field distribution given by Eq. (5.86) can be rewritten as:

$$E(x) = \left(\frac{J_C}{e_s V_s}\right) x \qquad (5.89)$$

Since the doping concentration in the N+ substrate is much larger than the majority carrier concentration in the drift region, the collector-base voltage can be assumed to be supported across a width W_J located primarily within the drift region. From Eq. (5.89), this width is related to the applied bias by:

$$W_J = \sqrt{\frac{2 e_s V_s V_{CB}}{J_C}} \qquad (5.90)$$

Thus, the current induced base width is given by:

$$W_{CIB} = W_N - \sqrt{\frac{2 e_s V_s V_{CB}}{J_C}} \qquad (5.91)$$

and the effective base width of the transistor at high current densities is given by:

$$W_B(J_B) = W_B + W_N - \sqrt{\frac{2 e_s V_s V_{CB}}{J_C}} \qquad (5.92)$$

The current gain of the bipolar transistor is reduced because the electrons injected from the emitter must traverse through this increase in the effective base width. An estimate of this phenomenon can be obtained by using Eq. (5.55) to derive an expression for the common emitter current gain:

$$\beta = \frac{2 L_{nB}^2}{W_B^2} \qquad (5.93)$$

By substituting for W_B from Eq. (5.91), it can be shown that:

$$\beta = \beta_{J=0}\left[1 - 2\left(\frac{W_D - \sqrt{2e_s V_s V_{CB}/J_C}}{W_B}\right)\right] \qquad (5.94)$$

under the assumption that the base widening is small compared with the physical base width (W_B). Here, $\beta_{J=0}$ is the current gain at low current levels. This expression indicates that the current gain decreases with increasing current density, while it increases with increasing collector-base bias voltage. A physical insight into this behavior can be obtained by considering the electric field profiles shown in Fig. 5.17. When the current density increases, the majority carrier density in the drift region increases leading to change in the slope of the electric field

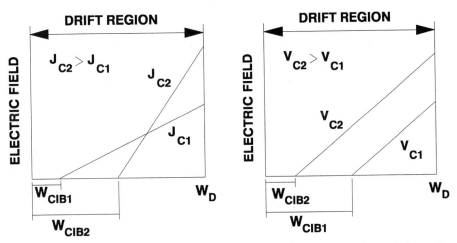

Fig. 5.17 Electric field profiles illustrating effect of (a) current density and (b) collector-base voltage up on the current induced base formation.

profile. This produces an increase in the width of the current induced base region. In contrast to this, when the collector-base voltage is increased, the area under the electric field profile increases, leading to a reduction in the width of the current induced base region.

It was previously demonstrated that the current gain of the bipolar transistor decreases when the current density is increased due to onset of high level injection in the base region (Rittner effect). The formation of the current induced base region (Kirk effect) occurs at even

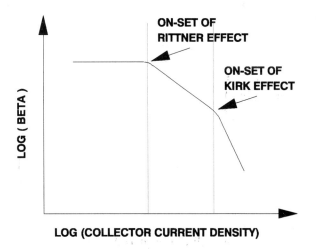

Fig. 5.18 Reduction in current gain of a bipolar transistor with increasing collector current density.

higher current densities leading to an enhancement in the degradation of current gain with increasing current density. This is illustrated in Fig. 5.18.

5.2.7 Emitter Current Crowding

In the previous sections, the current flow through the bipolar power transistor was assumed to occur uniformly through its entire emitter area. This forms the basis for the one-dimensional analysis used to derive all the expressions. In an actual power transistor, the

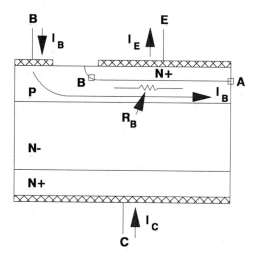

Fig. 5.19 Cross-section of power bipolar transistor illustrating interdigitation of emitter and base contacts.

emitter and base contacts must be interdigitated as illustrated in Fig. 5.19. When the transistor is biased into its on-state, a base drive current is applied at the base terminal. This current flows across the forward biased emitter-base junction resulting in the injection of electrons from the emitter. Due to the interdigitation of the contacts, in order to forward bias the emitter-base junction at point A, the base current must flow through the base resistance R_B. This produces a voltage drop between points A and B resulting in the forward bias across the emitter-base junction at point A being lower than that at point B. Consequently, the emitter injection at point A is smaller than that at point B and the emitter current distribution is not uniform across the width of the emitter N+ diffusion.

The resistance of the base region through which the base current flows is determined by the *pinch-resistance* formed by the superposition of the base and emitter diffusions. The integrated net P-type doping concentration in the base region below the emitter is much smaller than the P-type dopant diffused into the device. This can be seen from the doping profiles under the emitter shown in Fig. 5.20. Further, the depletion layers formed in the base region, at both

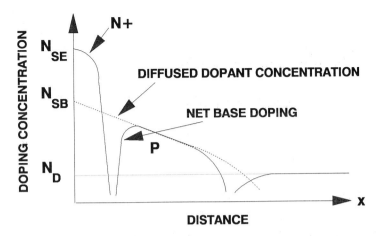

Fig. 5.20 Doping profiles under the emitter for a bipolar transistor.

the emitter-base and collector-base junctions, produce an increase in the base resistance, while any high level injection in the base produces a reduction in the base resistance. In general, the sheet resistance of the pinch-resistor can be substantial leading to a non-uniform emitter current density across its width.

In order to analyze the emitter current distribution, consider the device cross-section shown in Fig. 5.21 with an emitter half-width of W_E and a base width of W_B. It will be assumed that the emitter has a linear geometry with an emitter length of L_E orthogonal to the cross-section in the figure. The current distribution can be derived by considering an element of width dx located at a distance x from the edge of the emitter closest to the base contact (i.e. point B in the Fig. 5.19). The emitter current flowing through the segment dx in Fig. 5.21 is given by:

$$dI_E(x) = J_E L_E dx \qquad (5.95)$$

where J_E is the emitter current density at location x. The corresponding base current to support this emitter current flow is given by:

$$dI_B = (1 - \alpha) J_E L_E dx \qquad (5.96)$$

The voltage drop in the base across the segment dx produced by the flow of the base current $I_B(x)$ is given by:

$$dV_{BE}(x) = I_B(x) \frac{dx}{\sigma_B W_B L_E} \qquad (5.97)$$

where σ_B is the conductivity of the base region. The emitter current density at x can be related

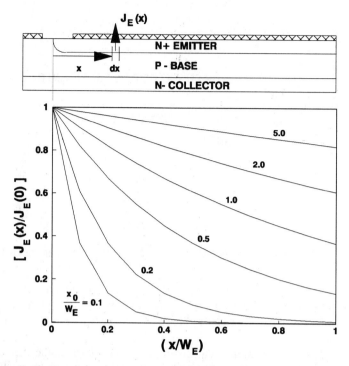

Fig. 5.21 Normalized current distribution along the width of the emitter in a power bipolar transistor.

to the emitter-base voltage at x by:

$$J_E(x) = J_0\, e^{qV_{BE}(x)/2kT} \tag{5.98}$$

under the assumption that high level injection conditions are applicable. Differentiation of this equation gives:

$$\frac{d\,J_E(x)}{d\,x} = \frac{q}{2\,k\,T}\,J_E(x)\,\frac{d\,V_{BE}(x)}{d\,x} \tag{5.99}$$

Using Eq. (5.97) in Eq. (5.99):

$$\frac{d\,J_E(x)}{d\,x} = \frac{q}{2\,k\,T}\,J_E(x)\,\frac{I_B(x)}{\sigma_B\,W_B\,L_E} \tag{5.100}$$

Under high level injection conditions in the base region, it can be assumed that the conductivity of the base region becomes proportional to the emitter current density:

Chapter 5 : BIPOLAR TRANSISTORS

$$\sigma_B = \sigma_0 \frac{J_E(x)}{J_{HL}} \qquad (5.101)$$

where J_{HL} is the emitter current density at which high level injection begins to occur. Using this expression in Eq. (5.100):

$$\frac{d J_E(x)}{d x} = \frac{q}{2 k T} J_{HL} \frac{I_B(x)}{\sigma_{B0} W_B L_E} \qquad (5.102)$$

Differentiation of this expression gives:

$$\frac{d^2 J_E(x)}{d x^2} = \frac{q}{2 k T} \frac{J_{HL}}{\sigma_{B0} W_B L_E} \frac{d I_B(x)}{d x} \qquad (5.103)$$

Combining Eqs. (5.96) and (5.103) gives:

$$\frac{d^2 J_E(x)}{d x^2} - \frac{q}{2 k T} \frac{J_{HL}}{\sigma_{B0} W_B} (1 - \alpha) J_E(x) = 0 \qquad (5.104)$$

The solution for this differential equation is:

$$J_E(x) = J_E(0) e^{(-x/x_0)} \qquad (5.105)$$

with a characteristic decay length for the emitter current density given by:

$$x_0 = \sqrt{\frac{2 k T W_B \sigma_B}{q J_{HL} (1 - \alpha)}} \qquad (5.106)$$

As an example, the characteristic decay length is about 200 microns for a device with a base width of 10 microns, an average base doping of 1×10^{17} per cm^3, and a common base current gain of 0.95, under the assumption that the onset of high level injection occurs at a current density of 10 amperes per cm^2. This allows fabrication of devices with emitter widths (which are twice W_E) of up to several hundred microns in size without severe current crowding at the edges of the emitter fingers. The normalized current distribution along the emitter width is shown in Fig. 5.21 for the case of different values of the normalized characteristic lengths. It can be seen that the current flow along the emitter becomes more uniform when the characteristic length is large when compared with the emitter width.

Based upon the above discussion, it can be concluded that the current carrying capability of the power bipolar transistor does not scale with an increase in the emitter area. Instead, due to current crowding at the edges of the emitter, the current scales with an increase in the emitter periphery. Consequently, it is common practice to interdigitate the emitter and base contacts into a finger geometry as shown in Fig. 5.22. Note that the base contact is placed on the outer

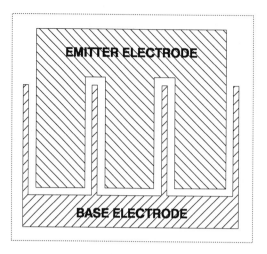

Fig. 5.22 Interdigitated emitter-base finger geometry for power bipolar transistors.

edges of the emitter contact on both sides of the device to obtain the benefits of current flow along these emitter edges. In spite of the need for interdigitation, the photolithography design rules for the fabrication of power bipolar transistors are not particularly demanding with state-of-the-art process technology because the size of the emitter width (W_E) is relatively large (hundreds of microns in size).

5.3 STATIC BLOCKING CHARACTERISTICS

The power bipolar transistor is expected to support high voltages during the forward blocking mode. This voltage is supported across the lightly doped collector drift region within the device. Devices with blocking voltages ranging from 100 to 1500 volts are used in applications. The power bipolar transistor is optimized to block high voltages for only one polarity of applied bias to the collector (positive bias for the case of the N-P-N transistor discussed in earlier sections). The device cannot support large voltages when the collector bias is reversed because in this case the base-collector junction becomes forward biased and the voltage is impressed upon the reverse biased emitter-base junction. The emitter-base junction can support only a relatively small reverse bias due to the high doping concentration on both sides of the junction. A typical breakdown voltage for this junction is about 20 volts. For this reason, the bipolar power transistor is used only in circuits operating from a DC power supply.

The voltage blocking characteristics of the bipolar power transistor in the forward active mode are discussed in this section. The ability of the device to support a large collector bias is dependent upon the impedance between the emitter and base terminals. If the base is open circuited, the blocking voltage is severely degraded due to enhancement of the leakage current

Chapter 5 : BIPOLAR TRANSISTORS

by the internal current gain of the transistor. These effects are analyzed in this section.

5.3.1 Open Emitter Breakdown Voltage

When the N-P-N bipolar transistor is operated with no connection to the emitter terminal (open emitter configuration) and a positive bias is applied to the collector terminal with respect to the base terminal, the device operates with the voltage supported across the base-collector junction. The breakdown voltage in this mode is referred to as the *open emitter breakdown voltage* (BV_{CBO}). In this mode of operation, the breakdown voltage of the device is dependent upon the properties of the N- collector drift region. The drift region doping and thickness are chosen based upon the avalanche breakdown considerations previously discussed in Chapter 3. A higher breakdown voltage can be achieved by reducing the drift region doping and increasing its thickness. However, it should be noted that this results in a larger collector drift region resistance. As discussed earlier in this chapter, a larger collector drift region resistance leads to an expansion of the quasi-saturation region, which results in an increase in the on-state power dissipation. It will also be shown in a subsequent section that a thicker drift region results in an increase in the storage time during turn-off.

5.3.2 Shorted Emitter-Base Breakdown Characteristics

When the N-P-N bipolar transistor is operated with the emitter terminal short-circuited to the base terminal and a positive bias is applied to the collector terminal with respect to the base terminal, the device operates with the voltage supported across the base-collector junction. In this mode of operation, the breakdown voltage of the transistor is dependent not only upon the properties of the N- collector drift region but also upon the current gain of the bipolar transistor.

This can be explained by considering the cross-section of the power bipolar transistor shown in Fig. 5.23 with the interdigitated emitter and base contacts. Upon the application of a positive bias to the collector terminal, the voltage is supported across the reverse biased base-collector junction. When the collector bias reaches the avalanche breakdown voltage of the base-collector junction (BV_{CBO}), enhanced current flow begins to occur across the base-collector junction due to impact ionization. This current flows into the base terminal, as illustrated in Fig. 5.23 producing a lateral voltage drop in the base region under the emitter diffusion. This voltage drop forward biases the emitter-base junction at the middle of the emitter finger, as indicated in Fig. 5.23. After the onset of the injection of electrons from the emitter, any base current flow is amplified by the gain (beta) of the bipolar transistor. Since the current levels under the blocking conditions are small, the transistor is operating at a high current gain. Thus, the emitter current becomes a dominant current. Beyond this collector current level, any increase in collector current promotes stronger injection from the emitter, resulting in a positive feed-back mechanism that produces a reduction in the collector voltage supported by the transistor with increasing collector current. A negative resistance characteristic is observed in this regime of operation, as shown in Fig. 5.24, until the voltage reaches the open base

Chapter 5 : BIPOLAR TRANSISTORS

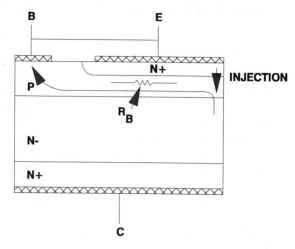

Fig. 5.23 Operation of power bipolar transistor with shorted base and emitter terminals.

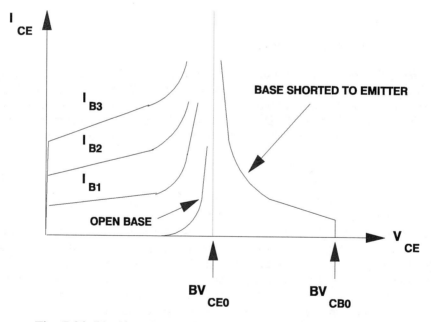

Fig. 5.24 Blocking characteristics of a power bipolar transistor.

breakdown voltage (discussed in the next section). It is therefore general practice to operate bipolar transistors only up to the open base breakdown voltage. This must be taken into consideration when designing the collector drift region doping and thickness.

5.3.3 Open-Base Breakdown Characteristics

When the N-P-N bipolar transistor is operated with no connection to the base terminal (open base configuration) and a positive bias is applied to the collector terminal with respect to the emitter terminal, the device operates with the voltage supported across the base-collector junction with a forward biased emitter-base junction. The breakdown voltage in this mode is referred to as the *open base breakdown voltage* (BV_{CEO}). In this configuration, the leakage current flowing across the base-collector junction must also flow across the emitter-base junction. Thus, the leakage current is amplified by the gain (beta) of the bipolar transistor. Due to the relatively low current levels in the blocking mode, the current gain of the bipolar transistor is high, resulting in significant enhancement in the leakage current.

The leakage current for an open base transistor was previously discussed in Chapter 3, where it was shown that:

$$I_E = I_C = \frac{I_L}{(1 - \alpha)} \tag{5.107}$$

where α is the common base current gain of the N-P-N transistor and I_L is the sum of the space-charge-generation and diffusion currents across the base-collector junction. The common base current gain is given by:

$$\alpha = \gamma_E \cdot \alpha_T \cdot M \tag{5.108}$$

where γ_E is the emitter injection efficiency, α_T is the base transport factor, and M is the avalanche multiplication factor. In the case of the open base transistor with a lightly doped base region that was discussed in Chapter 3, the current gain increases with applied bias primarily due to an increase in the base transport factor. In the case of the power bipolar transistor that is being considered in this chapter, the base doping is much larger than the doping concentration in the collector drift region. Consequently, as the collector bias is increased, the current gain increases primarily due to an increase in the multiplication factor with only a small increase in the base transport factor. The multiplication factor (M) can be empirically related to the collector bias by:

$$M = \frac{1}{[1 - (V_{CE}/BV_{CBO})^n]} \tag{5.109}$$

Using Eq. (5.108) in conjunction with Eq. (5.109), it is possible to trace the collector current-voltage characteristic as shown in Fig. 5.24 for the case of an open circuited base.

From Eq. (5.107), it can also be concluded that the collector current approaches infinity when the common base current gain approaches unity. This condition is satisfied when:

$$M = \frac{1}{\gamma_E \cdot \alpha_T} = \frac{1}{\alpha_0} \tag{5.110}$$

where α_0 is the common base current gain at low collector biases where the avalanche multiplication factor is equal to unity. Thus, the open base breakdown voltage (BV_{CB0}) can be obtained from the condition:

$$M(BV_{CE0}) = \frac{1}{[1 - (BV_{CE0}/BV_{CB0})^n]} = \frac{1}{\alpha_0} \qquad (5.111)$$

From this equation, it can be shown that:

$$BV_{CE0} = \frac{BV_{CB0}}{(\beta_0)^{1/n}} \qquad (5.112)$$

where β_0 is the common emitter current gain at low collector biases where the multiplication factor is unity. In general, the common emitter current gain at low current levels is large leading to an open base breakdown voltage that is substantially smaller than the open emitter breakdown voltage. As an example, if the common emitter current gain is 100, the open base breakdown voltage will be only about one-third of the open emitter breakdown voltage. As pointed out earlier, it is necessary to operate the power bipolar transistor below its open base breakdown voltage. Consequently, when designing a power bipolar transistor to operate up to any given breakdown voltage, it becomes necessary to use much higher resistivity collector drift regions than dictated by the parallel plane breakdown voltage analysis. The high collector drift region resistivity enlarges the quasi-saturation region leading to higher power dissipation during current conduction. It also promotes the onset of the Kirk effect (current induced base widening) at lower current densities leading to degradation of current gain at high current levels.

5.4 DYNAMIC SWITCHING CHARACTERISTICS

A primary application for the power bipolar transistor is as a switch that regulates the flow of energy from a power source to the load. This is commonly performed by rapidly switching the transistor between its on- and off-states. The efficiency with which this control of power flow is possible depends upon the power dissipated within the power transistor. Power dissipation in the bipolar power transistor occurs during its on-state due to a finite on-state voltage drop, during its off-state due to a finite leakage current, and during switching from both the off- to the on-state and from the on- to the off-state. The power dissipation in the off-state is usually much smaller than the other components. The on-state power dissipation depends upon whether the device is operated in its voltage saturation region or in the quasi-saturation region as discussed in a previous section of this chapter. A lower on-state voltage drop can be obtained by operating the device in the voltage saturation region as compared to the quasi-saturation region. However, it will be shown in this section that this results in an increase in the switching power losses.

5.4.1 Turn-on Transient

The switching of the bipolar power transistor from its off-state to its on-state is considered in this section. It will be assumed that a resistive load is connected between the transistor and the DC power supply of value V_{cc}. Whether the transistor is driven into quasi-saturation or into saturation depends upon the load resistance and the base drive current. Both of these cases are discussed below.

It will be assumed that the device is operating initially in the off-state with zero current.

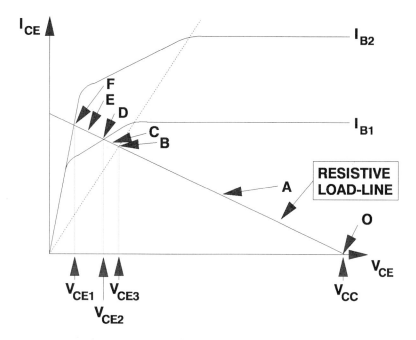

Fig. 5.25 Current-voltage locus during turn-on transient.

This corresponds to the operating point O on the output characteristics shown in Fig. 5.25. In the off-state, the collector drift region is depleted and the voltage is supported across the reverse biased base-collector junction.

A positive base drive current is applied at time t = 0. As the base drive current increases, the emitter current also increases. The electrons injected from the N+ emitter region diffuse through the P-base region and are collected by the reverse biased base-collector region. They then drift through the collector drift region at saturation velocity due to the high electric field in the drift region. This corresponds to an operating point A in Fig. 5.25. Note that a part of the collector power supply voltage is supported across the load resistor and the rest of the voltage is supported by the transistor. Thus, the voltage across the transistor decreases with time as illustrated in Fig. 5.26. This mode of operation occurs until the transistor enters the

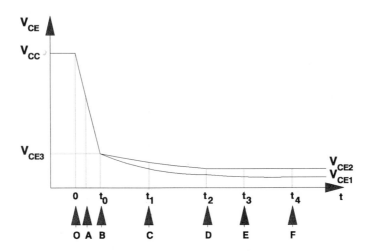

Fig. 5.26 Change in collector-emitter voltage across a bipolar transistor during turn-on.

quasi-saturation region. The intersection of the quasi-saturation resistance and the load line is indicated in Fig. 5.25 as point B and the time at which the transistor enters into the quasi-saturation region is indicated in Fig. 5.26 as t_0. At this time, the transistor is supporting a voltage V_{CE3}. Once the transistor enters the quasi-saturation region, the base-collector junction becomes forward biased and strong injection of holes occurs into the collector drift region resulting in modulation of its conductivity.

The carrier distribution within the collector drift region during the turn-on transient is shown in Fig. 5.27 for the period (for example at time t_1 corresponding to point C) after the transistor has entered into the quasi-saturation region. Note that the slope of the carrier concentration distribution is proportional to the collector current. Since the collector current does not change by a significant amount after the device enters the quasi-saturation region, the slope of the carrier concentration distribution also remains relatively unchanged. For a base drive I_{B1}, the load line intersects the output characteristics within the quasi-saturation region at point D. This corresponds to a time t_2 in the turn-on transient. Note that in this case the entire collector drift region is not modulated as indicated in Fig. 5.27. Beyond time t_2, the device reaches a steady-state operating point D with an on-state voltage drop of V_{CE2}.

If the base drive voltage applied during turn-on is increased to a value I_{B2} while maintaining the same load line, the device is driven into saturation. This is illustrated by the carrier distribution profiles in Fig. 5.27 marked t_3 for the operating point E and t_4 for the operating point F. After time t_4, the device reaches a steady-state operating point with an on-state voltage drop of V_{CE1}. It can be seen from the figures that the steady-state voltage drop in this case is smaller than for the case of base drive I_{B1} but the time taken to reach the steady-state operating point is longer and the total stored charge in the collector drift region is much larger.

During the time intervals $(t_3 - t_0)$ and $(t_4 - t_0)$, the power dissipation in the transistor is enhanced because the collector current has almost reached its steady-state value while the

Chapter 5 : BIPOLAR TRANSISTORS

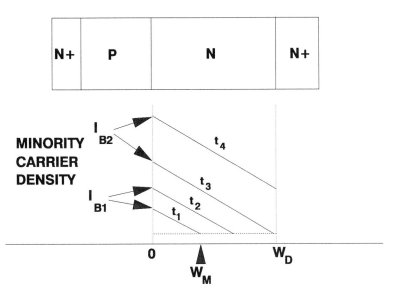

Fig. 5.27 Carrier distribution in the collector drift region during the turn-on transient.

collector voltage is larger than its steady-state value. It is therefore important to keep these intervals as small as possible. The shape of the voltage transient can be analyzed using the charge control principle. If the base current that provides for recombination in the collector drift region is defined as $J_{pC}(t)$ at time t, the charge per unit area in the collector drift region Q_c is given by:

$$\frac{dQ_c}{dt} = J_{pC}(t) - \frac{Q_c}{\tau_D} \qquad (5.113)$$

If the steady-state base drive current required to provide for recombination in the collector drift region is defined as J_{pC}, then the solution of this equation is:

$$Q(t) = J_{pC} \tau_D [1 - e^{-(t/\tau_D)}] \qquad (5.114)$$

According to this equation, the charge in the collector drift region increases exponentially with time.

The voltage drop across the collector drift region occurs primarily within the unmodulated region as discussed earlier. If the edge of the unmodulated region is defined as W_M, the voltage drop across the collector drift region is given by:

$$V_{CE}(t) = \frac{J_C}{q \mu_n N_D} (W_D - W_M) \qquad (5.115)$$

The position W_M of the edge of the modulated region is related to the total charge stored in the collector drift region by:

$$W_M = \frac{2 \, Q(t)}{q \, p(0)} \tag{5.116}$$

where p(0) is the carrier concentration at the base-collector junction. By relating the slope of the carrier distribution profile to the collector current density:

$$p(0) = \frac{J_C \, W_M}{2 \, q \, D_n} \tag{5.117}$$

Using this equation, it can be shown that:

$$W_M = 2 \, Q(t) \left(\frac{D_n}{J_C \, J_{pC} \, \tau_D} \right)^{1/2} \tag{5.118}$$

By using Eq. (5.114):

$$W_M = 2 \left(\frac{D_n \, \tau_D \, J_{pC}}{J_C} \right)^{1/2} [1 - e^{-(t/\tau_D)}] \tag{5.119}$$

Using this expression in Eq. (5.115), an equation describing the change in collector voltage with time is obtained:

$$V_{CE}(t) = \frac{J_C}{q \, \mu_n \, N_D} \left\{ W_D - 2 \left(\frac{D_n \, \tau_D \, J_{pC}}{J_C} \right)^{1/2} [1 - e^{-(t/\tau_D)}] \right\} \tag{5.120}$$

In deriving this expression, it has been assumed that the collector current density (J_C) is approximately constant because the transient occurs in the quasi-saturation region.

5.4.2 Turn-off Transient

In previous sections, it has been shown that high concentrations of electrons and holes are present within the base and collector regions during the on-state. This stored charge must be removed during the turn-off process to restore the device to its blocking state. One method for achieving this goal is to simply open circuit the base terminal (i.e., to stop providing the base drive current needed to keep the bipolar transistor in its on-state). The stored charge will then decay by recombination at a rate dictated by the lifetime in the base and collector regions. Since this process is relatively slow and produces turn-off times that are too long for applications, it is preferable to apply a reverse bias to the base-emitter junction to enhance the turn-off process. The reverse base current flow extracts a significant fraction of the stored charge, which enables faster turn-off. Due to the presence of the stored charge in the base region, the emitter-base junction must undergo a recovery akin to that of the P-i-N rectifier.

Consider the case of a bipolar power transistor being turned-off with a resistive load in

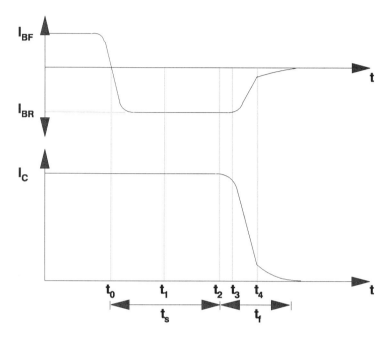

Fig. 5.28 Base and collector current waveforms during turn-off of a bipolar power transistor.

series with it. In Fig. 5.28, the base and collector current waveforms during turn-off are shown for a bipolar power transistor. It is assumed that the device is initially in its on-state with a base drive current of I_{BF} and a collector current of I_C. The carrier distribution within the base and collector regions in the on-state is indicated in Fig. 5.29 by the profile marked $t = t_0$.

At time t_0, the base current is reversed to a value I_{BR}. At this time, both the base-emitter and base-collector junctions are forward biased and the transistor cannot sustain a high voltage. The collector current then remains constant as determined by the power supply voltage and the load resistance. However, the emitter current is reversed, as indicated in the cross-section shown in Fig. 5.29. Thus, the slope of the carrier concentration profile in the base region also reverses at the emitter junction, as shown by the profile marked $t = t_1$. This situation continues until time t_2, beyond which an unmodulated portion of the collector drift region forms. Thus, after time t_2, the collector drift region begins to support voltage and the collector current begins to decrease. The time duration $(t_2 - t_0)$, during which the collector current remains constant, is referred to as the *storage time*.

An estimate of the storage time can be obtained by assuming that the stored charge is primarily resident in the collector drift region. This assumption is valid if the width of the

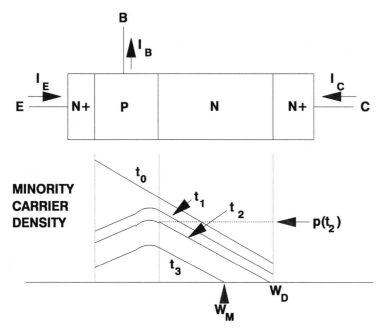

Fig. 5.29 Carrier distribution profiles in a power bipolar transistor during turn-off.

collector drift region is much larger than the base width of the bipolar transistor. The stored charge in the collector drift region just prior to the application of the reverse base drive current is given by:

$$Q_s = J_{pC} \tau_D \qquad (5.121)$$

where J_{pC} is the base current supplied for recombination in the collector drift region and τ_D is the lifetime in the drift region. At the end of the storage time, the stored charge remaining in the collector drift region is given by:

$$Q_s(t_2) = \frac{1}{2} p(t_2) W_D \qquad (5.122)$$

where $p(t_2)$ is the carrier concentration at the base-collector junction at time t_2. This carrier concentration is related to the collector current density by:

$$J_C = \frac{2 q D_n p(t_2)}{W_D} \qquad (5.123)$$

Thus, the charge extracted during the storage time is given by:

$$Q_s(t_2 - t_0) = \left[J_{pC}\tau_D - \frac{1}{4} \frac{J_C W_D^2}{q D_n} \right] \quad (5.124)$$

Since a constant reverse base current of magnitude J_{BR} flows during the storage period:

$$Q_s = J_{BR} t_s \quad (5.125)$$

Using these equations:

$$t_s = \frac{J_{pC}}{J_{BR}} \tau_D - \frac{1}{4} \frac{J_C}{J_{BR}} \frac{W_D^2}{q D_n} \quad (5.126)$$

The storage time is of the same order of magnitude as the high level recombination lifetime in the collector drift region. This value is typically about 10 microseconds.

After the storage phase, an unmodulated region begins to form within the collector drift region. This allows the transistor to support voltage, resulting in a reduction in collector current. Due to the presence of a high concentration of minority carriers at the base-collector junction, the junction is unable to support voltage at this time. Consequently, the rate of change of collector current is determined by the extraction of carriers by the reverse base drive current. This condition applies for the time t_3 in Fig. 5.28 and Fig. 5.29. When the minority carrier concentration at the base-collector junction falls to zero, it begins to support voltage with the formation of a depletion region. The collector voltage now rises rapidly. At this point in time, both the collector and base currents decrease rapidly. It should be noted that a relatively gradual decrease in current with time is usually observed after the rapid fall in collector current. This period, referred to as a *current tail*, is due to the removal of charge in the base and collector regions by recombination of minority carriers after the base-collector junction has become reverse biased. The rate of decay of the collector current during this time is determined by the minority carrier lifetime. Since the current changes by several orders of magnitude during this time, the minority carrier lifetime varies between its high level injection value and its low-level injection value.

5.5 SECOND BREAKDOWN CHARACTERISTICS

The current-voltage boundary within which a power bipolar transistor can be operated without destructive failure is defined as its *safe-operating-area or S-O-A*. At low current levels, this boundary is determined by the on-set of avalanche breakdown. At low voltage levels, this boundary is determined by the maximum current that the leads can handle without fusing. When both the current and voltage are simultaneously large, the device experiences a high instantaneous power dissipation. The safe operation of the device is then determined either by a thermal limitation or by an instability referred to as *second breakdown* to distinguish it from the previously discussed avalanche breakdown observed at low current levels. These limits on the S-O-A of bipolar power transistors are discussed below.

The second breakdown phenomenon is of particular importance to operation of the bipolar transistor with inductive loads. A typical example of the current-voltage locus observed

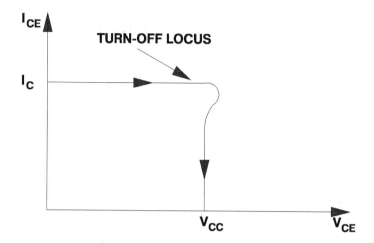

Fig. 5.30 Current-voltage locus during turn-off.

during the turn-off of the bipolar transistor is illustrated in Fig. 5.30. During turn-off, the collector voltage exceeds the DC power supply voltage. This *voltage over-shoot* is due to the presence of a high di/dt across any stray inductance in the circuit during turn-off. In the case of the turn-on of the power bipolar transistor, the current-voltage locus takes the form shown in Fig. 5.31 because of the reverse recovery of anti-parallel diodes. Note that, in this case, the current exceeds the steady-state value by a magnitude determined by the reverse recovery current of the rectifier. It is important to maintain these current-voltage loci within the safe-operating-area of the bipolar power transistor to ensure non-destructive operation.

5.5.1 Forward Biased Second Breakdown

The power bipolar transistor is prone to thermal runaway when biased in the forward active region of the output characteristics. This occurs because of the affinity for the formation of a local region in the emitter through which the current tends to constrict itself. In order to understand this effect, consider the transistor being driven with a voltage source (V_{BE}) applied between the base-emitter junction. As discussed in section 5.2.1, the collector current density is then given by:

Chapter 5 : BIPOLAR TRANSISTORS

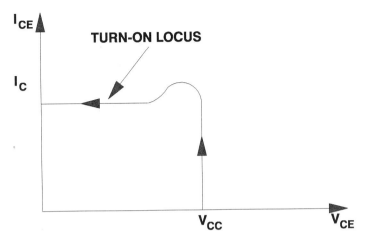

Fig. 5.31 Current-voltage locus during turn-on.

$$I_C = \frac{A\ q\ D_{nB}\ n_{0B}}{W_B}\ e^{(q\ V_{BE}/k\ T)} \qquad (5.127)$$

Consider a local region in the emitter of the power bipolar transistor. The power dissipation in this region is:

$$P_d = V_{CE}\ I_C \qquad (5.128)$$

If the current density in this region increases, then the local power dissipation also becomes larger resulting in an increase in the local temperature. This produces an increase in the minority carrier concentration in the local region as given by:

$$n_{0B} = \frac{n_i^2}{N_{AB}} = \frac{N_C\ N_V}{N_{AB}}\ e^{(-E_g/k\ T)} \qquad (5.129)$$

where N_C and N_V are the density of states in the conduction and valency bands, and N_{AB} is the acceptor doping concentration in the base region. This in turn results in an increase in the local current density which produces a further increase in the power dissipation and temperature. Thus, there is a positive feedback mechanism inherent in the collector current flow within the power bipolar transistor. This can create a high local current density with high temperatures that can result in melting the emitter metal producing destructive failure.

One method to prevent this localization of collector current flow is by placing a high resistance in series with the base terminal as illustrated in Fig. 5.32 so that the base drive current I_B is determined by the ratio of the base drive voltage V_{BE} and the base circuit resistance R_B. This corresponds to driving the bipolar power transistor with a current source. Under these

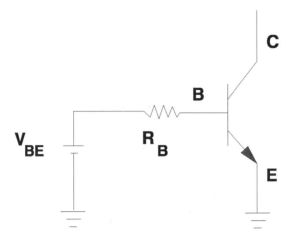

Fig. 5.32 Bipolar transistor with base drive resistance.

conditions, the base-emitter voltage V_{BE} is given by:

$$e^{(q V_{BE}/k T)} = \frac{L_{pE} I_B}{A \, q \, D_{pE} \, p_{0E}} \qquad (5.130)$$

Substituting this expression into Eq. (5.129):

$$I_C = \frac{D_{nB}}{D_{pE}} \frac{L_{pE}}{W_B} \frac{N_{DE}}{N_{AB}} \frac{n_{ieB}^2}{n_{ieE}^2} I_B \qquad (5.131)$$

Now the collector current becomes relatively insensitive to the local temperature because the temperature dependence of the intrinsic carrier concentrations in the base and emitter regions cancels out. Although this method produces a more stable and uniform current distribution in the bipolar power transistor, it is accompanied by the disadvantages of needing a high base drive voltage and a significant power loss in the base drive circuit within the series base drive resistance.

Another approach to increasing the uniformity of collector current flow with stable operation is by the addition of a resistance in series with the emitter. This method, illustrated in Fig. 5.33, is also referred to as *emitter ballasting*. The basic idea behind this approach is that if the emitter current increases at a local region, the voltage drop across the emitter ballast resistance also increases. This diverts the emitter current to other regions of the emitter promoting a more uniform current distribution. For this concept to work, it is not possible to use a lumped series emitter resistance because this would not result in promoting uniform current distribution within the transistor. Instead, it is necessary to distribute the emitter ballast resistance throughout the emitter fingers. The emitter ballast resistance must be relatively small

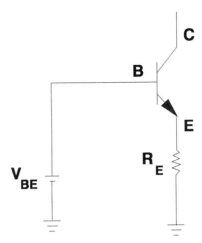

Fig. 5.33 Bipolar power transistor with emitter ballast resistance.

in value because the emitter current flows through it, producing an increase in the on-state voltage drop and power dissipation within the bipolar power transistor.

A simple method for integrating a distributed emitter ballast resistance is illustrated in Fig. 5.34, which shows the top view and cross-section of the bipolar power transistor with the emitter metallization restricted to the central portion of each emitter finger. The emitter ballast resistance is formed by utilizing the sheet resistance of the N$^+$ emitter diffusion as indicated by R_E. During the on-state, it was previously shown that the emitter current tends to crowd to the edges of the emitter closest to the base contact. As a consequence of this, the emitter current is forced to flow through the emitter ballast resistance formed in the N$^+$ emitter, which produces the desired voltage drop required to promote uniform current distribution. Although other methods to form the emitter ballast resistance have been proposed by using polysilicon layers, the approach shown in Fig. 5.34 is attractive because no additional processing is needed during device fabrication.

5.5.2 Reverse Biased Second Breakdown

The power bipolar transistor is susceptible to a different failure mechanism during turn-off. This limitation to its safe-operating-area is referred to as *reverse biased second breakdown or RBSOA* because the base drive current is reversed during turn-off. As discussed earlier, the reverse base drive current is needed to extract minority carriers (stored charge) in the base and collector drift regions. In the analysis in Section 5.4.2, a one-dimensional case was discussed. In an actual bipolar power transistor, it is necessary to take into account the finite emitter finger width during turn-off. As illustrated in Fig. 5.35, the reverse base drive current first extracts stored charge from the edge of the emitter closest to the base contact. As a consequence of this, the emitter current tends to constrict to the center of the emitter finger during turn-off. This

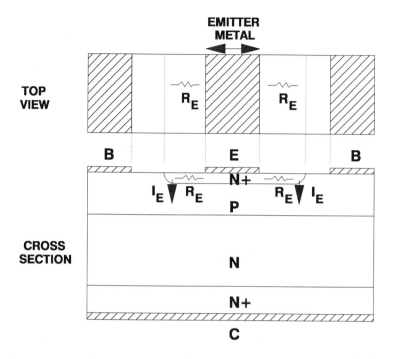

Fig. 5.34 Bipolar power transistor structure with emitter ballast resistance.

phenomenon is particularly important when the transistor is switching current with an inductive load. In this case, the total current flowing through the transistor tends to remain constant while the emitter conduction area decreases. The current density at the center of the emitter then increases drastically during the turn-off process.

In section 5.2.6, it was demonstrated that a base widening phenomenon occurs at high operating current densities and the peak electric field shifts from the base-collector junction to the junction between the N- drift region and the N+ substrate. During the turn-off process, the presence of a high collector voltage and the concomitant high current density at the center of the emitter finger creates a large electron concentration in the drift region moving at the saturated drift velocity. This carrier density is given by:

$$n = \frac{J_C}{q\, v_s} \qquad (5.132)$$

This carrier density becomes substantially greater than the background doping concentration in the drift region at high collector current densities. The electric field profile in the collector drift region then shifts as indicated in Fig. 5.36 for the case of current crowding. Notice that the peak electric field is increased by the current crowding. This can lead to the on-set of avalanche

Chapter 5 : BIPOLAR TRANSISTORS

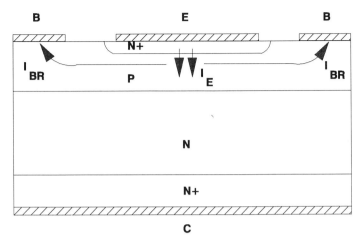

Fig. 5.35 Current constriction in a bipolar power transistor during turn-off.

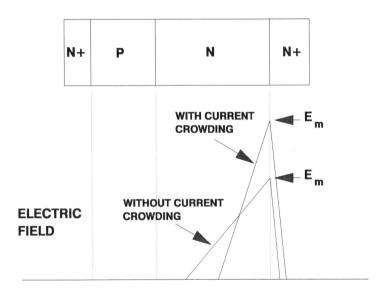

Fig. 5.36 Enhancement in electric field due to current crowding in the emitter finger.

breakdown at voltages well below that expected from the drift layer doping and thickness.

The breakdown voltage of the bipolar power transistor in the presence of a high collector current density can be solved by assuming that the electric field is determined by Eq. (5.86) with the mobile electron concentration much larger than the doping concentration in the drift region. Then the avalanche breakdown voltage under the presence of current flow is given by:

$$BV(J_C) = 5.34 \times 10^{13} \left(\frac{J_C}{q\,v_s}\right)^{-3/4} \qquad (5.133)$$

The collector breakdown voltage is determined by this value and the current gain of the bipolar transistor as discussed earlier in Section 5.3.3. Thus, the boundary of the safe-operating-area is given by:

$$BV_{SOA} = BV(J_C)\,(1 - \alpha)^{1/m} \qquad (5.134)$$

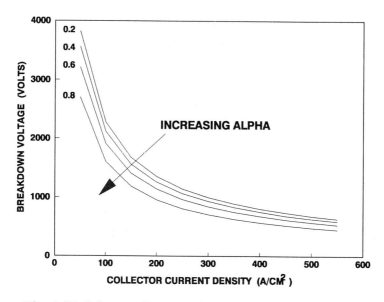

Fig. 5.37 Safe-operating-area of the bipolar power transistor.

As indicated in Fig. 5.37, the safe-operating-area is substantially reduced by the on-set of the avalanche breakdown at lower voltages when the collector current density increases. It should also be noted that the safe-operating-area is larger for smaller current gain values. A reduction in the current gain is not desirable due to the need to supply larger control signals during the on-state of the transistor. However, it is possible to resolve this conflict by making the emitter junction depth smaller in the center of the emitter than at its edges, as shown in Fig. 5.38. In this device structure, the on-state current gain is large due to the narrower base region with smaller base charge at the edges of the emitter where the emitter current tends to flow. At the same time, the current gain in the center of the emitter fingers is smaller due to the wider base region with greater base charge. Thus, it is possible to obtain a high current gain in the on-state and a low current gain during turn-off. This enables the bipolar power transistor to operate with a lower base drive current and superior safe-operating-area.

It is also possible to improve the safe-operating-area by changing the doping profile in

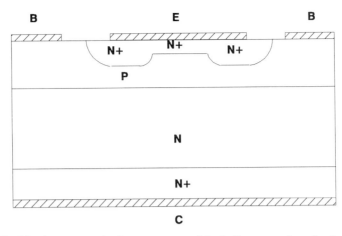

Fig. 5.38 Bipolar power device structure with shallower emitter in the center.

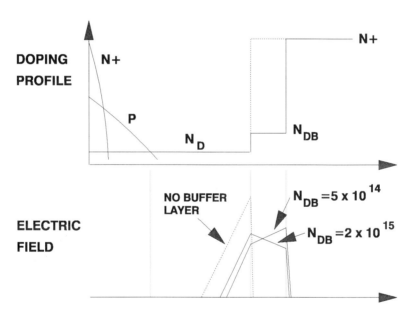

Fig. 5.39 Improved electric field distribution in the collector drift region with a buffer layer.

the collector drift region. If a portion of the collector drift region near the N+ substrate is doped at a higher concentration N_{DB}, the electric field distribution is altered from that shown in Fig. 5.36 to that shown in Fig. 5.39 for the case of two buffer layer doping concentrations and a current density of 1600 amperes per cm^2, which corresponds to a electron concentration in the drift region of 1×10^{15} per cm^3. The region with the higher doping concentration is called a *collector buffer layer*. This change in electric field distribution will occur only if the doping concentration in the buffer layer is comparable to the electron concentration in the drift region as determined by the collector current density [see Eq. (5.132)]. The reduction in the peak electric field by the presence of the buffer layer produces an improvement in the safe-operating area.

5.6 DARLINGTON POWER TRANSISTOR

From the discussion in the previous sections, it is apparent that the current gain of the power bipolar transistor becomes quite low at high collector current densities. An important method to increase the current gain is the use of another bipolar transistor to provide the base drive current for the transistor handling the load current. This circuit, illustrated in Fig. 5.40,

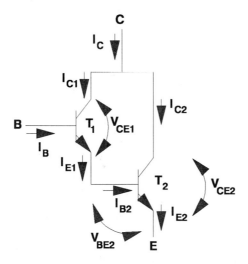

Fig. 5.40 The Darlington configuration for increasing the current gain.

is referred to as the *Darlington configuration*. The two transistors have a common collector connection and the emitter of the drive transistor (T_1) is connected to the base terminal of the output transistor (T_2). The base drive current (I_B) is supplied to the drive transistor. This turns-on transistor T_1 which then provides the base drive current for transistor T_2. The emitter current of transistor T_1, which is also the base current for transistor T_2 is given by:

$$I_{E1} = (1 + \beta_1) I_B = I_{B2} \tag{5.135}$$

and the collector current of transistor T_1 is given by:

Chapter 5 : BIPOLAR TRANSISTORS

$$I_{C1} = \beta_1 I_B \qquad (5.136)$$

where β_1 is the common emitter current gain of transistor T_1. If the common emitter current gain of transistor T_2 is β_2, then the collector current of transistor T_2 is given by:

$$I_{C2} = \beta_2 I_{B2} = \beta_2 (1 + \beta_1) I_B \qquad (5.137)$$

The total output current for the Darlington configuration is the sum of the collector currents flowing through both transistor T_1 and transistor T_2:

$$I_C = I_{C1} + I_{C2} = (\beta_1 + \beta_2 + \beta_1 \beta_2) I_B \qquad (5.138)$$

Thus, the net current gain of the Darlington configuration is approximately equal to the product of the current gains of the drive and output transistors:

$$\beta_D = (\beta_1 + \beta_2 + \beta_1 \beta_2) \approx \beta_1 \beta_2 \qquad (5.139)$$

This increase in the current gain occurs because part of the base drive current for the output transistor is derived from the collector via transistor T_1.

The Darlington configuration allows operation of the composite device with a higher current gain. However, in order for the current to flow via transistor T_1 through the base-emitter junction of transistor T_2, it is necessary to raise the potential on the collector terminal to above that across a forward biased diode. Consequently, the voltage drop across the Darlington configuration is substantially larger than that across a power bipolar transistor in saturation. If transistor T_1 is driven into saturation, then the voltage drop across the Darlington configuration is given by:

$$V_{CE,D} = V_{CE,SAT1} + V_{BE2} \qquad (5.140)$$

The high current gain of the Darlington configuration can lead to a high leakage current under blocking conditions. In order to prevent this and to allow removal of stored charge from transistor T_2, shunting resistances are placed across the emitter-base terminals of both transistors, as illustrated in Fig. 5.41. Since the base leakage current of transistor T_1 flows into the base of transistor T_2 after amplification by transistor T_1, the resistance R_2 must be made lower than the resistance R_1. These resistances raise the breakdown voltage of the Darlington configuration from the open-base value to the shorted base value.

The monolithic integration of the transistors, the shunting resistances, and the anti-parallel diode can be achieved, as illustrated in Fig. 5.42, by making use of the sheet resistance of the base region. The smaller resistance (R_2) is formed by using the sheet resistance of the base region without any N^+ emitter diffusion in this region, while the larger resistance (R_1) is formed by using the pinch-resistance formed in the base region with an N^+ emitter diffusion. The anti-parallel diode is formed by the base-collector junction in the region where the emitter E_2 metal contacts the base region. In a typical case, the area of the output transistor (T_2) is about 6 times

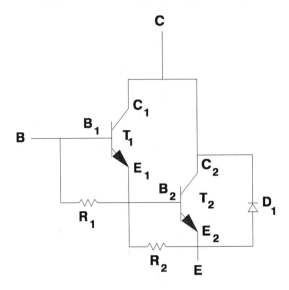

Fig. 5.41 Darlington transistor configuration with shunting resistors and anti-parallel diode.

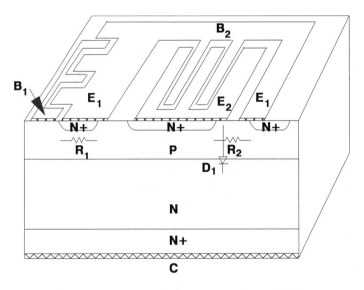

Fig. 5.42 Monolithic Darlington bipolar power transistor structure.

that of the drive transistor T_1.

5.7 TRENDS

The bipolar power transistor has been extensively used in power electronic systems since its initial commercialization in the 1950s. For these applications, devices have been developed with breakdown voltages up to 1500 volts and with current handling capability of several hundred amperes. However, as discussed in this chapter, the power bipolar transistor is a current controlled device with a relatively low current gain. Its application has required development of complex and expensive gate drive circuits. In addition, the devices are prone to failure due to the second breakdown phenomenon and cannot be easily paralleled because of the negative temperature coefficient for the on-state voltage drop. For these reasons, the bipolar power transistor has been replaced by the power MOSFET in applications where the operating voltages are below 200 volts and by the insulated gate bipolar transistor (IGBT) in applications where the operating voltages range from 200 volts to 1500 volts. The improvement in system performance achieved by the use of the MOS-gated devices in the 1980s has led to a lack of demand for further developments in bipolar power transistor technology.

REFERENCES

1. S.K. Ghandhi, "Semiconductor Power Devices," Wiley, New York (1977).

2. B.J. Baliga and D.Y. Chen, "Power Transistors: Device Design and Applications," IEEE Press, New York (1984).

3. D.J. Roulston, "Bipolar Semiconductor Devices," McGraw-Hill, New York (1990).

4. A. Blitcher, "Field Effect and Bipolar Power Transistor Physics," Academic Press, New York (1981).

5. W.J. Chudobiak, "The saturation characteristics of NPνN power transistors," IEEE Trans. Electron Devices, Vol. ED-17, pp. 843-852 (1970).

6. P.L. Hower, "Application of the charge-control model to high voltage power transistors," IEEE Trans. Electron Devices, Vol. ED-23, pp. 863-870 (1976).

7. E.J. McGrath and D.H. Navon, "Factors limiting current gain in power transistors," IEEE Trans. Electron Devices, Vol. ED-24, pp. 1255-1259 (1977).

8. R.J. Hauser, "The effects of distributed base potential on emitter current injection density and effective base resistance for stripe transistor geometries," IEEE Trans. Electron Devices, Vol. ED-11, pp. 238-242 (1964).

9. D. Navon and R.E. Lee, "Effect of non-uniform emitter current distribution on power transistor stability," Solid State Electronics, Vol. 13, pp. 981-991 (1970).

10. R.J. Whittier and D.A. Tremere, "Current gain and cut-off frequency fall-off at high currents," IEEE Trans. Electron Devices, Vol. ED-16, pp. 39-57 (1969).

11. C.T. Kirk, "A theory of transistor cutoff frequency falloff at high current densities," IEEE Trans. Electron Devices, Vol. ED-9, pp. 164-174 (1966).

12. H.A. Schafft, "Second breakdown - A comprehensive review," Proc. IRE, Vol. 55, pp. 1272-1288 (1967).

13. P.L. Hower and V.G.K. Reddi, "Avalanche breakdown in transistors," IEEE Trans. Electron Devices, Vol. ED-17, pp. 320-335 (1970).

14. R. Arnold and D. Zoroglu, "A quantitative study of emitter ballasting," IEEE Trans. Electron Devices, Vol. ED-21, pp. 385-391 (1974).

15. K. Owyang and P. Shafer, "A new power transistor structure for improved switching performance," IEEE Int. Electron Devices Meeting, pp. 667-670 (1978).

16. P.L. Hower, "Optimum design of power transistor switches," IEEE Trans. Electron Devices, Vol. ED-20, pp. 426-437 (1973).

17. C.F. Wheatley and W.G. Einthoven, "On the proportioning of chip area for multistage Darlington power transistors," IEEE Trans. Electron Devices, Vol. ED-23, pp. 870-878 (1976).

PROBLEMS

5.1 Consider an $N^+P\nu N^+$ bipolar power transistor with uniformly doped emitter, base, and collector drift regions. The emitter region has a doping concentration of 2×10^{19} per cm^3 and thickness of 10 microns. The base region has a doping concentration of 2×10^{17} per cm^3 and thickness of 10 microns. The drift region has a doping concentration of 2×10^{14} per cm^3 and thickness of 40 microns. The Shockley-Read-Hall (low-level, high-level, and space-charge- generation) lifetime in all the regions is 0.1 microseconds. Determine the emitter injection efficiency of the transistor excluding the Rittner effect.

5.2 Calculate the base transport factor for this bipolar transistor.

Chapter 5 : BIPOLAR TRANSISTORS

5.3 Determine the common-base and common-emitter current gains for this bipolar transistor.

5.4 Calculate the open emitter breakdown voltage for this bipolar transistor.

5.5 Calculate the open base breakdown voltage for this bipolar transistor.

5.6 What is the quasi-saturation resistance for this bipolar transistor?

5.7 Determine the value of the Rittner current density for this bipolar transistor.

5.8 What is the Early voltage for this bipolar transistor?

5.9 Determine the Kirk current density for this bipolar transistor at a collector bias of 200 volts.

5.10 Calculate the open-base leakage current density at room temperature for this bipolar transistor at a collector bias of 200 volts.

Chapter 6

POWER THYRISTORS

Power thyristors became commercially available in the 1950s and were rapidly accepted for power control applications. These devices offer both forward and reverse blocking capability of comparable magnitude. This makes them well suited for AC circuit applications. The basic thyristor structure can be triggered into its on-state by the application of a relatively small gate drive current. Once it enters into its on-state, the device remains in a self-sustaining mode of operation with a low on-state voltage drop even at high on-state current densities. Consequently, in both the on-state and off-state, no gate bias is required, thus making the device attractive to users. When used in an AC circuit, the device turns-off upon reversal of the anode voltage. A popular variation of this device structure consists of two thyristors integrated monolithically in an anti-parallel configuration. This device, called the *Triac*, is not only capable of blocking current flow for forward and reverse anode voltages but is also capable of conducting current flow in both directions upon application of a gate turn-on signal.

For DC circuit applications, it is preferable to be able to turn-off the current flow without reversal of the anode voltage. This has been achieved in a structure called the *Gate Turn-Off (GTO) Thyristor*. In the GTO, the turn-off is achieved by application of a large reverse gate drive current that extracts charge from the base region of the thyristor and disrupts the self-sustaining current conduction mechanism. A high degree of interdigitation between the emitter (cathode) and the gate is necessary to obtain turn-off.

The ratings of power thyristors have steadily grown over the years due to the demand for devices capable of controlling very high voltages and currents. The growth in the current ratings is traced in Fig. 6.1, while the growth in voltage blocking capability is traced in Fig. 6.2. The current handling capability of a power thyristor is directly proportional to its active area. The active area of power thyristors has been limited by the availability of silicon wafers with larger diameters. As silicon wafer pulling and zone refining technology improved, the power semiconductor industry has utilized the larger diameter wafers to increase the current handling capability.

The voltage blocking capability of power thyristors has been limited by the availability of silicon wafers with sufficiently high resistivity to obtain the desired avalanche breakdown voltage as discussed in chapter 3. The presence of oxygen in the silicon ingots pulled by the Czochralski process has been found to result in poor breakdown characteristics. Consequently, power thyristors have been manufactured by using float-zone silicon which has a much lower

Chapter 6 : POWER THYRISTORS 259

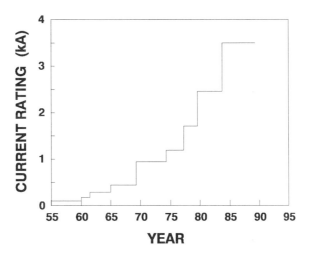

Fig. 6.1 Growth in the current handling capability of power thyristors.

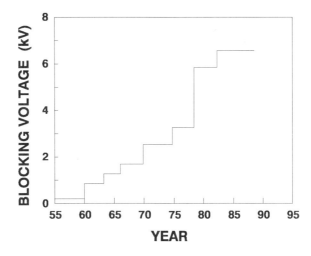

Fig. 6.2 Growth in the blocking voltage capability of power thyristors.

oxygen content. An important issue that decides the manufacturability of devices from this material has been the variation in the resistivity across the wafer. Large resistivity variations can produce poor blocking voltage as discussed in this chapter. The development of the neutron transmutation doping (NTD) process discussed in chapter 2 was motivated by the need to form uniformly doped silicon wafers with high resistivities for the manufacturing of power thyristors. With the development of the NTD process, a jump in the voltage blocking capability was obtained in the late 1970s, as shown in Fig. 6.2. Due to these developments, single power thyristors are available with blocking voltage capability of 6500 volts. It is impressive to note that these devices are made from a single wafer of silicon with a diameter of up to 12.5 cm, leading to a current handling capability of 2000 amperes. Thus, a single power thyristor is capable of controlling over 10 megawatts of power.

This chapter first introduces the basic four layer thyristor structure and then analyzes its forward and reverse blocking capability. The importance of cathode shorting to obtain good voltage blocking capability at high operating temperatures is brought out during this analysis. The forward conduction behavior and gate drive requirements are then discussed. Next, the switching behavior is analyzed, including plasma spreading effects as well as the use of an involute gate topology and amplifying gate structure to improve the turn-on. A discussion of the operation of the gate turn-off thyristor is provided at the end of the chapter, followed by design considerations to optimize its performance.

6.1 THYRISTOR STRUCTURE AND OPERATION

The structure of the power thyristor consists of four semiconductor layers of alternating types. One of these layers is lightly doped and serves as the drift region that supports a high voltage when the device is in its blocking state. The basic structure is illustrated in Fig. 6.3

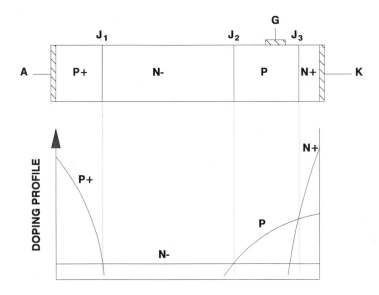

Fig. 6.3 Structure and doping profile for a power thyristor.

with its doping profile. The lightly doped N-type drift region serves as the base of a P-N-P transistor. The diffused P-base and the N$^+$ emitter regions on the cathode (K) side form a second (N-P-N) transistor in combination with the N- drift region. These transistors are coupled because they share the N- drift region. The equivalent circuit for the thyristor is shown in Fig. 6.4.

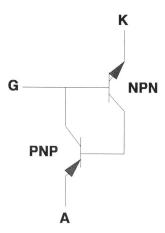

Fig. 6.4 Equivalent circuit for the power thyristor.

The power thyristor has the capability to support voltage in both directions. The structure shown in Fig. 6.3 contains three junctions (J_1, J_2, and J_3) in series. The cathode terminal is used as the reference, and the high voltages are applied to the anode terminal. When a negative bias is applied to the anode, junctions J_1 and J_3 are reverse biased while junction J_2 is forward biased. The junction J_3, formed between the N$^+$ emitter diffusion and the P-base diffusion, cannot support a high voltage due to the relatively high doping concentrations on both sides of the junction. Consequently, most of the negative anode voltage is supported across the reverse biased junction J_1. When a positive bias is applied to the anode, junction J_2 is reverse biased while junctions J_1 and J_3 are forward biased. Consequently, the positive anode voltage is supported across the reverse biased junction J_2. Since the voltage is supported across the same lightly doped N- drift region under both forward (positive anode bias) and reverse (negative anode bias) blocking conditions, the blocking capability of the power thyristor is nearly symmetric, as shown in Fig. 6.5.

It will be shown later in this chapter that the thyristor can be triggered into a current conduction mode by the application of a gate bias when the anode voltage is positive. This is indicated by the dashed lines in Fig. 6.5. This occurs due to the coupling of the two transistors within the thyristor structure. The gate bias serves to increase the current gain of one of these transistors. Once this happens, the two transistors can provide the base drive currents for each other. This allows on-state current conduction without the need for any gate drive. This mode of operation is referred to as *regenerative action*. The self-sustaining operation of the thyristor structure has the advantage that, unlike in the case of the power bipolar transistor, no gate drive

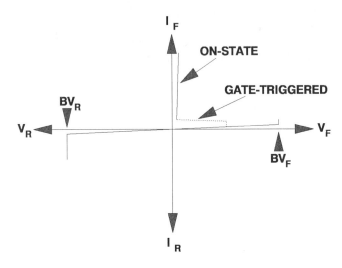

Fig. 6.5 Output characteristics of a power thyristor.

current is required to maintain it in its on-state. However, this action has the disadvantage that the device is difficult to turn-off. In the case of the conventional thyristor structure, the device is primarily used in AC circuits where the anode voltage periodically reverses. This switches the thyristor from its on-state to its reverse blocking state. Special thyristor structures, called *Gate Turn-Off (GTO) thyristors*, have been developed which enable switching off the regenerative action in the thyristor by the application of a gate drive current. These structures are suitable for DC circuits.

In the following sections, the different modes of operation of the thyristor are analyzed. This is followed by description of the special thyristor structures that enable gate controlled turn-off. In addition, the operation of a bi-directional current carrying device, called the *power Triac*, is described. These devices are extensively used for control of power flow in AC circuits operating at low frequencies.

6.2 STATIC BLOCKING CHARACTERISTICS

The blocking voltage capability of the power thyristor comprises one of its limits of performance. As discussed in the previous section, the power thyristor can support voltage for the case of either positive or negative voltage applied to the anode. This voltage is supported mainly across the lightly doped N-drift region. Although this results in nearly symmetric blocking voltage capability, it is important to understand the differences between the forward and reverse blocking capability. This is especially important when the operating temperature increases. In the following sections, it will be shown that the forward blocking capability is not as good as the reverse blocking capability. A design will then be discussed, based on periodically short-circuiting the N⁺ emitter to the P-base regions, which allows enhancement of the forward blocking capability to bring it closer to the reverse blocking capability.

6.2.1 Reverse Blocking

When a negative bias is applied to the anode of the thyristor structure shown in Fig. 6.3, the junctions J_1 and J_3 are reverse biased, while junction J_2 is forward biased. The breakdown voltage of junction J_3, formed between the N⁺ emitter diffusion and the P-base diffusion, is small due to the relatively high doping concentrations on both sides of the junction. This junction will typically have a breakdown voltage of less than 50 volts. Consequently, most of the negative anode voltage is supported across the reverse biased junction J_1. The reverse bias across junction J_1 results in the formation of a depletion regions that extends into the P⁺ anode and N-drift regions. The fabrication of power thyristors is performed by using aluminum as a dopant to create deep P-type diffusions with a highly graded doping profile at the junction (J_2). This allows supporting some (usually less than 10 percent) of the applied bias within the P⁺ anode region. Most of the anode bias is supported within the N-drift region. The breakdown voltage of the device is then dictated by the open-base transistor formed between junctions J_1 and J_2. The criteria for determining the breakdown voltage of an open-base transistor were discussed in Chapter 3 in section 3.7. The breakdown voltage is determined by the voltage at which the product of the current gain (α_{PNP}) and the multiplication factor (M) becomes equal to unity. Based upon this criterion, the doping concentration and thickness of the N-drift region must be optimized as illustrated in Fig. 3.39. It is important to obtain the desired reverse blocking capability while minimizing the thickness of the N-drift region because this results in the lowest on-state and switching power losses.

One of the problems encountered with power thyristors has been a degradation in the voltage blocking capability with increasing temperature. This has set a limit to their maximum operating temperature to about 125°C. An illustration of the reduction in reverse blocking capability with temperature is given in Fig. 6.6. The reverse blocking capability of a P-N junction diode is also shown in this figure for comparison. The reverse blocking capability of the diode increases monotonically with increasing temperature. This is due to a decrease in the impact ionization coefficients with increasing temperature by the enhanced scattering of the free carriers at elevated temperatures. This produces a reduction in the multiplication factor (M). In the case of the power thyristor, the reverse blocking capability also increases initially with temperature due to the same phenomenon. In this temperature range, the reverse blocking voltage of the thyristor is slightly below that of the diode due to the current gain of the PNP transistor. However, at higher temperatures, the minority carrier lifetime, and hence the diffusion length, increases producing a rapid increase in the current gain. This increase in the current gain is responsible for the observed reduction in the reverse blocking voltage of power thyristors at higher temperatures.

264   Chapter 6 : POWER THYRISTORS

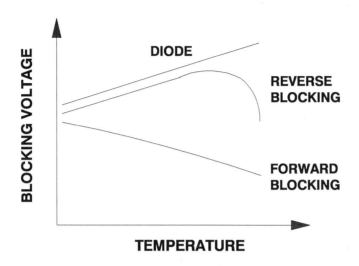

Fig. 6.6 Degradation in blocking voltage capability of power thyristors.

6.2.2 Forward Blocking

When a positive bias is applied to the anode of the thyristor structure shown in Fig. 6.3, junction J_2 is reverse biased while junctions J_1 and J_3 are forward biased. The junction J_2 serves as a common collector region for both the N^+-P-N bipolar transistor and the P^+-N-P bipolar transistor. Since this junction is reverse biased, both of these transistors are operating in their forward active regions. Any leakage current generated within the device is then amplified by the current gain of both of these transistors.

Consider the thyristor structure shown in Fig. 6.7 with the anode, cathode, and leakage currents indicated by the arrows. It will be assumed that there is no gate current flowing during the analysis. Then, the currents flowing at collector junction J_2 are: a component ($\alpha_{PNP} I_A$) related to the emitter current I_A of the P-N-P transistor at its collector; a component ($\alpha_{NPN} I_K$) related to the emitter current I_K of the N-P-N transistor at its collector; and the leakage current I_L. Since no gate current is assumed to be flowing, the application of Kirchhoff's law leads to:

$$I_A = \alpha_{PNP} I_A + \alpha_{NPN} I_K + I_L = I_K \qquad (6.1)$$

Based upon this equation:

$$I_A = \frac{I_L}{(1 - \alpha_{PNP} - \alpha_{NPN})} \qquad (6.2)$$

From this equation, it can be concluded that the anode current will go to infinity when the sum of the current gains of the NPN and PNP transistors becomes equal to unity:

Chapter 6 : POWER THYRISTORS

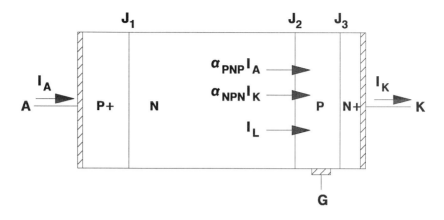

Fig. 6.7 Leakage current in a power thyristor structure.

$$\alpha_{PNP} + \alpha_{NPN} = 1 \tag{6.3}$$

At low temperatures, the current gains of the transistors are small. However, they increase with temperature due to an increase in the minority carrier lifetime and diffusion length. In addition, the current gains increase with increasing anode bias due to extension of the depletion regions on both sides of the junction J_2 which reduces the base width of the transistors through which the carriers must diffuse. The criterion derived above for the forward blocking voltage can be rewritten in the form:

$$M (\gamma \alpha_T)_{PNP} + M (\gamma \alpha_T)_{NPN} = 1 \tag{6.4}$$

where M is the multiplication coefficient applicable to both the transistors because they share the same collector junction. Thus, the multiplication coefficient that determines the forward blocking voltage is:

$$M_{FB} = \frac{1}{(\gamma \alpha_T)_{PNP} + (\gamma \alpha_T)_{NPN}} \tag{6.5}$$

In contrast, the multiplication coefficient that determines the reverse blocking voltage is:

$$M_{RB} = \frac{1}{(\gamma \alpha_T)_{PNP}} \tag{6.6}$$

This demonstrates that the multiplication factor that determines the forward breakdown voltage will always be less than the value that determines the reverse blocking capability, which in turn implies that the forward blocking capability will always be less than the reverse blocking capability.

The decrease in forward blocking voltage with increase in temperature is shown in Fig. 6.6 for comparison with the reverse blocking capability. It can be see that the forward blocking capability can severely limit the high temperature performance of a thyristor. An improvement in the high temperature forward blocking capability can be obtained by suppressing the gain of the upper N-P-N transistor at low currents corresponding to the leakage currents flowing during the forward blocking mode, while obtaining high gain values at current levels corresponding to the on-state mode. This is possible by using a thyristor structure in which the N$^+$ cathode is periodically short-circuited to the P-base region. This structure, referred to as

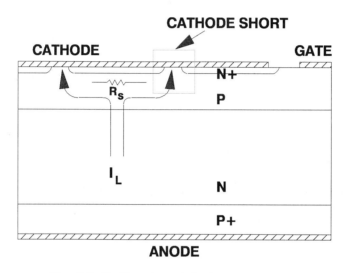

Fig. 6.8 Emitter-shorted thyristor structure.

an *emitter or cathode shorted structure*, is illustrated in Fig. 6.8. During the forward blocking mode, the leakage current (I_L) that flows into the P-base region is diverted directly into the contacts made by the cathode metal to the P-base regions at the cathode shorts. This diversion of the leakage current around the N+ emitter suppresses the gain of the N-P-N transistor and brings the forward blocking voltage closer to the reverse blocking voltage.

However, the leakage current generated below the N$^+$ emitter region flows laterally through the P-base region before reaching the metal contact at the shorts. This produces a voltage drop across the resistance R_S in the base region which forward biases the N$^+$ emitter/P-base junction. When the voltage drop increases sufficiently, the N$^+$ emitter begins to inject electrons into the P-base region and the current gain of the N-P-N transistor increases. The increase in current gain can be modeled by considering the N-P-N transistor to have a resistance shunting the emitter-base junction as illustrated in Fig. 6.9. The collector current I_C is related to the base-emitter voltage by:

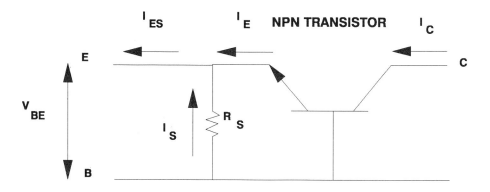

Fig. 6.9 Bipolar transistor with shunting resistance between emitter and base terminals.

$$I_C = I_o \, e^{\frac{q\,V_{BE}}{kT}} \tag{6.7}$$

where, using Eq. (5.21),:

$$I_o = \frac{A\,q\,D_n\,n_{0B}}{W_B} \tag{6.8}$$

The current I_{ES} flowing at the emitter terminal in the presence of the shunting resistance consists of the sum of the current flowing in the shunting resistance R_S and the emitter current I_E flowing within the N-P-N transistor. Thus:

$$I_{ES} = I_S + I_E = \frac{V_{BE}}{R_S} + \frac{I_o}{\alpha_{NPN}} \, e^{\frac{q\,V_{BE}}{kT}} \tag{6.9}$$

The current gain in the presence of the shunting resistance is given by:

$$\alpha_{NPN,S} = \frac{I_C}{I_{ES}} = \alpha_{NPN} \left[\frac{e^{q\,V_{BE}/kT}}{e^{qV_{BE}/kT} + \dfrac{V_{BE}\,\alpha_{NPN}}{I_o\,R_S}} \right] \tag{6.10}$$

This expression can be used to predict the change in current gain with increasing base-emitter voltage (V_{BE}). As the base-emitter voltage increases, most of the current initially flows through the shunting resistance as shown in Fig. 6.10. The net current gain of the transistor is then

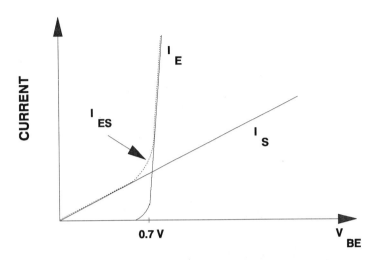

Fig. 6.10 Current distribution at shunted base-emitter junction.

close to zero. However, when the base-emitter voltage exceeds about 0.7 volts, the N^+/P junction becomes strongly forward biased and substantial current flows through the emitter of the N-P-N transistor.

This results in an increase in the current gain until it approaches the current gain of the transistor without the shunting resistance. The change in current gain then takes the form shown

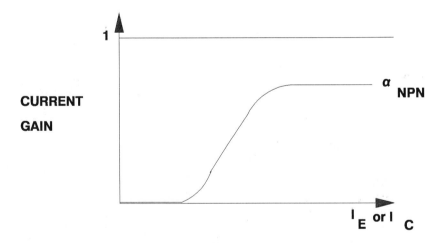

Fig. 6.11 Increase in current gain with current for a transistor with shunting resistance across base-emitter junction.

Chapter 6 : POWER THYRISTORS   269

in Fig. 6.11 as a function of the emitter or collector current. This behavior is ideal for obtaining good thyristor performance. The low current gain at small current levels, corresponding to the leakage currents in devices, promotes improved forward blocking capability at high temperatures. At the same time, the current gain at higher current levels, corresponding to on-state conditions, promotes regenerative action within the thyristor allowing self-sustaining operation with low on-state voltage drop.

In order to obtain an understanding of the influence of device structural parameters upon the performance of the cathode short, consider the case of a linear short geometry shown in

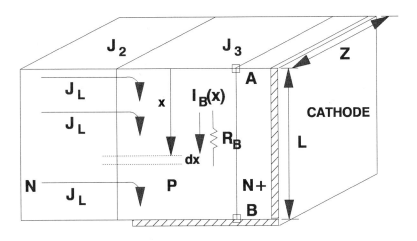

Fig. 6.12 Linear cathode short geometry for thyristors.

Fig. 6.12. It will be assumed that the device is in its forward blocking mode with junction J_2 reverse biased. Under these conditions, no current flows into the N^+ emitter region and all the leakage current can be assumed to flow laterally into the contact between the cathode metal and the P-base region at point B. Since the leakage current is created by space-charge-generation within the depletion region formed across junction J_2, a uniform current density J_L is collected across this junction. The current flowing in the P-base region ($I_B(x)$) will then increase from point A in the middle of the emitter finger to point B. The current dI_B collected within a small region dx along the length of the emitter finger is given by:

$$dI_B = J_L Z \, dx \qquad (6.11)$$

Then the base current flowing at point at a distance x from point A will be given by:

$$I_B(x) = \int_0^x J_L Z \, dx = J_L Z x \qquad (6.12)$$

In conjunction with the base resistance R_B, this base current flow produces a forward bias across the emitter-base junction. The voltage drop across the segment dx is given by:

$$dV_{BE}(x) = I_B(x) \frac{\rho_{SB}}{Z} dx \qquad (6.13)$$

where ρ_{SB} is the sheet resistance of the P-base region under the N$^+$ emitter. The total voltage drop created between point A and point B due to this base current flow is:

$$V_{BE}(A) = \int_0^L dV_{BE}(x) \, dx = J_L \, \rho_{SB} \frac{L^2}{2} \qquad (6.14)$$

This voltage drop forward biases the N$^+$ emitter/P-base junction at point A. When this forward bias becomes approximately equal to the junction potential (V_{bi}, which is approximately 0.7 volts), electron injection occurs at point A, resulting in an increase in the current gain. Thus, the maximum leakage current that can be tolerated before the thyristor is triggered from its forward blocking mode to its on-state is given by:

$$J_{Lm} = \frac{2 \, V_{bi}}{\rho_{SB} \, L^2} \qquad (6.15)$$

which corresponds to a leakage current of:

$$I_{Lm} = J_{Lm} \, Z \, L = \frac{2 V_{bi} \, Z}{\rho_{SB} \, L} \qquad (6.16)$$

From this equation, it can be concluded that, in order to obtain higher temperature operation corresponding to larger leakage currents, it is necessary to make the sheet resistance (ρ_{SB}) and the emitter length (L) between cathode shorts smaller.

In practical devices, it is necessary to form the cathode shorts in the form of a two dimensional array distributed uniformly over the top surface of the device. One such arrangement, shown in Fig. 6.13, is a square array with shorted regions of diameter d located with a spacing of D. For such an array, the maximum voltage drop under the N$^+$ emitter region due to a leakage current density J_L is given by:

$$V_{BE} = J_L \, \rho_{SB} \, A_S \qquad (6.17)$$

where the term A_S is given by:

$$A_S = \frac{1}{16} \left\{ d^2 + D^2 \left[2 \ln\left(\frac{D}{d}\right) - 1 \right] \right\} \qquad (6.18)$$

Note that this term is equal to $L^2/2$ for the case of the linear short treated earlier in this section.

It can be seen from Fig. 6.13 that there is an obvious reduction in the cathode area as a result of making the periodic cathode shorts. The physical fractional area consumed by the

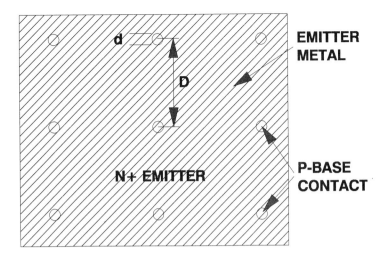

Fig. 6.13 A cathode shorting arrangement for thyristors.

contacts to the base region is given by:

$$F_S = \frac{\pi}{4} \left(\frac{d}{D}\right)^2 \tag{6.19}$$

However, the actual loss in the area of current conduction in the N⁺ emitter is greater than this because the region in the immediate vicinity of the short does not have sufficient forward bias to promote injection of electrons from the N+ emitter into the P-base region. Consequently, one of the compromises in designing the cathode short is the trade-off between improving the forward blocking capability and obtaining the lowest possible on-state voltage drop. It will be shown later in this chapter that another compromise must be made in designing the cathode shorts from the point of view of the gate drive current required to trigger the thyristor from the forward blocking mode to the on-state.

When the cathode shorting density is sufficient, all the leakage current can be assumed to flow into the cathode at the shorts. Then, the current gain of the upper N-P-N transistor becomes small at the leakage current levels. Consequently, the multiplication factor that governs the forward blocking capability becomes:

$$M_{FB} = \frac{1}{(\gamma\, \alpha_T)_{PNP} + (\gamma\, \alpha_T)_{NPN}} = \frac{1}{(\gamma \alpha_T)_{PNP}} \tag{6.20}$$

which is the same as for the case of reverse blocking, and the forward blocking capability with the presence of cathode-shorts becomes nearly equal to the reverse blocking capability.

6.3 FORWARD CONDUCTION CHARACTERISTICS

When the current gains of the two internal transistors within the thyristor structure become sufficiently large, the coupling of the transistors via their common collector junction produces a self-sustaining condition that leads to current flow with a low on-state voltage drop. In the self-sustaining mode of operation, the current flows through the thyristor without any gate voltage or current applied to the device. The self-sustaining operation is possible only when the sum of the current gains of the two transistors exceeds unity. In the case of devices with cathode shorts, the current gain of the upper N-P-N transistor decreases at lower current levels. This prevents self-sustained operation of the thyristor at lower current levels. The current level below which the device is unable to maintain self-sustaining current conduction is referred to as the *holding current*. The holding current is dependent upon the cathode shorting density.

6.3.1 On State Operation

When the thyristor is operating in the on-state, injection of carriers occurs across both junctions J_1 and J_3. A hole injected from the P^+ anode region into the N- drift region diffuses across the N- drift region and is collected by junction J_2. When the hole enters the P-base region it constitutes an additional majority carrier in the base of the N-P-N transistor. This results in the injection of an electron from the N^+ region into the P-base region. This electron now diffuses through the P-base region and is collected by junction J_2. When the electron enters the N- drift region, it provides the base drive current for the P-N-P transistor, promoting injection of a hole from the P^+ region. Thus, a positive feedback mechanism is inherent within the thyristor structure when both transistors operate at sufficient current gains.

When the thyristor operates in its on-state, the P-base/N- drift region junction is forced into a forward biased state with both transistors operating in the saturation mode. This can be demonstrated by using the following argument. The base current (holes in the N- drift region) for the P-N-P transistor supplied by the anode current I_A is given by $[(1 - \alpha_{PNP}) I_A]$. The electron current supplied to the N- drift region by the N^+ cathode is given by $[\alpha_{NPN} I_K]$. Since no gate drive current is assumed to be flowing and the current gains are close to unity:

$$(1 - \alpha_{PNP}) I_A < \alpha_{NPN} I_K \qquad (6.21)$$

This indicates that, in order to maintain charge neutrality in the N- drift region, holes must be injected from the P-base region. This condition is satisfied when junction J_2 becomes forward biased, putting the two transistors into their saturation region. Thus, in the on-state, the power thyristor operates with all three junctions under forward bias. The net voltage drop across the device consists of the voltage drop across forward biased junctions J_1 and J_3 minus the voltage drop across forward biased junction J_2. This is equivalent to the voltage drop across one forward biased diode.

In the on-state, the power thyristor can be treated as a P-i-N rectifier formed between the P^+ anode region and the N^+ cathode region. Since the cathode provides injection of electrons

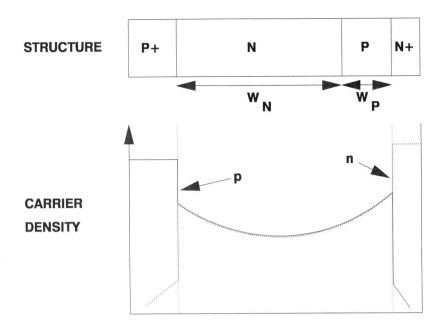

Fig. 6.14 Carrier distribution profile in a thyristor in the on-state.

through the P-base region into the N- drift region, it is sometimes referred to as a *remote emitter*. The carrier distribution in the thyristor in the on-state is illustrated in Fig. 6.14. In this diagram, it has been assumed that the carrier concentration has exceeded the doping in the P-base region. The effective width of the P-i-N diode is then the sum of the thickness of the P-base and N- drift regions. The on-state voltage drop can be calculated using the relationships derived in Chapter 4 on power rectifiers with:

$$2d = (W_N + W_P) \tag{6.22}$$

As in the case of the P-i-N rectifier, the effects of Auger recombination and carrier-carrier scattering, as well as end region recombination, must be taken into account during this analysis.

6.3.2 Gated Turn-on

As described in section 6.2, the thyristor can support a high voltage in the forward blocking mode with the voltage supported across the reverse biased junction J_2. The device can be triggered from the forward blocking state to the on-state by the application of a gate control signal. In order to analyze the thyristor current flow with a gate drive current I_G, consider the cross-section shown in Fig. 6.15. The currents at the blocking junction J_2 indicated in the cross-section are based upon the same considerations discussed earlier in Section 6.2. Using

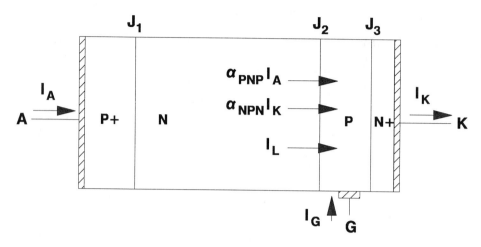

Fig. 6.15 Current flow in a thyristor with gate drive current.

Kirchhoff's law:

$$I_A = \alpha_{PNP} I_A + \alpha_{NPN} I_K + I_L \qquad (6.23)$$

and

$$I_K = I_G + I_A \qquad (6.24)$$

Combining these equations:

$$I_A = \frac{\alpha_{NPN} I_G + I_L}{(1 - \alpha_{PNP} - \alpha_{NPN})} \qquad (6.25)$$

Thus, the gate current serves to enhance the anode current by providing the base drive current for the upper N-P-N transistor.

In the case of the thyristor structure with cathode shorts, the gate drive current also serves the purpose of raising the current gain of the N-P-N transistor. In these devices, the gate contact is placed away from the cathode short as shown in Fig. 6.16. When a positive bias is applied to the gate electrode with reference to the cathode electrode, a gate drive current I_G flows from the gate contact to the ohmic contact formed between the cathode metal and the P-base region at the cathode shorts. This produces a voltage drop across the resistance of the P-base region under the N^+ emitter (pinch-resistance), that forward biases point A on the N^+ cathode with respect to P-base region. If a sufficiently large gate drive current I_G is applied, this forward bias will be sufficient to produce the injection of electrons from the N^+ region into the P-base region. This is also equivalent to raising the current gain of the N-P-N transistor. The thyristor self-sustaining action is then initiated by the gate drive current, allowing the device to

Chapter 6 : POWER THYRISTORS

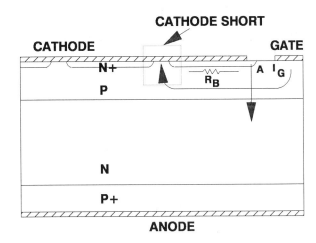

Fig. 6.16 Gate controlled triggering of a thyristor.

switch from its forward blocking mode to its on-state.

The gate drive current at which the thyristor will turn-on is dependent up on the cathode shorting design. In order to determine the influence of the short design on the gate triggering current, consider the case of the linear cathode short previously discussed in section 6.2.2. As mentioned in context with Fig. 6.16, the gate drive current flows from the gate contact to the cathode short producing a forward bias across the N^+ cathode/P-base junction. This is

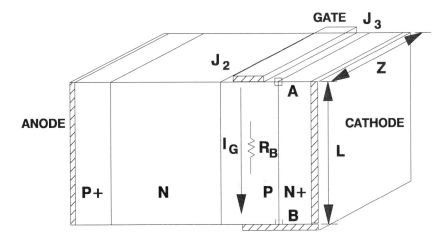

Fig. 6.17 Gate triggering in a linear short geometry.

illustrated in Fig. 6.17 for the case of the linear short geometry, where it has been assumed that the gate drive current flows from the gate contact to the cathode metal at the cathode short formed at the end of the cathode finger. This assumption is valid because the N$^+$ emitter does not conduct any current before the thyristor is turned-on. The condition required to turn-on the thyristor is a gate current at which the potential at point A rises to that required to forward bias the N$^+$/P junction (V_{bi}). Based upon this criterion:

$$I_G R_B = I_G \rho_{SB} \frac{L}{Z} = V_{bi} \qquad (6.26)$$

Thus, the gate triggering current is given by:

$$I_{GT} = \frac{V_{bi}}{\rho_{SB}} \frac{Z}{L} \qquad (6.27)$$

From this expression, it can be concluded that the gate drive current required to trigger the thyristor can be reduced by increasing the base sheet resistance (ρ_{SB}) and the emitter length (L) between the cathode shorts. This is in direct conflict with the requirements for improvement of high temperature forward blocking capability discussed in section 6.2.2. Thus, when designing thyristors, a compromise must be made between ensuring good high temperature blocking characteristics and obtaining ease of gate control.

In actual thyristors, the cathode shorting array must be located around the gate contact

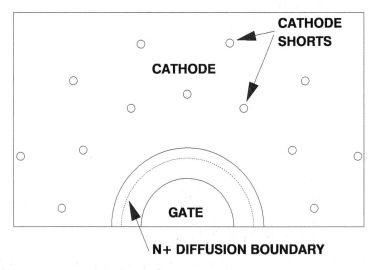

Fig. 6.18 Cathode short located around the gate contact in a thyristor.

as shown in Fig. 6.18. In this case, the gate triggering current is given by:

$$I_{GT} = \frac{V_{bi}}{\rho_{SB} A_S} \tag{6.28}$$

where A_S is the geometrical factor given earlier in Eq. (6.18). Once again, the same conflict between improving the forward blocking capability and reducing the gate triggering current needs to taken into account during the design.

6.3.3 Holding Current

As discussed earlier, the thyristor can maintain itself in the on-state by the regenerative action between the two coupled bipolar transistors within the structure. However, when the anode current is reduced without gate drive current, the device switches from its on-state to the

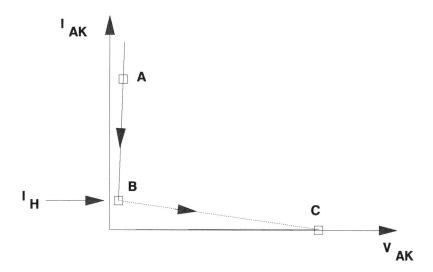

Fig. 6.19 Turn-off of the thyristor below its holding current.

forward blocking state as illustrated in Fig. 6.19. The anode current corresponding to point B where the thyristor switches from its on-state to its blocking state is referred to as the *holding current*. In order to maintain a large range of current over which the thyristor can operate in the on-state, it is desirable to obtain a low holding current in devices.

The value of the holding current is dependent upon the cathode shorting geometry because this determines the variation in the gain of the N-P-N transistor with anode current level. As the anode current is reduced, the current gain of the upper N-P-N transistor begins to decrease as shown in Fig. 6.11 due to the presence of the cathode shorts. Once the current gain falls below the value required to maintain the self-sustaining action, as defined by

Eq. (6.2), the thyristor turns-off and reverts to its forward blocking mode of operation.

In order to analyze the holding current, consider the case of a linear cathode short. It will be assumed that the device is initially in its on-state. Then, most of the current collected at the junction J_2 will flow directly into the N^+ emitter, with a fraction associated with recombination in the P-base region flowing to the cathode short. This current flow is illustrated

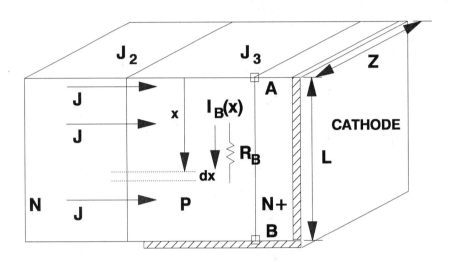

Fig. 6.20 Current distribution in a thyristor with linear short for holding current analysis.

in Fig. 6.20, where it is assumed that the current density at junction J_2 is uniform. Consider an elemental width dx located at a point x along the cathode finger. The base current component arising from the current J at the junction is given by:

$$dI_B = J_K (1 - \alpha_{NPN}) \, Z \, dx \quad (6.29)$$

The total base current at point x can then be obtained by integrating this equation between the limits 0 and x:

$$I_B = \int_0^x dI_B = J_K (1 - \alpha_{NPN}) \, Z \, x \quad (6.30)$$

This base current flows through the pinch resistance of the base region. The voltage drop produced by this current flow forward biases the junction J_3 resulting in the injection of electrons which is necessary to sustain the thyristor in its on-state. If the voltage drop produced by the base current flow is insufficient to bias the junction J_3 by at least a junction drop V_{bi}, then the thyristor will be unable to maintain its self-sustaining action and remain in its on-state. The voltage drop in the elemental section dx produced by the base current I_B at location x is given by:

$$dV(x) = I_B \frac{\rho_B}{W_B Z} dx = I_B \frac{\rho_{SB}}{Z} dx \qquad (6.31)$$

where ρ_B is the average resistivity of the P-base region. The potential at point A, where the forward bias across junction J_3 is the largest, can be obtained by integration of Eq. (6.31) across the emitter length (L):

$$V_A = \int_0^L dV(x) = \int_0^L I_B \frac{\rho_{SB}}{Z} dx \qquad (6.32)$$

Substituting for I_B from Eq. (6.30):

$$V_A = \int_0^L J_K (1 - \alpha_{NPN}) \rho_{SB} x \, dx = J_K (1 - \alpha_{NPN}) \rho_{SB} \frac{L^2}{2} \qquad (6.33)$$

Defining the holding current density (J_{HA}) as the anode current density at which the potential at point A becomes equal to V_{bi}, and making the approximation that the anode and cathode current densities are approximately equal, an expression for the holding current density is obtained:

$$J_{HA} = \frac{2 V_{bi}}{(1 - \alpha_{NPN}) \rho_{SB} L^2} \qquad (6.34)$$

The holding current for the thyristor is then given by:

$$I_{HA} = J_{HA} L Z = \frac{2 V_{bi} Z}{(1 - \alpha_{NPN}) \rho_{SB} L} \qquad (6.35)$$

From this expression, it can be concluded that the operating range of the thyristor in its on-state can be extended to lower current levels by increasing the sheet resistance of the base and the distance between the cathode shorts. As discussed in the previous section, this is in favor of reducing the gate drive current but is against obtaining improved forward blocking capability.

6.4 SWITCHING CHARACTERISTICS

In the previous sections, the static operating modes of the thyristor were discussed and the transition points between these modes were related to the device geometrical design parameters. In this section, the mechanisms involved during the transition between the blocking state and the on-state will first be discussed followed by analysis of the recovery of the thyristor from its on-state to the reverse blocking state when the potential at the anode is switched from positive to negative. Before performing this analysis, the ability of the thyristor to tolerate a high rate of change of anode voltage [dV/dt] will be analyzed because this is an important requirement that thyristors must satisfy when used in applications.

6.4.1 dV/dt Capability

High power thyristors are operated in circuits with large supply voltages. During circuit operation, the devices often experience a high rate of change in anode voltage with time. The ability of the thyristor to maintain its forward blocking state under the influence of a rapidly changing anode voltage is referred to as its *dV/dt capability*. When the voltage across the thyristor is rising rapidly, a current flows through the device because of the device capacitance. when this current flows into the P-base region, it can act as a gate drive current which can trigger the thyristor into its on-state. Since the thyristor is being triggered from its blocking state to its on-state, it can be expected that this characteristic of the thyristor is also determined by the cathode short design.

The analysis of the dV/dt capability can be performed by considering the case of the thyristor with a linear short design. This structure is shown in Fig. 6.21 with the junction

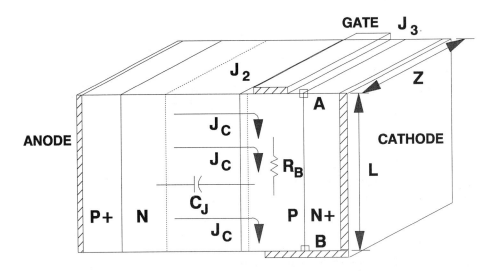

Fig. 6.21 Displacement current flow used to analyze dV/dt capability of thyristors.

depletion boundary and junction capacitance indicated. When a high dV/dt is applied to the anode, a displacement current flows across junction J_2 due to its capacitance C_J:

Chapter 6 : POWER THYRISTORS

$$J_C = C_J \left[\frac{dV}{dt}\right] \quad (6.36)$$

Note that C_J is the *capacitance per unit area* due to the depletion layer formed across junction J_2. The displacement current density can be assumed to be uniform across junction J_2. Since the thyristor is in its forward blocking mode, the current entering the P-base region flows to point B, where the cathode metal is connected to the P-base region. This produces a voltage drop across the P-base region that forward biases junction J_3 at point A. The current distribution under these conditions is identical to that used to perform the analysis of the forward blocking capability with a leakage current flowing across junction J_2. Consequently, the capacitive current at which the thyristor will be triggered into its on-state can be obtained from Eq. (6.15):

$$J_C = \frac{2 V_{bi}}{\rho_{SB} L^2} \quad (6.37)$$

By combining Eqs. (6.36 and 6.37), the dV/dt capability is obtained:

$$\left[\frac{dV}{dt}\right] = \frac{2 V_{bi}}{C_J \rho_{SB} L^2} \quad (6.38)$$

The junction capacitance (C_J) is a strong function of the anode voltage (V_A) and has its highest value at lower anode voltages because the depletion layer width increases with increasing reverse bias voltage across a P-N junction:

$$C_J = \frac{\epsilon_S}{W_D} = \sqrt{\frac{q \epsilon_S N_D}{2 V_A}} \quad (6.39)$$

where N_D is the doping concentration in the N- base region. Here, it has been assumed that the doping concentration in the P-base region is much greater than in the N- drift region so that the depletion region in the P-base region can be neglected. When this expression is substituted into Eq. (6.38), the dV/dt capability is obtained for the linear short geometry:

$$\left[\frac{dV}{dt}\right] = \frac{2 V_{bi}}{\rho_{SB} L^2} \sqrt{\frac{2 V_A}{q \epsilon_S N_D}} \quad (6.40)$$

From this expression, it can be concluded that the [dV/dt] capability can be increased by reducing the P-base sheet resistance and the N^+ emitter length between the cathode shorts. This is in direct conflict with obtaining low holding and gate drive currents. Thus, a compromise must be made during cathode short design among all these characteristics. It is worth pointing out that when performing the analysis of the [dV/dt] capability, a worst case analysis should be performed using low anode voltages.

6.4.2 Turn-on Transient

The thyristor can be triggered from its blocking state to its on-state by the application of a gate drive signal. As discussed in Section 6.3.2, the gate drive current flows from the gate contact to the locations where the cathode metal overlaps the P-base region at the cathode shorts. This produces a voltage drop in the P-base region that forward biases the N^+/P junction. Thus, injection of electrons from the N^+ emitter into the P-base region occurs when there is an adequate gate drive current. Before the thyristor can establish self-sustaining current conduction, the internal feedback loop between the transistors must be first completed. The time taken for this is referred to as the *delay time*. Once the feedback loop is established, the anode current rises rapidly. The rate of increase of the anode current is characterized by a *rise time* that is dependent on the establishment of the proper stored charges in the N-base and P-base regions. In a one-dimensional thyristor, the turn-on process is completed at this point in time. However, in actual devices, the turn-on occurs only in the vicinity of the gate contact. The thyristor conduction area then spreads across the rest of the cathode area. The time taken for the current to spread over the cathode area is characterized by a *turn-on spreading time*. These three fundamental phases observed during the turn-on of the thyristor are discussed in this section.

6.4.2.1 Delay time.

As mentioned above, the turn-on of the thyristor is initiated by the application of a gate drive current that forward biases the N^+ emitter, resulting in the injection of electrons into the P-base region. These electrons diffuse through the P-base region and are collected across the reverse biased junction J_2. The time taken for the electrons to cross the P-base region is referred to as the *base transit time* of the N-P-N transistor (t_p). The base transit time is given by:

$$t_p = \frac{W_{BP}^2}{2 D_n} \qquad (6.41)$$

where W_{BP} is the base width of the N-P-N transistor. The electrons are then rapidly swept through the deletion layer of junction J_2. Once the electrons enter the N-base region, the injection of holes occurs from the P^+/N junction J_1 in order to maintain charge neutrality. These holes now diffuse through the undepleted portion of the N-base region and are collected across the reverse biased junction J_2. The base transit time for holes in the N-base region (t_n) of the P-N-P transistor is given by:

$$t_n = \frac{W_{BN}^2}{2 D_p} \qquad (6.42)$$

where W_{BN} is the base width of the P-N-P transistor. Once the holes enter the depletion region across junction J_2, they are rapidly swept through it and enter the P-base region where they serve as additional base drive current. This sets up the internal current feedback loop that switches the thyristor into its on-state. At this point in time, the anode current begins to rise.

Based upon the above discussion, it is apparent that there is a delay between the application of the gate drive current pulse and a rise in the anode current. This period is therefore referred to as the *delay time*. The delay time is approximately equal to the sum of the transit times for the two transistors:

$$t_d = t_p + t_n = \frac{W_{BP}^2}{2 D_n} + \frac{W_{BN}^2}{2 D_p} \qquad (6.43)$$

Since the base width of the P-N-P transistor is much greater than that of the N-P-N transistor, the delay time is mostly associated with the transit time for holes in the N-base region. It is worth pointing out that the transit time is dependent upon the width of the undepleted portion of the N-base region and not its total width. Consequently, the transit time decreases with increasing forward blocking voltage. The transit time in the N-base region for high voltage thyristors is typically 15 microseconds.

6.4.2.2 Turn-on Rise Time. Once the regenerative action within the thyristor begins to occur, the minority carriers within the N-base and P-base regions increase with time until the device reaches its steady-state forward conduction mode of operation. During this time, the anode current is increasing. The rise time for the anode current can be determined by analysis of the growth of the stored charge within the two base regions of the thyristor. In this analysis, the increase in the stored charge with time is assumed to be provided by the base current for each transistor and minority carrier recombination is assumed to be negligible because the minority carrier lifetime is much longer than the rise time.

The base current for the P-N-P transistor is entirely derived from the injection of electrons from the N^+ cathode region. Consequently:

$$I_{B,PNP} = \alpha_{NPN} I_K(t) = \frac{dQ_N}{dt} \qquad (6.44)$$

where $I_K(t)$ is the cathode current flowing at time t and Q_N is the stored charge in N-base region of the P-N-P transistor. In the case of the N-P-N transistor, the base current consists of two components: namely, the holes provided by injection from the P^+ anode region and the gate drive current (I_G) supplied to turn-on the thyristor. Consequently:

$$I_{B,NPN} = \alpha_{PNP} I_A(t) + I_G = \frac{dQ_P}{dt} \qquad (6.45)$$

where Q_P is the stored charge in the P-base region of the N-P-N transistor. From the definition of the transit time, the current flowing through the P-base region is equal to the stored charge Q_P divided by the transit time t_P. Thus:

$$\alpha_{NPN} I_K(t) = \frac{dQ_N}{dt} = \frac{Q_P}{t_P} \qquad (6.46)$$

Differentiating this equation and substituting from Eq. (6.45):

$$\frac{d^2Q_N}{dt^2} = \frac{1}{t_P}\frac{dQ_P}{dt} = \frac{1}{t_P}(\alpha_{PNP} I_A + I_G) \tag{6.47}$$

From the definition of the transit time, the current flowing through the N-base region is equal to the stored charge Q_N divided by the transit time t_N. Thus:

$$\alpha_{PNP} I_A(t) = \frac{Q_N}{t_N} \tag{6.48}$$

Using this expression in Eq. (6.47):

$$\frac{d^2Q_N}{dt^2} = \frac{1}{t_P}\left(\frac{Q_N}{t_N} + I_G\right) \tag{6.49}$$

This is a second order differential equation that describes the change in the stored charge in the N-base region with time:

$$\frac{d^2Q_N}{dt^2} - \frac{Q_N}{t_N t_P} = \frac{I_G}{t_P} \tag{6.50}$$

This equation indicates that the stored charge will increase exponentially with a time constant given by the square root of the product of the transit time for the P-base and N-base regions. The stored charge grows until the anode current reaches its steady state value $I_{A,SS}$ after a rise time t_r. Based upon this, it can be shown that:

$$Q_N(t) = [(\alpha_{PNP} I_{A,SS} + I_G) t_N\, e^{(t-t_r)/\sqrt{t_N t_P}}] - I_G t_N \tag{6.51}$$

Based upon there being zero stored charge in the N-base region at the beginning of the rise time, this equation can be used to derive an expression for the rise time:

$$t_r = \sqrt{t_N t_P}\, \ln\left[\frac{\alpha_{PNP} I_{A,SS}}{I_G} + 1\right] \tag{6.52}$$

The rise time is typically 15 microseconds for high voltage thyristors. Substituting this expression in Eq. (6.51):

$$Q_N(t) = I_G t_N\, [e^{t/\sqrt{t_N t_P}} - 1] \tag{6.53}$$

By using Eq. (6.48), an equation for the change in anode current with time is then obtained:

$$I_A(t) = \frac{I_G}{\alpha_{PNP}}\, [e^{t/\sqrt{t_N t_P}} - 1] \tag{6.54}$$

The cathode current waveform can be obtained from Kirchhoff's law:

$$I_K(t) = I_A(t) + I_G \qquad (6.55)$$

Thus, the anode and cathode currents increase exponentially with time at a rate determined by the geometric mean of the transit time for the base regions.

An example of the waveform for the anode current as a function of time during the turn-on of the thyristor is shown in Fig. 6.22 with the delay time and the rise time indicated at the

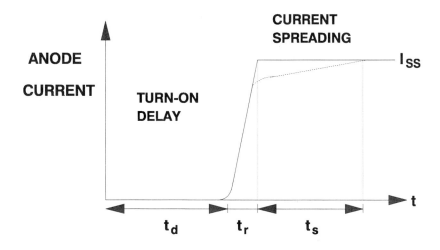

Fig. 6.22 Anode current waveform during turn-on of a thyristor.

bottom. Note that during the delay time, there is no anode current flow but that a cathode current flow equal to the gate current occurs. After the delay time, the anode and cathode currents rise rapidly to their steady-state values as shown by the solid line. This behavior is based upon current flow in a one-dimensional thyristor structure. In actual devices designed to handle high currents, the gate region is located in the center of a wafer with a large diameter. The turn-on then occurs first near the gate contact because the injection from the N^+ emitter is initiated in this region. Once this region turns-on, the current spreads to the rest of the cathode area because the region that has turned on supplies base drive current to the adjacent regions.

6.4.2.3 Turn-on Spreading Time. The spreading of the current over the cathode area has been found to be characterized by a propagation velocity called the *spreading velocity*. Measurements performed on the rate of spreading of the anode current across the area of the thyristor have been performed by measurements of the infra-red radiation emitted from the thyristor due to the recombination of holes and electrons in the region that has substantial current flow. A typical value for the spreading velocity is 5000 cm/sec. The spreading velocity (u_s) depends upon the thyristor design parameters and operating conditions.

The rate at which the current spreads from the turn-on region is determined by the rate of diffusion of the carriers from the on-region to the off-region. This process depends upon the concentration gradient, which is decided by the carrier concentration in the on-region, and the diffusion length. Using the average concentration of holes and electrons in the on-region given by Eq. (4.50) with a total i-region thickness (2d) of $(W_N + W_P)$, and assuming that the spreading velocity is proportional to the gradient of the carrier concentration:

$$u_s \propto \frac{dn_{av}}{dx} = \frac{n_{av}}{L_a} = \frac{J}{q(W_N + W_P)} \sqrt{\frac{\tau_{HL}}{D_a}} \qquad (6.56)$$

where J is the current density in the on-region. Based upon this equation, it can be anticipated that the spreading velocity will depend upon the anode current density, the high-level lifetime, and the base widths of the thyristor. This behavior has been empirically confirmed.

First, it has been found that:

$$u_s \propto J^{1/n} \qquad (6.57)$$

where J is the anode current density and the coefficient n is found to have a value between 2 and 3. This expression indicates that the spreading velocity increases with increasing anode current density. Thus, the current will be distributed more rapidly if the rate of rise of the anode current is more rapid. This is beneficial because it reduces the heat dissipation in the turn-on region around the gate contact.

Second, it has been found that:

$$u_s \propto \sqrt{\tau_{HL}} \qquad (6.58)$$

where τ_{HL} is the high level lifetime in the N-base region. This expression indicates that the spreading velocity increases with increasing lifetime in the N-base region. It is necessary to reduce the lifetime to decrease the turn-off time for the thyristor. These devices are referred to as *inverter grade thyristors*. The reduced lifetime produces a slower spread in the turn-on region. This is detrimental to the operation of the thyristor at higher frequencies.

Third, it has been found that:

$$u_s \propto \frac{1}{W_N} \qquad (6.59)$$

where W_N is the width of the N-base region. As discussed in Section 6.2 on static blocking characteristics, the width of the N-base region must be increased in order to support larger forward and reverse blocking voltages. Thus, the spreading velocity will be smaller for devices designed for higher blocking voltage capability. This indicates that there will be slower spreading of the current after initiation of the turn-on. This in turn implies a reduction in the [di/dt] capability of the thyristor with increasing breakdown voltage.

Chapter 6 : POWER THYRISTORS 287

6.4.2.4 Involute Gate Structure. Based upon the above discussion of current spreading within the thyristor after turn-on, it can be concluded that the best thyristor design will contain an interdigitated gate and cathode structure with all regions of the cathode of equal width located equidistant from the gate edge. In the case of high current thyristors fabricated from an entire wafer, such a design is possible by using the *involute pattern*. An involute is generated when a string is tightly unwound from a cylinder.

Consider the case of two involutes generated from a cylinder with radius r, as illustrated

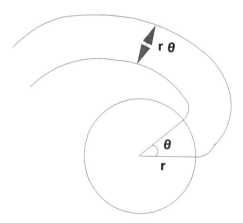

Fig. 6.23 Illustration of uniform cathode fingers generated using involutes.

in Fig. 6.23, with starting points located away from each other by an angle θ. It can be shown that these involutes will be always spaced equidistant from each other. The space between the involutes is (r θ). Thus, if these lines represent the gate electrodes, the cathode fingers formed with the involute topology will be uniform in width. This will allow distribution of the cathode current within the thyristor in the most efficient manner. The involute structure produces a large increase in the gate edge. This has the advantage of greatly increasing the di/dt capability of the thyristor because the turn-on is distributed over a large region of the device. However, this increase in gate width requires a corresponding increase in the gate drive current. Further, the interdigitation of the gate and cathode fingers requires higher resolution lithographic techniques. which increases the fabrication cost. Another drawback of this geometry is that a substantial surface area is consumed by the gate contact and fingers, which reduces the current handling capability of the thyristor. For these reasons, the involute design is used only for devices designed for operation in inverter circuits operating at higher frequencies.

6.4.2.5 Amplifying Gate Structure. As discussed in the context of the involute gate structure, a substantial gate drive current is required to turn-on a thyristor that operates with a high [di/dt] capability. In order to reduce the gate drive current while maintaining the [di/dt] capability, a thyristor structure has been developed that utilizes the anode current flow after initiation of turn-on to provide gate drive current to the rest of the thyristor. This method for triggering the thyristor is called an *amplifying gate* because the gate current is amplified during the turn-on process.

The top view and cross-section of the amplifying gate thyristor structure is illustrated in

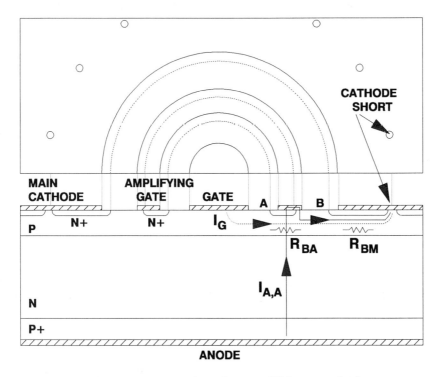

Fig. 6.24 Top view and cross-section of an amplifying gate thyristor structure.

Fig. 6.24 with a central circular gate region. The amplifying gate is located between the gate and the main cathode region of the thyristor. It should be noted that the cathode metal for the amplifying gate thyristor makes contact with the P-base region at the edges located away from the gate contact. The main thyristor region is assumed to contain cathode shorts in order to improve the blocking characteristics. All the diffusions and metal patterning steps for the amplifying gate thyristor are performed simultaneously with those for the main thyristor.

When a gate drive current is applied, it flows from the gate contact to the first row of cathode shorts located in the main cathode region, as indicated by the dashed line in Fig. 6.24. This produces a voltage drop in the P-base region that forward biases the N^+ emitter of both the amplifying gate region and the main thyristor. The thyristor with amplifying gate is designed so that the forward bias at the N^+ emitter of the amplifying thyristor region is greater than that for the main thyristor region. Then, the injection of electrons occurs at point A due to the gate drive current. This turns-on the thyristor formed in the amplifying gate region. The resulting anode current is indicated by the solid line. Since no external connection is made to the cathode metal in the amplifying thyristor portion of the structure, the current that flows into the N^+ emitter in the amplifying thyristor must flow via the short to the P-base region to the cathode short in the main thyristor. As the anode current in the amplifying thyristor region increases, it produces an increase in the forward bias on the N^+ emitter of the main thyristor at point B until it begins to inject electrons. This turns-on the main thyristor. Thus, the gate triggering current is used to turn-on only the amplifying gate thyristor and the current used to trigger the main thyristor is derived from the anode current. Although a simple annular gate and amplifying thyristor design is illustrated in the figure, the amplifying gate can be used in conjunction with the involute cathode design with the amplifying gate thyristor supplying the gate drive current to the cathode fingers of the main thyristor.

An equivalent circuit for the amplifying gate thyristor structure is shown in Fig. 6.25

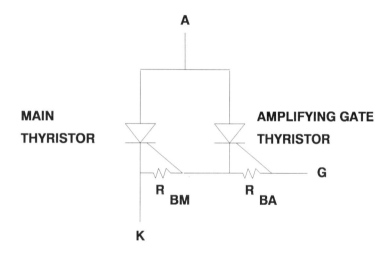

Fig. 6.25 Equivalent circuit for the amplifying gate thyristor.

with the amplifying gate region shown as a separate thyristor. Note that the cathode current of the amplifying gate thyristor serves as the gate current for the main thyristor. The resistances of the P-base region under the N^+ emitter in the amplifying gate region and the main thyristor region which determine the forward bias across the N^+/P junction are shown as R_{BM} and R_{BA}. These resistances are also indicated in the cross-section shown in Fig. 6.24 for the amplifying gate thyristor. The relative magnitudes of these resistances determines whether the gate drive current triggers the amplifying gate thyristor prior to the main thyristor.

In order to ensure that the gate drive current triggers the amplifying gate thyristor and not the main thyristor, it is necessary to analyze the voltage drop under the N^+ emitter for the amplifying gate and main thyristors when the gate drive current is applied. In the case of a concentric or annular gate and amplifying gate thyristor design, the voltage drop can be

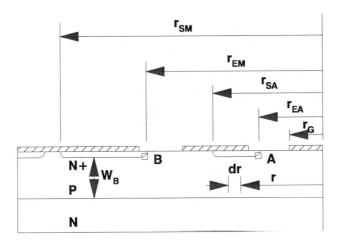

Fig. 6.26 Amplifying gate thyristor structure used for analysis of triggering conditions.

determined by taking into account the radial spreading of the gate current. A cross-section of the upper portion of the thyristor with an amplifying gate region is shown in Fig. 6.26 with the radii of the N+ emitter and cathode metal defined. Consider a segment of the P-base region of length dr located at radius r. The resistance of this segment through which the gate current flows is given by:

$$R(r) = \frac{\rho_B \, dr}{2 \pi r W_B} = \frac{\rho_{SB} \, dr}{2 \pi r} \qquad (6.60)$$

where ρ_B is the resistivity of the P-base region, W_B is the thickness of the P-base region under the N+ emitter, and ρ_{SB} is the sheet resistance of P-base region under the N+ emitter. The voltage drop due to a gate current I_G flowing through this segment dr is:

$$dV(r) = I_G R(r) = \frac{I_G \rho_{SB} \, dr}{2 \pi r} \qquad (6.61)$$

This expression can now be used to determine the voltage drop below the N+ emitter in the amplifying gate region and in the main thyristor region. In the case of the amplifying gate thyristor region, the voltage drop at point A is given by:

$$V_A = \int_{r_{EA}}^{r_{SA}} dV(r) = \frac{I_G \rho_{SB}}{2 \pi} \ln\left(\frac{r_{SA}}{r_{EA}}\right) \qquad (6.62)$$

Similarly, in the case of the main thyristor region, the voltage drop at point B is given by:

$$V_B = \int_{r_{EM}}^{r_{SM}} dV(r) = \frac{I_G \, \rho_{SB}}{2\pi} \ln\left(\frac{r_{SM}}{r_{EM}}\right) \tag{6.63}$$

In order to ensure that the amplifying gate thyristor and not the main thyristor is triggered by the gate drive current, it is necessary to satisfy the criterion that V_B is less than V_A. Based upon the above equations, this criterion is satisfied when:

$$\frac{r_{SA}}{r_{EA}} < \frac{r_{SM}}{r_{EM}} \tag{6.64}$$

This must be achieved by proper choice of the dimensions for the N^+ emitter diffusions and metal patterns when designing the amplifying gate thyristor structure.

Using the expression for the potential at point A, it is possible to obtain the gate triggering current for the amplifying gate thyristor structure. The amplifying gate thyristor region will begin to conduct current when the potential at point A becomes sufficient to produce the injection of electrons from the N^+ emitter into the P-base region. Thus, the gate triggering current is given by:

$$I_{GT} = \frac{2\pi \, V_{bi}}{\rho_{SB} \ln(r_{SA}/r_{EA})} \tag{6.65}$$

This equation is useful for determination of the radii for the N^+ emitter and the cathode metal in the amplifying gate region to obtain any desired value for the gate drive current.

6.4.3 Turn-off Transient

As discussed earlier in Section 6.3.3, a power thyristor will switch from its on-state to the forward blocking state when the anode current is reduced below the holding current. This behavior is not usually used in power circuits for turning off the thyristor. Power thyristors are usually switched off by reversal of the anode voltage. This occurs naturally in circuits operating from an AC power source. The control of the power delivered to a load via the thyristor is determined by the duration for which the thyristor remains in its on-state. A typical set of waveforms for the power source voltage, the voltage across the thyristor, and the anode current flowing through the thyristor are shown in Fig. 6.27, where the thyristor has been switched on at time t_0. The thyristor remains in its on-state during the period from t_0 to T/2. At time T/2, the anode voltage reverses. The thyristor is now required to support the reverse voltage across the P^+/N junction (junction J_1 in Fig. 6.3). However, since the thyristor is conducting current just prior to time T/2, there is a large stored charge in the N-base region. This charge must be removed before the junction can become reverse biased and support voltage. This process is analogous to that for the reverse recovery of a P-i-N rectifier as discussed in Chapter 4. However, the additional junctions in the thyristor must be taken into account to understand the reverse recovery waveforms.

292   Chapter 6 : POWER THYRISTORS

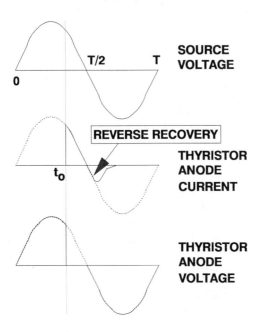

Fig. 6.27 Current and voltage waveforms for a thyristor operated with an AC power source.

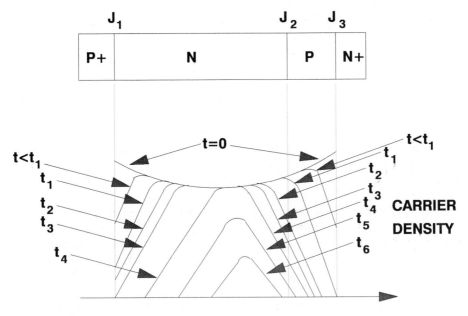

Fig. 6.28 Carrier distribution profiles in a thyristor during its reverse recovery.

Consider the case of a thyristor operated with a resistive load R_L, as shown in Fig. 6.28, with the carrier distribution illustrated for various points in time during the turn-off process. If the device is turned-off from its on-state with an operating current of I_F with a constant reverse anode bias of V_R, then the reverse current will initially be given by:

$$I_R = \frac{V_R}{R_L} \qquad (6.66)$$

When this reverse current flows, the carrier distribution shifts to the curve for $t < t_1$. It should be noted that the slope of the carrier distribution at junctions J_1 and J_3 is proportional to the anode current. Thus, the slope reverses at time $t = 0$ because of the change in the direction for the anode current through the thyristor. The reverse current remains constant until time t_1 is reached when the carrier concentration at junction J_3 goes to zero. This carrier distribution profile is shown in Fig. 6.28 as the case $t = t_1$. At this time, junction J_3 becomes reverse biased and begins to support voltage. The maximum voltage that this junction can support is limited to its breakdown voltage. Due to the relatively high doping concentrations in the N^+ emitter region and the P-base region, the breakdown voltage of junction J_3 (BV_{J3}) is small (less than 50 volts) when compared to the forward blocking voltage of the thyristor. Consequently, when the anode voltage increases to BV_{J3}, this junction breaks down and the anode current continues to flow with a magnitude I_{R1} given by:

$$I_{R1} = \frac{V_R - BV_{J3}}{R_L} \qquad (6.67)$$

Since BV_{J3} is small when compared with V_R, the change in anode current is small. The carrier distribution when the thyristor is operating in this mode is shown by the line marked t_2 in Fig. 6.28. The anode current I_{R1} continues to extract the stored charge in the base regions of the thyristor until the carrier concentration becomes equal to zero at junction J_1 at time t_3. The carrier distribution at this time is shown in Fig. 6.28 by the line marked t_3.

Once the carrier concentration at junction J_1 becomes equal to zero, it begins to support voltage and a depletion layer extends from the junction into the N-base region. Since junction J_1 can support a high voltage, the anode voltage now increases until it reaches the reverse bias voltage V_R. This is accompanied by a decrease in the anode current. The carrier distribution during this time interval is shown in Fig. 6.28 by the line marked t_4. As the voltage increases, the depletion layer extends into the N-base region. This results in the excess carrier density being confined to a portion of the N-base region located in the proximity of junction J_2. When the anode voltage has become approximately equal to the applied reverse bias V_R, the charge distribution is shown by the line marked t_5. Beyond this time, the depletion width remains relatively constant because the anode voltage has almost reached its steady-state value. The excess carriers remaining within the device are removed by recombination, and the carrier profile during this time is shown by the line marked t_6. There is a slow decay in the anode current during this time interval, which is referred to as the *recombination tail*.

The anode voltage and current waveforms during the reverse recovery of the thyristor are shown in Fig. 6.29. The various time intervals for which the carrier distributions were

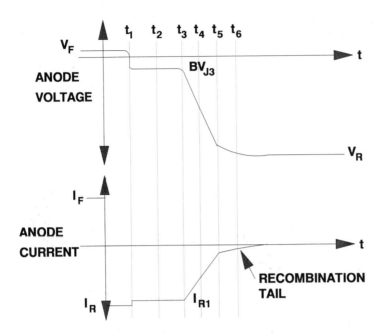

Fig. 6.29 Anode voltage and current waveforms for a thyristor during reverse recovery.

shown in Fig. 6.28 are indicated on the x-axis. Note the plateau in the anode voltage during the time interval from t_3 to t_1 and the corresponding shoulder in the reverse anode current. The reverse recovery of the thyristor differs from the reverse recovery of a P-i-N rectifier due to this additional time interval created by the presence of the additional reverse biased junction (J_3) in the thyristor that does not exist in the rectifier.

In the case of the thyristor operated in an AC circuit, the reverse recovery is accompanied by an approximately triangular reverse current pulse followed by a current tail associated with the recombination of the stored charge. This current waveform was shown in Fig. 6.27 for the case of a reverse recovery time that is short when compared with the half period for the sinusoidal AC voltage source. However, if the operating frequency is increased such that the half period of the sinusoid becomes comparable to the turn-off time of the thyristor, the anode voltage applied to the thyristor switches from negative to positive before all the stored charge within the thyristor has been removed. The anode voltage and current waveforms for this case are illustrated in Fig. 6.30. Under these conditions, the thyristor enters its forward blocking regime of operation while there are excess carriers still present within the N-base region. These carriers are now collected by the blocking junction J_2 and flow in the P-base region to the cathode shorts. If the resulting base current is sufficiently large, it can trigger the thyristor into its on-state. It is worth pointing out that, for a sinusoidal voltage source, at the time T when the anode voltage reverses polarity, the device is also subjected to the highest [dV/dt]. Thus, there is a significant displacement current flowing at the same time that the remanent stored charge is being extracted via the P-base region. This promotes the triggering

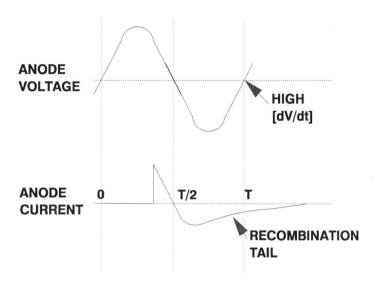

Fig. 6.30 Thyristor waveforms for the case of a half-period comparable to the reverse recovery time.

of the thyristor at time T. If the thyristor is triggered to its on-state at time T, then it will conduct current throughout the half-period during which the anode voltage is positive. Consequently, once the triggering occurs due to the stored charge, all gate control is lost. The highest operating frequency for a thyristor is therefore limited by the reverse recovery time. Typically, high power thyristors are operated at frequencies of less than 1000 Hz.

6.5 LIGHT TRIGGERED THYRISTORS

In the previous sections, it was shown that the thyristor can be triggered from its forward blocking state to its on-state by the application of a gate drive current. In high power systems, such as power transmission and distribution networks, the system operating voltages are very large (typically over 100,000 volts). Since these voltages far exceed the blocking capability of power thyristors, it becomes necessary to operate the thyristors in a series string. One of the problems encountered in this arrangement is the large potential difference between the gates of the thyristors. Isolation of the gate drive (control) circuit from the high voltages is difficult. One elegant method for addressing this problem has been developed by using optical fibers to transmit the control signal to the thyristors. The voltage isolation for the control circuit is obtained via the fiber link. However, this requires the conversion of the optical signal into an electrical current pulse that can be applied to the gate of the thyristor to trigger it. Although this can be accomplished by using opto-couplers and local power supplies to deliver the gate current, this is found to be an expensive solution. Instead, thyristors have been developed that can be directly triggered via an optical signal. These devices are referred to as *light triggered thyristors*.

The basic principle of operation of a light triggered thyristor is based upon the generation of hole-electron pairs in silicon by the incident light. These carriers are used to create a gate drive current within the structure. The light sources that are attractive for this application are either gallium-arsenide LEDs operating at a wavelength of about 0.9 microns. The output power of such LEDs is relatively small, which greatly limits the amount of available light generated current. For this reason, lasers are sometimes preferred. Even in this case, the amount of current available to trigger the thyristor is relatively small. Consequently, it is necessary to incorporate an amplifying thyristor region in the light triggered thyristor. This is also very important from the point of view of increasing the [di/dt] capability of the thyristor. This can be demonstrated by using Eq. (6.54) derived in Section 6.4.2 for the anode current variation with time during turn-on under an applied gate drive current I_G:

$$\frac{dI_A}{dt} = \frac{I_G}{\alpha_{PNP}} \frac{1}{\sqrt{t_N t_P}} e^{t/\sqrt{t_N t_P}} \qquad (6.68)$$

Thus, the anode [di/dt] is directly proportional to the applied gate drive current I_G. Consequently, light triggered thyristors can have a very poor [di/dt] capability unless an amplifying gate is incorporated with the light sensitive region.

The light triggered thyristor structure with an amplifying gate region is shown in Fig. 6.31 with a central circular gate region for optical triggering. Note that the gate region contains an N^+ emitter region so that a thyristor is formed in this portion of the structure. The top metal pattern is designed to leave a window in the center of the device for the incident light used to trigger the device. When the thyristor is in its forward blocking mode and a pulse of light is incident in the gate region, hole-electron pairs are created in the P-base and N-base regions, resulting in a photo-generated current I_P. This current flows from the central gate region to the cathode shorts, as indicated in Fig. 6.31 by the dashed line, producing a voltage drop across the resistance (R_{BG}) of the P-base region under the N^+ emitter in the gate region. This voltage drop forward biases the N^+ emitter at point C. If the bias is sufficiently large, electron injection occurs at point C, which triggers the thyristor in the gate region into its on-state. This current flows from the gate region to the cathode shorts, as indicated in Fig. 6.31 by the solid line marked I_G. The current I_G is similar to the gate drive current used in the earlier analysis for the amplifying gate thyristor. It serves to turn-on the adjacent amplifying gate thyristor, which then turns-on the main thyristor. Since the gate current I_G is derived from the anode, the magnitude of the optically generated current I_P can be small. This enables the design of a light triggered thyristor with a high optical sensitivity while obtaining [di/dt] capability comparable to that of the conventional thyristor.

An analysis of the design of the gate region for the light triggered thyristor can be performed by assuming that the incident light creates a uniform photo-generated current density J_P in the gate region. For the case of a circular gate region, the hole current collected within a radius r is given by:

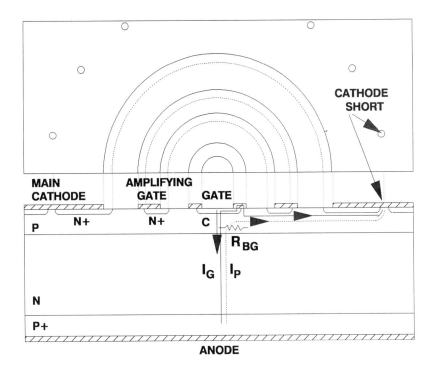

Fig. 6.31 Light triggered thyristor structure with amplifying gate region.

$$I_B(r) = \pi \, r^2 \, J_P \qquad (6.69)$$

When this current flows through a narrow segment of the P-base region of radius dr, it produces a voltage across a resistance dR given by:

$$dV(r) = I_B(r) \, dR(r) = \pi \, r^2 \, J_P \, \frac{\rho_{SB}}{2 \, \pi \, r} \, dr = \frac{J_P \, \rho_{SB}}{2} \, r \, dr \qquad (6.70)$$

This produces a forward bias across the N$^+$/P junction at point C:

$$V_C = \int_0^{r_{SG}} dV(r) = \frac{J_P \, \rho_{SB} \, r_{SG}^2}{4} \qquad (6.71)$$

Using the criterion that the thyristor will get triggered by the photo-current when this voltage drop is equal to V_{bi}, an expression for the photo-current J_{PT} required to trigger the thyristor can be derived:

$$J_{PT} = \frac{4 \, V_{bi}}{\rho_{SB} \, r_{SG}^2} \tag{6.72}$$

Thus, the triggering sensitivity can be controlled by adjusting the radius r_{SG} and the sheet resistance of the P-base region. It is worth pointing out that if a large radius r_{SG} and high sheet resistance are used to reduce the photo-current requirements, the [dV/dt] capability and high temperature blocking capability will be reduced.

6.6 THYRISTOR SELF-PROTECTION

Thyristors can be destroyed under adverse circuit operating conditions. Two of these conditions occur when there is a voltage surge created by switching high currents through stray lead inductances and when the thyristor is triggered in a local region by a high [dV/dt] applied to the device especially at high operating temperatures. It is possible to incorporate protective measures during the design of the thyristor to allow it to survive these conditions by controlled triggering of the device into its on-state. The basic design philosophy is to ensure turn-on of the thyristor at its gate region even when the triggering is due to the adverse circuit operating conditions. When this is achieved, the device turns-on as if it were being gated by the gate drive circuit. This ensures failure proof turn-on. The thyristor is then not prone to destructive failure during the adverse operating conditions. Thyristor designs for self-protection against over-voltage and high [dV/dt] conditions are described below.

6.6.1 Breakdown Protection

The basic approach to obtaining protection against conditions where the anode voltage exceeds the forward blocking capability is by ensuring that the avalanche current generated by the anode bias flows through the gate region and triggers the thyristor into its on-state. One such design is illustrated in Fig. 6.32, where a trench has been formed under the gate contact. The trench can be formed by etching the silicon in a portion of the gate region. The depth of the trench is selected to allow the depletion layer in the P-base region during the forward blocking mode of operation to reach-through to the metal contact. Due to the metal interface, a high leakage current is generated in the depletion layer. This leakage current flows from the gate region to the cathode short, as indicated in Fig. 6.32 by the dashed line. This current flow is similar to that during gate controlled turn-on. It serves to trigger the amplifying gate region, which in turn triggers the main thyristor.

Another design for providing the self-protection for the thyristor against excessive anode bias in the forward blocking mode is shown in Fig. 6.33. In this design, an N$^+$ emitter region is incorporated under the trench region formed in the gate contact area. This creates a thyristor below the gate contact. When the anode voltage increases beyond a certain value, the depletion

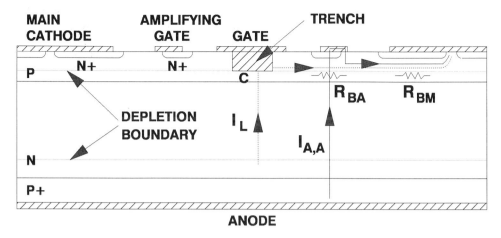

Fig. 6.32 Thyristor design for protection against avalanche breakdown during forward blocking.

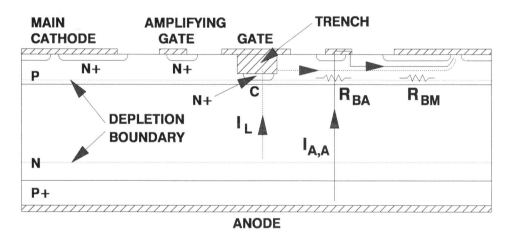

Fig. 6.33 Thyristor self-protection against high anode bias during forward blocking.

layer in the P-base region reaches-through to the N+ emitter. This results in the injection of electrons into the N-base region which triggers the turn-on of the thyristor under the gate contact. The current in this thyristor flows to the cathode short and acts like a gate drive current for the amplifying gate region. This method of providing self-protection for the thyristor is less sensitive to the operating temperature because it does not depend up on the leakage current, which is a very strong function of temperature.

6.6.2 [dV/dt] Turn-on Protection

Although it is desirable to design the thyristor so that it is not triggered into the on-state when a high [dV/dt] is applied in its forward blocking mode, this is difficult to ensure at elevated temperatures due to the significant leakage current flow. The turn-on of the thyristor under a high [dV/dt] condition can occur at any position under the main cathode due to inhomogeneities in resistivity and minority carrier lifetimes. When the device turns-on, a large anode current flows through a local region of the cathode. This produces excessive power dissipation leading to destructive failure.

The basic approach to protect a thyristor against destructive failure during a high [dV/dt] condition is based upon ensuring that the turn-on occurs in the gate region. This requires the design of the gate region in such a manner that a thyristor is triggered on under the gate contact when the [dV/dt] exceeds a defined limit. A thyristor structure for achieving this goal is shown

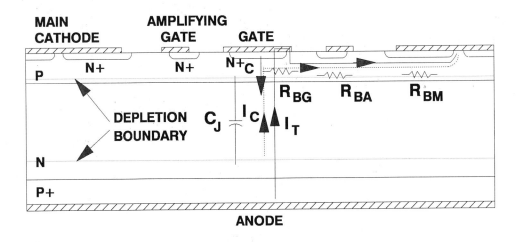

Fig. 6.34 Thyristor structure with protection against high [dV/dt] conditions.

in Fig. 6.34, where an N^+ emitter has been incorporated under the gate contact. The integration of the N^+ region under the gate creates a thyristor in this region. The dimensions of the N^+ region are chosen such that the [dV/dt] current triggers this thyristor into its on-state prior to the rest of the thyristor. The current in this thyristor flows into the cathode short acting like a gate drive current. This triggers the amplifying gate region, which in turn triggers the main thyristor.

A design criterion for ensuring the turn-on in the gate region under high [dV/dt] conditions can be derived for the case of a circular device geometry. Consider the device cross-section shown in Fig. 6.35 with a gate region containing an N^+ emitter region with a radius of r_{SG}. When a high [dV/dt] is applied to the anode under forward blocking conditions, a

Chapter 6 : POWER THYRISTORS

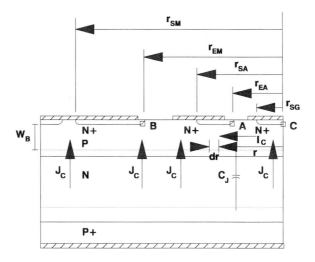

Fig. 6.35 Structure of thyristor with protection against high [dV/dt] conditions used for analysis of design criteria.

displacement current flows due to the junction capacitance C_J. A uniform current density J_C flows at the reverse biased junction J_2 under these conditions. The current collected by the junction flows through the P-base region to the first set of anode shorts. This produces a voltage drop in the P-base region which forward biases the N$^+$ emitter of the gate region, the amplifying gate region, and the main thyristor. The maximum forward bias at the N$^+$/P-base junction occurs at point C for the gate region, at point A for the amplifying gate region, and at point B at the main cathode. In order to ensure that the turn-on under a high [dV/dt] condition occurs at the main gate region, it is necessary to ensure that the forward bias at point C exceeds that at point A and B.

In order to determine the voltage drop at points A, B, and C under a high [dV/dt] condition, consider a location in the P-base region with radius r and radial length dr as shown in Fig. 6.35. The resistance of this segment is given by:

$$R(r) = \frac{\rho_B \, dr}{2 \pi r W_B} = \frac{\rho_{SB} \, dr}{2 \pi r} \quad (6.73)$$

where ρ_{SB} is the sheet resistance of the P-base region under the N$^+$ emitter. The displacement current in the P-base region flowing through the segment of length dr at radius r consists of the current collected by junction J_2 up to radius r. Thus:

$$I_C(r) = \pi r^2 J_C \quad (6.74)$$

The voltage drop across segment dr produced by this current is obtained by combining Eq. (6.73) and Eq. (6.74):

$$dV(r) = I_C(r) \, R(r) = \frac{1}{2} J_C \rho_{SB} \, r \, dr \qquad (6.75)$$

The forward bias at point C can be obtained by integrating the voltage drop from $r = 0$ to $r = r_{SG}$, where r_{SG} is the radius of the edge of the N^+ emitter region under the gate contact. Thus:

$$V_C = \int_0^{r_{SG}} dV(r) = \int_0^{r_{SG}} \frac{1}{2} J_C \rho_{SB} \, r \, dr = \frac{1}{4} J_C \rho_{SB} \, r_{SG}^2 \qquad (6.76)$$

Similarly, the forward bias at point A in the amplifying gate region is given by:

$$V_A = \int_{r_{EA}}^{r_{SA}} \frac{1}{2} J_C \rho_{SB} \, r \, dr = \frac{1}{4} J_C \rho_{SB} \, (r_{SA}^2 - r_{EA}^2) \qquad (6.77)$$

and the forward bias at point B in the main thyristor region is given by:

$$V_B = \int_{r_{EM}}^{r_{SM}} \frac{1}{2} J_C \rho_{SB} \, r \, dr = \frac{1}{4} J_C \rho_{SB} \, (r_{SM}^2 - r_{EM}^2) \qquad (6.78)$$

Based upon the criterion that the forward bias at point C must be larger than that at point A, which in turn should be larger than that at point B, the design requirements for fail-safe turn-on under high [dV/dt] conditions are given by:

$$r_{SG}^2 > (r_{SA}^2 - r_{EA}^2) > (r_{SM}^2 - r_{EM}^2) \qquad (6.79)$$

When designing a thyristor with self-protection against failure during [dV/dt] induced turn-on, these criteria should be combined with that given in section 6.4.2.5 for the amplifying gate region as defined by Eq. (6.64).

6.7 GATE TURN-OFF THYRISTORS

The power thyristor structures discussed in the preceding sections of this chapter are designed for operation in AC circuits. These devices can be turned-on from the forward blocking state and turn-off when the anode voltage reverses. They are not suitable for applications with DC power sources. Although the bipolar power transistor is suitable for applications with DC power sources, its maximum blocking voltage is limited by a degradation in its current gain and on-state voltage drop. This has motivated the development of power thyristors that can be switched from the on-state to the off-state under gate bias control. Such devices are referred to as *Gate Turn-Off Thyristors, GTO Thyristors, or GTOs*.

6.7.1 Turn-off Criterion

In a gate turn-off thyristor, gate current is applied in the opposite polarity to that used

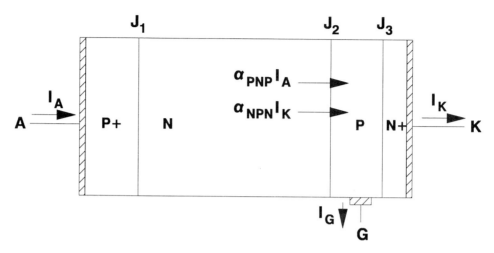

Fig. 6.36 Thyristor structure with reverse gate drive current.

to turn-on the thyristor, as shown in Fig. 6.36 with the currents at junction J_2. In the on-state, the holes injected from the anode of the thyristor serve as the base drive current for the N-P-N transistor. The turn-off gate current diverts the holes from the P-base region into the gate circuit. This reduces the current gain of the N-P-N transistor, and the thyristor is no longer able to maintain its self-sustaining mode of operation.

The base drive current required to maintain current conduction in the N-P-N transistor is $[(1 - \alpha_{NPN}) I_K]$. The base drive current available to the N-P-N transistor with a reverse gate I_G is $[(\alpha_{PNP} I_A) - I_G]$. Thus, the condition for obtaining the turn-off of the thyristor by application of the reverse gate current is given by:

$$\alpha_{PNP} I_A - I_G < (1 - \alpha_{NPN}) I_K \qquad (6.80)$$

Using Kirchhoff's law:

$$I_K = I_A - I_G \qquad (6.81)$$

Using these two equations, the condition for turn-off of the thyristor can be written as:

$$I_G > \frac{(\alpha_{PNP} + \alpha_{NPN} - 1)}{\alpha_{NPN}} I_A \qquad (6.82)$$

If the ratio of the anode current to the gate current is defined as the *turn-off gain (β)*, then from Eq. (6.82), the maximum turn-off gain is obtained as:

$$\beta_m = \frac{\alpha_{NPN}}{(\alpha_{PNP} + \alpha_{NPN} - 1)} \qquad (6.83)$$

A large value for the turn-off gain is desirable in order to reduce the complexity and cost of the gate drive circuit. From Eq. (6.83), it can be concluded that a high turn-off gain is achievable by making the gain of the N-P-N transistor close to unity and the gain of the P-N-P transistor small. A high current gain for the N-P-N transistor is obtained in gate turn-off thyristors by not shorting the N^+ emitter to the P-base region, i.e. no cathode shorts are used. A low gain for the P-N-P transistor is obtained by using a combination of methods. The first is a reduction in the minority carrier lifetime in the N-base region by lifetime control techniques. This method results in a reduction in the base transport factor for the P-N-P transistor. The second method is to form *anode shorts* (a short between the P+ anode region and the N-base region. This method reduces the injection efficiency of the anode junction. The third method is to include a highly doped N-type region at the anode junction. This region is called a *buffer layer*. This layer reduces the injection efficiency of the anode junction. When these methods are used to obtain acceptable values for the turn-off gain (typically in the range of 2 to 10 for a GTO), they produce an increase in the forward voltage drop.

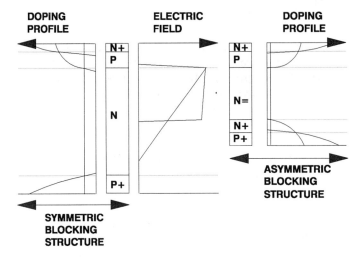

Fig. 6.37 Comparison of doping and electric field profiles for buffer layer structure with conventional thyristor structure.

Since the gate turn-off thyristor is designed for operation in DC circuits, it is required to exhibit only a high forward blocking capability, and its reverse blocking capability does not have to be large as in the case of the power thyristors. It is possible to improve the on-state voltage drop for the gate turn-off thyristor by taking advantage of the buffer layer to obtain the desired forward blocking capability with a reduced thickness for the N-base region. A comparison of the electric field distribution for the thyristor with a buffer layer is made in Fig. 6.37 with a conventional thyristor that has a uniformly doped N-base region. In this figure the doping profiles for the devices are also shown. Note that the doping concentration of the N= region of the thyristor with the buffer layer is low and that its width has been chosen so that its forward blocking voltage is the same as that for the conventional thyristor. The low doping concentration of the N= region alters the electric field profile to that for a punch-through diode discussed in Chapter 3, while the breakdown voltage for the conventional thyristor is determined by open-base breakdown, which requires an undepleted portion in the N-base region at breakdown. This results in a net N-base width (sum of the width of the N= and N- buffer layer) for the thyristor with a buffer layer that is approximately half that for the conventional thyristor. For the same lifetime in the N-base region, this produces a reduction in the on-state voltage drop because of a smaller [d/L] ratio as discussed in Chapter 4 on power rectifiers. A typical forward voltage drop for the gate turn-off thyristor with a buffer layer is about 3 volts.

6.7.2 Turn-off Time

The turn-off process in a GTO occurs in three phases. In the first phase, stored charge in the P-base region is extracted until the junction J_2 becomes reverse biased across its entire

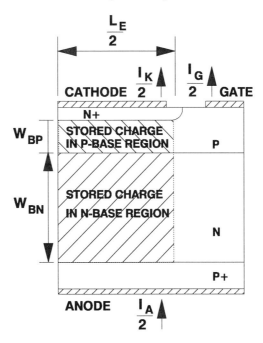

Fig. 6.38 GTO structure used for one-dimensional storage time analysis.

area. During this time interval, the cathode current flows through a part of the N⁺ emitter that remains in its on-state. The cathode current then remains approximately unchanged during this time interval. In the second phase of the turn-off process, the cathode current decreases and the anode voltage rises rapidly to the supply voltage. During this time interval, the depletion layer across junction J_2 expands in the N-base region. This sweeps out stored charge from the N-base region. In the third phase of the turn-off process, the remaining stored charge in the N-base region is removed by recombination. This produces a current tail in the anode current waveform. These three phases are discussed below.

6.7.2.1 Storage Time Analysis. The turn-off time for the gate turn-off thyristor is dominated by a time delay between the application of the reverse gate current and a decrease in the anode current indicating the on-set of device turn-off. This time interval is referred to as the *storage time*. During the storage phase, the gate current extracts charge from the P-base region. An estimate of the storage time can be obtained by charge control analysis performed using the GTO structure shown in Fig. 6.38. In this analysis, it will be assumed that the reverse gate current I_G extracts the stored charge in the P-base region under the N⁺ emitter in a time interval t_s, which is the storage time. If the average free carrier density in the P-base and N-base regions during the on-state is defined as n_a, then:

$$I_G t_s = Q_{SB} = q n_a W_{BP} L_E Z \qquad (6.84)$$

where L_E is the length of the N⁺ emitter, Z is the width of the emitter, and W_{BP} is the thickness of the P-base region. The average free carrier concentration in the P-base and N-base regions is related to the anode current flowing in the on-state by:

$$n_a = \frac{I_A \tau_{HL}}{q L_E Z (W_{BP} + W_{BN})} \qquad (6.85)$$

where W_{BN} is the thickness of the N-base region and τ_{HL} is the high level lifetime. Combining these two equations:

$$t_s = \tau_{HL} \left(\frac{W_{BP}}{W_{BP} + W_{BN}} \right) \frac{I_A}{I_G} = \tau_{HL} \beta \left(\frac{W_{BP}}{W_{BP} + W_{BN}} \right) \qquad (6.86)$$

This expression indicates that the storage time can be reduced by reducing the lifetime and operating the GTO at a lower turn-off gain. However, it is worth pointing out that a reduced lifetime leads to a larger on-state voltage drop, and operating at a lower current gain involves larger gate drive currents, which makes the control circuit more expensive.

The above discussion is valid for a one-dimensional thyristor where the gate current can extract charge uniformly from the base regions of the thyristor. In actual devices, the gate and cathode regions must be interdigitated. In this case, when the reverse gate current is applied, the stored charge is first removed from the regions below the N⁺ emitter that are in proximity to the gate contact. After this region turns-off, the gate current removes charge from the segment a little further removed from the gate contact until this portion under the N⁺ emitter

Chapter 6 : POWER THYRISTORS 307

turns-off. This process proceeds until all portions under the N⁺ emitter are turned-off. Consequently, there is a constriction of the cathode current towards the center of the N⁺ emitter during turn-off.

In order to analyze the gate turn-off thyristor with current constriction towards the center of the N⁺ emitter fingers, consider the cross-section of the device shown in Fig. 6.39 with a

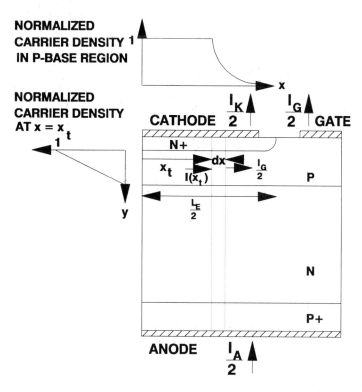

Fig. 6.39 Gate turn-off thyristor structure with current constriction during turn-off.

portion of the N⁺ emitter turned-off during the storage time interval. Note that this cross-section has been taken through the center of an N⁺ emitter finger and the gate contact. Consequently, the currents flowing in the cathode and gate contacts are $I_K/2$ and $I_G/2$, respectively. A constant reverse gate current is assumed to be flowing during the storage phase. This condition is valid if the gate circuit inductance is small. (Often, the GTO gate drive circuit consists of a voltage source in series with an inductance. In this case, the gate current increases linearly with time during the storage phase.)

It will be assumed that a portion of the thyristor under the N⁺ emitter finger from $x = x_t$ to $x = L_E/2$ has been turned-off by extraction of charge by the applied reverse gate current $I_G/2$ while the remainder of the thyristor from $x = 0$ to $x = x_t$ is still operating in the on-state. Now consider a segment dx within the P-base region at the boundary between the portion of the thyristor that is in the on-state and in the off-state. It will be assumed that the junction J_2 has

emitter edge to the N-base edge as shown on the left hand side of Fig. 6.39, if recombination in the P-base region can be neglected and because the carrier concentration at the P-base/N-base junction is zero. The electron concentration in the P-base region within the region where the thyristor is operating in its on-state will be independent of x, while the electron concentration will decay exponentially from the on-region ($x = x_t$) towards the right hand side with a decay length equal to the diffusion length for electrons (L_n) in the P-base region. Based upon these considerations, the variation of the electron concentration can be written as:

$$n(x,y) = n_0 \left(1 - \frac{y}{W_{BP}}\right) e^{-(x - x_t)/L_n} \qquad (6.87)$$

for the portion of the N$^+$ emitter located on the right hand side of $x = x_t$. Note that this equation is not applicable for the portion of the N$^+$ emitter located on the left hand side of $x = x_t$ because the thyristor is in its on-state in this portion and junction J_2 is forward biased in this portion.

Based upon the triangular electron concentration profile in the P-base region in the y-direction, the hole charge stored in the segment dx is given by:

$$dQ = \frac{1}{2} q\, n_0\, W_{BP}\, Z\, dx_t \qquad (6.88)$$

if the variation of electron concentration with x is neglected due to the small value for dx. If recombination is neglected, the magnitude of this stored charge is determined by the difference between the current entering the segment from the left hand side [$I(x_t)$] and the current leaving the segment from the right hand side [$I_G/2$]:

$$\frac{dQ}{dt} = I(x_t) - \frac{I_G}{2} \qquad (6.89)$$

The current $I(x_t)$ entering the segment from the left hand side is determined by the diffusion of carriers:

$$I(x_t) = -\int_0^{W_{BP}} Z\, J_x\, dy \qquad (6.90)$$

with

$$J_x = q\, D_n\, \left.\frac{\delta n}{\delta x}\right|_{x=x_t} = \frac{q\, D_n\, n_0}{L_n}\left(1 - \frac{y}{W_{BP}}\right) \qquad (6.91)$$

Combining these equations:

$$I(x_t) = \frac{q\, n_0\, D_n\, Z\, W_{BP}}{2\, L_n} \qquad (6.92)$$

Chapter 6 : POWER THYRISTORS

$$I(x_t) = \frac{q \, n_0 \, D_n \, Z \, W_{BP}}{2 \, L_n} \quad (6.92)$$

The cathode current flowing at this point in time during the turn-off process consists of the component from x = 0 to x = x_t where the thyristor is operating in its on-state and of the component from x = x_t to x = $L_E/2$ where the thyristor has been turned off but current continues to flow due to the diffusion of holes from the on region into this region. In the on-region from x = 0 to x = x_t, the current density can be assumed to be uniform with a value:

$$J_{Kon} = \frac{q \, (W_{BP} + W_{BN}) \, n_0}{\tau_{HL}} \quad (6.93)$$

based upon application of the charge control principle to the charge within the P-base and N-base regions with the approximation that the average charge density n_a in the base regions is equal to carrier concentration at junction J_3. The cathode current density flowing in the region from x = x_t to x = $L_E/2$ where the thyristor has been turned off is given by:

$$J_{Koff} = q \, D_n \, \frac{\delta n}{\delta y} = q \, D_n \, \frac{n_0}{W_{BP}} \, e^{-(x - x_t)/L_n} \quad (6.94)$$

based upon the linear distribution of electron concentration within the P-base region in this portion of the device. The cathode current is obtained by integration of the cathode current density over both segments:

$$\frac{I_K}{2} = Z \left[\int_0^{x_t} J_{Kon} \, dx + \int_{x_t}^{L_E/2} J_{Koff} \, dx \right] \quad (6.95)$$

which leads to:

$$\frac{I_K}{2} = \frac{qZ(W_{BP} + W_{BN}) n_0 x_t}{\tau_{HL}} + \frac{qZD_n n_0 L_n}{W_{BP}} \left[1 - e^{(x_t - L_E/2)/L_n} \right] \quad (6.96)$$

Using the relationship $L_n^2 = \tau_{HL} \, D_n$ and assuming that the diffusion length L_n is short compared with the distances x_t and L_E, this equation can be rewritten as:

$$\frac{I_K}{2} = \frac{q \, Z \, D_n \, n_0 \, (W_{BP} + W_{BN}) \, x_t}{L_n^2} + \frac{q \, Z \, D_n \, n_0 \, L_n}{W_{BP}} \quad (6.97)$$

From this expression, the electron concentration is obtained:

$$n_0 = \frac{I_K}{2 \, q \, Z \, D_n} \, \frac{L_n^2 \, W_{BP}}{[(W_{BP} + W_{BN}) \, W_{BP} \, x_t + L_n^3]} \quad (6.98)$$

From Eq. (6.88):

$$\frac{dQ}{dt} = \frac{1}{2} q n_0 W_{BP} Z \frac{dx_t}{dt} \qquad (6.99)$$

Using Eq. (6.98) to substitute for n_0 in Eq. (6.99):

$$\frac{dQ}{dt} = \frac{1}{4} \frac{I_K \tau_{HL} W_{BP}^2}{[(W_{BP} + W_{BN}) W_{BP} x_t + L_n^3]} \frac{dx_t}{dt} \qquad (6.100)$$

Similarly, using Eq. (6.98) to substitute for n_0 in Eq. (6.92):

$$I(x_t) = \frac{1}{4} \frac{I_K L_n W_{BP}^2}{[(W_{BP} + W_{BN}) W_{BP} x_t + L_n^3]} \qquad (6.101)$$

Combining Eq. (6.100) and Eq. (6.101) with Eq. (6.89):

$$\frac{dx_t}{dt} = \frac{D_n}{L_n} - \frac{2}{(\beta - 1)} \frac{[(W_{BP} + W_{BN}) W_{BP} x_t + L_n^3]}{\tau_{HL} W_{BP}^2} \qquad (6.102)$$

because

$$\frac{I_G}{I_K} = \frac{1}{(\beta - 1)} \qquad (6.103)$$

This expression can be rewritten as:

$$\frac{dx_t}{\left\{ \dfrac{D_n}{L_n} - \dfrac{2}{(\beta - 1)} \dfrac{[(W_{BP} + W_{BN}) W_{BP} x_t + L_n^3]}{\tau_{HL} W_{BP}^2} \right\}} = dt \qquad (6.104)$$

The storage time (t_s) is the time taken for the on-region to constrict from the entire N$^+$ emitter width of $L_E/2$ to a very small value of width (δ) at the center of the N$^+$ emitter. Note that the current constriction is not assumed to occur to a zero width because this would imply an infinite cathode current density at the end of the storage time. By integration of the above equation:

$$t_s = \int_{L_E/2}^{\delta} \left\{ \frac{D_n}{L_n} - \frac{2}{(\beta-1)} \frac{[(W_{BP}+W_{BN}) W_{BP} x_t + L_n^3]}{\tau_{HL} W_{BP}^2} \right\}^{-1} dx_t \qquad (6.105)$$

which gives:

Chapter 6 : POWER THYRISTORS

$$t_s = \frac{(\beta - 1)}{2} \frac{\tau_{HL} W_{BP}}{(W_{BP} + W_{BN})} \ln\left\{ \frac{\frac{D_n}{L_n} - \frac{2}{(\beta-1)\tau_{HL}W_{BP}^2}\left[L_n^3 + (W_{BP}+W_{BN})W_{BP}\frac{L_E}{2}\right]}{\frac{D_n}{L_n} - \frac{2}{(\beta-1)\tau_{HL}W_{BP}^2}\left[L_n^3 + (W_{BP}+W_{BN})W_{BP}\delta\right]} \right\}$$

(6.106)

From this expression, it can be concluded that the storage time can be reduced by performing the turn-off with a lower current gain, in other words by increasing the reverse gate drive current. However, this increases the cost and complexity of the gate drive circuit. It can also be reduced by decreasing the high level injection lifetime (τ_{HL}). However, this increases the on-state voltage drop, which increases the power loss in the GTO during current conduction. The storage time can also be reduced by making the width of the P-base region (W_{BP}) as small as possible. However, this is limited by reach-through breakdown problems. Thus, a careful compromise between the various device parameters is necessary during GTO design.

6.7.2.2 Maximum Turn-Off Gain. It is advantageous to use the smallest possible gate drive current in order to reduce the complexity and cost of the gate control circuit. The *maximum turn-off gain* (β_{max}) is defined as the condition when the GTO turns-off with the lowest possible gate drive current. This corresponds to the condition when the storage time becomes infinite, indicating failure to turn-off the GTO if the gate current is reduced any further. From Eq. (6.106), the condition for infinite storage time is:

$$\frac{D_n}{L_n} = \frac{2}{(\beta_{max} - 1)\tau_{HL}W_{BP}^2}\left[L_n^3 + (W_{BP} + W_{BN})W_{BP}\delta\right] \quad (6.107)$$

From this equation, the maximum turn-off gain is obtained:

$$\beta_{max} = 1 + \frac{2 L_n}{D_n \tau_{HL} W_{BP}^2}\left[L_n^3 + (W_{BP} + W_{BN})W_{BP}\delta\right] \quad (6.108)$$

If the on-region at the end of the storage time interval (δ) is small compared with the diffusion length, then the maximum turn-off gain is given by:

$$\beta_{max} = 1 + 2 \frac{L_n^2}{W_{BP}^2} \quad (6.109)$$

From this expression, it can be concluded that it is necessary to make the width of P-base region as small as possible and to make the diffusion length for electrons as large as possible. These requirements are in conflict with obtaining a high forward blocking voltage and fast turn-off time.

6.7.2.3 Linear ramp gate drive.
As mentioned earlier, it is customary to drive the gate using a linearly increasing gate current with time by using an inductance in the gate circuit. In this case, Eq. (6.89) must be rewritten as:

$$\frac{dQ}{dt} = I(x_t) - \frac{1}{2} R t \qquad (6.110)$$

where R is the ramp rate for the gate current. Using the same procedure as in the case of a constant gate current, the rate of movement of the boundary x_t can be derived:

$$\frac{dx_t}{dt} = \frac{D_n}{L_n} - \frac{2 R t}{I_K} \frac{[(W_{BP} + W_{BN}) W_{BP} x_t + L_n^3]}{\tau_{HL} W_{BP}^2} \qquad (6.111)$$

If the assumption is made that the anode and cathode currents are approximately constant during the storage phase (which is true if the turn-off gain is large), an expression for the storage time can be derived:

$$t_s = \frac{W_{BP}^2}{2 L_n^2} \frac{I_K}{R} \qquad (6.112)$$

which indicates that the storage time can be reduced by increasing the gate current ramp rate R in proportion to the cathode current.

6.7.2.4 Fall Time Analysis.
At the end of the storage time interval, junction J_2 becomes reverse biased across the entire area under the N^+ emitter region. Although this junction is designed to support a large voltage, this requires the establishment of a depletion region in the N-base region. Since the N-base region still contains a high concentration of holes and electrons at the end of the storage time interval, these carriers must be swept out during the formation of the depletion layer.

The analysis of the fall time can be performed for the case of a device turning off with a resistive load. A schematic of this circuit is shown in Fig. 6.40 with R_L as the load resistance and V_S as the DC supply voltage. A one-dimensional cross-section of the GTO is shown in this figure because the sweep out of the stored charge can be assumed to occur uniformly across the reverse biased junction J_2. At an instant of time t during the fall time, the depletion layer in the N-base region is assumed to be located at a distance x(t) from junction J_2. At this point in time, the anode current and voltage are $I_A(t)$ and $V_A(t)$, respectively. During the fall time, it will be assumed that there is an average carrier concentration p_a within the neutral N-base region. Then, the anode current is given by:

$$I_A(t) = q A p_a \frac{dx}{dt} \qquad (6.113)$$

where A is the cross-sectional area of junction J_2. The depletion layer width [x(t)] at time t is related to the anode voltage [$V_A(t)$] at time t by:

Chapter 6 : POWER THYRISTORS

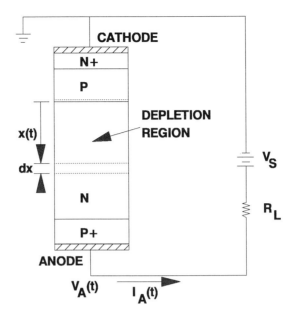

Fig. 6.40 Circuit for analysis of fall time in a GTO.

$$x(t) = \sqrt{\frac{2\,\epsilon_s\,V(t)}{q\,N_D}} \qquad (6.114)$$

The rate of change in the depletion layer width is then given by:

$$\frac{dx}{dt} = \sqrt{\frac{\epsilon_s}{2\,q\,N_D\,V(t)}}\,\frac{dV}{dt} \qquad (6.115)$$

Substituting this expression in Eq. (6.113) gives:

$$I_A(t) = A\,p_a\sqrt{\frac{q\,\epsilon_s}{2\,N_D\,V(t)}}\,\frac{dV}{dt} \qquad (6.116)$$

Application of Ohms law to the circuit gives:

$$V_s - V_A(t) = I_A(t)\,R_L = R_L\,A\,p_a\sqrt{\frac{q\,\epsilon_s}{2\,N_D\,V(t)}}\,\frac{dV}{dt} \qquad (6.117)$$

The solution for this equation with the boundary condition of an initial anode current $I_A(0)$ is given by:

$$I_A(t) = I_A(0) \, \text{sech}^2 \left[\frac{I_A(0)}{A \, p_a} \sqrt{\frac{N_D}{2 \, q \, \epsilon_s \, V_s}} \, t \right] \qquad (6.118)$$

The *fall time* (t_f) is defined as the time taken for the anode current to decrease from 90 percent of its initial value to 10 percent of its initial value. From Eq. (6.118), the fall time is obtained as:

$$t_f = \frac{1.82 \, A \, p_a}{I_A(0)} \sqrt{\frac{2 \, q \, \epsilon_s \, V_s}{N_D}} \qquad (6.119)$$

If recombination in the middle region of the thyristor is dominant during current conduction in the on-state, the initial anode current $I_A(0)$ is related to the average carrier concentration by:

$$I_A(0) = \frac{q \, A \, p_a \, (W_{BP} + W_{BN})}{\tau_{HL}} \qquad (6.120)$$

Combining this equation with Eq. (6.112):

$$t_f = \frac{1.82 \, \tau_{HL}}{(W_{BP} + W_{BN})} \sqrt{\frac{2 \, \epsilon_s \, V_s}{q \, N_D}} \qquad (6.121)$$

From this equation, it can be concluded that the fall time can be reduced by decreasing the high level lifetime. However, this is accompanied by an increase in the on-state voltage drop and power dissipation. It should also be noted that the fall time increases with increase in the operating supply voltage (V_s) because of the need to establish a larger depletion layer width during this time interval. However, the turn-off time is independent of the initial anode current $[I_A(0)]$. This is true only if the middle region recombination is dominant.

If the end region recombination becomes dominant, as observed in thyristors and rectifiers at higher on-state current densities, the average carrier concentration in the middle region is related to the anode current by:

$$p_a = n_{ie} \sqrt{\frac{I_A(0)}{I_S}} \qquad (6.122)$$

where I_S is the saturation current for the injecting junctions (see section 4.2). Substituting for p_a in Eq. (6.119) gives:

$$t_f = 1.82 \; A \; n_{ie} \sqrt{\frac{2 \; q \; e_s \; V_s}{N_D \; I_A(0) \; I_s}} \qquad (6.123)$$

In this case, the fall time decreases with an increase in the initial anode current. This behavior is commonly observed when the anode current is increased to higher values.

6.7.2.5 Current Tail: At the end of the second phase of the turn-off process, the depletion layer has reached its maximum value as determined by the DC power supply voltage. The depletion layer width at this time cannot extend through the N-base region because of reach-through breakdown considerations. Consequently, some stored charge is still left after the end of the fall time. The third phase of the turn-off process consists of the removal of the remnant stored charge in the N-base region. This process occurs by recombination of the free carriers via deep levels in the band gap.

The gate and anode current waveforms for the case analyzed in the above sections is

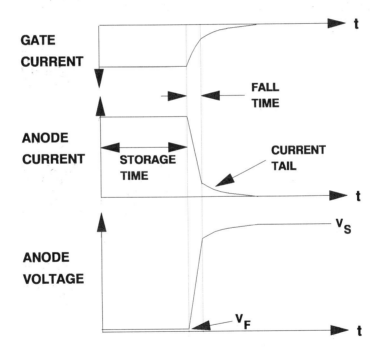

Fig. 6.41 Current and Voltage waveforms during turn-off for a GTO.

shown in Fig. 6.41 together with the anode voltage waveform. The storage time is usually the longest time interval and it sets one limit to the maximum operating frequency of the GTO. However, during the storage phase, the voltage drop across the GTO is small and the power dissipated in the device during this period is similar to that in the on-state. In contrast, the anode voltage rises rapidly during the fall time. This results in a substantial power dissipation during the switching transient. Further, during the anode current tail time, the anode voltage is nearly equal to the DC supply voltage. This results in significant power dissipation during the third phase of the turn-off process. It is therefore important to reduce the current tail time by using lifetime control processes such as gold doping or electron irradiation. The degree to which the lifetime can be reduced is limited by the concomitant increase in the on-state voltage. Thus, there is a trade off between the on-state and switching power losses. Another method to reduce the anode current tail is by using anode shorting.

6.7.3 Anode Shorted GTO Structure

A gate turn-off thyristor with its anode shorted to its N-base region is shown in

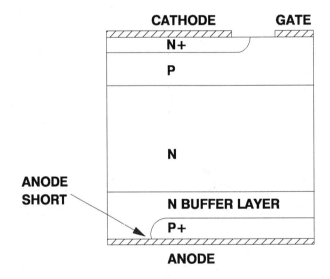

Fig. 6.42 Anode Shorted GTO Structure.

Fig. 6.42. The shorting of the P$^+$ anode region to the N-base region is performed by patterning the P$^+$ anode diffusion on the back surface of the wafers. Since anode shorting destroys any reverse blocking capability, this method is usually used in conjunction with the buffer layer in order to reduce the thickness of the N-base region so as to improve the on-state and switching behavior of the GTO. The anode short greatly reduces the anode current tail. This is because, during the turn-off process, the excess electrons remaining in the N-base region beyond the edge of the depletion layer can be extracted via the ohmic contact to the N-base region. It is worth noting that the anode shorting reduces the injection efficiency of the P-N-P transistor, which is detrimental to obtaining a low on-state voltage drop. Thus, this method also requires a trade-off between the on-state power dissipation and the power dissipation during switching.

Although the removal of the excess electrons in the N-base region can be accomplished by placing the anode short at any random location with reference to the cathode and gate regions, it is common practice to align the anode short as shown in Fig. 6.42 to lie below the center of the N^+ emitter fingers. This design is important for improvement of the turn-off performance. As discussed earlier in section 6.7.2.1, during the storage phase, the on-region of the thyristor is squeezed towards the center of the N^+ emitter fingers by the extraction of stored charge by the reverse gate current. Placement of the anode short at the center of the N^+ emitter results in a low current gain for the P-N-P transistor towards the end of the storage phase. This is beneficial to the turn-off process because the holding current of the thyristor can exceed the operating current level in this region. The alignment of the anode short to the cathode fingers requires special photolithographic tools for pattern recognition on both sides of the wafer or by simultaneous masking of the front and back surfaces.

6.7.4 Maximum Controllable Current

An important parameter for the gate turn-off thyristor is the *maximum controllable current*, which is defined as the highest anode current that can be turned-off under gate control. A larger maximum controllable current allows the GTO to be operated over a wider range of anode current levels with controlled turn-off. This enables the GTO to handle transient anode currents and the reverse recovery currents from anti-parallel diodes during turn-on. This current

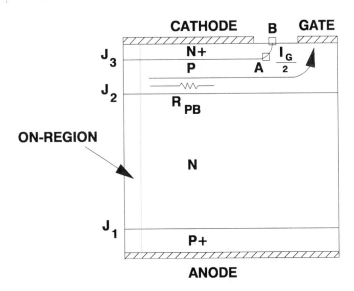

Fig. 6.43 GTO structure with point of breakdown indicated during the turn-off process.

is limited by the size of the N+ emitter fingers and the doping profile in the P-base region.

6.7.4.1 Constant gate current case. Consider the GTO cross-section shown in Fig. 6.43 operating at the end of the storage phase. At this time, the gate current is extracting charge from the center of the cathode finger. This current must flow under the entire length of the N$^+$ emitter region, producing a voltage drop in the P-base region. This voltage drop reverse biases the N$^+$ emitter/P-base junction (J$_3$) with the maximum reverse bias at point A nearest to the gate contact. Due to the relatively high doping concentration in the P-base region, this voltage drop can exceed the breakdown voltage of junction J$_3$ at high gate drive currents. The maximum gate drive current is therefore related to the breakdown voltage of junction J$_3$ by:

$$\frac{I_{GM}}{2} R_{PB} = BV_{J3} \tag{6.124}$$

where R_{PB} is the resistance of the P-base region under the N$^+$ emitter given by:

$$R_{PB} = \frac{\rho_{SB} L_E}{2 Z} \tag{6.125}$$

Any applied gate current that exceeds this value will not enable turn-off of higher anode currents because the impact ionization at point A will generate a current that shunts the gate current from the path for extracting the stored charge in the center of the N$^+$ emitter. The maximum controllable anode current occurs when the device is operated with the above maximum gate current at the limit of turn-off [i.e., under the conditions for maximum turn-off current gain given by Eq. (6.109)]. Thus:

$$I_{AM} = \beta_{max} I_{GM} \tag{6.126}$$

Using Eqs. (6.109), (6.124), (6.125), and (6.126):

$$I_{AM} = 4 \left(1 + \frac{2 L_n^2}{W_{BP}^2} \right) \frac{Z \, BV_{J3}}{\rho_{SB} L_E} \tag{6.127}$$

From this equation, it can be concluded that the highest value for the maximum controllable current can be obtained by maximizing the ratio of the breakdown voltage of the N$^+$ emitter/P-base junction (J$_3$) to the sheet resistance of the P-base region. Unfortunately, an increase in the breakdown voltage of junction J$_3$ requires reducing the P-base doping concentration which results in an increase in the sheet resistance of the P-base region. This creates a conflict during device design. This conflict can be resolved by considering the case of a uniformly doped P-base region with a doping concentration N_{AP} and width W_{BP}. The sheet resistance of the P-base region is then given by:

$$\rho_{SB} = \frac{\rho_{BP}}{W_{BP}} = \frac{1}{q \, \mu_p \, N_{AP} \, W_{BP}} \tag{6.128}$$

while the breakdown voltage of junction J$_3$ is given by:

Chapter 6 : POWER THYRISTORS   319

$$BV_{J3} = 5.34 \times 10^{13} \, N_{AP}^{-3/4} \tag{6.129}$$

Substituting these expressions into Eq. (6.127) gives:

$$I_{Am} = 3.42 \times 10^{-5} \left(1 + \frac{2 \, L_n^2}{W_{BP}^2}\right) \frac{\mu_P \, N_{AP}^{1/4} \, W_{BP} \, Z}{L_E} \tag{6.130}$$

The maximum controllable cathode current density is then given by:

$$J_{KM} = 3.42 \times 10^{-5} \left(1 + \frac{2 \, L_n^2}{W_{BP}^2}\right) \frac{\mu_P \, N_{AP}^{1/4} \, W_{BP}}{L_E^2} \tag{6.131}$$

From this equation, it can be concluded that the maximum controllable anode current increases when the P-base doping concentration is increased. It is clear from this expression that it becomes important to reduce the length of the N^+ emitter to obtain large values for the maximum controllable current.

6.7.4.2 Linear gate ramp case. The storage time expression derived for the linearly increasing gate drive current case does not indicate the existence of any maximum controllable current because the gate current increases with time, allowing any magnitude of anode current to be turned-off eventually. However, in practice, a maximum controllable current limit is observed even when turning off the GTO with a linearly increasing gate current. The reason for this is the increase in the absolute value for the gate current with time. At the end of the storage phase, the gate current has a value:

$$I_G(t_s) = R \, t_s \tag{6.132}$$

This current produces a voltage drop across the resistance of the base region under the N^+ emitter. When this voltage drop becomes equal to the breakdown voltage of the N^+ emitter/P-base junction, any further increase in the gate current flows across junction J_3 via avalanche breakdown. This does not allow any further extraction of the stored charge located at the middle of the N^+ emitter finger. Based upon this limitation, the maximum gate current that allows extraction of stored charge is given by:

$$I_{GM}(t_s) = \frac{BV_{EB}}{R_B} \tag{6.133}$$

where BV_{EB} is the breakdown voltage of the N^+ emitter/P-base junction and R_B is the base resistance between the center of the N^+ emitter and the gate contact. Combining this expression with Eq. (6.112), the maximum controllable current is obtained:

$$I_{KM} = \frac{2\ BV_{EB}}{R_B} \frac{L_n^2}{W_{BP}^2} \qquad (6.134)$$

The maximum controllable cathode current density can be obtained by combining this equation with Eqs. (6.128) and (6.129) for a uniformly doped P-base region:

$$J_{KM} = 3.42 \times 10^{-5} \frac{L_n^2}{W_{BP}\ L_E^2} \mu_P\ N_{AP}^{1/4} \qquad (6.135)$$

As in the case of the constant gate current drive, the maximum controllable current can be increased by increasing the P-base doping concentration and reducing the length of the N$^+$ emitter.

6.7.4.3 Gate-cathode breakdown voltage. One of the problems with the GTO structure shown in Fig. 6.43 with an N$^+$ emitter formed by using planar diffusion is its relatively poor breakdown voltage. This arises due to two reasons. Firstly, the P-base region is usually formed

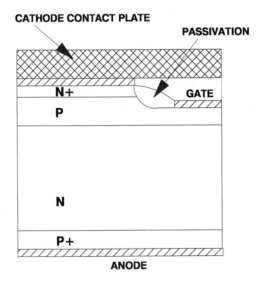

Fig. 6.44 Mesa etched GTO structure with recessed gate regions.

by diffusion of aluminum or gallium from the top surface, which results in a higher doping concentration at point B than that under the N$^+$ emitter. Consequently, the breakdown voltage of junction J$_3$ becomes determined by the high surface concentration of the P-base diffusion while the P-base sheet is determined by the lower doping concentration below the N$^+$ emitter region. This makes the ratio of the breakdown voltage (BV$_{J3}$) to the sheet resistance (ρ_{SB})

smaller than for the case of a homogeneously doped P-base region. Secondly, the junction curvature of the planar diffused junction shown in Fig. 6.43, reduces the breakdown voltage to that for a cylindrical junction. Since the depth of the N^+ diffusion is small, the breakdown voltage of junction J_3 is reduced.

Both of these problems can be overcome by using a mesa etched structure shown in Fig. 6.44 with the gate region recessed from the top surface. The mesa etch is performed after diffusion of the N^+ emitter region. Note that no mask is now required to pattern the N^+ emitter because this is achieved by the mesa etching step. The mesa etch raises the breakdown voltage of junction J_3 because it reduces the maximum doping concentration at the junction to that below the N^+ emitter and it produces a bevel at the junction with an angle close to 90°. A further advantage of the mesa etched GTO structure is that the gate regions are recessed from the top surface. This allows placement of a large contact plate on the top of the device to make contact to all the cathode fingers.

6.7.5 GTO Layout

It is important in a GTO to ensure uniform turn-off of all cathode regions simultaneously. If this does not occur, the anode current can preferentially constrict to a few cathode fingers while the rest of the device turns-off. In this case, the local cathode current density can exceed the maximum controllable current density and the device will fail to turn-off. Uniform turn-off in a GTO is achievable by making a uniform array of cathode fingers. In such an array, all cathode fingers must be identical in width and length. Since the GTOs are designed to carry high currents, it is customary to fabricate a single device from each silicon wafer. The cathode

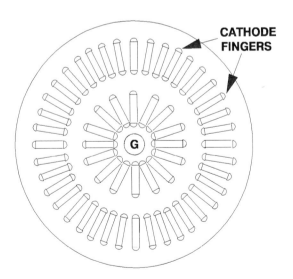

Fig. 6.45 GTO layout with uniform cathode fingers in concentric arrays.

fingers are then placed in concentric circles around a central gate contact. An example of such a GTO layout is shown in Fig. 6.45.

6.7.6 GTO Ratings

Gate turn-off thyristors have been developed with very high blocking voltage and current handling capability. Single devices can be fabricated from wafers with diameters of over 100 mm. The blocking voltages are typically 4500 volts with a maximum controllable current of over 1000 amperes. Such devices are suitable for applications requiring control of high power levels from a DC power source. Typical applications for GTOs are in the area of traction drives, such as those used in electric street cars and electric locomotives. These devices are designed using the asymmetric blocking structure with a buffer layer to optimize the trade-off between the on-state voltage drop, the turn-off time, and the turn-off gain. This trade-off between the device parameters results in typical on-state voltage drops of 3 volts, a turn-off time of about 5 microseconds, and a turn-off gain of between 5 and 10. Consequently, substantial gate control currents are necessary to turn-off the GTO. This makes the cost, size, and power loss of the gate control circuit large. It is also necessary to use large snubbers to protect the GTO during the switching process because of its poor safe-operating-area. These limitations have encouraged the development of alternate device structures where a MOS-gate is used to control the turn-off process in thyristors. These structures are discussed in Chapter 9.

6.8 TRIACS

A commonly encountered application for power semiconductor devices is the control of power delivered to loads from an AC power source. A prime example of this is home appliances that are operated directly from the AC outlet in residential neighborhoods. The power thyristor described in Section 6.1 can be used for such applications because of its forward and reverse blocking capability. However, this device can be triggered into its on-state only when the anode voltage is positive with reference to the cathode terminal. This restricts the on-state current flow to only one-half cycle of the AC input waveform, as discussed in Section 6.4.3. It is more efficient to deliver power to the load during both half-cycles of the input waveform. The anode current waveform for this case is shown in Fig. 6.46 for comparison with that for a thyristor previously shown in Fig. 6.27. The device is assumed to be triggered on during the positive half-cycle of the anode voltage at time t_1 and triggered on during the negative half-cycle of the anode voltage at time t_2. It is assumed that the device turns-off instantaneously at the voltage zero cross-over point.

This requires a solid state device that can be triggered into its on-state when the anode voltage is either positive or negative with reference to the cathode terminal. The *Triac (Triode AC Switch)* is such a device with bi-directional voltage blocking capability and bi-directional current conduction capability. The output characteristics of this device are illustrated in

Chapter 6 : POWER THYRISTORS 323

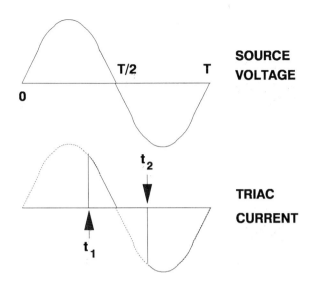

Fig. 6.46 Current and Voltage waveforms for a Triac operated with an AC power source.

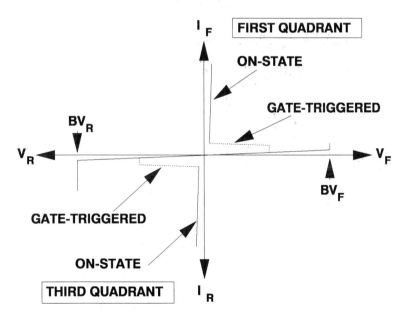

Fig. 6.47 Output characteristics of a Triac.

Fig. 6.47. The unique characteristics of this device are an approximately equal forward and reverse breakdown voltage, and gate triggered current conduction in both the first and third quadrant with a low on-state voltage drop.

One option for obtaining the above characteristics is to connect two thyristors in a back-

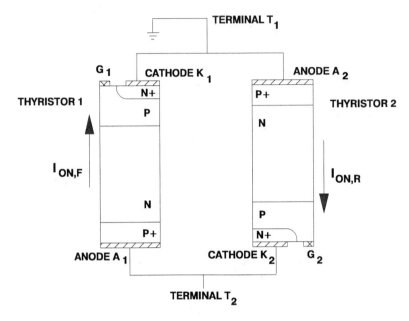

Fig. 6.48 AC switch formed by connecting two thyristors in a back-to-back configuration.

to-back circuit configuration, as shown in Fig. 6.48. In this circuit configuration, the terminal T_1 is used as the reference (ground) terminal and the terminal T_2 is connected to the load and AC power source. When terminal T_2 has a positive potential (during the positive half-cycle of the input voltage waveform), thyristor 1 is operating in its forward blocking mode while thyristor 2 is operating in its reverse blocking mode. During the positive half-cycle of the applied voltage, thyristor 1 can therefore be triggered into its on-state allowing current conduction in the direction $I_{ON,F}$ as shown in Fig. 6.48. The triggering of this thyristor can be performed by using a gate control current applied to its gate G_1 with reference to its cathode K_1. Since cathode K_1 is connected to reference terminal T_1, this gate control signal can be provided from a circuit which is operating at close to ground potential.

When the voltage applied to terminal T_2 reverses from positive to negative, thyristor 1 turns-off via its reverse recovery process and begins to operate in its reverse blocking mode. At the same time, thyristor 2 begins to operate in its forward blocking mode. Consequently, thyristor 2 can be triggered into its on-state by the application of a gate current. This gate control signal must be applied to its gate terminal G_2 with reference to its cathode terminal K_2. Now, the on-state current for thyristor 2 flows in the direction shown in Fig. 6.48 by $I_{ON,R}$. This enables the desired gate controlled current conduction in the reverse direction. However, the triggering of the current flow in the reverse direction requires application of a gate control signal to a gate terminal G_2 that is referenced to a cathode terminal K_2 connected to terminal T_2 whose potential can be very large. Thus, the operating voltage of the gate control circuit must be shifted to that of terminal T_2. This is an undesirable complexity because it adds cost to the power electronics. It is therefore important not only to obtain bi-directional blocking voltage and bi-directional current conduction capability, but also to enable gate control of the device with a single gate referenced to only one of the device terminals (the ground terminal).

6.8.1 Triac Structure and Operation

These features have been incorporated in the triac structure shown in Fig. 6.49 by the

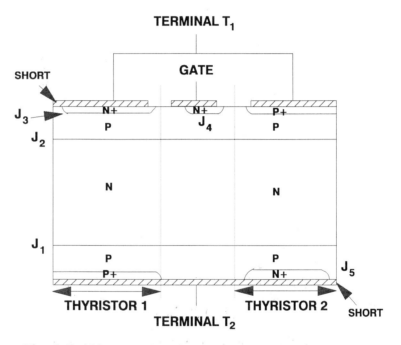

Fig. 6.49 Triac structure with single gate control electrode.

addition of a semiconductor layer to the basic thyristor structure. The triac structure is basically an integration of the two back-to-back thyristors shown in Fig. 6.48 into a single (monolithic) structure. This is achieved by replacing the P^+ anode region of thyristor 1 with a second P-base region and by the diffusion of an additional N^+ emitter region on the back surface of the wafer. This results in the formation of an additional junction (J_5) in the structure. Note that in this structure, in the off-state, the voltage is supported across the same wide N-base region. However, the on-state current in the forward and reverse directions flows through two separate areas within the structure. Thus, the area of the triac is approximately twice that for a single thyristor of the same current rating. It is also worth pointing out that a unique gate structure has been developed which allows triggering the device into its on-state during operation in either the positive or negative quadrant. This gate structure contains an N^+ emitter region whose function is described below. In the following discussion, it will be assumed that terminal T_1 is the reference or ground terminal and that all biases are applied with reference to this electrode.

6.8.1.1 Operating Mode 1: Consider the case of the triac operating with terminal T_2 at a positive potential. In this case, junctions J_1 and J_3 will be forward biased, while junction J_2 and J_5 will be reverse biased. Thyristor 1 will then be operating in its forward blocking mode while thyristor 2 will be operating in its reverse blocking mode. If a positive bias is applied to the gate electrode, a gate current will flow from this electrode to the cathode shorts as shown in

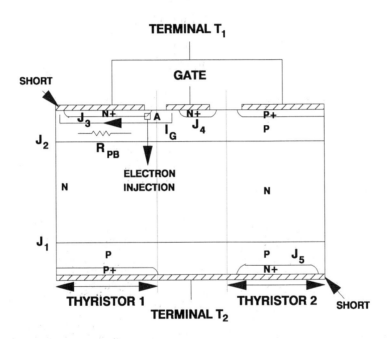

Fig. 6.50 Triac operating in first quadrant with positive gate bias for triggering.

Fig. 6.50. This produces a voltage drop across the resistance (R_{PB}) in the P-base region under the N^+ emitter of thyristor 1, which will forward bias the N^+ emitter/P-base junction at point A. If a sufficient gate control current is supplied, the voltage drop in the P-base region will be enough to create the injection of electrons from the N^+ emitter at point A. This will trigger the thyristor 1 into its on-state. Note that this method of triggering is the same as that used in single thyristors.

6.8.1.2 Operating Mode 2: The triac can also be triggered into the on-state when the potential

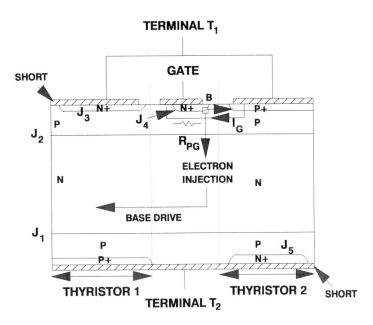

Fig. 6.51 Triac operating in first quadrant with negative gate bias for triggering.

applied to terminal T_2 is positive by the application of a negative bias to the gate electrode. In this case, the gate control current will flow in the opposite direction to that for mode 1 as illustrated in Fig. 6.51. This gate current flow will produce a voltage drop in the resistance (R_{PG}) of the P-base region under the N^+ emitter located under the gate electrode. This voltage drop will allow the potential on the N^+ emitter region under the gate electrode at point B to become negative with respect to terminal T_1. If the gate control current is sufficient, the N^+ emitter region will inject electrons at point B. Some of these electrons will be collected across the reverse biased junction J_2. When these electrons enter the N-base region, they will serve as the base drive current for the P-N-P transistor within thyristor 1. This will trigger thyristor 1 into its on-state.

6.8.1.3 Operating Mode 3: An important feature of the triac structure is that it can be triggered into its on-state when the potential at terminal T_2 is negative by using the same gate electrode. Consider the case of the triac with a negative voltage applied to terminal T_2. The junctions J_1 and J_3 will be reverse biased while junctions J_2 and J_5 will be forward biased. The applied bias will then be supported across the wide N-base region. Thus thyristor 1 will be operating in its reverse blocking mode while thyristor 2 will be operating in its forward blocking mode. If a positive gate voltage is applied to the triac, a gate control current will flow from the gate electrode as shown in Fig. 6.52 to the cathode short. This will produce a voltage drop in the resistance (R_{PB}) of the P-base region under the N^+ emitter of thyristor 1, which will forward bias the N^+ emitter/P-base junction J_3 at point A. If sufficient gate control current is supplied,

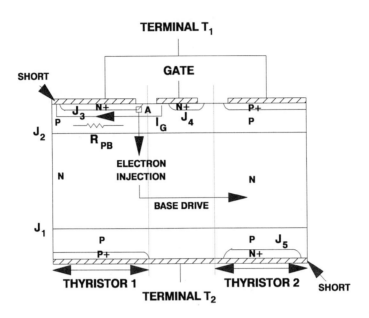

Fig. 6.52 Triac operating in third quadrant with positive gate bias for triggering.

the N⁺ emitter will inject electrons into the P-base region at point A. Although the junction J_2 is forward biased, it will collect the electrons. When these electrons enter the N-base region they will provide the base drive current for the P-N-P transistor within thyristor 2. This will trigger thyristor 2 into its on-state. Since the electrons that initiate the triggering process are derived from a region located away from the N-base region, this method of triggering is referred to as *remote gate control*.

6.8.1.4 Operating Mode 4: The triac can be triggered into its on-state while operating in the third quadrant by the application of a negative voltage to the gate electrode. In this case, the gate control current will flow in the opposite direction as illustrated in Fig. 6.53 to that for mode 3. This gate current flow will produce a voltage drop in the resistance (R_{PG}) of the P-base region under the N+ emitter located under the gate electrode. This will forward bias the N⁺ emitter region under the gate electrode at point B. If the gate control current is sufficient, the N⁺ emitter region will inject electrons into the P-base region at point B. These electrons will be collected by junction J_2 even though it is forward biased. When these electrons enter the N-base region, they will serve as the base drive current for the P-N-P transistor within thyristor 2. This will trigger thyristor 2 into its on-state.

Thus, the triac has the unique features of having both forward and reverse blocking capability, of being able to conduct current in both directions, and of being controlled by a gate signal applied to a single gate electrode referenced to one of its output terminals. This makes the triac a very attractive device for a variety of applications where the power is delivered from

Chapter 6 : POWER THYRISTORS

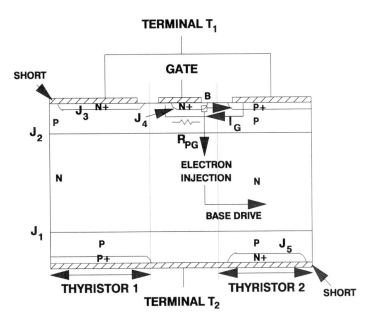

Fig. 6.53 Triac operating in third quadrant with negative gate bias for triggering.

an AC source. However, the integration of the two thyristors in a back-to-back configuration in a monolithic implementation degrades the [dV/dt] capability of the device and curtails its maximum operating frequency. These limitations are discussed below.

6.8.2 [dV/dt] Limitations

Consider the triac operating in its on-state in the first quadrant where the voltage on terminal T_2 is positive. When the source potential reverses from positive to negative, the thyristor 1 undergoes its reverse recovery process. Simultaneously, a high [dV/dt] is applied to thyristor 2 which enters its forward blocking mode. Since a large density of carriers are present in the N-base region of thyristor 1 at the instant when the voltage at terminal T_2 reverses, these carriers can diffuse as shown in Fig. 6.54 from the thyristor 1 region into the thyristor 2 area. As the depletion region across junction J_1 builds up, the holes in the N-base within thyristor 2 are collected by the depletion region extension and flow into the P-base region of thyristor 2. When this hole current flows through the P-base region to its contact at point D, it produces a voltage drop across the resistance (R_{PB}) in the P-base region under the N+ emitter of thyristor 2. This forward biases the N+ emitter/P-base junction at point C. As the operating frequency of the AC source is increased, the [dV/dt] during the switching process also increases. Further, there is less time for the recombination of the stored charge in the N-base region of thyristor 1 before the reverse voltage on terminal T_2 reaches a large value. This results in the

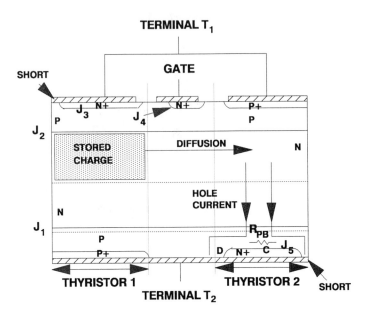

Fig. 6.54 Carrier flow in N-base region from thyristor 1 to thyristor 2 within a triac.

triggering of the thyristor into its on-state without the application of the gate control signal. Thus, the maximum operating frequency of the triac is substantially lower than that for power thyristors.

One method of increasing the operating frequency of triacs is to perform lifetime reduction. This reduces the stored charge during the reverse recovery process and consequently allows operation of the device at higher frequencies. However, a reduction in the lifetime results in a higher on-state voltage drop that in turn reduces the efficiency of the power control system. This problem can be mitigated by performing selective lifetime reduction, as shown in Fig. 6.55, of the region between thyristor 1 and thyristor 2. The stored charge in thyristor 1 is now unable to diffuse into the N-base region of thyristor 2 during the reversal of the voltage on terminal T_2 because any charge entering the low lifetime zone will be removed rapidly by recombination. A convenient method for obtaining such a low lifetime zone is by using electron irradiation to produce the recombination centers in the low lifetime zone with a tungsten mask to block the electron irradiation from the rest of the triac area. Thus, both thyristor 1 and thyristor 2 can have low on-state voltage drops while the [dV/dt] capability and operating frequency are improved.

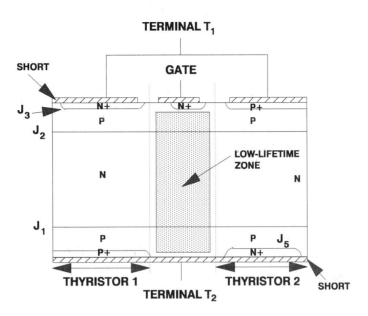

Fig. 6.55 Triac structure with reduced lifetime between thyristor 1 and thyristor 2.

6.9 TRENDS

The commercial availability of power thyristors earmarked the conversion of power electronics from vacuum tube based systems to solid-state power convertors. The demand for devices with higher power handling capability has fostered the development of devices with ever increasing voltage blocking capability and higher current handling capability. This demand can be expected to continue into the future. However, the large gate control current required for the gate turn-off thyristors places a serious limitation to the size, weight, and cost of systems. This has created a strong interest in the development of thyristor structures that can be both turned-on and turned-off with a voltage applied to the gate of an MOS structure incorporated within the thyristor structure. These structures are discussed in Chapter 9. When these developments are successful, it is anticipated that gate turn-off thyristors will be replaced by MOS-gated thyristors.

REFERENCES

1. S.K. Ghandhi, "Semiconductor Power Devices," Wiley, New York (1977).

2. B.J. Baliga and D.Y. Chen, "Power Transistors: Device Design and Applications," IEEE Press, New York (1984).

3. A. Blitcher, "Thyristor Physics," Springer-Verlag, New York (1976).

4. J.L. Moll, M. Tannenbaum, J.M. Goldey, and N. Holonyak, "P-N-P-N transistor switches," Proc. IEEE, Vol. 44, pp. 1174-1182 (1956).

5. A. Herlet, "The maximum blocking capability of silicon thyristors," Solid State Electronics, Vol. 8, pp. 655-671 (1965).

6. J. Cornu and A. Jaecklin, "Processes at turn-on of thyristors," Solid State Electronics, Vol. 18, pp. 683-689 (1975).

7. R.L. Longini and J. Melngailis, "Gated turn-on of four layer switch," IEEE Trans. Electron Devices, Vol. ED-10, pp. 178-185 (1963).

8. W.H. Dodson and R.L. Longini, "Probed determination of turn-on spread of large area thyristors," IEEE Trans. Electron Devices, Vol. ED-13, pp. 478-484 (1966).

9. H. J. Ruhl, "Spreading velocity of the active area boundary in a thyristor," IEEE Trans. Electron Devices, Vol. ED-17, pp. 672-680 (1970).

10. Y. Yamasaki, "Experimental observation of the lateral plasma propagation in a thyristor," IEEE Trans. Electron Devices, Vol. ED-22, pp. 65-68 (1975).

11. H.F. Storm and J.G. St. Clair, "An involute gate-emitter configuration for thyristors," IEEE Trans. Electron Devices, Vol. ED-21, pp. 520-522 (1974).

12. F.E. Gentry and J. Moyson, "The amplifying gate thyristor," IEEE Int. Electron Devices Meeting, Paper 19.1 (1968).

13. P.S. Raderecht, "A review of the shorted emitter principle," Int. J. Electronics, Vol. 31, pp. 541-564 (1971).

14. P. Voss, "A thyristor protected against di/dt failure at breakover turn-on," Solid State Electronics, Vol. 17, pp. 655-661 (1974).

15. A. Munoz-Yague and P. Leturcq, "Optimum design of thyristor gate-emitter geometry," IEEE Trans. Electron Devices, Vol. ED-23, pp. 917-924 (1976).

16. A. Herlet and K. Raithel, "Forward characteristics of thyristors in the fired state," Solid State Electronics, Vol. 9, pp. 1089-1105 (1966).

17. R.A. Kokosa, "The potential and carrier distributions of a P-N-P-N device in the on-state," Proc. IEEE, Vol. 55, pp. 1389-1400 (1967).

18. D. Silber, "Progress in light activated power thyristors," IEEE Trans. Electron Devices, Vol. ED-23, pp. 899-904 (1976).

19. E.D. Wolley, "Gate turn-off in P-N-P-N devices," IEEE Trans. Electron Devices, Vol. ED-13, pp. 590-597 (1966).

20. M. Azuma and M. Kurata, "GTO thyristors," Proc. IEEE, Vol. 76, pp. 419-427 (1988).

21. F.E. Gentry, R.I. Scace, and J.K. Flowers, "Bidirectional triode P-N-P-N switches," Proc. IEEE, Vol. 53, pp. 355-369 (1965).

22. J.F. Essom, "Bidirectional triode thyristor applied voltage rate effect following conduction," Proc. IEEE, Vol. 55, pp. 1312-1317 (1967).

PROBLEMS

6.1 Consider a N^+PN-P^+ power thyristor with uniformly doped semiconductor layers. The N^+ emitter region has a doping concentration of 2×10^{19} per cm^3 and thickness of 10 microns. The P-base region has a doping concentration of 2×10^{17} per cm^3 and thickness of 20 microns. The N-base region has a doping concentration of 5×10^{13} per cm^3 and thickness of 300 microns. The P^+ anode region has a doping concentration of 1×10^{19} per cm^3 and thickness of 10 microns. The Shockley-Read-Hall lifetime in all the regions is 2 microseconds. What is the on-state current flowing in the thyristor at a forward voltage drop of 1.5 volts?

6.2 Determine the gate current required to trigger this thyristor if a linear cathode shorting geometry is used with an N^+ emitter length of 0.5 cm between the shorts and its width is 1 cm.

6.3 Determine the holding current of this thyristor if a linear cathode shorting geometry is used with an N^+ emitter length of 0.5 cm between the shorts and its width is 1 cm.

6.4 Determine the [dV/dt] capability of this thyristor if a linear cathode shorting geometry is used with an N^+ emitter length of 0.5 cm between the shorts and its width is 1 cm when it is blocking 100 volts.

6.5 Determine the leakage current density in this thyristor at room temperature when a negative bias of 500 volts is applied to the anode without the cathode shorts.

6.6 Determine the leakage current density in this thyristor at room temperature when a positive bias of 500 volts is applied to the anode without the cathode shorts.

6.7 Determine the maximum length of the N^+ emitter between the cathode shorts required to make the leakage current under forward blocking conditions equal to that under reverse blocking conditions.

6.8 Consider an amplifying gate thyristor with annular gate geometry and uniformly doped semiconductor layers. The N^+ emitter region has a doping concentration of 1×10^{19} per cm^3 and thickness of 5 microns. The P-base region has a doping concentration of 1×10^{17} per cm^3 and thickness of 15 microns. The N-base region has a doping concentration of 3×10^{13} per cm^3 and thickness of 300 microns. The P^+ anode region has a doping concentration of 1×10^{19} per cm^3 and thickness of 10 microns. The Shockley-Read-Hall lifetime in all the regions is 2 microseconds. The gate current required to turn-on the device is 100 milliamperes. In order to bond a gate lead, the diameter of the gate pad must be 4 times the wire diameter of 20 mils. The device fabrication process requires a separation between the metal regions of 10 mils and a separation between the N^+ emitter edge and the metal of 5 mils. In order to ensure turn-on via the amplifying gate region, design the main thyristor so that it will not turn-on until the gate current is twice as large as that required to turn-on the amplifying gate thyristor.

6.9 Consider a gate turn-off thyristor with linear cathode geometry and uniformly doped semiconductor layers. The N^+ emitter region has a doping concentration of 1×10^{19} per cm^3 and thickness of 10 microns. The P-base region has a doping concentration of 1×10^{17} per cm^3 and thickness of 15 microns. The N-base region has a doping concentration of 5×10^{13} per cm^3 and thickness of 300 microns. The P^+ anode region has a doping concentration of 1×10^{19} per cm^3 and thickness of 10 microns. The Shockley-Read-Hall lifetime in all the regions, except the N-base region, is 1 microsecond, while that in the N-base region is 10 microseconds. Calculate the maximum turn-off gain of the device based upon the storage time analysis with a constant gate drive current if the N^+ emitter has a length of 750 microns and a width of 1 cm.

6.10 What is the maximum controllable current for this gate turn-off thyristor assuming the N^+ emitter/P-base junction has parallel plane breakdown voltage? Compare this value with the on-state current assuming that the on-state current density is 200 amperes per cm^3.

Chapter 7

POWER MOSFET

Prior to the development of the power metal-oxide-semiconductor field-effect-transistors (MOSFETs), the only device available for high speed, medium power applications was the power bipolar transistor. The power bipolar transistor was first developed in the early 1950s, and its technology has matured to a high degree allowing the fabrication of devices with current handling capability of several hundred amperes and blocking voltages of 600 volts. Despite the attractive power ratings achieved for bipolar transistors, there exist several fundamental drawbacks in their operating characteristics. First of all, the bipolar transistor is a current controlled device. A large base drive current, typically one-fifth to one-tenth of the collector current, is required to maintain them in the on-state. Even larger reverse base drive currents are necessary for obtaining high speed turn-off. These characteristics make the base drive circuitry complex and expensive. The bipolar transistor is also vulnerable to a second breakdown failure mode under the simultaneous application of a high current and voltage to the device as commonly required in inductive power circuits. Further, it is difficult to parallel these devices. The forward voltage drop in bipolar transistors decreases with increasing temperature. This promotes diversion of the current to a single device unless emitter ballasting schemes are utilized.

In order to address these performance limitations experienced with power bipolar transistors, the power MOSFET was developed in the 1970s. These devices evolved from MOS integrated circuit technology. In the power MOSFET, the control signal is applied to a metal gate electrode that is separated from the semiconductor surface by an intervening insulator (typically silicon dioxide). The control signal required is essentially a bias voltage with no significant steady-state gate current flow in either the on-state or the off-state. Even during the switching of the devices between these states, the gate current is small at typical operating frequencies (less than 100 kHz) because it serves only to charge and discharge the input gate capacitance. The high input impedance is a primary feature of the power MOSFET that greatly simplifies its gate drive circuitry.

In contrast with the power bipolar transistor, the power MOSFET is a unipolar device. Current conduction occurs via transport of majority carriers in the drift region without the presence of minority carrier injection required for bipolar transistor operation. No delays are observed due to storage or recombination of minority carriers in power MOSFETs during turn-off. Their inherent switching speed is orders of magnitude faster than for bipolar transistors. This feature is particularly attractive in circuits operating at high frequencies where switching power losses are dominant. Power MOSFETs have also been found to display an excellent safe-operating-area (i.e., they can withstand the simultaneous application of high current and voltage, for a short duration, without undergoing destructive failure due to second breakdown. Further, these devices can be easily paralleled because the forward voltage drop of power MOSFETs increases with increasing temperature. This feature promotes an even current distribution between paralleled devices. These characteristics of power MOSFETs make them important candidates for many applications. They are being used in audio/rf circuits and in high frequency inverters such as for switch mode power supplies. Other applications for these devices are for lamp ballasts and motor control circuits.

7.1 BASIC STRUCTURE AND OPERATION

The operation of the power MOSFET relies upon the formation of a conductive layer at the surface of the semiconductor. The modulation of the charge at the surface of a semiconductor by using the bias on a metal plate with an intervening insulating layer was first demonstrated in 1948. Using this principle, the first surface field effect transistor was fabricated in silicon with thermally grown silicon dioxide as the insulator. Since then, the insulated gate field effect transistor (IGFET) has been applied to the fabrication of a large variety of integrated circuits. Until recently, the development of discrete devices has followed the basic concept of the lateral channel structure used in these earlier applications. Such devices have the drain, gate, and source terminals on the same surface of the silicon wafer. Although this feature makes them well suited for integration, it is not optimum for achieving a high power rating. The vertical channel structure, with source and drain on opposite surfaces of the wafer, is more suitable for a power device because more area is available for the source region and because the electric field crowding at the gate is reduced.

Three discrete vertical channel power MOSFET structures have been explored. The first structure was the vertical channel V-MOS power FET, shown in Fig. 7.1, whose name is derived from the V-shaped groove within which the gate is located. This structure can be fabricated by first performing an unpatterned P-base diffusion followed by the N^+ source region diffusion. A V-shaped groove extending through these diffusions is then formed by using a preferential etch. The gate electrode is then deposited and patterned so as to overlap the N^+ source and extend into the groove beyond the bottom of the P-base region. The channel region for this structure is formed along the walls of the V-groove. Although the V-MOSFET was the first commercial structure, it was superseded by the DMOSFET because of stability problems during manufacturing and a high local electric field at the tip of the V-groove.

A cross-section of the DMOSFET structure is provided in Fig. 7.2. This DMOS structure is fabricated by using planar diffusion technology with a refractory gate, such as polysilicon, as a mask. In these devices, the P-base region and the N^+ source regions are diffused through a common window defined by the edge of the polysilicon gate. The name for this device is derived from this *double-diffusion process*. The P-base region is driven-in deeper

Chapter 7 : POWER MOSFET   337

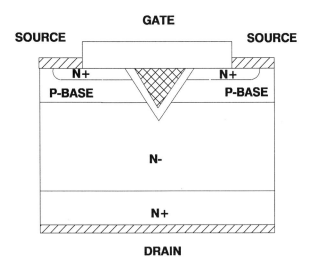

Fig. 7.1 The VMOSFET Structure.

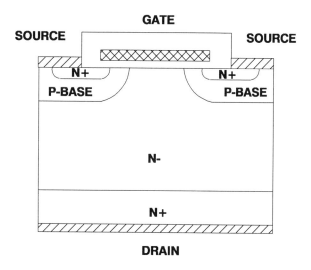

Fig. 7.2 The DMOSFET structure.

than the N⁺ source. The difference in the lateral diffusion between the P-base and N⁺ source regions defines the surface channel region. This has been the most commercially successful structure.

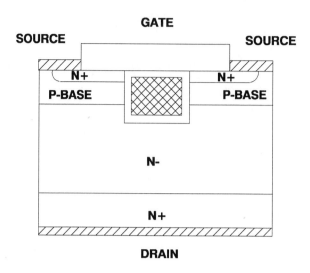

Fig. 7.3 The UMOSFET structure.

The third power MOSFET structure that has been explored is the U-MOSFET structure shown in Fig. 7.3. The name for this structure is derived from the U-shaped groove formed in the gate region by using reactive ion etching. The fabrication of this structure can be performed by following the same sequence as described earlier for the VMOSFET structure with the V-groove replaced by the U-groove. The U-groove structure has a higher channel density than either the VMOS or DMOS structures which allows significant reduction in the on-resistance of the device. The technology for the fabrication of this structure was derived from the trench etching techniques developed for the storage capacitor in memories. The commercialization of the UMOSFET occurred in the 1990s.

In all three power MOSFET device structures, the P-N junction between the P-base region and the N-drift region provides the forward blocking capability. Note that the P-base region is connected to the source metal by a break in the N^+ source diffusion. This is important to establishing a fixed potential to the P-base region during device operation. If the gate electrode is externally shorted to the source, the surface of the P-base region under the gate (i.e., the channel region) remains unmodulated at a carrier concentration determined by the doping level. When a positive drain voltage is now applied, it reverse biases the P-base/N-drift region junction. This junction supports the drain voltage by the extension of a depletion layer on both sides. Due to the higher doping level of the P-base region, the depletion layer extends primarily into the N-drift region. Its doping concentration and width must be chosen in accordance with the criteria established in Chapter 3 for avalanche breakdown of P-N junctions. A higher drain blocking voltage capability requires a lower drift region doping and a larger width. It is important to connect the gate electrode to the source to establish its potential at the lowest point during the forward blocking state. If the gate is left floating, its potential can rise via capacitive coupling to the drain potential. This induces modulation of the channel region, which can produce an undesirable current flow at drain voltages well below the avalanche breakdown limit. Thus, power MOSFETs will not support large drain voltages unless the gate is grounded (i.e., connected to the source during forward blocking).

In order to carry current from drain-to-source in the power MOSFET, it is essential to form a conductive path extending between the N^+ source regions and the N-drift region. This can be accomplished by applying a positive bias to the gate electrode. The gate bias modulates the conductivity of the channel region by the strong electric field created normal to the semiconductor surface through the oxide layer. For a typical gate oxide thickness of 1,000 °A and gate drive voltage of 10 volts, an electric field of 10^6 volts per cm is created in the oxide. The gate induced electric field attracts electrons to the surface of the P-base region under the gate. This field strength is sufficient to create a surface electron concentration that overcomes the P-base doping. The resulting surface electron layer in the channel provides a conductive path between the N^+ source regions and the drift region. The application of a positive drain voltage now results in current flow between drain and source via the N-drift region and the channel. This current flow is controlled by the resistance of these regions. Note that the current flow occurs solely by transport of majority carriers (electrons for n-channel devices) along a resistive path comprising the channel and drift regions. No minority carrier transport is involved for the power MOSFET during current conduction in the on-state.

In order to switch the power MOSFET into the off-state, the gate bias voltage must be reduced to zero by externally shorting the gate electrode to the source electrode. When the gate voltage is removed, the electrons are no longer attracted to the channel and the conductive path from drain to source is broken. The power MOSFET then switches rapidly from the on-state to the off-state without any delays arising from minority carrier storage and recombination as experienced in bipolar devices. The turn-off time is controlled by the rate of removal of the charge on the gate electrode because this charge determines the conductivity of the channel. Turn-off times of under 100 nanoseconds can be achieved with a moderate gate drive current flow arising from the discharge of the input gate capacitance of the device.

All the power MOSFET structures illustrated in the figures contain a parasitic N^+-P-N-N^+ vertical structure. This parasitic bipolar transistor must be kept inactive during all modes of operation of the power MOSFET. To accomplish this, the P-base region is shorted to the N^+ emitter region by the source metallization as shown in the cross-section of the devices. The short between the P-base and N^+ emitter can be provided within each cell of the device structure as illustrated in the figures or occasionally at selected locations. The latter approach makes the fabrication of the cells much easier because it eliminates an alignment step necessary to form the short within the cells. However, the resistance of the P-base region between the shorts can become large and any lateral current flow in the P-base, due to capacitive currents arising at high applied [dV/dt] to the drain, can lead to forward biasing of the N^+/P junction at locations remote from the shorts. Forward biasing of the N^+/P junction activates the parasitic bipolar transistor and leads to the initiation of minority carrier transport. This not only can slow down the switching of the power MOSFET but also can lead to second breakdown. Due to the high [dV/dt]'s observed in the high frequency applications for which the power MOSFET is particularly well suited, it is common practice to form the short in every cell and minimize the length of the N^+ source region from the edge of the channel to the short.

7.2 OUTPUT CHARACTERISTICS

The power MOSFET is capable of blocking voltage only in one quadrant. For n-channel structures, the devices are operated with a positive voltage applied to the drain. When the gate electrode is shorted to the source, the device can support a large drain voltage across the P-base/N-drift region junction. The forward blocking capability is shown in Fig. 7.4 by the

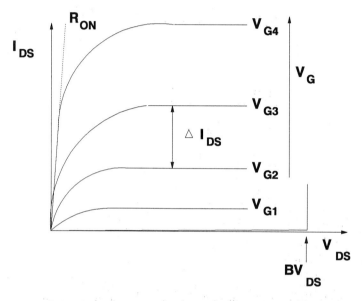

Fig. 7.4 Output characteristics of the power MOSFET.

lowest trace. Although a finite leakage current is illustrated in the figure, it is very small except at high operating temperatures. The maximum forward blocking voltage is determined by the avalanche breakdown voltage of the P-base/N-drift layer junction. The breakdown voltage is dependent not only up on the device termination but is affected by the internal cell structure of the devices.

When a positive gate bias is applied, the channel becomes conductive. At low drain voltages, the current flow is essentially resistive with the on-resistance determined by a combination of the channel and drift region resistances. The channel resistance decreases with increasing gate bias while the drift region resistance remains constant. The total on-resistance decreases with increasing gate bias until it approaches a constant value. At large gate bias voltages, the channel resistance becomes small compared with the drift region resistance and the device on-resistance becomes independent of gate bias. The on-resistance is an important power MOSFET parameter. It is a measure of the current handling capability of the device because it determines the power dissipation during current conduction. The on-resistance is defined as the slope of the output characteristic in the linear region at low drain voltages. At high drain voltages, the resistance of the power MOSFET increases. Ultimately, the current saturates at high drain voltages as shown in Fig. 7.4. The current saturation in power MOSFETs can be used to provide a current limiting function in power circuits as long as the power dissipation in the devices is kept within reasonable limits.

An important device parameter for power MOSFETs is the transconductance. It is defined by:

$$g_m = \left[\frac{\Delta I_{DS}}{\Delta V_{GS}}\right]_{V_{DS}} \tag{7.1}$$

where $\Delta V_{GS} = (V_{G3}-V_{G2})$ as illustrated in Fig. 7.4. A large transconductance is desirable to obtain a high current handling capability with low gate drive voltage, and for achieving high frequency response. In the saturated current region of operation, the output characteristics are controlled by the gate induced channel characteristics. The transconductance is, therefore, determined by the design of the channel and gate structure.

Although the power MOSFET was originally intended for operation in only one quadrant as illustrated in Fig. 7.4 for n-channel devices, it can be operated with negative drain voltages as a synchronous rectifier. When a negative drain voltage is applied to the devices illustrated in the figures with their gates connected to the source, the P-base/N-drift layer junction becomes forward biased. When the drain voltage exceeds a knee voltage (V_K) of approximately 0.7 volts at room temperature, the P-N junction will inject minority carriers and begin to conduct current

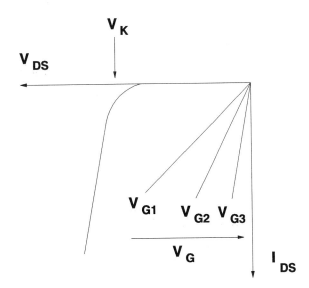

Fig. 7.5 Power MOSFET operated in the third quadrant.

as illustrated in Fig. 7.5. In this quadrant, if a positive gate voltage is applied to the device in order to create a channel, an alternate path for current flow between drain and source is created. This current flow occurs via majority carriers and is limited by the resistance of the channel and drift region. If these resistances are small, the power MOSFET characteristics will look like those shown in Fig. 7.5. In this figure, the device is shown to carry current with a voltage drop well below the knee point (V_K). Under these operating conditions, the P-N junction will not inject any significant concentration of minority carriers. Consequently, the power MOSFET will exhibit a very low forward drop, significantly lower than for a P-N junction rectifier, and retain its high switching speed. For these reasons, it has been suggested that the power MOSFET be used as a replacement for diodes in low voltage (< 5 volts) switch mode power supplies. To accomplish this function, it is essential to provide a gate signal to the power MOSFET in synchronism with the supply voltage to maintain it in the low forward drop on-state, as well as the blocking state, at the appropriate times. The device is therefore called a *synchronous rectifier*. To be used as a synchronous rectifier, the power MOSFET must exhibit an extremely low on-resistance if it is to be capable of handling significant drain currents. For example, a device capable of switching 50 amperes with a forward drop of less than 0.25 volts must have an on-resistance of less than 5 milliohms. To achieve this, very large devices are required with breakdown voltages below 50 volts. The large input capacitance of such devices creates difficulties when providing the synchronous gate drive signal at high operating frequencies.

7.3 STATIC BLOCKING CHARACTERISTICS

As in the case of other power semiconductor devices, the most important distinguishing feature of a power MOSFET is its relatively high blocking voltage capability. In the power MOSFET, the ability to block current flow at high voltages is obtained by supporting the voltage across a reverse biased P-N junction. However, the presence of the MOS-gate structure introduces unique design problems that must be overcome to sustain high voltages without rupturing the thin gate oxide. In addition, the power MOSFET structure contains a parasitic bipolar transistor. The blocking voltage capability can be limited by the activation of this transistor by reach-through of the depletion layer across the base of the transistor. In addition, it has been found that the breakdown voltage is affected by the junction curvature within the DMOS cell, and by the high electric field created at the points at the bottom of the V-groove in the VMOS structure, and by that created at the corners of the trench gate in the UMOS structure. These issues are analyzed in this section of the chapter.

7.3.1 Parasitic Bipolar Transistor

In the forward blocking mode, the gate electrode of the power MOSFET is externally shorted to the source. Under these conditions, no surface channel forms under the gate at the surface of the P-base region. When a positive drain voltage is applied, the P-base/N-drift layer junction becomes reverse biased and supports the drain voltage. The breakdown voltage of this junction is not given by the simple parallel-plane analysis described in Chapter 3 due to several reasons. Firstly, the power MOSFET structure consists of a N^+-P-N transistor where the P-base region is shorted at selected points to the N^+ emitter by the source metallization. Despite the shorting of the N^+ emitter and P-base at some locations, this structure will conduct current as soon as the depletion layer in the P-base reaches-through to the N^+ emitter because the N^+ emitter then becomes forward biased and injects electrons into the P-base region. The reach-through breakdown condition is determined by the doping profiles used for device fabrication.

In the forward blocking mode, the drain voltage is supported across the P-base/N- drift region junction. From the previous discussion on abrupt junction breakdown, it may be concluded that the depletion layer extends primarily on the lightly doped N- drift region. This situation does not apply in the case of the power MOSFET. Here, the depletion layer of the P-base/N-drift region junction extends on both sides. This occurs because the P-base doping concentration must be maintained at a relatively low value in order to be able to invert the surface under the gate during the operation of the power MOSFET in its on-state.

7.3.2 Doping Profile

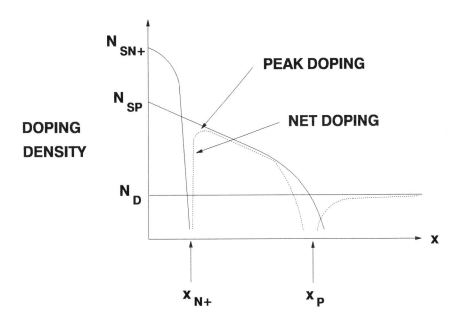

Fig. 7.6 Typical doping profile for the power MOSFET.

A typical diffusion profile for the device is shown in Fig. 7.6. The solid lines indicate the dopant distributions while the dashed lines show the resulting carrier concentration profiles which differ from the dopant profiles due to compensation effects. The surface concentration (N_{SP}) of the P-base diffusion and the N^+ emitter depth combine to determine the peak doping (N_{AP}) in the P-base indicated by the arrow in Fig. 7.6. The peak P-base doping is an important parameter for the doping profile because it controls the threshold voltage of the power MOSFET, i.e., the minimum gate voltage required to induce a surface channel. For a typical threshold voltage of 2 to 3 volts for an n-channel power MOSFET, the peak base concentration (N_{AP}) is about 1×10^{17} per cm^3.

The channel length is another important design parameter in power MOSFETs. It has a strong influence on the on-resistance and the transconductance. The channel length is determined by the difference in the depths of the P-base and N^+ emitter diffusions, i.e., ($x_P - x_{N+}$). (It should be noted that, although a vertical doping profile is measured to characterize devices, the lateral profile under the gate determines the channel properties.) The need to maintain a low peak base concentration and a narrow base width to achieve good on-state characteristics can adversely impact the forward blocking capability. Despite the shorting of the N^+ emitter to the P-base by the source metal, a parasitic N^+-P-N bipolar transistor exists in the power MOSFET. When the P-base/N-drift layer junction is reverse biased, the depletion layer in the P-base can extend to the N^+ emitter/P-base junction and cause premature reach-through breakdown. It is important to design the P-base diffusion profile so that sufficient charge is resident in the P-base to prevent reach-through of the depletion layer to the N^+ emitter.

The depletion width extension on the diffused side of a P-N junction has been calculated at breakdown using numerical techniques. In Fig. 7.7, the change in the depletion width on the

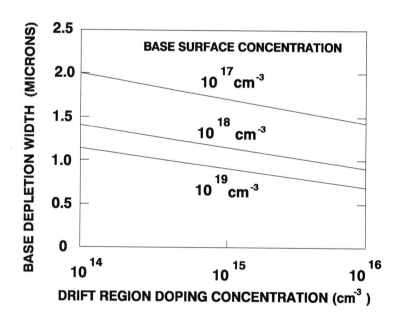

Fig. 7.7 Depletion layer extension in the P-base region during forward blocking.

diffused side of the junction is provided as a function of the background doping level and surface concentration of the diffusion. As the drift layer concentration decreases, the junction profile becomes more graded causing an increase in the depletion width on the diffused side. It should be noted that when the surface concentration is below 10^{18} per cm^3, the depletion width on the diffused side can extend well over 1 micron especially for low drift layer concentrations. Care must be taken during device design to prevent reach-through breakdown in these cases. From the data in Fig. 7.7, it can also be concluded that the fabrication of devices with channel lengths of less than 1 micron is difficult to achieve unless the device is required to support only low voltages, i.e., for devices with high drift layer concentrations.

7.3.3 Cell Structure.

The breakdown voltage of the power MOSFET can be determined by either breakdown

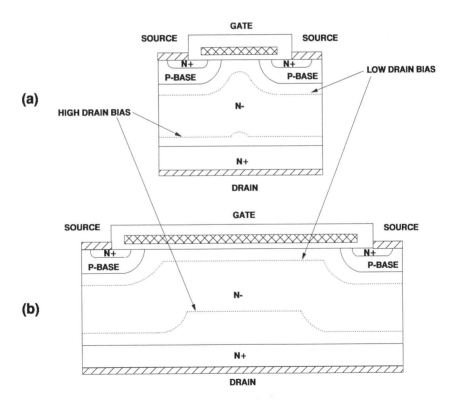

Fig. 7.8 Comparison of depletion layer shape for power MOSFETs with (a) narrow and (b) wide polysilicon gate electrodes.

at the edges of the device or by breakdown within the MOS cell structure. The edge breakdown is determined by the edge termination. Since the on-resistance of the power MOSFET increases very rapidly with increasing breakdown voltage as shown later in this chapter, it is important to use an edge termination that approaches the ideal breakdown voltage. With such terminations, the breakdown voltage can shift to the MOS cell structure. It is therefore important to consider the effect of the MOS cell structure on the breakdown voltage when performing its optimization.

7.3.3.1 DMOS Cell Structure: In order to achieve a large channel width for good on-state characteristics, the power MOSFETs are fabricated with a repetitive pattern of small cells. In the DMOS cell structure shown in Fig. 7.2 the P-base/N-drift layer is brought to the surface between the individual cells. Consequently, each cell contains a planar junction edge. However, the polysilicon gate electrode acts as a field plate that reduces the electric field at the junction edges. During forward blocking the depletion layer extends out from the junction as well as from the gate overlap region between the cells. The depletion layer profile at low and high drain voltages is shown for two examples of a very small cell spacing and a large cell spacing in Fig. 7.8. In the case of the DMOS design with small cell spacing, the depletion layer curvature becomes small at high drain voltages. These devices will breakdown at the edge termination. However, if the cell spacing is large, the curvature of the depletion layer at the edges of the P-base regions is significant resulting in lower breakdown voltage as was discussed for the case of cylindrical or spherical junctions in Chapter 3.

Numerical analysis of the effect of cell spacing upon the breakdown voltage has been performed for the DMOS structure. As an example, the calculated breakdown voltage is shown

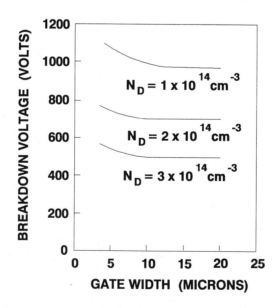

Fig. 7.9 Effect of polysilicon gate width on the breakdown voltage of the DMOS cell.

as a function of the gate width in Fig. 7.9 for several drift region doping levels. It can be seen that the breakdown voltage increases when the cell spacing decreases below 10 microns. Note that the rise in the breakdown voltage occurs at larger cell spacing for the case of lower drift layer doping concentrations. This is consistent with the larger depletion layer spreading at lower doping levels which tends to alleviate the junction curvature. In general a 15 percent increase in the breakdown voltage can be obtained by bringing the cells close together. However, the cells cannot be placed arbitrarily close together because of an increase in the on-resistance as discussed later in this chapter.

7.3.3.2 V-Groove Corner: In the case of the VMOS structure shown in Fig. 7.1 the depletion layer from the P-base/N-drift layer junction expands into the drift region from either side of the V-groove. The depletion layer in the N-drift layer extends well beyond the tip of the V-groove

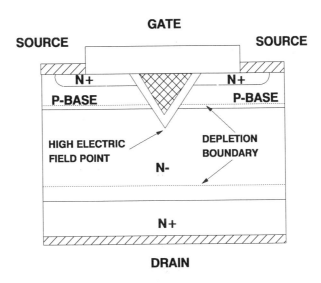

Fig. 7.10 High electric field created at VMOS tip.

as illustrated in Fig. 7.10. A higher electric field is created at the vicinity of the sharp tip of the V-groove. The higher local electric fields at the tip of the groove causes avalanche breakdown along a path near the tip, prior to avalanche breakdown in the parallel plane portion of the junction.

In order to alleviate the increase in electric field at the sharp pointed end of the V-groove, a truncated V-groove structure has been developed. The truncation of the groove is achieved by using a wider window during etching of the groove and stopping the etch before a sharp point is created. This creates a wide drain overlap region and eliminates the sharply pointed end of the groove. In spite of this, a discontinuity in the semiconductor surface still remains. The impact of this discontinuity upon the breakdown voltage has been analyzed by numerical techniques.. As the drain overlap distance, i.e., the distance over which the gate lies over the drift region, gets shorter the structure approaches the V-groove case and the breakdown voltage decreases. In comparison, the breakdown voltage of the DMOS structure increases slightly as the drain overlap gets smaller. These results apply when the gate electrode completely extends between the adjacent cells over the entire drain overlap region.

When no overlap of the gate electrode is provided, i.e., if the gate extends only to the edge of the junction between the P-base and N-drift regions, the breakdown voltage is found to rapidly decrease with increasing spacing between the cells for both the DMOS and VMOS structures. For this reason, it is advisable to let the gate electrode extend completely between the cells during power MOSFET design. An exception to this rule occurs when designing for maximizing the frequency response. The overlap of the gate electrode over the drain region adds significantly to the input capacitance which degrades the frequency response. By minimizing the overlap of the gate over the drain and keeping the separation between cells small, it is possible to achieve a reasonable compromise between achieving high breakdown voltage and high frequency response. Another approach to achieving this goal is to use a thicker oxide in the region between the cells.

The above results of numerical analysis of the breakdown of V-MOS structures indicates that these devices will exhibit a lower avalanche breakdown voltage when compared with DMOS devices for the same epitaxial layer doping level and thickness. This has serious consequences upon the minimum on-resistance achievable by using the VMOS structure. The higher specific on-resistance for a given avalanche breakdown voltage for the VMOS structure has resulted in most manufacturers converting to the DMOS structure for commercial fabrication of power MOSFETs.

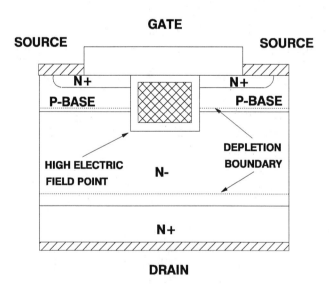

Fig. 7.11 High electric field in the power UMOSFET structure.

7.3.3.3 UMOS Structure:
In the case of the UMOS structure shown in Fig. 7.3, a high electric field can occur at the corners of the U-groove because it extends below the edge of the P-base/N-drift junction. In order to turn-on the UMOSFET, the channel must extend from the N^+ source to the N- drift region. This is achieved when the trench depth exceeds the depth of the P-base region. However, if the trench depth is much greater than the P-base/N- drift region junction, the electric field at the corners of the trench increases resulting in a reduction in the cell breakdown voltage as indicated in Fig. 7.11. In addition, the electric field increases when the spacing between the trench region is increased. For these reasons, it is preferable to keep the trench depth to just below the P-base depth and to make the spacing between the trenches as small as possible within the constraints of the process technology. Fortunately, this is consistent with minimizing the on-resistance of the power MOSFET as discussed in the next section.

7.4 FORWARD CONDUCTION CHARACTERISTICS

Current flow in a power MOSFET during forward conduction is achieved by the

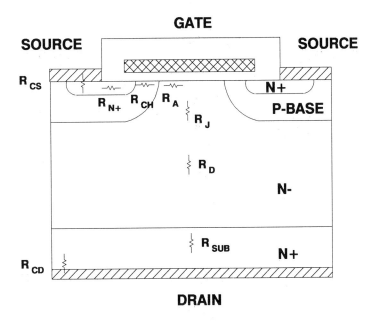

Fig. 7.12 Components of resistance within the DMOSFET structure during on-state current flow.

application of a positive gate bias voltage for n-channel devices in order to create a conductive path across the P-base region underneath the gate. The current flow is limited by the total resistance between the source and drain. This resistance consists of many components, as illustrated in the DMOS cross-section in Fig. 7.12, which combine to determine the on-state voltage drop when the device is carrying current. The resistances of the N$^+$ emitter (R_{N+}) and substrate (R_S) regions are generally negligible for high voltage power MOSFETs that have high drift region resistance. They become quite important when the drift and channel resistance become small as in the case of low ($<$ 100 volts) breakdown voltage devices. The channel (R_{CH}) and accumulation layer (R_A) resistances are determined by the conductivity of the thin surface layer induced by the gate bias. These resistances are a function of the charge in the surface layer and the electron mobility near the surface. The analysis of the charge at the surface will be treated in this section. The electron mobility in surface inversion and accumulation layers was treated in Chapter 2.

In addition to these resistances, the drift layer contributes two more components to the total on-resistance. The portion of the drift region that comes to the upper surface between the cells contributes a resistance R_J that is enhanced at higher drain voltages due to the pinch-off action of depletion layers extending from adjacent P-base regions. This phenomenon has been termed the *JFET action*. Finally, the main body of the drift region contributes a large series resistance (R_D) especially for high voltage devices. The analysis of each of these components of the on-resistance is provided in this section.

7.4.1 MOS Surface Physics

In this section, the charge created at the surface of the semiconductor due to the gate bias applied to the metal-oxide semiconductor structure is analyzed. For this analysis, a one dimensional structure is assumed with no electric field applied parallel to the surface. To perform the analysis, a P-type semiconductor region is assumed. The treatment is applicable to the analysis of current transport in the n-channel power MOSFET. A similar analysis can be performed for an N-type semiconductor region by appropriate changes in the polarity of the voltage, electric field, and free carrier charge. In the analysis presented here, the oxide layer is assumed to be a perfect insulator that does not allow the transport of any charge between the gate and the semiconductor.

7.4.1.1 Flatband: The energy band diagram for an ideal MOS structure with a P-type semiconductor when no potential is applied to the metal electrode is shown in Fig. 7.13. Here, an ideal MOS structure is defined as one that satisfies the following conditions: (a) the insulator has infinite resistivity, (b) charge can exist only in the semiconductor and on the metal electrode, and (c) there is no energy difference between the work function of the metal and the semiconductor. Under these conditions, there is no band bending in the absence of a gate bias and the band structure assumes the form shown in the figure. This energy band diagram is known as the *flatband condition*. In this figure, Φ_M is the metal work function, χ and χ_o are the semiconductor and oxide electron affinities, Φ_B is the barrier height between the metal and the oxide, and Ψ_B is the potential difference between the intrinsic and Fermi levels in the

Chapter 7 : POWER MOSFET

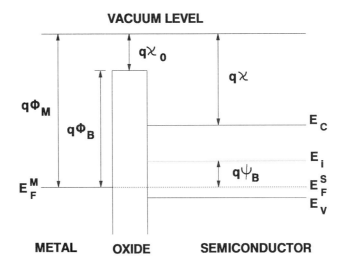

Fig. 7.13 MOS structure under flat band conditions.

semiconductor. Under flatband conditions:

$$q\Phi_m = q\chi + \frac{E_g}{2} + q\Psi_B = q\phi_B + q\chi_o \qquad (7.2)$$

7.4.1.2 Accumulation: When a negative bias is applied to the metal electrode, holes are attracted to the surface and the valency band is bent closer to the Fermi level as shown in Fig. 7.14. This condition of excess majority carriers at the surface is referred to as *accumulation*. Since the semiconductor charge is located at the oxide interface, all the applied bias is supported across the oxide. The charge in the semiconductor then increases in proportion to the applied bias on the metal electrode. This condition is applicable to the drift region below the gate electrode in the power MOSFET structure.

7.4.1.3 Depletion: In contrast, when a positive bias is applied to the metal electrode, the holes are repelled from the surface. At small negative biases, this results in the bending of the valence band away from the Fermi level resulting in the formation of a surface depletion layer as shown in Fig. 7.15. The applied bias is now supported partially in the semiconductor depletion layer and partly across the oxide. The width of the depletion layer depends up on the doping concentration. and the surface potential, i.e. the semiconductor potential below the oxide. This condition is applicable to the P-base region of the power MOSFET when the gate bias is below the threshold voltage. Since there is no mobile charge in the semiconductor under these conditions, no surface conduction can be induced.

7.4.1.4 Inversion: At larger negative biases on the metal, the band bending increases and the

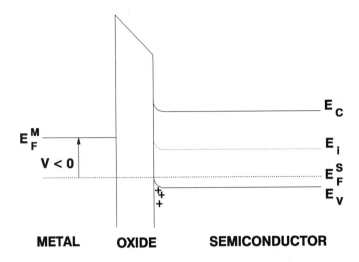

Fig. 7.14 Energy band diagram for the MOS structure with negative gate bias to form an accumulation layer.

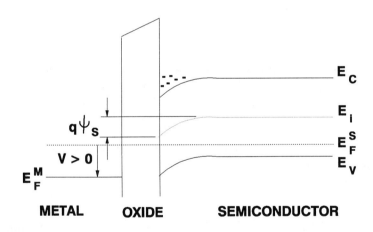

Fig. 7.15 Energy band diagram for the MOS structure with positive gate bias to form a depletion layer in the semiconductor.

intrinsic level crosses the Fermi level as shown in Fig. 7.16. At this point, the semiconductor surface becomes N-type while the bulk substrate is P-type. Since the surface of the semiconductor contains free electrons, it is now possible to utilize this layer to induce a current conduction path along the surface. When the Fermi level at the surface lies close to the intrinsic

Chapter 7 : POWER MOSFET

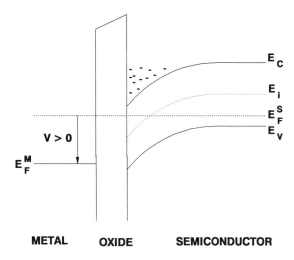

Fig. 7.16 Energy band diagram for MOS structure with positive bias applied to gate to create an inversion layer.

level, the density of electrons is relatively small. This condition is referred to as *weak inversion*. However, the density of electrons at the surface exceeds the density of holes (majority carriers) in the bulk when the band bending is large. This condition is referred to as *strong inversion*.

7.4.1.5 Surface Charge: The charge in the inversion layer plays a key role in determining current transport in MOSFET devices. In order to determine this charge, the potential distribution [$\psi(x)$] in the semiconductor must be first determined. The electron and hole concentrations can then be obtained from:

$$n_p = n_{p0} \, e^{q\Psi/kT} \tag{7.3}$$

and

$$p_p = p_{p0} \, e^{-(q\Psi/kT)} \tag{7.4}$$

where n_{p0} and p_{p0} are the equilibrium electron and hole concentrations in the bulk. The potential distribution $\psi(x)$ can be derived by solving Poisson's equation:

$$\frac{d^2\Psi}{dx^2} = -\frac{\rho(x)}{\epsilon_s} \tag{7.5}$$

with the charge density given by:

$$\rho(x) = q(N_D^+ - N_A^- + p_P - n_P) \tag{7.6}$$

with the boundary condition that charge neutrality must exist in the bulk of the semiconductor far from the surface. In the bulk, charge neutrality requires

$$(N_D^+ - N_A^-) = (n_{PO} - p_{PO}) \tag{7.7}$$

Using Eq. (7.3), (7.4) and (7.5) in Eq. (7.6), Poisson's equation can be rewritten as:

$$\frac{d^2\Psi}{dx^2} = -\frac{q}{\epsilon_s}\left[p_{PO}(e^{-(q\Psi/kT)} - 1) - n_{PO}(e^{(q\Psi/kT)} - 1)\right] \tag{7.8}$$

Integrating this equation from the bulk toward the surface gives the electric field distribution:

$$E(x) = -\frac{d\Psi}{dx} = \frac{\sqrt{2}\,kT}{q\,L_D} F\left(\frac{q\Psi}{kT}, \frac{n_{PO}}{p_{PO}}\right) \tag{7.9}$$

where

$$L_D = \sqrt{\frac{kT\,\epsilon_s}{q^2\,p_{PO}}} \tag{7.10}$$

is called the *extrinsic Debye length for holes* and where

$$F\left(\frac{q\Psi}{kT}, \frac{n_{PO}}{p_{PO}}\right) = \left\{\left[e^{-(q\Psi/kT)} + \left(\frac{q\Psi}{kT}\right) - 1\right] + \frac{n_{PO}}{p_{PO}}\left[e^{(q\Psi/kT)} - \left(\frac{q\Psi}{kT}\right) - 1\right]\right\}^{1/2} \tag{7.11}$$

The space charge required per unit area to create this electric field can be obtained from Gauss's law:

$$Q_S = -\epsilon_s E_s \tag{7.12}$$

where E_s is the surface electric field. If the surface potential is called ψ_s, then the surface charge can be obtained from Eq. (7.9) and (7.12):

$$Q_S = \frac{\sqrt{2}\,\epsilon_s\,kT}{q\,L_D} F\left(\frac{q\Psi_s}{kT}, \frac{n_{PO}}{p_{PO}}\right) \tag{7.13}$$

The variation of the surface charge with shift in the surface potential is shown in Fig. 7.17. For negative values of the surface potential, the bands are bent upwards near the surface as shown in Fig. 7.14, creating a positively charged surface accumulation layer. In this regime of operation, the charge in the accumulation layer varies exponentially with the surface potential because the first term on the right-hand side of Eq. (7.11) becomes dominant:

$$Q_S = \frac{\sqrt{2}\,\epsilon_s\,kT}{q\,L_D}\,e^{(q|\Psi_s|/2kT)} \tag{7.14}$$

where ψ_S is the magnitude of the surface potential. As the magnitude of ψ_S decreases the accumulation layer charge decreases exponentially until it becomes equal to zero at the flat-band condition shown in Fig. 7.13. Note that under accumulation conditions, the electric field is essentially confined to the oxide and the entire gate voltage is supported by the oxide. The

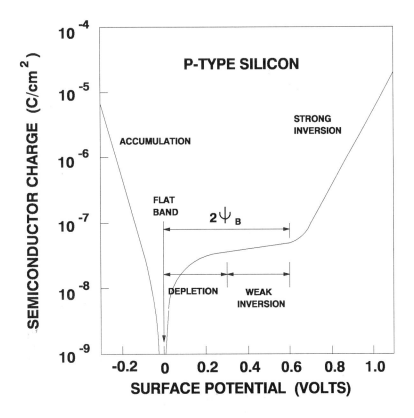

Fig. 7.17 Variation in the surface charge with change in surface potential.

variation of the accumulation layer charge with surface potential is shown in Fig. 7.17.

When the surface potential increases from zero in the positive direction, the bands are bent downwards as illustrated in Fig. 7.15. As long as the surface potential ψ_S remains below the bulk value (ψ_B), the intrinsic level E_i does not cross the Fermi level E_F. The surface is then still P-type and contains a depletion layer. The negative charge in the depletion layer can also be calculated from Eq. (7.13). In this case, the second term of the function F becomes dominant and the surface charge is given by:

$$Q_S = \frac{\epsilon_s}{L_D}\sqrt{2\frac{kT}{q}\Psi_s} \qquad (7.15)$$

Using Eq. (7.10) for L_D,:

$$Q_S = \sqrt{2\,\epsilon_s\, q\, p_{PO}\, \Psi_s} \qquad (7.16)$$

This expression indicates that the surface charge increases gradually as the square root of the surface potential, as illustrated in Fig. 7.17.

When the surface potential (ψ_s) exceeds the bulk value (ψ_B), the intrinsic level E_i crosses the Fermi level. At this point, a mobile negative charge begins to form at the surface. The concentration of this charge is small as long as ψ_s is less than $2\psi_B$. This regime is called *weak inversion*. In this region of operation, the surface charge continues to rise as the square root of the surface potential. When the surface potential exceeds $2\psi_B$, the surface charge begins to rise rapidly as shown on the right-hand side of Fig. 7.17. This occurs because the fourth term in Eq. (7.11) for the function F now becomes dominant. This regime of operation is known as *strong inversion*. It occurs when:

$$\Psi_s > 2\,\Psi_B = 2\,\frac{kT}{q}\ln\left(\frac{N_A}{n_i}\right) \qquad (7.17)$$

The surface charge under strong inversion can be derived from Eq. (7.13) by assuming that the fourth term in Eq. (7.11) is dominant:

$$Q_S = \sqrt{2\,\frac{n_{PO}}{p_{PO}}\,\frac{\epsilon_s\, kT}{q\, L_D}}\, e^{q\Psi_s/2kT} \qquad (7.18)$$

Substituting for L_D from Eq. (7.10) gives:

$$Q_S = \sqrt{2\,\epsilon_s\, n_{PO}}\, e^{q\Psi_s/2kT} \qquad (7.19)$$

In the strong inversion region, the surface charge again increases exponentially with surface potential. This charge is an important quantity because it determines the conductivity of the channel in power MOSFETs.

In the depletion region of operation, the applied gate voltage is shared across the gate oxide and the semiconductor depletion layer. With the onset of strong inversion, any further increase in the gate voltage is dropped across the oxide and the voltage across the semiconductor depletion layer remains essentially constant. As a consequence, the depletion layer width in the semiconductor expands until strong inversion occurs and then remains constant with further increase in gate voltage. The maximum depletion layer width is approximately given by:

$$W_m = \sqrt{\frac{2\,\epsilon_s}{q\,N_A}\,(2\,\Psi_B)} \qquad (7.20)$$

Using Eq. (7.17) for (ψ_B) gives :

$$W_m = \sqrt{\frac{4\,\epsilon_s\,k\,T}{q^2\,N_A}\,\ln\!\left(\frac{N_A}{n_i}\right)} \qquad (7.21)$$

A plot of the variation in the maximum depletion width with background doping concentration

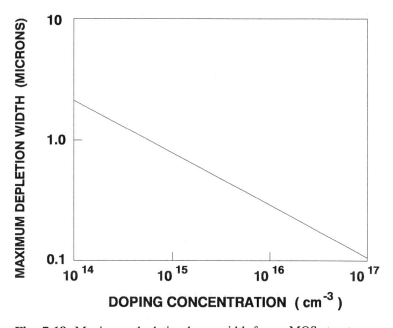

Fig. 7.18 Maximum depletion layer width for an MOS structure.

is provided in Fig. 7.18. It is worth pointing out that the maximum depletion layer width for the MOS structure is considerably smaller than the maximum depletion layer width in the semiconductor determined by avalanche breakdown.

7.4.2 Threshold Voltage

The voltage on the gate electrode at which strong inversion begins to occur in the MOS structure is an important design parameter for power MOSFETs because it determines the minimum gate bias required to induce an N-type conductance in the channel. This voltage is called the *threshold voltage*. For proper device operation, its value cannot be too large or too small. If the threshold voltage is large, a high gate bias voltage will be needed to turn-on the power MOSFET. This imposes problems with the design of the gate drive circuitry. It is also important that the threshold voltage not be too low. Due to the existence of charge in the gate oxide, it is possible for the threshold voltage to be negative for n-channel power MOSFETs. This is an unacceptable condition because a conductive channel will now exist at zero gate bias voltage, i.e., the devices will exhibit normally-on characteristics. Even if the threshold voltage is above zero for an n-channel power MOSFET, its value should not be too low because the device can then be inadvertently triggered into conduction either by noise signals at the gate terminal or by the gate voltage being pulled up during high speed switching. Typical power MOSFET threshold voltages are designed to range between 2 and 3 volts.

In the absence of any difference in the work function of the metal and the semiconductor, the applied gate voltage and the semiconductor surface potential are related by:

$$V_G = V_{ox} + \Psi_s \tag{7.22}$$

where V_{ox} is the voltage across the oxide. The voltage across the oxide can be related to the surface charge by:

$$V_{ox} = E_{ox} t_{ox} = \frac{Q_S}{\epsilon_{ox}} t_{ox} = \frac{Q_S}{C_{ox}} \tag{7.23}$$

At the point of transition into the strong inversion condition, the gate bias voltage is equal to the threshold voltage:

$$V_T = \frac{Q_S}{C_{ox}} + 2\Psi_B \tag{7.24}$$

Using Eq. (7.16) for Q_s at the point of transition into strong inversion:

$$V_T = \frac{\sqrt{4 \epsilon_s k T N_A \ln(N_A/n_i)}}{(\epsilon_{ox}/t_{ox})} + \frac{2kT}{q} \ln\left(\frac{N_A}{n_i}\right) \tag{7.25}$$

From this expression, it can be concluded that the threshold voltage will increase linearly with gate oxide thickness and approximately as the square root of the semiconductor doping. The calculated threshold voltage for silicon MOS structures with gate oxide thicknesses ranging from 100 °A to 10,000 °A are provided in Fig. 7.19 for doping levels ranging from 10^{14} to 10^{18} per cm^3. These curves apply to a uniformly doped semiconductor and do not take into account the presence of any charge in the oxide.

In actual metal-oxide-semiconductor structures, the threshold voltage is altered due to several factors.: (a) an unequal work function for the metal and the semiconductor. If barrier height between silicon-dioxide and the metal is ϕ_B, the difference between the metal and the

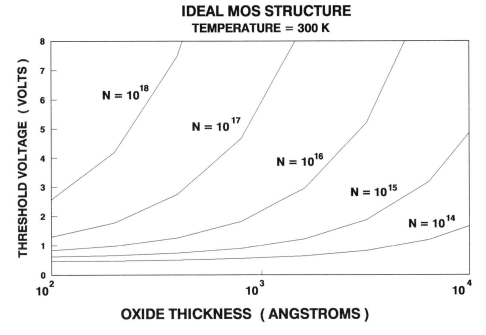

Fig. 7.19 Dependence of threshold voltage on oxide thickness and doping concentration.

semiconductor work function can be obtained as:

$$q\,\Phi_{ms} = q\,\Phi_B + q\,\chi_0 - \left(q\,\chi + \frac{E_g}{2} + q\,\Psi_B\right) \quad (7.26)$$

(b) the presence of fixed surface charge (Q_{fc}) at the oxide-silicon interface; (c) the presence of mobile ions in the oxide with charge Q_I; and (d) the presence of charged surface states at the oxide-silicon interface with charge Q_{SS}. All of these charges cause a shift in the threshold voltage:

$$V_T = \Phi_{ms} + \frac{Q_S}{C_{ox}} + 2\,\Psi_B - \left(\frac{Q_{SS} + Q_I + Q_{FC}}{C_{ox}}\right) \quad (7.27)$$

The fixed surface charge Q_{fc} is located within 20 °A of the oxide-silicon interface and cannot be charged or discharged over a wide range of surface potentials such as those normally encountered in device operation. The density of this charge is dependent upon the oxide growth conditions and its subsequent heat treatment. The presence of excess silicon or a deficiency of oxygen has been postulated as the origin of the fixed charge but the influence of impurities cannot be excluded.

The presence of mobile ions (charge Q_I) has been related to the shift in the threshold voltage under bias stressing at elevated temperatures or under long periods. Since the mobility of alkali ions decreases with increasing atomic weight, sodium ions are the most prominent mobile species in the oxide. The concentration of these ions is dependent upon the cleanliness of the wafers during processing, and the purity of the gate metal.

Interface states (charge Q_{SS}) are energy levels in the silicon band gap that are located close to the surface which can charge and discharge with changes in the surface potential. The origin of interface states on a clean free surface in vacuum has been shown to arise from unsaturated bonds at the surface. The interface state density at the silicon- silicon dioxide interface is usually several orders of magnitude lower (10^{10} to 10^{12} per cm^2) than on the free surface. The origin of these surface states is not completely understood but the effect of processing upon their density has been extensively studied. This will be discussed in the section on device fabrication.

The dependence of the threshold voltage of an aluminum gate MOS capacitor fabricated with 1,000 °A gate oxide layer grown on P-type silicon is provided in Fig. 7.20 as a function

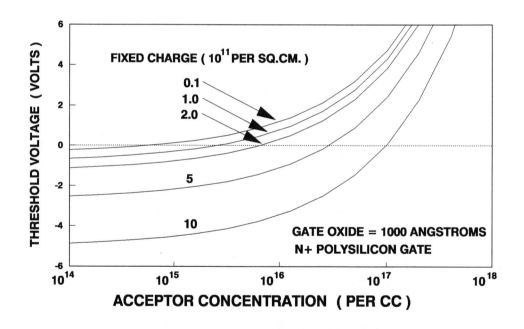

Fig. 7.20 Threshold voltage for the case of a polysilicon gate MOS capacitor with 1000 angstrom gate oxide.

of the doping level. For a typical peak channel doping on the order of 5×10^{16} atoms per cc and a charge density of 10^{11} per cm^2, a threshold voltage of 2 to 3 volts can be obtained with an oxide thickness of 1,000 °A. This threshold voltage is convenient for controlling the device with integrated drive circuits.

It is important that channels not form in regions of the device which are not in the main current flow path. An example is the region under the gate bonding pad. It is typical to place this bonding pad over a thick field oxide. The threshold voltage for inverting the surface for

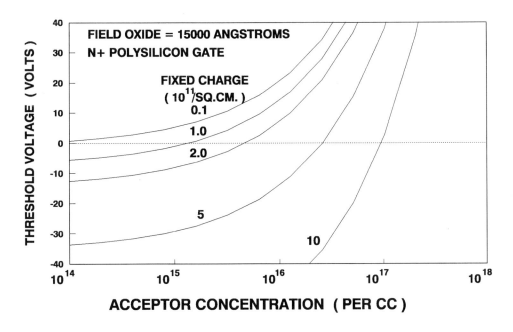

Fig. 7.21 Threshold voltage for an MOS capacitor with polysilicon gate over 15000 field oxide.

an oxide thickness of 15,000 °A is shown in Fig. 7.21. The 15,000 °A oxide thickness represents the thick field oxide in the inactive portions of the device where the influence of the gate bias must be minimized. In these figures, curves are shown for various values of the charge in the oxide which may arise from either fixed positive interface charge or positive ions at the oxide-silicon interface or from positively charged surface states. For this doping level and interface charge, the threshold voltage of the field oxide is over 30 volts which provides sufficient margin for over-driving the gate without inverting the regions outside the device active area.

As indicated by Eq. (7.27), the threshold voltage for inversion in an MOS capacitor is affected by the barrier height between the metal and the oxide. The threshold voltage of a device can be tailored to only a very limited extent by choosing the proper metal, and it is not common to form a metal gate structure. A greater control over the threshold voltage can be achieved by substituting polycrystalline silicon for the metal. By doping the polycrystalline silicon either N- or P-type, the work function difference in the polysilicon-SiO$_2$-Si system can be varied. Measurements have shown that the Fermi level in heavily doped polycrystalline silicon corresponds exactly to that in monocrystalline silicon. Based upon this, the work function difference between N-type polysilicon and P-type monocrystalline silicon is given by the expression :

$$q \, \Phi_{PS-Si} = -\frac{kT}{q} \ln\left(\frac{N_{DPS} \, N_A}{n_i^2}\right) \quad (7.28)$$

where N_A is the substrate doping concentration and N_{DPS} is the donor concentration in the polycrystalline silicon.

Most power MOSFETs are fabricated by using heavily doped polycrystalline silicon as the gate electrode. The use of polycrystalline silicon as the gate electrode provides a refractory gate material that is compatible with silicon device processing and allows the fabrication of large area devices with a two level electrode process as illustrated in Fig. 7.2. The conductivity of polycrystalline silicon is an order of magnitude lower than for metals. In the case of power MOSFETs designed for operation at very high frequencies, the RC gate charging time constant becomes too large for polycrystalline silicon gate electrodes. For these devices, aluminum has been used as the gate electrode, with source and gate metal being interdigitated on the wafer surface. With the development of refractory gate silicide or polycide technology for integrated circuits, this technology has been applied to the fabrication of power MOSFETs for very high frequency applications, such as switch mode power supplies with two level electrode structures. It is worth pointing out that the polycide gate electrode can be treated as a polysilicon electrode for calculation of the threshold voltage because the polysilicon region of the polycide gate is at the oxide interface.

7.4.3 Channel Resistance

Consider a lateral MOS field effect transistor having a P-type base region with N$^+$ source and drain located on either side as shown in Fig. 7.22. The current flow between source and drain is controlled by the electron charge available for transport in the surface inversion layer of the P-base as well as the surface mobility of these electrons. According to the MOS analysis, when the surface potential (ψ_S) exceeds twice the bulk potential (ψ_B), a strong inversion layer begins to form. Since the band bending is small beyond this point, the inversion layer charge available for current conduction is given by:

$$Q_n = C_{ox} \, (V_G - V_T) \quad (7.29)$$

Thus, the channel resistance at low drain voltages, where the voltage drop along the channel is negligible, is given by :

$$R_{ch} = \frac{L}{Z \, \mu_{ns} \, C_{ox} \, (V_G - V_T)} \quad (7.30)$$

Chapter 7 : POWER MOSFET 363

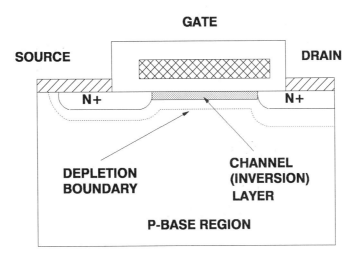

Fig. 7.22 Lateral MOSFET structure used for analysis of channel resistance.

where Z and L are the width and length of the channel and μ_{ns} is the surface mobility of electrons. The surface mobility, and its dependence upon doping, temperature and surface orientation, is discussed extensively in Chapter 2 because of its importance in determining the conductance of the channel in the MOS field effect transistor.

As the drain current increases, the voltage drop along the channel between drain and

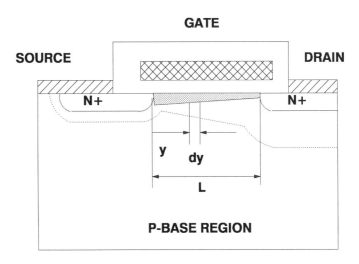

Fig. 7.23 Channel charge distribution in a lateral MOSFET under drain current flow.

source becomes significant. The positive drain potential opposes the gate bias voltage and reduces the surface potential in the channel. The channel charge near the drain end is then reduced by the voltage drop along the channel as illustrated in Fig. 7.23. In this figure, the coordinates used for the analysis of the voltage drop across the MOSFET are also indicated.

When the drain voltage becomes equal to $(V_G - V_T)$, the charge in the channel at the drain end becomes equal to zero. This condition, called *channel pinch-off*, is illustrated in Fig 7.24.

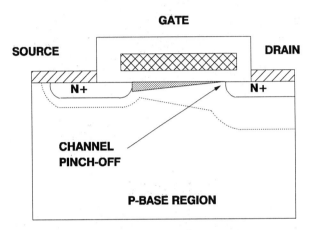

Fig. 7.24 Channel charge distribution in a lateral MOSFET under pinch-off conditions.

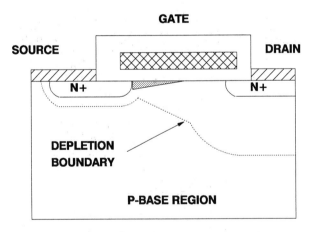

Fig. 7.25 Lateral MOSFET biased beyond the pinch-off point.

At this point, the drain current saturates. Further increase in drain voltage results in no further increase in drain current. The drain voltage is now supported across an extension of the depletion layer under the gate as shown by Fig. 7.25, which results in a reduction in the effective channel length.

The channel resistance can be derived as a function of the gate and drain voltages under the following assumptions: (a) the gate structure is an ideal MOS structure as discussed earlier, (b) the free carrier mobility is a constant independent of the electric field strength, (c) the base region is uniformly doped, (d) the current transport occurs exclusively by drift, (e) the leakage current is negligible, and (f) the longitudinal electric field along the surface is small compared with the transverse electric field resulting from the gate bias. The last assumption is referred to as the *gradual channel approximation*.

The resistance (dR) of an elemental segment (dy) of the channel is dependent upon the inversion layer charge per unit area and the mobility of the free carriers:

$$dR = \frac{dy}{Z \, \mu_{ns} \, Q_n(y)} \qquad (7.31)$$

The charge in the inversion layer not only depends upon the gate voltage but also depends upon the drain current because of the potential drop along the channel:

$$Q_n(y) = C_{ox} [V_G - V_T - V(y)] \qquad (7.32)$$

where V(y) is the voltage drop along the channel. The voltage drop in segment dy is then given by:

$$dV = I_D \, dR \qquad (7.33)$$

From these expressions,

$$\int_o^L I_D \, dy = -Z \, \mu_{ns} \, C_{ox} \int_0^{V_D} (V_G - V_T - V) \, dV \qquad (7.34)$$

Using the condition that the drain current (I_D) must remain constant throughout the channel allows the integrations to be performed:

$$I_D = \frac{\mu_{ns} \, C_{ox} \, Z}{2 \, L} [2(V_G - V_T) V_D - V_D^2] \qquad (7.35)$$

According to this equation, when the drain voltage is small, the drain current increases linearly with drain voltage:

$$I_D = \mu_{ns} \, C_{ox} \, \frac{Z}{L} (V_G - V_T) V_D \qquad (7.36)$$

In this linear region, the channel resistance is given by:

$$R_{CH} = \frac{L}{Z \mu_{ns} C_{ox} (V_G - V_T)} \qquad (7.37)$$

which agrees with the expression given in Eq. (7.30) representing a homogeneous inversion layer extending between drain and source. In this linear regime, the transconductance can be obtained by differentiating Eq. (7.36) with respect to the gate voltage:

$$g_m = \frac{dI_D}{dV_G} = \mu_{ns} C_{ox} \frac{Z}{L} V_D \qquad (7.38)$$

As the drain voltage and current increase, the second term in Eq. (7.35) becomes increasingly important and causes the drain current to saturate. Physically, this corresponds to the reduction in the channel inversion layer charge near the drain with increasing drain voltage. Ultimately, the inversion layer charge at the drain end of the channel becomes zero and the drain current saturates. The drain voltage at which current saturation occurs is the channel pinch-off point:

$$V_{DS} = (V_G - V_T) \qquad (7.39)$$

Substituting for V_{DS} in Eq. (7.35), the saturated drain current is obtained:

$$I_{DS} = \frac{\mu_{ns} C_{ox} Z}{2 L} (V_G - V_T)^2 \qquad (7.40)$$

The saturated drain current is an important parameter because it determines the maximum current that the channel will support. In actual devices, the saturated drain current will be lower than that indicated by Eq. (7.40) because the surface mobility (μ_{ns}) is a function of the longitudinal electric field.

The transconductance of the device in the saturated current region of operation can be obtained by differentiating Eq. (7.40) with respect to the gate voltage:

$$g_{ms} = \frac{dI_D}{dV_G} = \mu_{ns} C_{ox} \frac{Z}{L} (V_G - V_T) \qquad (7.41)$$

The basic features of the current-voltage characteristics of a MOSFET as described by Eq. (7.35) are shown in Fig. 7.26.

According to the above analysis, the drain current will saturate above a certain drain voltage. In actual practice, the drain current does not remain constant in the saturation region because when the drain voltage increases beyond V_{DS}, the length of the channel is reduced as illustrated in Fig. 7.25. This decreases the effective channel length and produces a finite drain output conductance. For devices with higher P-base doping levels, the channel shrinkage (ΔL) can be approximated by treating the gate-drain depletion region like an N^+/P step junction:

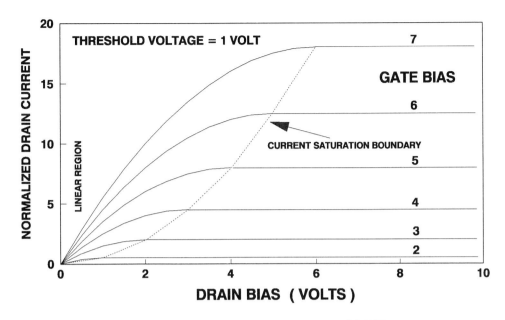

Fig. 7.26 Output characteristics of the lateral MOSFET structure.

$$\Delta L = \sqrt{\frac{2\,\epsilon_s}{q\,N_A}\,(V_D - V_{DS})} \qquad (7.42)$$

The drain output resistance can then be obtained by accounting for the change in channel length with increasing drain voltage. A fairly good agreement between the calculated drain output conductance with this expression and the measured output conductance of n-channel devices has been reported. Better results can only be obtained by resorting to a two dimensional numerical analysis of current transport in the devices.

7.4.4 DMOSFET Specific On-Resistance

The on-resistance of a power MOSFET is the total resistance between the source and drain terminals in the on-state. The on-resistance is an important device parameter because it determines the maximum current rating. The power dissipation in the power MOSFET during current conduction is given by:

$$P_D = I_D\,V_D = I_D^2\,R_{on} \qquad (7.43)$$

Expressed in terms of the chip area (A):

$$\frac{P_D}{A} = J_D^2 \, R_{on,sp} \qquad (7.44)$$

where (P_D/A) is the power dissipation per unit area; J_D is the on-state current density; and $R_{on,sp}$ is the *specific on-resistance*, defined as the on-resistance per unit area. These expressions are based upon the assumption that the power MOSFET is operated in its linear region at a relatively small drain bias during current conduction. The maximum power dissipation per unit area is determined by the maximum allowable junction temperature and the thermal impedance. For typical power packages, the maximum power dissipation per unit area is in the range of 100 watts per cm^2. The maximum operating current density will then vary inversely as the square root of the specific on-resistance.

The specific on-resistance of the power MOSFET is determined by the resistance components illustrated in Fig. 7.12 for the DMOS structure. Thus:

$$R_{on} = R_{N+} + R_{CH} + R_A + R_J + R_D + R_S \qquad (7.45)$$

where R_{N+} is the contribution from the N$^+$ source diffusion, R_{CH} is the channel resistance, R_A is the accumulation layer resistance, R_J is the contribution from the drift region between the P-base regions, R_D is the drift region resistance, and R_S is the substrate resistance. Additional resistances can arise from a non-ideal contact between the source/drain metal and the N$^+$ semiconductor regions as well as from the leads used to connect the device to the package. Each of these contributions is described below.

7.4.4.1 Substrate Resistance: The contribution from the substrate is generally negligible for high voltage power MOSFETs. However, in the case of low voltage devices with breakdown voltages below 50 volts, it can contribute significantly to the on-resistance. This is especially true because the substrate must be thick in order to impart adequate strength to the wafers during device fabrication. It can be assumed that the current density is uniform within the substrate because of rapid current spreading at the drift region interface. The specific resistance contributed by the substrate is then given by:

$$R_{SB,sp} = \rho_{SB} \, t_{SB} \qquad (7.46)$$

where ρ_{SB} is the resistivity of the substrate and t_{SB} is its thickness. In the case of a typical antimony doped substrate with thickness of 20 mils and resistivity of 0.01 ohm-cm, the substrate resistance per unit area is 5 x 10^{-4} ohm-cm^2. This value is comparable to the ideal specific resistance of the drift region for a 50 volt device. The substrate resistance can be reduced by using 0.001 ohm-cm arsenic doped substrates and by lapping the substrate to reduce its thickness after fabrication of the device structure on the top surface.

7.4.4.2 Source Resistance: To analyze the contribution of the N$^+$ source resistance, consider the DMOS cell structure shown in Fig. 7.27. In this cell, 2m is the cell diffusion window and

Chapter 7 : POWER MOSFET

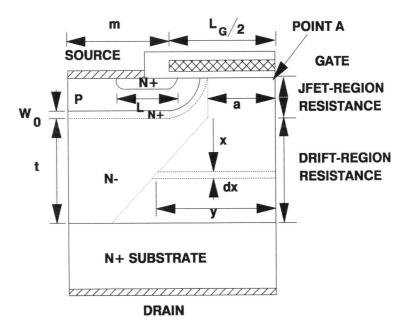

Fig. 7.27 Cross-section of DMOS cell used for analysis of specific on-resistance.

L_G is the length of the gate electrode between the adjacent cells. The cell diffusion window (2m) is determined by the photolithographic design tolerances used for device fabrication. These tolerances also determine the length (L_{N+}) of the N$^+$ source region. If a linear cell is considered with a width Z perpendicular to the cross-section shown in Fig. 7.27, the resistance per cm^2 due to the N$^+$ source region is given by :

$$R_{N+,sp} = \frac{1}{2} \rho_{SN+} L_{N+} (L_G + 2m) \qquad (7.47)$$

where ρ_{SN+} is the sheet resistance of the N$^+$ diffusion. In a typical device, ρ_{SN+} = 10 ohms per square, L_{N+} = 10 microns and the cell repeat spacing (L_G + 2m) = 40 microns. The resulting specific resistance of the N$^+$ emitter regions (2 x 10^{-5} ohm-cm^2) is negligible compared with all other resistances in the structure.

7.4.4.3 Channel Resistance: The channel resistance was discussed in the previous section for the basic MOSFET structure. The contribution from the channel depends upon the ratio (L_{CH}/Z), the gate oxide thickness (via C_{ox}), and the gate drive voltage (V_G). The contribution from the channel can be minimized by making the channel length (L_{CH}) small and keeping its width (Z) large. This requires a high cell density and good control over the P-base and N$^+$ emitter diffusion profiles to keep the channel short without causing reach-through breakdown as discussed earlier in Section 7.3. To calculate the contribution from the channel, consider the

DMOS cell structure shown in Fig. 7.27. The channel resistance per cm² for the linear cell structure is given by:

$$R_{CH,sp} = \frac{L_{CH}(L_G + 2m)}{2\mu_{ns}C_{ox}(V_G - V_T)} \quad (7.48)$$

For a typical device with a channel length (L_{CH}) of 2 microns, a gate oxide thickness of 1000 °A, and a gate drive voltage of 10 volts, the channel resistance per cm² is found to be 2.5×10^{-3} ohm-cm². Note that the channel resistance decreases when the cell repeat spacing is reduced. This occurs because of an increase in the channel density (i.e., the channel width per cm² of active cell area). The channel resistance can also be reduced by decreasing the gate oxide thickness while maintaining the same gate drive voltage.

7.4.4.4 Accumulation Layer Resistance: The resistance of the accumulation layer (R_A) accounts for the current spreading from the channel into the JFET region. The accumulation layer resistance is dependent upon the charge in the accumulation layer and the mobility for free carriers at the accumulated surface. For the linear cell geometry, the accumulation layer resistance per cm² is:

$$R_A = \frac{K(L_G - 2x_P)(L_G + 2m)}{2\mu_{nA}C_{ox}(V_G - V_T)} \quad (7.49)$$

Here, the current flowing through the channel is assumed to flow from the edge of the P-base region to the center of the polysilicon gate (point A) via the accumulation layer. However, in actuality, the current spreads from the channel into the JFET region. The factor K is introduced to account for the two-dimensional nature of the current flow from the channel into the JFET region via the accumulation layer. Good agreement with experimental results has been observed for K = 0.6, which implies that the effective resistance to the drain current flow is 60 percent of the total accumulation region resistance.

For the above example, with L_G = 20 microns and x_P = 3 microns, the accumulation resistance per cm² is found to be 6×10^{-3} ohm-cm². The accumulation layer resistance can be reduced by decreasing the length (L_G) of the gate electrode between the cells. However, this has an adverse effect upon the JFET resistance (R_J).

7.4.4.5 JFET Region Resistance: The resistance of the drift region between the P-base diffusions is referred to as the *JFET resistance* because the current flow resembles that in a junction field effect transistor with the P-base regions acting as the gate regions. This resistance contribution can be calculated easily if the effect of the voltage drop along the vertical direction on the depletion region is neglected. Under the assumption that the current is flowing uniformly down from the accumulation layer into the JFET region, the resistance of the JFET region becomes that of a semiconductor region with a cross-section:

$$A_{JFET} = aZ = \left(\frac{L_G}{2} - x_P\right)Z \quad (7.50)$$

where Z is the width of the cell orthogonal to the cross-section. The JFET contribution to the specific resistance is then given by:

$$R_J = \frac{\rho_D (L_G + 2m)(x_P + W_0)}{(L_G - 2x_P - 2W_0)} \quad (7.51)$$

In the case of high voltage power MOSFETs, the drift region doping must be small to obtain the desired breakdown voltage. The depletion layer extension (W_0) can then be a significant fraction of the gate length (L_G) leading to a large JFET resistance contribution. This problem can be solved by increasing the gate length. However, this leads to a poor channel density and to reduced cell breakdown voltages. It is therefore preferable to increase the doping concentration in the JFET region while maintaining a lower doping concentration in the drift region to obtain the desired breakdown voltage. The maximum doping concentration in the JFET region must be kept below about 5×10^{16} per cm^{-3} to avoid a high local electric field and to prevent significant alteration of the channel doping.

7.4.4.6 Drift Region Resistance: In this analysis, the drift region is assumed to begin below the bottom of the P-base diffusion as indicated in Fig. 7.27. The current spreads from the JFET region into the drift region as shown by the dotted lines in the figure. Many possible models for the current spreading into the drift region can be proposed. One such model that allows a reasonably accurate estimation of the drift region spreading resistance, is based on the current spreading from a cross-section of ($a = L_G - 2x_P$) at a 45 degree angle. The cross-section for the current flow then increases with depth through the drift region. At a depth x below the P-base region, the cross-section for current flow is $y = (a + x)Z$, where Z is the width orthogonal to the cross-section. Consequently, the drift region spreading resistance is given by:

$$R_D = \int_0^t \frac{\rho_D}{(a + x) Z} dx = \frac{\rho_D}{Z} \ln\left(\frac{a + t}{a}\right) \quad (7.52)$$

Using this equation, the specific resistance contribution for the drift region is obtained as:

$$R_{D, sp} = \frac{\rho_D (L_G + 2m)}{2} \ln\left(\frac{a + t}{a}\right) \quad (7.53)$$

It is important to note that even if ideal breakdown voltage is assumed to be achievable at the device edge termination and within the DMOS cell structure, the specific resistance of the drift region is not equal to the ideal specific on-resistance because of the effect of the current spreading from the JFET region into the drift region. This effect can be appreciated by taking the ratio of the drift region specific resistance given in Eq. (7.53) to the ideal specific on-resistance:

$$\frac{R_{D, sp}}{R_{D, ideal}} = \frac{(L_G + 2m)}{2 t} \ln\left(\frac{a + t}{a}\right) \quad (7.54)$$

This expression indicates that the deviation of the drift region specific resistance from the ideal value becomes worse as the cell size becomes large when compared with the drift region thickness. The reason is that the drain current tends to flow only under the gate region, creating a significant dead space under the P-base region. When the cell size is reduced by reducing the diffusion window (2m), the dead space becomes smaller, allowing the drift region contribution to become closer to the ideal value.

The above expression for the drift region specific resistance was derived under the assumption that the 45° current spreading does not lead to any overlap in the current flow paths. In the case of higher breakdown voltage devices with larger drift region thickness, the current flow paths will overlap. The drift region resistance must then be modeled as the sum of a region where the cross-section increases with depth and a second region with uniform cross-section equal to the cell width. This leads to a drift region specific resistance given by:

$$R_{D,sp} = \frac{\rho_D (L_G + 2m)}{2} \ln\left(\frac{L_G + 2m}{L_G - 2x_P - 2W_0}\right) + \rho_D (t - m - x_P - W_0) \tag{7.55}$$

7.4.4.7 Contact Resistance:

It is important to include the effect of a finite contact resistance when analyzing the DMOSFET resistance. This is particularly important because the area of the contact to the source region is a small fraction of the total cell area. This tends to amplify the contact resistance contribution. This does not occur on the drain side because the contact covers the entire back surface of the device. Thus, the drain contact resistance contribution to the specific resistance is given by:

$$R_{CD,sp} = \rho_C \tag{7.56}$$

where ρ_C is the specific contact resistivity.

However, in the case of the source region, the contact area depends upon the area where the source metal and the source N^+ region overlap. This overlap is defined by the edges of the contact window and the mask used to pattern the N^+ diffusion. If the resulting contact area is A_{CS}, the contribution to the specific resistance by the source contact is given by:

$$R_{CS,sp} = \frac{A_{cell}}{A_{CS}} \rho_C \tag{7.57}$$

This expression indicates that if the process design rules result in a small source contact area, it can lead to a large contribution to the specific resistance. As an example, if the area of the source contact is 5 percent of the cell area and the specific contact resistance is 1×10^{-5} ohm-cm^2, the contribution of the source contact resistance to the specific on-resistance will be 2×10^{-4} ohm-cm^2, which is significant when compared with the ideal specific on-resistance. For this reason, methods have been developed to reduce the contact resistivity for the source contact by replacing the aluminum contact with silicides. This is particularly important when scaling down the polysilicon window because it leads to a smaller contact area for the source region.

7.4.5 Ideal Specific On-Resistance

Consider the ideal case where the resistances of the N$^+$ emitter, N$^+$ substrate, n-channel region, accumulation region, and JFET region are negligible. The specific on-resistance of the power MOSFET will then be determined by the drift region alone. In addition, if it is assumed that the current flows uniformly through the drift region without current spreading effects, the resistance of the drift region is referred to as the *ideal specific on-resistance* for the power MOSFET. This is the resistance of a drift region with the doping concentration and thickness required to support the desired breakdown voltage. Using the equations derived for the doping concentration and thickness of the drift region in Chapter 3 on breakdown voltage:

$$R_{on,sp} = \frac{W_D}{q \mu_n N_D} = 5.93 \times 10^{-9} (BV_{PP})^{2.5} \qquad (7.58)$$

for n-channel devices and

$$R_{on,sp} = \frac{W_D}{q \mu_p N_A} = 1.63 \times 10^{-8} (BV_{PP})^{2.5} \qquad (7.59)$$

for p-channel devices. In deriving the numerical values in these equations, it has been assumed that the mobility is that for a relatively low doping concentration. The ideal specific on-resistance for a p-channel MOSFET is higher than that for the n-channel MOSFET due to lower mobility for holes than electrons in silicon. Since the drift region doping concentration is below 1×10^{16} per cm^3, the mobility is relatively independent of the doping concentration. Consequently, the ideal specific on-resistance of the p-channel MOSFET is about 2.5 times larger than that for the n-channel MOSFET.

7.4.6 DMOS Cell Optimization

In the above analysis of the specific on-resistance of the power DMOSFET structure, it was found that the contributions to the on-resistance from various terms are dependent upon the device geometrical design parameters. It was also pointed out that the dominant components of the on-resistance are the channel resistance, the accumulation layer resistance, the JFET region resistance, and the drift region resistance. When the gate length (L_G) is small, the JFET and drift region resistances become large due to the small width (a) through which the current must flow into the channel. At the same time, the accumulation layer resistance becomes small because of the shorter path along the surface and the channel resistances become small because of a reduction in the cell pitch that is equivalent to an increase in the channel density. The opposite trends occur when the gate length is increased. This implies that there is an optimum gate length at which the total specific on-resistance has a minimum value. For this reason, it is necessary to perform a calculation of the specific on-resistance as a function of the polysilicon gate length during the optimization of the design of a power DMOSFET structure.

The impact of changing the cell geometry on the on-resistance can best be illustrated by

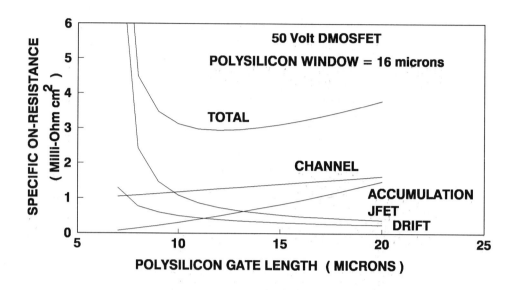

Fig. 7.28 Optimization of gate length for a power DMOSFET structure with breakdown voltage of 50 volts.

considering some specific cases. First, consider the case of a relatively low voltage (50 volt) power DMOSFET structure. For this analysis, it will be assumed that ideal breakdown voltage can be obtained at the edge termination and within the DMOS cell. Under these assumptions, the drift region has a thickness of 2.5 microns and resistivity of 0.4 ohm-cm. It will be assumed that the DMOS process creates a channel length of 2 microns based up on a P-base depth of 3 microns and an N^+ source depth of 1 micron. The gate oxide thickness will be assumed to be 1000 °A with an inversion layer mobility of 500 cm^2/V s and an accumulation layer mobility of 1000 cm^2/V s. The polysilicon window will be assumed to be 16 microns with a N^+ mask of 8 microns and a contact window of 12 microns. These windows define the source contact area. The specific on-resistance of the contacts to the source and drain will be assumed to be 1×10^{-6} ohm-cm^2.

The impact of increasing the polysilicon gate length (L_G) upon the various components of the on-resistance is provided in Fig. 7.28 together with the total specific on-resistance. It can be seen that the channel and accumulation layer resistances increase as the gate length (L_G) increases. Concurrently, the resistances of the JFET and drift regions decreases because of an increase in the cross-sectional area for the current flow. A well defined minimum in the total on-resistance is observed at an optimum gate length of 12 microns. The minimum specific on-resistance for this example is found to be 3 milliohm-cm^2. In comparison, the ideal specific on-resistance for a breakdown voltage of 50 volts is 0.1 milliohm-cm^2. Thus, the device specific on-resistance deviates from the ideal value by a large factor of 30 times. It is worth pointing out that the channel resistance at the optimum gate length is significantly larger than all the other components. This indicates that improvements in the performance of the low breakdown voltage power DMOSFETs can be obtained (a) by increasing the channel density, (b) by reducing the channel length, and (c) by reducing the gate oxide thickness.

It is instructive to perform a similar analysis for optimization of the specific on-resistance of the DMOSFET for a higher breakdown voltage. For this analysis, consider a device designed with a breakdown voltage of 500 volts. As in the 50 volt case, it will be assumed that ideal breakdown voltage is attainable at the edge termination and within the DMOS cell. Then, the drift region must have a thickness of 37 microns and a resistivity of 1.0 ohm-cm. Apart from these values, all other device parameters used in the calculations will be assumed to be the same as those used for the 50 volt case. The impact of increasing the polysilicon gate length (L_G) upon the various components of the on-resistance for this case is provided in Fig. 7.29 together

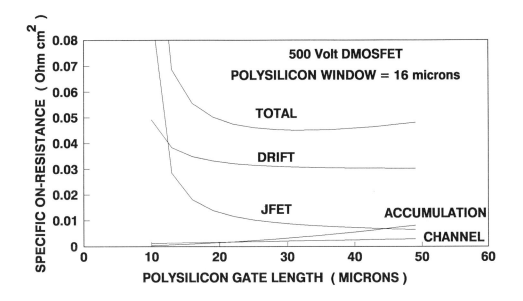

Fig. 7.29 Optimization of the gate length for a power DMOSFET structure with breakdown voltage of 500 volts.

with the total specific on-resistance. In this case, the minimum specific on-resistance occurs at a polysilicon gate length of 30 microns, which is significantly larger than that for the 50 volt case. The minimum specific on-resistance has a value of 45 milliohm-cm^2. In comparison, the ideal specific on-resistance for a breakdown voltage of 500 volts is 33 milliohm-cm^2. Thus, the device specific on-resistance deviates from the ideal value by a relatively small factor of 1.36 times. The reason for this DMOS specific resistance approaching the ideal specific on-resistance

376   Chapter 7 : POWER MOSFET

is that in this case the drift region resistance at the optimum gate length is dominant and the other components are relatively much smaller.

From these two examples, it can be concluded that it is possible to approach the ideal specific on-resistance for DMOSFETs with higher breakdown voltages when compared to those with lower breakdown voltages. This is illustrated in Fig. 7.30, where the specific on-resistance

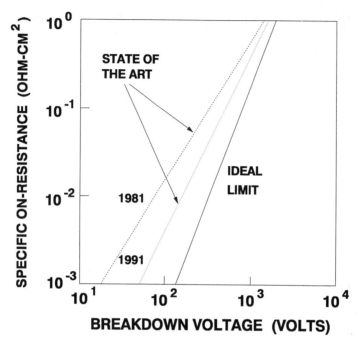

Fig. 7.30 Specific on-resistance of the power DMOSFET structure as a function of breakdown voltage.

is plotted as a function of the breakdown voltage. The ideal specific on-resistance is also shown in the figure for comparison. The state-of-the-art line for devices of 1981 vintage is representative of a process design rule that requires a 16 micron polysilicon window. With advancements in lithographic tools used to fabricate these devices, it has been possible to shrink the polysilicon window size.

The impact of reducing the polysilicon window size can be demonstrated by performing the optimization of the polysilicon gate length for a 50 volt DMOSFET structure with a polysilicon window of 8 microns. Apart from this change, all other device parameters will be assumed to be the same as those used in the earlier analysis. The calculated contributions to the specific on-resistance from the various components are given in Fig. 7.31 together with the total specific on-resistance. In comparison to the 50 volt design with the 16 micron polysilicon window, the minimum total specific on-resistance now occurs at a slightly smaller gate length. More importantly, the minimum value for the specific on-resistance is reduced from 3 milliohm-

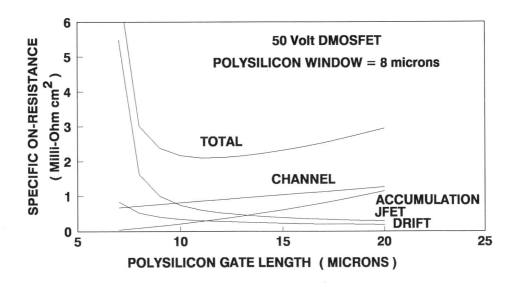

Fig. 7.31 Optimization of the on-resistance of the 50 volt DMOSFET with 8 micron polysilicon window.

cm² to about 2 milliohm-cm². This improvement is due to an increase in the channel density as well as better current spreading under the P-base region, i.e. more effective utilization of the area under the P-base region. The 8 micron polysilicon gate window is representative of the 1991 vintage state-of-the-art devices in Fig. 7.30. In these devices, it is customary to also use a thinner gate oxide and a smaller channel length to reduce the specific on-resistance even further. Typical devices with 50 volt breakdown voltage have a specific on-resistance of 1 milliohm-cm². From the plots in Fig. 7.30, it can be clearly seen that silicon power MOSFETs with high breakdown voltages approach the ideal performance limit, while improvements in the MOSFET structure are required for approaching ideal performance for the devices with lower breakdown voltages.

7.4.7 UMOSFET On-Resistance

In the previous section, it was demonstrated that the specific on-resistance of silicon power DMOSFETs is significantly higher than the ideal specific on-resistance. This deviation is caused by the resistance contributions from the channel region, the accumulation layer, and the JFET regions in the DMOS structure. The UMOSFET structure was proposed to reduce the resistance contributions from these regions. The basic UMOSFET structure is shown in Fig. 7.32. In this structure, no JFET region exists. Instead the current spreads out from an

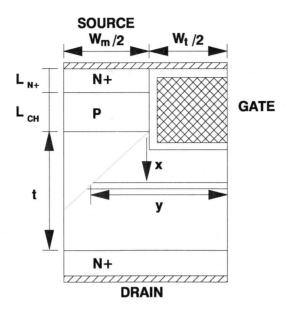

Fig. 7.32 Current flow path in the UMOSFET structure used for on-resistance analysis.

accumulation layer formed on the surface of the U-groove as illustrated in Fig. 7.32. The UMOSFET resistance then consists mainly of the channel resistance and the drift region resistance per unit area. In addition, the resistances from the contacts to the source and drain, and the resistances of the N⁺ source and N⁺ substrate must be accounted for in a complete analysis.

For the UMOSFET structure, the channel resistance is given by:

$$R_{ch,sp} = \frac{L_{ch}(W_m + W_t)}{2\mu_{ns} C_{ox}(V_G - V_T)} \qquad (7.60)$$

The UMOS structure can be fabricated with narrow mesa and trench regions because of the absence of the JFET region. The cell pitch in the UMOS structure is determined by the process and the lithographic design rules. It is important to use a UMOS process that allows self-aligned fabrication of the trench region and the contact to the source region. In combination with good lithographic tools, it is possible to reduce the UMOS cell size to less than 6 microns. Thus, the cell size for the UMOSFET can be made much smaller than for the DMOS cell structure. This allows obtaining a much higher channel density than in a DMOS structure. For a typical gate oxide thickness of 1000 °A and a channel length of 1 micron, the specific on-resistance of the channel is calculated to be less than 0.2 milliohm-cm².

Using an approach similar to that for the DMOS structure to account for the change in cross-sectional area for current flow (y in Fig. 7.32), an expression for the drift region spreading resistance can be derived:

$$R_D = \rho_D \left\{ \left[\left(\frac{W_m + W_t}{2} \right) \ln\left(\frac{W_m + W_t}{W_t} \right) \right] + \left(t_D - \frac{W_m}{2} \right) \right\} \quad (7.61)$$

where the first term is associated with the portion of the drift region where the current spreads at 45° while the second term is associated with the portion of the drift region where the area of cross-section is equal to the cell area. Unlike the DMOS structure, an overlap of the current spreading occurs even for low breakdown voltage designs for the UMOS structure because of the very small half-width for the mesa region. As an example, in the case of a 50 volt device fabricated with mesa and trench widths of 3 microns, the drift region contribution is calculated to be 0.15 milliohm-cm². Thus, the spreading resistance for the drift region in the UMOS structure is only 1.5 times greater than the ideal specific on-resistance.

The contribution to the specific on-resistance arising from the current flowing through the N^+ source region is given by:

$$R_{N+,sp} = \rho_{N+} \, t_{N+} \left(\frac{W_m + W_t}{W_m} \right) \quad (7.62)$$

where ρ_{N+} is the resistivity of the N^+ region and t_{N+} is its thickness. In deriving this expression, it has been assumed that the current flows vertically through this region. Due to the high doping concentration and small thickness of the N^+ source region, the contribution from this resistance is usually negligible.

The contribution from the N^+ substrate is given by:

$$R_{SB,sp} = \rho_{SB} \, t_{SB} \quad (7.63)$$

where ρ_{SB} and t_{SB} are the resistivity and thickness of the substrate. In the case of an arsenic doped substrate with resistivity of 0.002 ohm-cm and a thickness of 500 microns, this resistance contribution is 0.1 milliohm-cm². It is important to reduce this value by thinning the substrate because this contribution is comparable to the channel and drift region resistances for the UMOSFET structure.

The contribution from the source contact resistance can be obtained by taking into account the source contact area. In the UMOS structure, it is possible to obtain a relatively large contact area as shown in Fig. 7.32 if the contact to the P-base region is provided at selected locations orthogonal to the device cross-section. In this case, the source contact resistance is given by:

$$R_{CS,sp} = \rho_C \left(\frac{W_m + W_t}{W_m} \right) \quad (7.64)$$

When equal widths are used for the mesa and trench regions, the specific contact resistance contributed by the source contact is only twice the contact resistivity. This contribution is small when compared with the other components if the specific contact resistivity is below 1 x 10^{-5}

ohm-cm^2. This conclusion is also applicable to the drain contact resistance.

Unlike the DMOS structure, there is no optimum design for the UMOS cell. In this case, it is beneficial to reduce the mesa and trench widths as much as possible. As these dimensions become smaller, the channel resistance contribution becomes smaller due to an increase in the channel density. In addition, the spreading resistance of the drift region approaches the ideal specific on-resistance when the mesa width (W_m) becomes smaller. For the example of a UMOSFET with a breakdown voltage of 50 volts fabricated using mesa and trench widths of 3 microns, the total specific on-resistance is 0.35 milliohm-cm^2. Thus, the UMOSFET structures allows obtaining specific on-resistances approaching the ideal value.

7.4.8 Optimum Doping Profile

The above analysis of the on-resistance for the DMOS and UMOS structures is based upon a homogeneously doped drift region. This is appropriate because all power MOSFETs are manufactured using uniformly doped epitaxial layers. It has been proposed that a lower on-resistance can be obtained by using a non-uniform epitaxial doping profile to obtain the same breakdown voltage. Consider a doping profile $N(x)$ that will minimize the on-resistance of the drift layer:

$$R_{on} = \int \frac{1}{q \mu N(x)} \, dx \tag{7.65}$$

under the condition that the electric field distribution satisfy the breakdown voltage requirement:

$$BV = \int E \, dx \tag{7.66}$$

Using Poisson's equation to relate the doping and electric field,

$$\frac{dE}{dx} = \frac{q N_D}{\epsilon_s} \tag{7.67}$$

These equations can be solved to obtain the optimum doping profile:

$$N_D(x) = \frac{\epsilon_s E_c^2}{3 q BV \sqrt{1 - (2 E_c x / 3 BV)}} \tag{7.68}$$

According to this equation, the doping concentration must increase from the surface towards the interface between the drift layer and the N$^+$ substrate. Using this doping profile, the specific on-resistance is given by:

$$R_{on,min} = \frac{3 BV^2}{\epsilon_s \mu_n E_c^3} \tag{7.69}$$

7.5 FREQUENCY RESPONSE

The power MOSFET is inherently capable of operation at high frequencies due to the absence of minority carrier transport. Two limits to operation at high frequencies are set by the transit time across the drift region and the rate of charging of the input gate capacitance. In a power MOSFET, the input capacitance is relatively large due to the large active area required to carry substantial drain currents. Consequently, the frequency response is usually limited by the charging and discharging of the input capacitance. In addition to the gate-to-source capacitance, a significant gate-to-drain capacitance must be included in the analysis due to the overlap of the gate electrode over the drift region. This capacitance is amplified by the Miller effect into an equivalent input gate capacitance of :

$$C_M = (1 + g_m R_L) \; C_{GD} \qquad (7.70)$$

The total input capacitance is :

$$C_{INPUT} = (C_{GS} + C_M) \qquad (7.71)$$

The maximum frequency of operation, as defined by the frequency at which the input current becomes equal to the load current, is then given by:

$$f_m = \frac{g_m}{2 \pi \; C_{INPUT}} \qquad (7.72)$$

To analyze the frequency response, it is necessary to evaluate each of the components of the input capacitance arising in the power MOSFET structure. Consider the conventional DMOS structure shown in Fig. 7.33 with the gate electrode extending between adjacent DMOS cells. The input gate-to-source capacitance for this structure contains several components: (a) the capacitance C_{N+} arising from the overlap of the gate electrode over the N$^+$ source region, (b) the capacitance C_P arising from the MOS structure created by the gate electrode over the P-base region, and (c) the capacitance C_o arising from running the source metal over the gate electrode. The total gate-to-source capacitance is given by:

$$C_{GS} = C_{N+} + C_P + C_o \qquad (7.73)$$

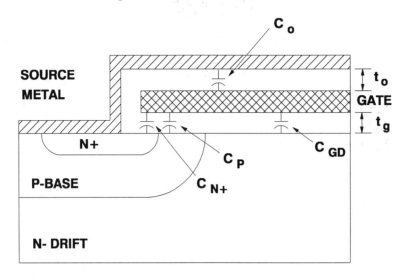

Fig. 7.33 DMOS cell structure indicating various capacitances.

The capacitance between the source and gate electrodes, which is referred to as the *overlap capacitance* (C_o) is determined by the dielectric constant and thickness of the intervening insulator:

$$C_o = \frac{e_I A_o}{t_o} \tag{7.74}$$

where A_o is the area of the overlap between the source and gate electrodes. To reduce this capacitance, a thick insulator layer is generally used during device fabrication. The other components of the capacitance require analysis of the MOS structure.

7.5.1 MOS Capacitance

Consider the basic MOS structure shown in Fig. 7.34. This structure contains two capacitances in series, namely, the oxide capacitance and the semiconductor capacitance. Thus:

$$\frac{1}{C_G} = \frac{1}{C_{ox}} + \frac{1}{C_S} \tag{7.75}$$

The oxide capacitance per unit area is determined by the thickness of the gate oxide:

$$C_{ox} = \frac{e_{ox}}{t_{ox}} \tag{7.76}$$

Chapter 7 : POWER MOSFET 383

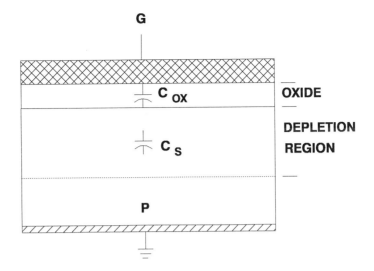

Fig. 7.34 Basic MOS structure used for capacitance analysis.

The semiconductor capacitance is determined by the width of the depletion layer. For the case of a P-type substrate, if a negative gate bias is applied, an accumulation layer will form at the surface. Any changes in the gate voltage cause a corresponding change in the accumulation layer charge with a response time corresponding to the dielectric relaxation time. The capacitance is then equal to the oxide capacitance :

$$C_A = C_{ox} \qquad (7.77)$$

When a positive gate bias is applied, a depletion layer forms in the semiconductor. The capacitance of the depletion layer is related to the space charge in the semiconductor:

$$C_S = \frac{dQ_S}{d\Psi_S} \qquad (7.78)$$

Using Eq. (7.13), it can be shown that:

$$C_S = \frac{\epsilon_s}{\sqrt{2}\, L_D} \frac{\{1 - e^{-(q\Psi_s/kT)} + (n_{p0}/p_{p0})[e^{(q\Psi_s/kT)} - 1]\}}{F(q\Psi_s/kT,\, n_{p0}/p_{p0})} \qquad (7.79)$$

The MOS gate capacitance will decrease with increasing positive gate voltage due to widening of the depletion layer. Beyond a certain voltage, a surface inversion layer will form. Once the inversion layer forms, the depletion layer reaches its maximum value and the semiconductor capacitance attains its lowest value. The resulting C-V curve is illustrated in Fig. 7.35 by the solid line. This curve is called the high frequency response curve (typically measured at one megahertz) because it is assumed that the inversion layer charge cannot follow

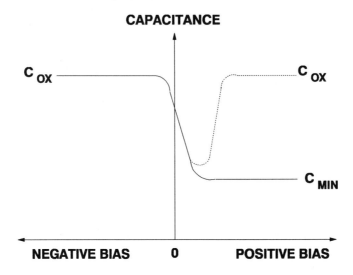

Fig. 7.35 C-V curves for an MOS capacitor with P-type semiconductor.

the AC signal used to measure the capacitance. If either the measurement frequency or the semiconductor lifetime is reduced until the inversion layer charge can respond to the AC signal, the capacitance will become equal to the oxide capacitance, as illustrated by the dashed line.

In the case of the power MOSFET, this type of C-V curve is not observed because the inversion layer formation in the P-base region is accomplished by transport of electrons from the N^+ source region into the inversion layer. Consequently, even at high operating frequencies, the inversion layer charge can respond to the applied high frequency signal by the rapid transfer of electrons from the N^+ emitter into the inversion layer, and the measured input gate C-V curve of power MOSFETs follows the dashed line rather than the solid line shown in Fig. 7.35.

7.5.2 Input Capacitance

Based upon the above analysis of the MOS capacitance for the gate structure, the various components of the input capacitance can be determined as follows. In the case of the heavily doped N^+ emitter region, the capacitance is determined by the gate oxide thickness:

$$C_{N+} = \frac{\varepsilon_{ox} A_{N+O}}{t_{ox}} \qquad (7.80)$$

where A_{N+O} is the area of overlap of the gate electrode over the N^+ emitter. For the DMOS structure with a channel width Z:

$$A_{N+O} = x_{N+} Z \qquad (7.81)$$

where x_{N+} is the depth of the N^+ emitter diffusion. It is assumed here that the lateral diffusion of the emitter is equal to its junction depth and that the gate fringing capacitance can be neglected. When taken together, these are reasonable approximations because they compensate for each other.

The capacitance of the gate over the P-base (C_P) is dependent upon the gate bias. It goes through a minimum with increasing gate bias as shown in Fig. 7.35 by the dashed lines. This capacitance can be reduced by keeping the channel length small. It represents a fundamental contribution to the input capacitance of the power MOSFET.

The gate-to-drain capacitance also varies with gate and drain voltage. It has a high value during the on-state because the surface of the drift region is under accumulation. As the drain voltage increases and the device supports high voltages, the gate-to-drain capacitance decreases. Due to the amplification of the gate-to-drain capacitance by the Miller effect, this capacitance can severely reduce the frequency response. It is therefore important to reduce its magnitude. The gate-to-drain overlap capacitance can be drastically reduced by altering the gate structure

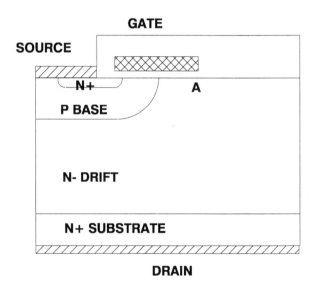

Fig. 7.36 DMOSFET structure with reduced gate-drain capacitance.

by eliminating the gate overlap over the drift region as shown in Fig. 7.36. In this structure, the source electrode is also confined to the diffusion window to eliminate the capacitance (C_o). A significant improvement in the high speed switching performance of power MOSFETs has been reported by taking these measures during device design and fabrication. However, it is important to note that eliminating the gate-to-drain overlap creates a high electric field at the edge of the gate. This can reduce the cell breakdown voltage to below that for the edge termination. In addition, it adversely impacts the on-resistance because of an increase in the resistance between the channel and the drift region because an accumulation layer is no longer formed over a portion of the surface between the base regions.

An improvement in the cell breakdown voltage can be obtained with a reduced drain-gate capacitance by incorporation of a shallow P-type diffusion as shown in Fig. 7.37 in the portion

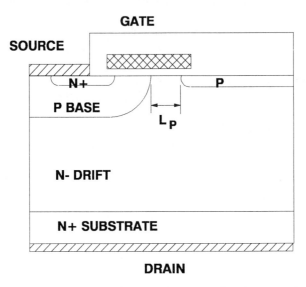

Fig. 7.37 DMOSFET structure with a shallow P-region between the P-base regions.

where the gate electrode has been interrupted. This P-type region behaves like a guard ring for the electrode, reducing the electric field crowding at its edge. However, it also enhances the JFET action within the cell during current flow. Thus, a reduction in capacitance can be achieved at the drawback of an increase in on-resistance. An optimization must be performed for each breakdown voltage design by varying the length L_P to obtain the best compromise between reducing the input capacitance and increasing the on-resistance.

7.5.3 Gate Series Resistance

In power MOSFETs, a series gate resistance is always present arising either from the resistance of the gate electrode internal to the device or from the gate drive circuit. The frequency response limited by the RC charging time constant of the input gate circuit is given by:

$$f_{INPUT} = \frac{1}{2\pi C_{INPUT} R_G} \quad (7.82)$$

In the case of conventional polysilicon gate power MOSFETs, the sheet resistance of the gate electrode is typically over 10 ohms per square. This has been found to seriously lower the frequency response. By changing to a molybdenum gate structure, an increase in frequency response by an order of magnitude has been reported, demonstrating that this is the limiting factor to high frequency performance. Power MOSFETs capable of delivering 100 watts at 900 MHz have been developed by using a molybdenum silicide gate process.

7.5.4 Maximum Operating Frequency

A *maximum operating frequency* for the power MOSFET can be defined as the frequency at which the input gate current becomes equal to the output drain current. For the case of an AC input waveform, the input current for the MOSFET is given by:

$$i_{INPUT} = 2 \pi f C_{INPUT} v_G \qquad (7.83)$$

where v_G is the input AC voltage. The output drain current of the MOSFET is given by:

$$i_{OUTPUT} = g_m v_G \qquad (7.84)$$

Based upon the definition, these currents are equal at the maximum operating frequency (f_m). This leads to:

$$f_m = \frac{g_m}{2 \pi C_{INPUT}} \qquad (7.85)$$

This limit to the operating frequency is generally avoided during operation of power MOSFETs because of the high power dissipation in the gate circuit.

7.6 SWITCHING PERFORMANCE

Due to the inherent high speed turn-on and turn-off capability of power MOSFETs, they are often used as power switches. When used in this manner, the devices are maintained in either the on-state or forward blocking state for most of the time and they must rapidly switch between these modes of operation. The power dissipation is determined by the on-resistance during the conducting state and the leakage current in the forward blocking state. These parameters have been discussed earlier. In this section, the parameters that govern the transition between the on and off-states are analyzed. In addition, high speed switching is accompanied with a very high rate of change of the drain voltage, which can cause an undesirable turn-on of the parasitic bipolar transistor in the power MOSFET. This can limit the switching speed and safe-operating-area of power MOSFETs.

A simple analytical procedure for predicting the switching transients experienced with power MOSFETs can be achieved by using the circuit model shown in Fig. 7.38. Due to the interaction between the device and the circuit, it is necessary to consider a specific type of load for analysis. Here a clamped, inductive load (L_l) is considered with a steady state current I_L flowing through it. The inductance L_S is the stray circuit inductance not clamped by the diode (D).

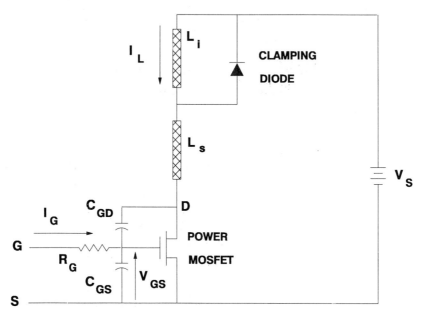

Fig. 7.38 Circuit used for analysis of switching times during turn-on and turn-off of a power MOSFET.

7.6.1 Turn-on Transient Analysis

First, consider the case where the power MOSFET is in its off-state and the load current is '*circulating*' (flowing) through the diode. Since the diode is in its on-state, all the supply voltage is supported by the power MOSFET under these conditions. Thus, the initial conditions for the power MOSFET are : $V_G = 0$; $I_D = 0$; and $V_D = V_S$. At time $t = 0$, a step voltage (V_{GA}) is applied at the gate terminal G.

7.6.1.1 Time interval t_1: The gate voltage (V_{GS}) of the power MOSFET that controls the drain current flow is determined by the voltage across the gate-to-source capacitance (C_{GS}). This voltage is governed by the charging of the capacitances C_{GS} and C_{GD} via resistor R_G. As long as the gate voltage (V_{GS}) remains below the threshold voltage (V_T), no drain current will flow. The time taken for the gate voltage (V_{GS}) to reach the threshold voltage (V_T) represents a turn-on delay period. During this period, the input capacitance is simply ($C_{GS} + C_{GD}$) because the Miller effect comes into effect only when the transistor operates in its active region. The gate voltage then rises exponentially with time:

Chapter 7 : POWER MOSFET

$$V_{GS}(t) = V_{GA}\{1 - e^{-[t/R_G(C_{GS} + C_{GD})]}\} \quad (7.86)$$

The *turn-on delay time*, defined as the time taken for the gate voltage to rise to the threshold voltage, obtained from this expression is given by:

$$t_1 = R_G (C_{GS} + C_{GD}) \ln\left[\frac{1}{1 - (V_T/V_{GA})}\right] \quad (7.87)$$

7.6.1.2 Time interval t_2: Beyond the turn-on delay time, drain current flow begins to occur. If the load current I_L is assumed to remain constant during the switching interval because of the high load inductance, the current from the diode transfers to the power MOSFET. For simplifying this analysis, it will be assumed that the diode does not exhibit any reverse recovery current. In addition, it will be assumed that the power MOSFET has a linear transfer

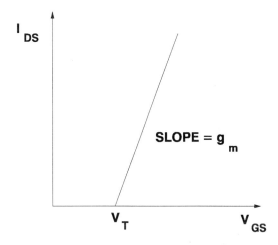

Fig. 7.39 Linear transfer characteristic for power MOSFET used in the switching time analysis.

characteristics as shown in Fig. 7.39. In this case, the drain current will increase in proportion to the gate voltage. Since the device is now in its active region, the Miller effect will determine the gate-to-drain capacitance being charged by the gate circuit. If the stray inductance L_S is small, the Miller effect is small and the gate voltage will continue to rise exponentially as described by Eq. (7.86). The drain current will then take the form:

$$I_D(t) = g_m (V_{GS} - V_T) = g_m \{V_{GA}[1 - e^{-t/R_G(C_{GD} + C_{GS})}] - V_T\} \quad (7.88)$$

This waveform is illustrated in Fig. 7.40 with the MOSFET drain and gate voltage waveforms.

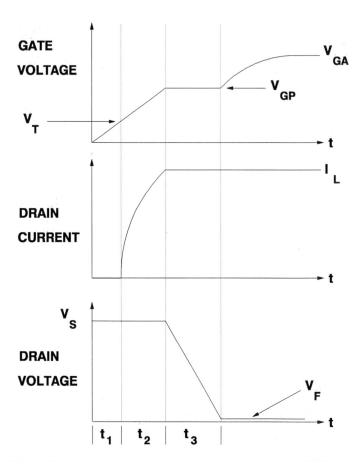

Fig. 7.40 Turn-on transient waveforms for a power MOSFET.

Since the diode is forward biased during this time interval, the power MOSFET must support the entire supply voltage (V_S) during this time and the drain voltage will remain constant. This time period will end when the drain current reaches the load current (I_L), i.e., when all the circulating diode current is transferred to the power MOSFET. Based up on this, an expression for the time interval t_2 is obtained:

$$t_2 = R_G \, (C_{GD} + C_{GS}) \, \ln\left[\frac{g_m \, V_{GA}}{g_m(V_{GA} - V_T) - I_L}\right] \quad (7.89)$$

7.6.1.3 Time interval t_3: At the end of time interval t_2, all the load current has transferred from the diode to the power MOSFET. Beyond this point in time, the diode enters its reverse blocking mode allowing the drain voltage to fall from V_S to the on-state voltage drop of the power MOSFET. Since the drain current is now constant, based up on the transfer characteristics, the gate voltage during this time must also remain constant. This value of the gate voltage is given by:

$$V_{GS} = V_T + \frac{I_L}{g_m} = V_{GP} \qquad (7.90)$$

Since the gate voltage (V_{GS}) is constant, all the input current flows into the Miller capacitance (C_{GD}) during this period. The input current is given by:

$$I_G = \frac{V_{GA} - V_{GP}}{R_G} = \frac{1}{R_G}\left[V_{GA} - \left(V_T + \frac{I_L}{g_m}\right)\right] \qquad (7.91)$$

Since this gate current charges the gate-drain capacitance C_{GD} and since the gate-source voltage (V_{GS}) is constant during this time interval:

$$\frac{dV_{GD}}{dt} = \frac{dV_{DS}}{dt} = \frac{I_G}{C_{GD}} = \frac{V_{GA} - (V_T + I_L/g_m)}{R_G C_{GD}} \qquad (7.92)$$

Integration of this equation yields:

$$V_D(t) = V_L - \left[\frac{V_{GA} - (V_T + I_L/g_m)}{R_G C_{GD}}\right] t \qquad (7.93)$$

Thus, the drain voltage decreases linearly with time during the time interval t_3. At the end of time interval t_3, the drain voltage reaches its steady-state value, which is the on-state voltage drop (V_F) of the power MOSFET. Using this criterion, an expression for the time interval t_3 is obtained:

$$t_3 = \frac{(V_S - V_F) R_G C_{GD}}{[V_G - (V_T + I_L/g_m)]} \qquad (7.94)$$

The gate voltage continues to rise beyond time t_3 until it reaches V_{GA}, but this has no influence on the drain current and voltage because they have reached their steady-state values.

The preceding analysis is valid only when the ratio (L_S/R_G) is small. Other cases can be treated in a similar manner. From the turn-on waveforms in Fig. 7.40, it can be concluded that a high power dissipation occurs during the intervals t_2 and t_3, where high current and voltage are sustained by the device simultaneously. In addition, the device current-voltage locus must stay within its safe-operating area during this period. The power loss can be reduced by keeping these time intervals short. This is possible by reduction of the gate series resistance (R_G) and the drain-gate capacitance (C_{GD}). Although the series resistance of the gate drive circuit can be reduced, this has the disadvantage of increasing the cost of the drive circuit. It is therefore important to use the methods discussed in the earlier section to reduce the drain-gate capacitance.

7.6.2 Turn-off Transient Analysis

The analysis of the turn-off transient for the power MOSFET in the circuit shown in Fig. 7.38 can be performed in a similar manner to the turn-on analysis. In this case, the power MOSFET is assumed to be initially in its on-state with a gate bias of V_{GA} applied to the gate terminal. Thus, the initial conditions for the power MOSFET are : $V_G = V_{GA}$; $I_D = I_L$; and $V_D = V_F$. Since all the load current is flowing through the power MOSFET, there is no current flowing through the diode and it is in its reverse blocking state. At time $t = 0$, the applied gate voltage is abruptly reduced to zero at terminal G.

7.6.2.1 Time interval t_4: The voltage (V_{GS}) then decreases exponentially with time due to

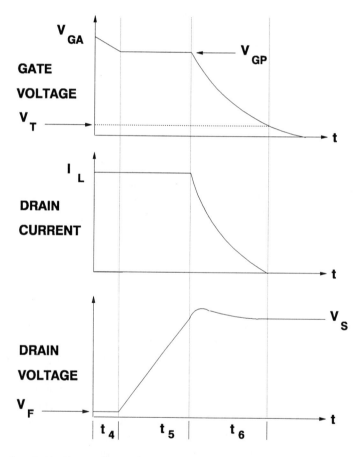

Fig. 7.41 Turn-off transient waveforms for a power MOSFET.

discharging of the gate capacitance via the gate series resistance R_G as shown in Fig. 7.41. During this period:

$$V_{GS} = V_{GA}\, e^{-t/R_G\,(C_{GS} + C_{GD})} \qquad (7.95)$$

No changes in the drain current or voltage occur until the gate voltage reaches the magnitude required to saturate the drain current at a value equal to the load current (I_L). Thus, at the end of time interval t_4:

$$V_{GS}(t_4) = V_T + \frac{I_L}{g_m} = V_{GA}\, e^{-t_4/R_G\,(C_{GS} + C_{GD})} \qquad (7.96)$$

From this equation, an expression for the *turn-off delay time* (t_4) is obtained:

$$t_4 = R_G\,(C_{GS} + C_{GD})\, \ln\!\left[\frac{V_{GA}}{V_T + (I_L/g_m)}\right] \qquad (7.97)$$

7.6.2.2 Time interval t_5: Beyond this time, the drain current will remain at I_L while the drain voltage begins to rise towards V_S because the load current cannot be diverted from the MOSFET into the diode until V_D exceeds V_S. As long as the drain current is constant, the gate voltage (V_{GS}) also remains constant and the gate current (I_G) discharges the gate-drain capacitance (C_{GD}). During this period,

$$V_{GS} = V_T + \frac{I_L}{g_m} \qquad (7.98)$$

and

$$I_G = \frac{V_{GS}}{R_G} = \frac{V_T + (I_L/g_m)}{R_G} \qquad (7.99)$$

Since this current discharges the gate-drain capacitance (C_{GD}) while the gate potential remains constant:

$$\frac{dV_{DS}}{dt} = \frac{dV_{DG}}{dt} = \frac{I_G}{C_{GD}} \qquad (7.100)$$

Integration of this equation provides the drain voltage waveform:

$$V_{DS}(t) = V_{on} + \frac{1}{R_G\, C_{GD}}\!\left(V_T + \frac{I_L}{g_m}\right) t \qquad (7.101)$$

According to this equation, the drain voltage rises linearly with time during this period from the on-state voltage drop (V_F). At the end of the time interval t_5, the drain voltage reaches the supply voltage (V_S). The duration of the period t_5 can therefore be obtained by equating $V_{DS}(t)$ to V_S at time t_5:

$$t_5 = \frac{R_G C_{GD} (V_L - V_{on})}{(I_L/g_m) + V_T} \tag{7.102}$$

7.6.2.3 Time interval t_6:
At the end of the previous time interval, the drain voltage of the power MOSFET has risen to the supply voltage. At this point in time the freewheeling diode D will turn-on. To achieve this, it is essential that the MOSFET drain voltage exceed the load supply voltage (V_S). If the stray inductance L_S is small, the overshoot in the drain voltage will be small. However, if L_S is large and the rate of change of the drain current becomes large, the voltage on the MOSFET drain can exceed its breakdown voltage. If the stray inductance is small, the drain voltage can be assumed to remain relatively constant. The gate voltage will then continue to decrease exponentially:

$$V_{GS}(t) = \left(V_T + \frac{I_L}{g_m}\right) e^{-t/R_G(C_{GS} + C_{GD})} \tag{7.103}$$

The drain current will follow this change in the gate voltage:

$$I_D(t) = (I_L + g_m V_T) e^{-t/R_G(C_{GS} + C_{GD})} - g_m V_T \tag{7.104}$$

This period will extend until the gate voltage reaches the threshold voltage (V_T) and the drain current is reduced to zero. Based upon this criterion, the time interval for this period is obtained:

$$t_6 = R_G (C_{GS} + C_{GD}) \ln\left(\frac{I_L}{g_m V_T} + 1\right) \tag{7.105}$$

After this interval, the gate voltage will continue to decay exponentially to zero. Since the drain current and voltage are at their steady-state values (in the off-state), this will have no influence upon them.

The composite waveforms corresponding to this discussion of turn-off are shown in Fig. 7.41. These waveforms apply to the case of a small ratio (L_S/R_G). Other cases can be treated in a similar manner. The power dissipation during the turn-off transient occurs primarily during the periods t_5 and t_6, where both the current and voltage are large. Once again, the current-voltage locus must remain within the safe-operating-area of the device. The drain voltage overshoot during period t_6 must especially be noted. Its magnitude depends on the size of the stray inductance. If the stray inductance is large, the MOSFET can be forced into avalanche breakdown, causing destructive failure. The analysis presented here also indicates the need to keep the gate-to-drain capacitance small. This can be achieved by either breaking the

Chapter 7 : POWER MOSFET

gate electrode as shown in Fig. 7.36 or by using a thick oxide in the portion where the drift region comes to the surface. This structure is sometimes referred to as a *terraced gate structure*.

7.6.3 [dV/dt] Capability

As discussed in the case of thyristors, it has been found that power MOSFETs can also be forced into current conduction under a high [dV/dt] applied to the drain. In certain cases, this will lead to destructive failure of the devices. The various mechanisms that can lead to [dV/dt] induced turn-on are discussed here, and device design approaches to raising the [dV/dt] capability are provided.

To analyze the [dV/dt] induced turn-on in power MOSFETs, consider the equivalent

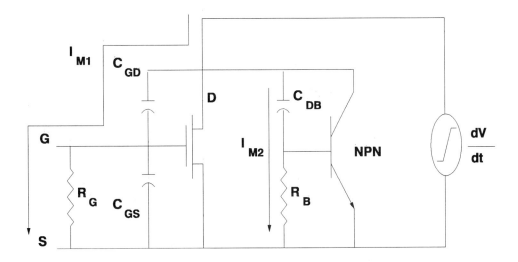

Fig. 7.42 Power MOSFET equivalent circuit used for analysis of [dV/dt] capability.

circuit of the device shown in Fig. 7.42 with a ramp applied between drain and source. In addition to the device capacitances, the equivalent circuit shows the parasitic N-P-N bipolar transistor with a base-emitter shunting resistance R_B. This shunting resistance is included in the device structure by source metal overlapping both the N^+ emitter and the P-base regions. Using this equivalent circuit, two modes for [dV/dt] induced turn-on can be projected and analyzed.

7.6.3.1 Mode 1: For the first mode for [dV/dt] induced turn-on, consider the current (I_{M1}) flowing via the gate-to-drain capacitance (C_{GD}) into the gate circuit resistance R_G. If the drain voltage is much larger than the gate voltage, the voltage drop across the gate resistance will be approximately given by :

$$V_{GS} = I_{M1} R_G = R_G C_{GD} \left[\frac{dV}{dt}\right] \quad (7.106)$$

If the gate voltage (V_{GS}) exceeds the threshold voltage of the power MOSFET, the device will be forced into current conduction. This can produce a high power dissipation in the device, leading to destructive failure. Using the criterion that the gate bias must remain below the threshold voltage, the [dV/dt] capability set by this mode is given by:

$$\left[\frac{dV}{dt}\right] = \frac{V_T}{R_G C_{GD}} \quad (7.107)$$

Based upon this equation, a higher [dV/dt] capability can be obtained by using a very low impedance gate drive circuit and by raising the threshold voltage. These options are not attractive because a lower drive impedance is more expensive and because a higher threshold voltage leads to a larger on-resistance. Since the threshold voltage decreases with increasing temperature, this mode of turn-on can become aggravated with increasing power dissipation within the device. In general, this mode of turn-on is non-destructive because the gate voltage does not rise much above the threshold voltage and the device current is limited by the high device resistance.

7.6.3.2 Mode 2: This mode of [dV/dt] induced turn-on occurs due to the existence of the parasitic bipolar transistor. At high rates of change of the drain voltage, a current (I_{M2}) flows via the capacitance (C_{DB}) into the base shorting resistance (R_B) as shown in Fig. 7.42. If this current is sufficient for the base-emitter junction of the bipolar transistor to become forward biased, it will turn-on the transistor. For low values of R_B, when the emitter junction is inactive, the breakdown voltage of the transistor, and hence the power MOSFET, approaches the collector-base breakdown voltage (BV_{CBO}). But under a high [dV/dt] applied to the device and with large values of R_B, the emitter-base junction becomes strongly forward biased. The breakdown voltage then collapses to the level of the open-base transistor breakdown voltage (BV_{CEO}), which is about 60 percent of the collector-base breakdown voltage (BV_{CBO}). If the applied drain voltage is greater than BV_{CEO}, the device will go into avalanche breakdown or may be destroyed by second breakdown if the drain current is not externally limited.

The [dV/dt] induced turn-on in this mode is dependent upon the internal device structure. Consider the DMOS structure shown in Fig. 7.43 with one end of the N^+ emitter shorted to the P-base region. The applied [dV/dt] creates a displacement current flow in the capacitance C_{DB}. This current must flow laterally in the P-base region to the source metal. It produces a voltage drop along resistance R_B, which forward biases edge A of the N^+ emitter. When edge A of the N^+ emitter becomes forward biased by the lateral current flow, the bipolar transistor will turn-on, precipitating further current flow. An estimate of the [dV/dt] at which this will occur can be obtained by assuming that the bipolar transistor will turn-on at an emitter-base forward bias voltage of V_{bi}:

Chapter 7 : POWER MOSFET   397

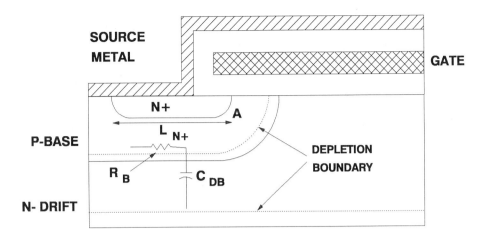

Fig. 7.43 Internal resistance and capacitance of power MOSFET used for [dV/dt] analysis.

$$\left[\frac{dV}{dt}\right] = \frac{V_{bi}}{R_B \, C_{DB}} \qquad (7.108)$$

where R_B is a distributed base resistance. To obtain a high [dV/dt] capability, it is important to keep the resistance R_B small. This can be achieved by increasing the doping level of the P-base and by keeping the length (L_{N+}) of the N⁺ emitter as small as possible within the constraints of the lithography used for device fabrication. It should be noted that the P-base sheet resistance increases with increasing drain voltage due to the extension of the depletion layer. This lowers the [dV/dt] capability. Further, increasing the temperature will lower the voltage (V_{bi}) at which the emitter will begin to inject. The P-base resistance will also increase with temperature due to a reduction in the mobility. These factors will cause a reduction in the [dV/dt] capability as temperature increases. Experimental confirmation of this mode of [dV/dt] induced breakdown has been obtained on special structures, where the P-base was not shorted to the N⁺ emitter, allowing the use of large external base resistance (R_B) to make the effect pronounced. In the case of practical power MOSFETs designed for operation at high switching speeds, the [dV/dt] capability is over 10,000 volts per microsecond. Consequently, these devices are immune to the failure modes discussed in this section.

7.7 SAFE OPERATING AREA

The safe-operating-area defines the limits of operation of the device. It is well known that the maximum current at low drain voltages is limited by power dissipation if the leads are sufficiently rated to prevent fusing. The maximum voltage at low drain currents is determined by the avalanche breakdown phenomena as discussed in Section 7.3. Under the simultaneous application of high current and voltage, the device may be susceptible to destructive failure even if the duration of the transient is small to prevent excessive power dissipation. This failure mode has been referred to as second breakdown.

7.7.1 Bipolar Second Breakdown

The term *second breakdown* refers to a sudden reduction in the blocking voltage capability when the drain current increases. This phenomenon has been observed in power MOSFETs. It originates from the presence of the parasitic bipolar transistor in the device structure. When the drain voltage is increased to near the avalanche breakdown voltage, current flows into the P-base region in addition to the normal current flow within the channel inversion layer. The avalanche current collected within the P-base region flows laterally along the P-base region to its contact. The voltage drop along the P-base region forward biases the edge of the N$^+$ emitter furthest from the base contact. When the forward bias on the emitter exceeds 0.6 to 0.7 volts, it begins to inject carriers. The parasitic bipolar transistor is no longer capable of supporting the P-base/N-drift layer breakdown voltage (BV_{CBO}). Its breakdown voltage is instead reduced to BV_{CEO}, which is typically 60 percent of BV_{CBO}.

To analyze second breakdown in power MOSFETs, consider the equivalent circuit of the

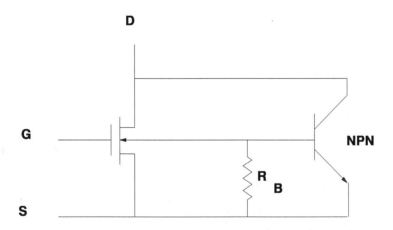

Fig. 7.44 Power MOSFET equivalent circuit showing the parasitic bipolar transistor.

device with the parasitic bipolar transistor as shown in Fig. 7.44. A base-to-emitter resistor (R_B) is shown in this figure corresponding to the lateral resistance in the P-base region illustrated in

Chapter 7 : POWER MOSFET

Fig. 7.45. The current flow in the device takes two paths - one via the MOS channel and the other via the active bipolar transistor. These currents must satisfy the following conditions:

$$I_D = I_C + I_M \tag{7.109}$$

$$I_S = I_E + I_M \tag{7.110}$$

and

$$I_B = I_C - I_E \tag{7.111}$$

In addition, the emitter and collector currents are related by the gain of the bipolar transistor:

$$I_C = \gamma_E \alpha_T M I_E \tag{7.112}$$

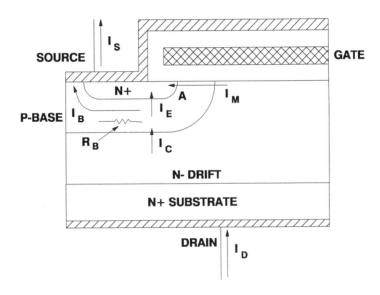

Fig. 7.45 Current paths in power MOSFET with activated parasitic bipolar transistor.

where γ_E is the injection efficiency, α_T is the base transport factor and M is the avalanche multiplication factor. The first two parameters can be assumed to be equal to unity for a typical power MOSFET structure. Further, during second breakdown, the emitter current is caused by the forward bias at point A. It can therefore be related to the potential at A by:

$$I_E = I_0 \, e^{(qV_A/kT)} \tag{7.113}$$

where V_A is the forward bias caused by the lateral base current (I_B). The voltage drop due to the lateral base current flow is determined by the base resistance (R_B):

$$V_A = R_B I_B \qquad (7.114)$$

Combining these equations gives:

$$I_E = I_0 \exp\left[\frac{q R_B}{k T} (M - 1) I_E\right] \qquad (7.115)$$

If the first order expansion of the exponential is used to evaluate the initiation of the second breakdown effect, it can be shown that:

$$I_E = \frac{I_0}{[1 - (q R_B/k T)(M - 1) I_0]} \qquad (7.116)$$

The multiplication factor (M) is related to the drain voltage by:

$$M = \frac{1}{[1 - (V_D/BV)^n]} \qquad (7.117)$$

As the drain voltage increases, the emitter current, and hence the source current can rise catastrophically in accordance with Eq. (7.116). The voltage at which this will occur can be obtained from the above equations:

$$V_{D,SB} = \frac{BV}{[1 + q R_B I_0/k T]^{1/n}} \qquad (7.118)$$

A reduction of the second breakdown voltage is predicted with increasing base resistance (R_B) by this analysis. For this reason, it is desirable to include a deep P+ diffusion in the center of the DMOS cell to reduce the resistance R_B without altering the threshold voltage.

7.7.2 MOS Second Breakdown

Another phenomenon that causes second breakdown occurs because of the effect of the lateral voltage drop in the P-base region upon the channel current. This effect is called the *body bias effect*. In this case, it is necessary to define a body bias coefficient:

$$\gamma = \frac{\Delta I_D}{\Delta V_B} \qquad (7.119)$$

where

Chapter 7 : POWER MOSFET   401

$$V_B = R_B I_B \quad (7.120)$$

The source current is then given by:

$$I_S = I_M + \gamma V_B = I_M + \gamma R_B I_B \quad (7.121)$$

At high drain voltages, the large electric field causes avalanche multiplication of the channel current. The base current is now given by:

$$I_B = I_D - I_S = (M - 1) I_S \quad (7.122)$$

Using Eq. (7.121):

$$I_B = (M - 1) (I_M + \gamma R_B I_B) \quad (7.123)$$

and from this equation:

$$I_B = \frac{(M - 1) I_M}{1 - \gamma R_B (M - 1)} \quad (7.124)$$

and

$$I_S = \frac{I_M}{1 - \gamma R_B (M - 1)} \quad (7.125)$$

which is of the same form as Eq. (7.116) derived earlier. As the drain voltage increases, the multiplication factor increases and causes a catastrophic increase in the source current. The drain voltage at which this second breakdown occurs is given by:

$$V_{D,SB} = \frac{BV}{(1 + \gamma R_B)^{1/n}} \quad (7.126)$$

Once again, a reduction of the second breakdown voltage occurs with increasing base resistance. Consequently, this second breakdown mechanism can also be suppressed by the incorporation of a deep P^+ diffusion within the DMOS cell.

As temperature increases, the breakdown voltage (BV) and the coefficient (n) will increase. This will tend to raise the second breakdown voltage. However, the P-base resistance (R_B) also increases with temperature due to a reduction in the mobility. This has a compensating effect. It has been found that the second breakdown voltage and current are only weak functions of temperature. In addition, p-channel power MOSFETs exhibit a smaller collapse in voltage when compared with n-channel devices due to the lower N-base resistance and the lower hole impact ionization coefficient.

Due to the rugged design with very efficient shorting of the emitter-base junction of the

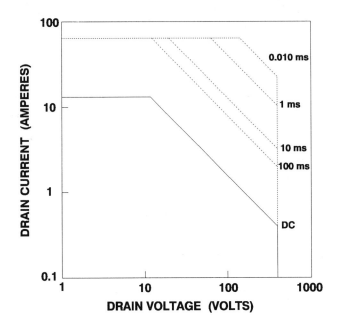

Fig. 7.46 Typical power MOSFET safe operating area limits. The times indicated represent the duration of the stress.

parasitic bipolar transistor, modern commercial power MOSFETs are rated with excellent safe-operating-area. As an example, the safe-operating-area of a typical 500 volt power MOSFET is shown in Fig. 7.46. The curves shown in this figure are simply based upon a power dissipation which will keep the junction temperature below 150°C. The ability of a power MOSFET to operate along these constant power contours substantiates the absence of second breakdown, by adequate shorting of the internal parasitic bipolar transistor, in modern commercially available devices.

7.8 INTEGRAL DIODE

Some power switching circuits require reverse current flow through the active power switching devices. Even though the power switching devices do not have to exhibit any reverse blocking capability, in these applications they must be capable of reverse conduction. Some examples of these types of circuits are DC to AC inverters for adjustable speed motor drives, switch mode power supplies, and DC to DC choppers for motor speed control with regenerative braking. When using bipolar transistors, it is necessary to use an antiparallel diode across the transistor to conduct the reverse current. This diode must exhibit good reverse recovery characteristics with a small reverse recovery charge to keep the power dissipation low and a reverse recovery waveform which does not contain an abrupt change in current (i.e., snap-off). Any abrupt current changes create very high transient voltages which can damage the power devices.

When the power MOSFET is used in place of the bipolar transistor as the power switching device, the possibility of utilizing the reverse conducting diode inherent in the structure becomes attractive because it eliminates the complexity and cost of adding the external

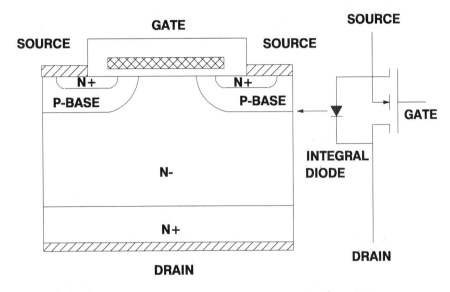

Fig. 7.47 Power MOSFET structure with its integral diode indicated.

diode. The existence of the integral reverse conducting diode is illustrated in Fig. 7.47. The reverse conducting diode is formed across the P-base/N-drift layer junction with its anode current flowing via the source contact to the P-base region. It should be noted that the applications under consideration here require high voltage operation. The power MOSFET forward drop in these cases exceeds the junction knee voltage discussed earlier with reference to Fig. 7.5 and it is not practical to gate the power MOSFET to utilize it as a synchronous rectifier.

In the as-fabricated form, the lifetime in the N-drift region of power MOSFETs is generally high. The integral diode can be treated as a P-i-N rectifier between the source and drain terminals. It will conduct current very efficiently with a forward voltage drop in the range of 1 volt due to the injection of a high concentration of minority carriers into the drift region. The theory of current conduction in a P-i-N rectifier has been discussed in Chapter 4. In the case of the power MOSFET integral diodes, the maximum current carried by the diode must match the power MOSFET rated current because in these applications, the diode current in one transistor invariably flows through another transistor in the circuit. Due to the high on-resistance of high voltage power MOSFETs, their forward conduction current density is

404 Chapter 7 : POWER MOSFET

limited to below 100 Amperes per cm^2. At these relatively low current densities for the integral diode, its forward voltage drop will be low (about 1 volt). It can, therefore, easily sustain the current requirements of the intended applications.

The primary difficulty with using the integral diodes of power MOSFETs is with their reverse recovery characteristics. Due to the high lifetime in the drift region of as-fabricated devices, the integral diodes exhibit very slow reverse recovery with large reverse recovery charge. In these diodes, a very large reverse recovery current (I_P) is observed. The peak reverse recovery current will increase with increasing di/dt (i.e., with higher switching speed). Unfortunately, the peak reverse current (I_P) flows through the transistors in the circuit imposing power dissipation and stress upon them. In addition, some diodes can go through the reverse recovery with an abrupt change in current. This behavior, called *snappy recovery* associated with a high rate of change of current, generates a very high voltage ($L_s$di/dt) across the stray inductance (L_S) which was discussed in Section 7.6 (see Fig. 7.38). This voltage appears across the power MOSFET terminals and can take the device beyond its safe-operating-area.

7.8.1 Switching Speed

Improvement of the reverse recovery characteristics of the integral diode in power

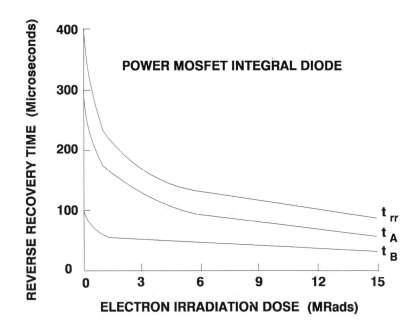

Fig. 7.48 Improvement of reverse recovery time for the integral diode in a power MOSFET by electron irradiation.

MOSFETs was first accomplished by using electron irradiation. Electron irradiation can be used to introduce recombination centers into the drift region as discussed in Chapter 2. During reverse recovery, the resulting lifetime reduction can greatly reduce the reverse recovery charge and the reverse recovery time. The improvement that can be achieved by using the electron irradiation process is illustrated in Fig. 7.48. An important feature of the process is the excellent control over the reverse recovery characteristics obtained by the ability to choose the radiation dose precisely. The reverse recovery time can be tailored from 350 nanoseconds for the as-fabricated devices to less than 100 nanoseconds at a dose of 16 Mrads. Since even good discrete P-i-N rectifiers exhibit a reverse recovery time of over 200 nanoseconds, the radiation process can improve the power MOSFET integral diode to the level that it can replace the external flyback diode. Another important observation with regard to the electron irradiation process is that the ratio of the time intervals for the two segments (t_A/t_B) shown in Fig. 4.29 remains approximately constant at a value of less than two at all radiation doses. This is an indication that no snap recovery phenomenon occurs as the diode speed is increased by the electron irradiation.

It should be noted that the electron irradiation process introduces a positive charge in the gate oxide. This causes a reduction in the threshold voltage of n-channel devices and an increase in the threshold voltage of p-channel devices. It has been observed that the threshold voltage can become negative for n-channel devices after electron irradiation, causing a drastic reduction in the forward blocking capability. Fortunately, the oxide charge due to electron irradiation can be annealed out at relatively low temperatures (150°C to 200°C). At these temperatures, the recombination centers in the bulk that improve the reverse recovery of the integral diode are stable. Other power MOSFET parameters remain essentially unaffected by the electron irradiation process. Consequently, it is an excellent method for obtaining power MOSFETs with high quality integral diodes because electron irradiation, and subsequent annealing, can be performed after complete device fabrication in a highly controlled manner.

The speed of the integral diode of power MOSFETs can also be improved by gold or platinum doping. A problem observed when using this process is an increase in the on-resistance when higher gold or platinum diffusion temperatures are used because the solid solubility of gold and platinum increases with temperature. This causes a greater compensation due to deep levels arising from the increase in gold or platinum concentration. The compensation becomes apparent when the concentration of the deep level becomes comparable to the background doping level. Since the capture cross-section for the platinum deep level responsible for minority carrier lifetime reduction is larger than for gold, it produces less compensation for a given reverse recovery speed. Experimental comparison of gold and platinum doping indicates that platinum doping will provide a 50 percent lower reverse recovery time when compared with gold for the same on-resistance. It is worth pointing out that the gold or platinum doping processes must be performed during device fabrication prior to contact metallization. This provides much less control over the minority carrier lifetime compared with electron irradiation, leading to a wider spread in device characteristics. Further, the platinum and gold atoms tend to segregate to regions of the device with strain such as the N^+ substrate and under the gate oxide. This phenomenon can severely degrade the MOS characteristics. The absence of the segregation effect in the case of electron irradiation makes the compensation negligible.

7.8.2 Parasitic Bipolar Transistor

A potential problem with the use of the integral diode in the power MOSFET arises from the existence of the bipolar transistor in the structure. The equivalent circuit for the integral

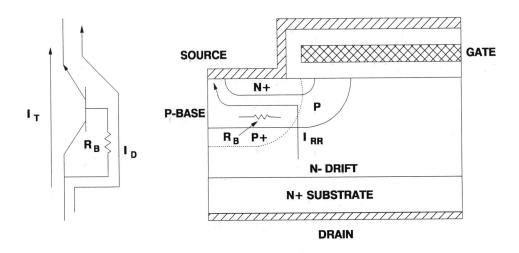

Fig. 7.49 Parasitic bipolar transistor activation when integral diode is utilized.

diode contains this bipolar transistor as shown in Fig. 7.49. The primary diode current flow path is shown by I_D in this figure. This current flow produces a voltage drop across the base resistance (R_B) which forward biases the emitter-base junction of the bipolar transistor. If the base resistance is not sufficiently low, the bipolar transistor will be activated and a significant current can flow along path (I_T) because even in the inverse mode the bipolar transistor has some gain. When the diode in the structure goes through its reverse recovery, the high voltage across the transistor can result in its going into second breakdown leading to catastrophic failure. This problem can be avoided by the addition of a deep P^+ diffusion in the DMOS cell indicated by the dashed lines in Fig. 7.49. This allows a reduction of R_B without altering the threshold voltage.

7.9 HIGH TEMPERATURE PERFORMANCE

Power MOSFETs exhibit very good high temperature operating characteristics. Devices are commercially available with peak junction temperature ratings as high as 200°C. The ability of a power MOSFET to operate at elevated temperatures is related to the absence of minority carrier injection. The effect of increasing temperature upon various device parameters is discussed in this section.

7.9.1 On-Resistance

An important merit of the power MOSFET is an increase in its on-resistance with increasing temperature. Although this may appear at first sight to be an undesirable feature due to increase in power dissipation, it imparts an important benefit in terms of device stability and paralleling. Local variations in drift region resistivity are unavoidable during the fabrication of power devices. Some locations inside the device will contain a lower on-resistance compared with the rest of the device. This promotes localization of current along the paths of least resistance. Localization of current produces additional heating. Fortunately, in the case of the power MOSFET, the mobility for holes and electrons decreases with temperature, causing the local resistivity to increase. This will tend to homogenize the current distribution and prevent thermal runaway. Further, when power MOSFETs are connected in parallel, the increase in their on-resistance with temperature promotes good current sharing without the need for external ballasting used in the case of bipolar transistors.

A typical example of the increase in the on-resistance of a power MOSFET with

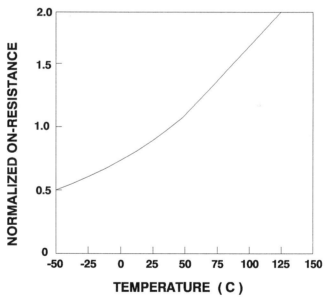

Fig. 7.50 Increase in the on-resistance with temperature for a power MOSFET.

temperature is provided in Fig. 7.50. This curve closely follows the mobility change with temperature discussed in Chapter 2. A general expression that can be used to predict the variation of the on-resistance of p- and n-channel power MOSFETs is:

$$R_{on}(T) = R_{on}(25°C) \left(\frac{T}{300}\right)^{2.3} \quad (7.127)$$

where T is the absolute temperature.

7.9.2 Transconductance

The reduction in the mobility with increasing temperature adversely affects the transconductance of power MOSFETs. A typical example of the variation of the

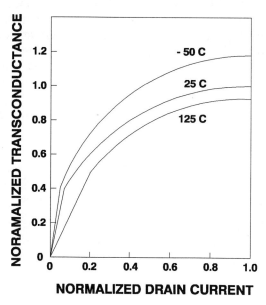

Fig. 7.51 Decrease in the transconductance of a power MOSFET with temperature.

transconductance is provided in Fig. 7.51. The transconductance will follow the mobility variation because the other terms that determine its value are relatively unaffected by temperature. Consequently, the variation in the transconductance with temperature can be predicted by using an expression similar to that for the on-resistance:

$$g_m(T) = g_m(25°C) \left(\frac{T}{300}\right)^{-2.3} \quad (7.128)$$

Care must be taken during device design to provide sufficient channel periphery to allow for the degradation in the transconductance if high temperature operation is anticipated.

7.9.3 Threshold Voltage

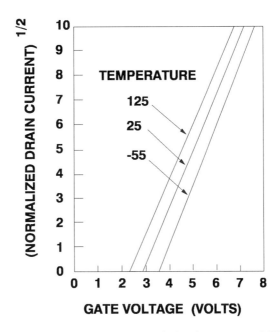

Fig. 7.52 Typical transfer characteristics for a power MOSFET.

In Section 7.4.3, it was shown that the drain current of a power MOSFET will vary as the square of the gate drive voltage ($V_G - V_T$) in the saturated current regime. A plot of the square root of the drain current as a function of the gate voltage can, therefore, be used to obtain a measurement of the threshold voltage. A typical example of this transfer characteristic is shown in Fig. 7.52. The measured drain current indeed varies as the square of the gate voltage resulting in the straight lines. The intercept of these lines at $I_D = 0$ provides the threshold voltage. It can be seen that the threshold voltage decreases with increasing temperature.

In the absence of a work function difference and oxide charge, it was shown in Section 7.4.2, that the threshold voltage is given by:

$$V_T = \frac{\sqrt{4\,\epsilon_s\,q\,N_A\,\psi_B}}{C_{ox}} + 2\,\psi_B \tag{7.129}$$

Differentiating this expression with respect to temperature:

$$\frac{dV_T}{dT} = \frac{d\psi_B}{dT}\left[\frac{1}{C_{ox}}\sqrt{\frac{\epsilon_s\,q\,N_A}{\psi_B}} + 2\right] \tag{7.130}$$

This expression is valid even in the presence of a work function difference and oxide charge because these parameters are not strongly affected by the temperature. The bulk potential (Ψ_B) varies with temperature because the energy gap changes with temperature:

$$\frac{d\Psi_B}{dT} = \frac{1}{T}\left[\frac{E_g(T=0)}{2\,q} - |\Psi_B(T)|\right] \quad (7.131)$$

From these expressions, the rate of variation of the threshold voltage with temperature can be calculated as a function of the background doping level and oxide thickness. Note that the threshold voltage will vary at a higher rate for thicker gate oxides and higher substrate doping levels. During the design and fabrication of power MOSFETs, it is important to provide an adequate room temperature threshold voltage that will allow operation at high temperatures with sufficient margin to ensure noise immunity and protection from inadvertent turn-on when a high [dV/dt] is applied.

7.10 DEVICE STRUCTURES AND TECHNOLOGY

Two fundamentally different processes have evolved for the fabrication of discrete vertical channel power MOSFETs. One of these processes relies upon the planar diffusion of the P-base and N^+ emitter from a common window defined by the refractory gate electrode to form the DMOS structure shown in Fig. 7.2, while the other process utilizes reactive ion etching of the silicon surface to form the U-groove structure shown in Fig. 7.3. Due to the poor performance of the V-groove power MOSFET structure shown in Fig. 7.1, its process will not be discussed. The discussion here will be confined to the case of n-channel devices. The fabrication of complementary (p-channel) devices is similar.

7.10.1 DMOS (Planar) Structure

For the fabrication of the planar power MOSFET structure shown in Fig. 7.2, the wafer orientation can be chosen to provide the highest surface mobility in order to achieve a low channel contribution to the on-resistance. For this reason, these devices are fabricated using (100) oriented wafers. Due to the extremely rapid increase in on-resistance with increasing breakdown voltage, power MOSFETs are confined to blocking voltages of less than 1,000 volts. The ideal depletion width for these devices is less than 100 microns. They must be fabricated by growing lightly doped epitaxial layers (typically 5 to 50 microns thick) on heavily doped substrates. In the case of n-channel devices, the N^+ substrates are antimony doped with a resistivity of 0.01 ohm-cm. Antimony is chosen as the dopant because its autodoping effects are small during epitaxial growth. For low on-resistance devices, arsenic doped substrates with a resistivity of 0.001 ohm-cm are preferred.

After epitaxial growth, a thick field oxide is grown. This oxide is patterned to create the desired planar diffused edge termination. Sometimes the processing of the diffusion for the edge termination is combined with the DMOS cell fabrication. Following the termination diffusion, the device active area (central region where the DMOS cells are to be fabricated) is opened up. The gate oxide is then grown to a typical thickness of 1,000 °A. Recently device manufacturers have been scaling down the gate oxide thickness to obtain an increase in the transconductance so as to make it possible to drive the power MOSFETs at voltages compatible with low voltage logic circuits.

A layer of heavily doped polysilicon (or other refractory gate material) is now deposited. After oxidizing the polysilicon, a photomasking step is used to pattern the polysilicon to the surface of the gate oxide. A cross-section of a small segment of the active region is shown at

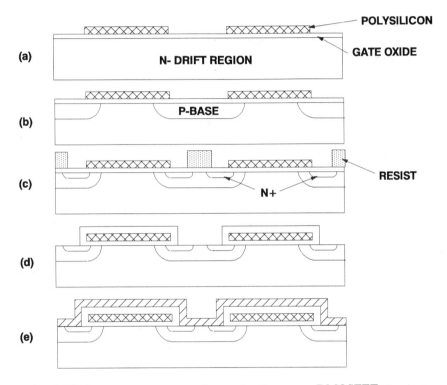

Fig. 7.53 Process sequence used to make the power DMOSFET structure.

this point in the process in Fig. 7.53(a). At this point in the process, the cell window has been defined. Boron is now implanted through the gate oxide to a dose (typically 10^{14} per cm^2) that is necessary to obtain the desired threshold voltage in conjunction with the N$^+$ emitter diffusion. The boron is driven-in with a wet oxidation step to obtain the structure shown in Fig. 7.53(b). An oxide island is now patterned at the center of each cell window to mask the N$^+$ emitter diffusion. This oxide island leaves a portion of the P-base accessible at the surface in each cell allowing good contacts to be subsequently made to it.

The patterning of the oxide island is one of the critical alignment steps during the DMOS process. If the oxide island is misaligned with respect to the edges of the polysilicon in the cell window, the length of the N$^+$ emitter will increase, causing degradation of the [dV/dt] and safe-operating-area. After the oxide island is formed, the N$^+$ emitter is diffused either by using phosphorus predeposition or by ion implantation. The latter process provides greater control over the emitter profile and threshold voltage. A typical dose is 1×10^{15} per cm^2. The device cross-section at this point is shown in Fig. 7.53(c). After the N$^+$ emitter is diffused, a thick layer of oxide is deposited by using low pressure chemical vapor deposition (LPCVD). This provides a conformal insulating film over the top and sides of the polysilicon gate electrode.

The oxide layer is now patterned to form the contact windows as illustrated in Fig. 7.53(d). This is another critical photolithographic step in the DMOS process because misalignment can either cause poor contact to the P-base region or lead to an overlap of the contact with the polysilicon at the cell edges. This overlap will produce a short between source and gate. The final step in the device process consists of metallization to create the structure shown in Fig. 7.53(e). The DMOS process described here provides the fundamental basis for the fabrication of modern power MOSFETs. Many variations of the process have been explored in an attempt either to eliminate some of the critical alignment steps or to provide self-alignment of the N$^+$ emitter and contact windows.

7.10.2 UMOS Structure

As in the case of the DMOS process, the UMOS process begins with the growth of a lightly doped N-type epitaxial layer on an N$^+$ substrate. After the growth and patterning of a thick field oxide, a boron diffusion is performed to create a P-base region across the entire active area of the device and to form the multiple field ring edge termination. The oxide layer grown during the P-base diffusion step is next patterned to define windows for the source region, and an N$^+$ implant is performed followed by its drive-in step. A cross-section of the active region at this point in the process is shown in Fig. 7.54(a). Note that, in this process, the N$^+$ source region extends across the entire top surface of the mesa region. The contact to the P-base region is made at selected locations along the length of the fingers orthogonal to the cross-section shown in Fig. 7.54.

A silicon nitride layer is now deposited on the top surface with a thin oxide buffer layer. This nitride layer is used later in the process to allow selective oxidation of the trench sidewalls. A thick oxide layer is deposited on top of the nitride layer to act as a mask during the reactive ion etching step used to form the trenches. This oxide/nitride stack is patterned using the third mask to define windows where the trenches are to be formed. Reactive ion etching is performed using conditions that favor the formation of vertically walled trenches. An example of such an etch is SF$_6$ mixed with oxygen. This method has been shown to not only produce vertical sidewalls but to also create a rounded bottom for the trench. This reduces the high electric field normally observed at the corners of the trenches. A cross-section of the active region at this point in the process is shown in Fig. 7.54(b).

A polysilicon layer is now deposited on the wafer using a conformal coating process. The polysilicon will then fill the U-groove formed by the trench etch step and will also cover

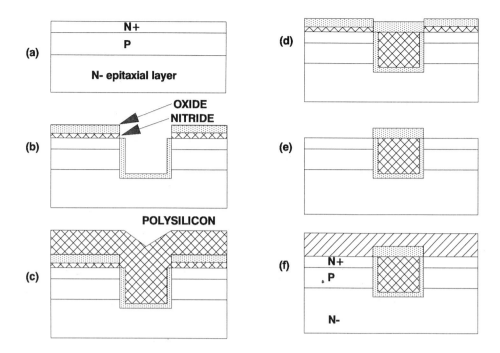

Fig. 7.54 Process sequence used to create the power UMOSFET structure.

the top surface. A cross-section of the active region at this point in the process is shown in Fig. 7.54(c). A polysilicon planarization process is now used to etch away the polysilicon on the surface above the mesa regions where the source contact must be made. There are several planarization methods described in the literature. After planarization of the polysilicon, an oxide layer is grown to act as the insulator between the gate and source electrodes. A cross-section of the active region at this point in the process is shown in Fig. 7.54(d).

The silicon nitride layer is now stripped. This can be performed, without removing the oxide layer on top of the polysilicon gate regions, by using hot phosphoric acid. This exposes the mesa surface to enable making contacts to the source region. Since no mask layer is needed to define the source contact in the mesa regions, this process produces self-aligned contacts which enables minimizing the mesa and trench widths. This is important for obtaining the lowest possible specific on-resistance for the power UMOSFET. A mask is now used to pattern a contact window for the polysilicon gate. A cross-section of the active region at this point in the process is shown in Fig. 7.54(e). The source contact metal is now evaporated and patterned to complete the device fabrication sequence. A cross-section of the active region at this point in the process is shown in Fig. 7.54(f).

As in the case of the DMOS process, the UMOS process described here is intended to only provide a fundamental sequence for device fabrication. Many innovative techniques to simplify the process and reduce the alignment problems have been attempted. Using some of these processes, it is possible to form a short between with P-base region and the N^+ source region within the mesa region. This provides the lowest shunting resistance for the parasitic bipolar transistor.

7.10.3 Gate Oxide Fabrication

Since the electrical characteristics of the MOS field effect transistor are modulated by the application of an electric field across the gate oxide, the oxide properties play an important role in determining the device characteristics. A large amount of effort has consequently been expended in the semiconductor industry to develop an understanding of oxidation and annealing conditions upon the oxide properties. The important process criteria for the growth of the gate oxide are provided in this section.

The silicon dioxide that is used as the insulating barrier between the gate metal and the channel can be grown by subjecting the silicon surface to an oxygen ambient at elevated temperatures. The ambients that have been most commonly used are either dry oxygen or oxygen containing water vapor. Lower interface state densities have been reported when the oxidation ambient contains water vapor. However, the slower oxidation rate in dry oxygen allows greater control over the gate oxide thickness. The higher interface density in this case can be reduced by increasing the oxidation temperature and subsequently annealing the oxide as described below.

Before the effects of annealing upon the oxide-silicon interface are considered, it is worth pointing out that the lowest fixed interface charge and surface state density have been observed for the (100) plane. In the case of the devices made using the anisotropic etching technology, the gate oxide must be formed on a (111) surface since this surface is exposed by the preferential etch. In contrast, the devices made by planar diffusion techniques (DMOS) can be fabricated from (100) oriented wafers in order to obtain a lower fixed interface charge and surface state density.

Low temperature annealing of the oxide films either in a mixture of hydrogen and nitrogen or in a nitrogen ambient containing water vapor has proven to be very effective in decreasing the interface state density. A 1,000°C anneal in dry nitrogen must be performed prior to the low temperature annealing to achieve a minimum interface state density. It is believed that, during the subsequent low temperature anneal, the hydrogen in the oxide diffuses to the oxide-silicon interface and ties up the "dangling" bonds, reducing the surface state density. The optimum annealing temperature is about 400°C, with a low surface state density being observed after 60 minutes of annealing time. A similar annealing of the interface states can be accomplished after the aluminum gate metallization has been applied. A faster annealing rate has been observed in this case. This is believed to arise from the generation of hydrogen at the aluminum by the reduction of water vapor in the oxide. This process has the drawback of the possible introduction of mobile ions into the silicon dioxide from the aluminum, which can lead to an instability in the threshold voltage.

A drift in the threshold voltage due to the migration of mobile ionic species in the oxide has been one of the most serious problems in MOS gated field effect transistors. This is particularly true in the case of the devices made by the preferential etching of the V-groove with a potassium hydroxide-isopropanol solution. In this case, the complete removal of potassium ions from the silicon surface prior to the growth of the gate oxide is imperative. The high mobility of alkali ions in silicon dioxide at relatively low temperatures (in the range of 200°C to 300°C) has been conclusively demonstrated. The migration of these alkali ions can be suppressed by either doping the oxide with phosphorus or capping it with a layer of silicon nitride. However, polarization effects in the phosphosilicate glass films and the trapping of charge at the oxide-nitride interface have prevented the application of these techniques to the gate area of power MOSFET devices.

A more promising process innovation has been the addition of halogens to the gas stream prior to and during oxidation. The exposure of the oxidation tube to a mixture of HCl and dry O_2 has been found to be effective in reducing the mobile ion contamination of oxides grown subsequently in the tube. The mobile ion contamination in the oxide has also been found to decrease with increasing HCl concentration in the gas stream during oxidation. However, it is important to keep in mind that HCl will etch silicon at the oxidation temperature if its concentration exceeds a few percent. Typically, an HCl concentration of about one percent by volume in the gas stream is effective in suppressing ionic contamination. The addition of chlorine during oxidation has also been reported to lower the interface state density and the fixed oxide charge density, as well as to improve the dielectric breakdown strength of the oxide. It is interesting to note that these improvements are observed only during dry oxidation and that the addition of HCl during oxidation with water vapor has been found to have no effect on the oxide properties.

In addition to the effect of the oxide thickness upon the gate threshold voltage, the growth of the oxide over the channel can also affect the threshold voltage by the redistribution of the boron in the channel during oxidation. Experiments conducted on the thermal oxidation of silicon doped with gallium, boron, and indium have found that the surface concentration of these impurities decreases after oxidation. This occurs because these impurities segregate into the oxide, thus depleting the silicon surface. Impurity segregation can seriously alter the diffusion profile of boron in the P-base region of both the DMOS and UMOS devices and thus influence the threshold voltage as well as the penetration of the depletion layer into the P-base region when the device is blocking current flow. Several models have been developed to allow the calculation of the diffusion profile of impurities with inclusion of the impurity redistribution during oxidation. These models have also been applied to the calculation of the effect of oxidation upon the threshold voltage of MOS field effect transistors. In addition, the strain at the surface under the polysilicon gate of the DMOS structure alters the lateral diffusion of impurities. It has been found that the boron diffusion is retarded while the phosphorus diffusion is enhanced. This effect can severely alter the channel length of both n- and p-channel DMOS power MOSFETs.

7.10.4 Cell Topology

In previous discussions of power MOSFETs in this chapter, cross-sections of devices were shown without consideration of the cell layout on the surface. For the VMOS device structure, the anisotropic etching of the grooves can be accomplished only when the grooves are oriented along the <110> directions on the (100) wafer surface. This constraint allows only two design options for the surface topology of VMOS devices, namely, long stripes or rectangular cells with V-grooves running perpendicular to each other. Even the latter option is impractical due to the poor contour at the corners where the V-grooves intersect at right angles. V-groove power MOSFET cells are therefore confined to long fingers with interdigitation of source and gate.

In contrast, the DMOS and UMOS structures allow any conceivable cell topology as long as it meets all technological constraints such as alignment tolerances. The cell windows that are commonly used for the design of power MOSFETs have linear, square, circular, hexagonal, or

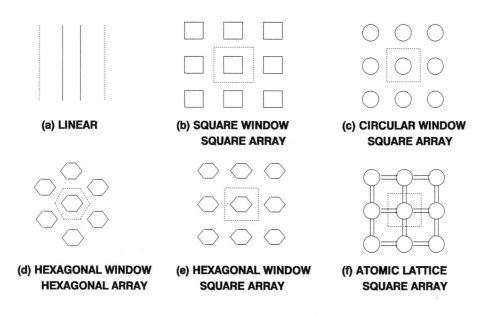

Fig. 7.55 Cell topologies used to make power DMOSFETs.

the atomic-lattice-layout (A-L-L) shapes as illustrated in Fig. 7.55. These windows can be located in either a square cell or hexagonal cell pattern. The impact of the layout of these cells upon the resistance has been analyzed under the assumption that they all have the same drift region doping concentration. It has been shown that the on-resistances of all cellular designs are equal if the size of the cell window and the ratio of the area of the cell window to the total cell area are kept the same. However, it has been found that the drift region doping concentration must be adjusted based upon the electric field crowding within the cell. When this is taken into account, the A-L-L design has been found to be superior to the other designs. The A-L-L design also has a lower drain overlap capacitance, which is favorable for operation at

Chapter 7 : POWER MOSFET   417

high frequencies.

7.11 SILICON CARBIDE POWER MOSFETS

As discussed earlier in this chapter, the specific on-resistance of the silicon power DMOSFET and UMOSFET structures approaches the ideal specific on-resistance of the drift region. Unfortunately, the ideal specific on-resistance for silicon, given by Eq. (7.58), increases rapidly with the breakdown voltage. For breakdown voltages above 200 volts, the ideal specific on-resistance for the n-channel silicon MOSFET becomes greater than 1×10^{-2} ohm-cm^2. This implies that the on-state voltage drop will exceed 10 volts for an on-state current density of 100 amperes per cm^2, resulting a very high power dissipation within the devices. Although it is possible to reduce the on-state power loss by decreasing the on-state current density, this is undesirable because of the corresponding increase in device area which leads to higher chip cost.

From a fundamental physics viewpoint, a lower specific on-resistance for the drift layer is achievable if the semiconductor has a higher breakdown electric field strength. Using the basic equation derived in Chapter 3 for an abrupt parallel plane junction, it can be shown that for obtaining a desired breakdown voltage (BV), the drift region must have a doping concentration N_D of:

$$N_D = \frac{\epsilon_s E_c^2}{2 q BV} \quad (7.132)$$

and a thickness W_D of:

$$W_D = \frac{2 BV}{E_c} \quad (7.133)$$

where E_c is the critical electric field at which avalanche breakdown occurs in the semiconductor. The specific on-resistance of the drift region, which was also defined as the ideal specific on-resistance, is then given by:

$$R_{on-sp(ideal)} = \frac{W_D}{q \mu_n N_D} = \frac{4 BV^2}{\epsilon_s E_c^3 \mu_n} \quad (7.134)$$

Thus, the ideal specific on-resistance decreases inversely proportional to the mobility and as the cube of the breakdown electric field strength. The denominator ($\epsilon_s E_c^3 \mu_n$) in Eq. (7.134) has been referred to as *Baliga's figure of merit for unipolar power devices*.

By using the known material properties of semiconductors, it is possible to select those that will exhibit a lower ideal specific on-resistance when compared with silicon by using this expression. It has been found that the most promising semiconductors are gallium arsenide, whose Baliga's figure-of-merit is 12.7 times larger than silicon, and silicon carbide whose

Baliga's figure-of-merit is 200 times larger than silicon. Although some research has been performed on the fabrication of vertical power MESFETs from gallium arsenide, this material has been found to be difficult to work with due to the dissociation of the compound during processing. In contrast, silicon carbide offers a much larger improvement in the ideal specific on-resistance and is stable even at extremely high temperatures.

In terms of the fabrication of silicon carbide based power MOSFETs, there are several outstanding problems that must be overcome. First, the diffusion rates for impurities in silicon carbide are orders of magnitude lower than for silicon. This precludes the possibility for fabrication of the DMOS cell structure using polysilicon as the gate electrode. The UMOS gate structure is more amenable to fabrication if the P-base and N^+ source regions are epitaxially grown. As discussed earlier in this chapter, the UMOS gate structure also offers a substantial increase in the channel density, which reduces the channel resistance contribution. In spite of this, calculations of the specific on-resistance of silicon carbide power UMOSFETs indicate very high contributions from the channel resistance resulting from the low electron inversion layer mobilities reported in the literature. This indicates that silicon carbide power MOSFETs would not be competitive with silicon devices unless the breakdown voltages exceed 1000 volts.

This problem can be overcome by using a new power switch configuration, called the *Baliga-pair*, which, as shown in Fig. 7.56, consists of a silicon power MOSFET connected in series with the source region of a silicon carbide high voltage power MESFET (which has a

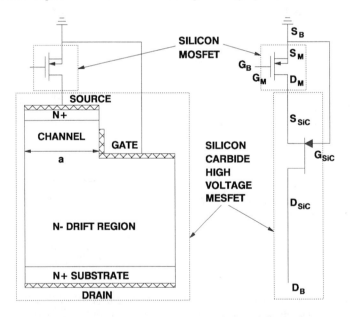

Fig. 7.56 The *Baliga-pair* power switch configuration.

Schottky barrier gate structure) or JFET (which has a P-N junction gate structure). Although a vertical channel trench-gate MESFET structure is illustrated in this figure, a lateral device could instead be used. It is important to note that the gate region of the silicon carbide

MESFET is connected to the reference terminal (or source region of the silicon power MOSFET) and that the composite switch is controlled by a signal applied to the gate of the silicon power MOSFET. The basic operating principles of this switch are discussed below.

If the half-width 'a' of the MESFET channel is designed to be larger than the zero bias depletion width of the MESFET gate structure, the MESFET will behave as a depletion-mode (or normally-on) device structure. If an increasing positive bias is applied to the drain D_B of the Baliga-pair with the gate G_B shorted to the source S_B, the voltage will be initially supported by the silicon MOSFET because the MESFET channel is not depleted. This will result in an increase in the potential of the source region S_{SiC} of the silicon carbide MESFET. Since the gate G_{SiC} of the silicon carbide MESFET is at zero potential, this will produce a reverse bias across the gate-source junction of the MESFET. As the voltage applied to the drain D_B is increased, this reverse bias will produce a pinch-off of the MESFET channel by the extension of a depletion region from the gate contact. Once the MESFET channel pinches-off, any further increase in the voltage applied to the drain D_B will be supported by the extension of a depletion region in the drift region of the silicon carbide MESFET. It has been shown by two-dimensional numerical simulations that, once the channel is pinched-off, the potential at the drain D_M of the MOSFET remains relatively constant and independent of the voltage applied to the drain D_B of the composite switch. Since a channel pinch-off voltage of less than 25 volts can be easily designed, this implies that a silicon power MOSFET with relatively low breakdown voltage can be used in conjunction with a high voltage silicon carbide MESFET to form the Baliga-pair.

This is important from the point of view of obtaining a low total on-state voltage drop for the composite switch. In order to turn on the Baliga-pair, a positive gate bias is applied to the gate G_M of the silicon power MOSFET, which also serves as the gate G_B of the composite switch. This switches the silicon MOSFET to its highly conductive state. When a positive voltage is applied to the drain D_B, current can now flow through the undepleted MESFET channel and the silicon MOSFET. The simulations have demonstrated that the specific on-resistance of the silicon carbide MESFET is very close to the ideal specific on-resistance for the drift region because of a uniform current distribution in the drift region. The resistance contribution from the MESFET channel increases the specific on-resistance by less than 25 percent because the current is transported in the bulk and not along a surface. Thus, the Baliga-pair is projected to have on-state voltage drops of only 0.1 volt when the MESFET is designed to block up to 1000 volts.

The Baliga-pair has several other important attributes. The first is an excellent forward biased safe operating area (FBSOA). This is achieved by simply reducing the gate bias applied to the switch until it approaches the threshold voltage for the silicon power MOSFET. In this case, when a voltage is applied to the drain D_B, the MOSFET operates in its current saturation regime. This limits the current flowing through the composite switch. When the voltage applied to the drain D_B is increased, the voltage across the MOSFET increases until the channel of the MESFET is pinched-off, allowing high voltages to be sustained with a current flow dictated by the MOSFET channel. The numerical simulations indicate a square FBSOA for the Baliga-pair because no minority carrier transport is involved.

The absence of minority carrier transport in the Baliga-pair is also important in obtaining a high switching speed. Since both the silicon MOSFET and the silicon carbide MESFET are unipolar devices, the turn-off time for the composite switch is determined by the charging and discharging time constants for the silicon MOSFET. Well optimized silicon power MOSFETs can be designed at the required low breakdown voltages, resulting in a very high switching speed for the Baliga-pair. This is attractive for the reduction of power losses in medium/high voltage power electronics systems operating at high frequencies.

Another attribute of the Baliga-pair is that it incorporates an excellent integral flyback diode. In the case of the silicon power MOSFET, it was shown in Section 7.8 that the junction between the P-base region and the N-drift region can be utilized as a reverse conducting (flyback) diode. However, this diode operates with the injection of minority carriers into the drift region, which compromises the switching speed and power losses in the devices. In the case of the silicon carbide MOSFET structure, there is an additional disadvantage that the potential required for the injection of minority carriers is much larger (typically 3 volts) when compared with silicon (typically 1 volt) due to its larger energy band gap. This results in a severe increase in the power losses for the flyback rectifier. In contrast to this, in the case of the Baliga-pair, the application of a negative bias to the drain D_B forward biases the Schottky barrier gate structure. As discussed in Chapter 4 on power rectifiers, the silicon carbide Schottky barrier rectifier has been demonstrated to have excellent on-state and switching characteristics because it is a unipolar device. Thus, the Baliga-pair also contains an excellent flyback diode if implemented by using a high voltage silicon carbide MESFET structure.

The basic operation of the Baliga-pair has been verified by using silicon power MOSFETs and silicon high voltage vertical channel JFETs. The implementation of this switch with silicon carbide devices awaits the successful fabrication of silicon carbide MESFET structures with high breakdown voltages.

7.12 TRENDS

Power MOSFETs have the following attributes: (a) a voltage controlled characteristic with low input gate power, (b) a stable negative temperature coefficient for the on-resistance, (c) a wide safe-operating-area, and (d) a high switching speed. Due to these features, when first introduced, the devices were expected to replace power bipolar transistors in many applications. This has occurred for systems with low operating voltages. Examples are switch-mode power supplies, computer peripherals, and automotive electronics. This displacement has not occurred for systems operating at high voltages due to the high on-resistance of power MOSFETs designed to support voltages above 200 volts. With rapid progress along the learning curve, the manufacturing cost of power MOSFETs now rivals that for the bipolar transistors. Since the on-resistance of the silicon power MOSFETs is already close to the ideal value, progress in increasing their power handling capability can be achieved only by increasing the area of the device. This is expected to occur by a pace dictated by improvements in process technology that lead to a reduction in defect density. Recently, it has been theoretically shown that replacement of silicon with silicon carbide would lead to a 200 fold reduction in the specific on-resistance. This offers a tremendous opportunity for improvement of device characteristics if the technological problems with manufacturing silicon carbide devices can be overcome.

REFERENCES

1. W. Shockley and G.L. Pearson, "Modulation of conductance of thin films of semiconductors by surface charges," Phys. Rev., Vol. 24, pp. 232-233 (1948).

2. D. Kahng and M.M. Atalla, IRE-AIEE Solid State Device Research Conference (1960).

3. M.N. Darwish and K. Board, "Optimization of breakdown voltage and on-resistance of VDMOS transistors," IEEE Trans. Electron Devices, Vol. ED-31, pp. 1769-1773 (1984).

4. V.A.K. Temple and P.V. Gray, "Theoretical comparison of DMOS and VMOS structures for voltage and on-resistance," IEEE Int. Electron Devices Meeting Digest, Abs. 4.5, pp. 88-92 (1979).

5. C.G.B. Garrett and W.H. Brattain, "Physical theory of semiconductor surfaces," Phys. Rev., Vol. 99, pp. 376-387 (1955).

6. V.G.K. Reddi and C.T. Sah, "Source to drain resistance beyond pinch-off in metal-oxide-semiconductor transistors (MOST)," IEEE Trans. Electron Devices, Vol. ED-12, pp. 139-141 (1965).

7. S.C. Sun and J.D. Plummer, "Modelling of the on-resistance of LDMOS, VDMOS, and VMOS power transistors," IEEE Trans. Electron Devices, Vol. ED-27, pp. 356-367 (1980).

8. C. Hu, "A parametric study of power MOSFETs," IEEE Power Electronics Specialists Conference Record, pp. 385-395 (1979).

9. B.J. Baliga, "Switching lots of watts at high speed," IEEE Spectrum, Vol. 18, pp. 42-48 (1981).

10. C. Hu, "Optimum doping profile for minimum ohmic resistance and high breakdown voltage," IEEE Trans. Electron Devices, Vol. ED-26, pp. 243-244 (1979).

11. H. Ikeda, K. Ashikawa, and K. Urita, "Power MOSFETs for medium-wave and short-wave transmitters," IEEE Trans. Electron Devices, Vol. ED-27, pp. 330-334 (1980).

12. H. Esaki and O. Ishikawa, "A 900 MHz, 100 watt, VD-MOSFET with silicide gate self-aligned channel," IEEE Int. Electron Devices Meeting Digest, Abstract 16.6, pp. 447-449 (1984).

13. K. Shenai, et al, "A novel high frequency power FET structure fabricated using LPCVD WSi$_2$ gate and LPCVD W source contact technology," IEEE Int. Electron Devices Meeting Digest, Abstr. 34.4, pp. 809-912 (1988).

14. T. Syau, P. Venkatraman, and B.J. Baliga, "Extended trench gate power UMOSFET structure with ultra-low specific on-resistance," Electronics Letters, Vol. 28, pp. 865-867 (1992).

15. B.J. Baliga, T. Syau, and P. Venkatraman, "The accumulation mode field effect transistor -a ultra-low on-resistance MOSFET," IEEE Electron Device Letters, Vol. EDL-13, pp. 427-429 (1992).

16. S. Clemente and B.R. Pelly, "Understanding the power MOSFET switching performance," Proc. IEEE Industrial Applications Society Meeting, Abs. 32.B, pp. 763-776 (1981).

17. R. Severns, "dV/dt effects in MOSFETs and bipolar junction transistor switches," IEEE Power Electronics Specialists Conference Record, pp. 258-264 (1981).

18. D.S. Kuo, C. Hu, and M.H. Chi, "dV/dt breakdown in power MOSFETs," IEEE-Electron Device Letters, Vol. EDL-4, pp. 1-2 (1983).

19. C. Hu and M.H. Chi, "Second breakdown of vertical power MOSFETs," IEEE Trans. Electron Devices, Vol. ED-29, pp. 1287-1293 (1982).

20. B.R. Pelly, "Power MOSFETs - A status review," International Power Electronics Conference Record, pp. 19-32 (1983).

21. B.J. Baliga and J.P. Walden, "Improving the reverse recovery of power MOSFET integral diodes by electron irradiation," Solid State Electronics, Vol. 26, pp. 1133-1141 (1983).

22. Y. Ohata, "New MOSFETs for high power switching applications," Proc. Powercon 11, Abs. C5, pp. 1-11 (1984).

23. R. Wang, J. Dunkley, T.A. DeMassa, and L.F. Jelsma, "Threshold voltage variations with temperature in MOS transistors," IEEE Trans. Electron Devices, Vol. ED-18, pp. 386-388 (1971).

24. Y. Shimada, K. Kato, and T. Sakai, "High efficiency MOS-FET rectifier device," Proc. IEEE Power Electronics Specialists Conference, pp. 129-136 (1983).

25. P.V. Gray and D.M. Brown, "Density of SiO$_2$-Si interface states," Appl. Phys. Letters, Vol. 8, pp. 31-33 (1966).

26. E. Arnold, J. Ladell, and G. Abowitz, "Crystallographic symmetry of surface state density in thermally oxidized silicon," Appl. Phys. Letters, Vol. 13, pp. 413-416 (1968).

27. Y.T. Yeow, D.R. Lamb, and S.D. Brotherton, "An investigation of the influence of low-temperature annealing treatments on the interface state density at the Si-SiO$_2$ interface," J. Phys. D: Appl. Phys., Vol. 8, pp. 1495-1506 (1975).

28. E. Yon, W.H. Ko, and A.B. Kuper, "Sodium distribution in thermal oxide on silicon by radiochemical and MOS analysis," IEEE Trans. Electron Devices, Vol. ED-13, pp. 276-280 (1966).

29. S.R. Hofstein, "Stabilization of MOS devices," Solid State Electronics, Vol. 10, pp. 657-670 (1967).

30. E.H. Snow and B.E. Deal, "Polarization phenomena and other properties of phosphosilicate glass films on silicon," J. Electrochemical Society, Vol. 113, pp. 263-269 (1966).

31. R.J. Kriegler, Y.C. Cheng, and D.R. Colton, "The effect of HCl and Cl$_2$ on the thermal oxidation of silicon," J. Electrochemical Society, Vol. 119, pp. 388-392 (1972).

32. M.C. Chen and J.W. Hile, "Oxide charge reduction by chemical gettering with trichloroethylene during thermal oxidation of silicon," J. Electrochemical Society, Vol. 119, pp. 223-225 (1972).

33. C.M. Osburn, "Dielectric breakdown properties of SiO$_2$ films grown in halogen and hydrogen containing environments," J. Electrochemical Society, Vol. 121, pp. 809-815 (1974).

34. A.S. Grove, O. Leistiko, and C.T. Sah, "Redistribution of acceptor and donor Impurities during thermal oxidation of silicon," J. Appl. Phys., Vol. 35, pp. 2695-2701 (1964).

35. W.G. Allen and C. Atkinson, "Comparison of models for redistribution of dopants in silicon during thermal oxidation," Solid State Electronics, Vol. 16, pp. 1283-1287 (1973).

36. H.G. Lee, J.D. Sansbury, R.W. Dutton, and J.L. Moll, "Modelling and measurement of surface impurity profiles of laterally diffused regions," IEEE J. Solid State Circuits, Vol. SC-13, pp. 445-461 (1978).

37. C. Hu, M.H. Chi, and V.M. Patel, "Optimum design of power MOSFET's," IEEE Trans. Electron Devices, Vol. ED-31, pp. 1693-1700 (1984).

38. B.J. Baliga, "Semiconductors for high voltage vertical channel field effect transistors," J. Applied Physics, Vol. 53, pp. 1759-1764 (1982).

39. B.J. Baliga, "Power semiconductor device figure of merit for high frequency applications," IEEE Electron Device Letters, Vol. EDL-10, pp. 455-457 (1989).

40. M. Bhatnagar and B.J. Baliga, "Comparison of 6H-SiC, 3C-SiC, and Si for power devices," IEEE Trans. Electron Devices, Vol. ED-40, pp. 645-655 (1993).

PROBLEMS

7.1 Determine the threshold voltage of a power MOSFET with a P-base region that is uniformly doped at a concentration of 1×10^{17} per cm^3 when a gate oxide of 1000 °A is used with a highly doped N-type polysilicon gate electrode.

7.2 Calculate the effect of a positive fixed charge density of 1×10^{11} and 1×10^{12} per cm^2 on the threshold voltage of this device.

7.3 The power MOSFET of Problem 7.1 is fabricated with a P-base region of 3 microns in depth and an N^+ source region of 1 micron in depth. Determine the channel width required to obtain a channel resistance of 1 milli-ohm for a gate drive voltage of 15 volts.

7.4 Determine the drift layer doping concentration and thickness required to obtain a breakdown voltage of 100 volts if ideal parallel plane breakdown is achieved at the edge termination.

7.5 Calculate the ideal specific on-resistance for this drift layer.

7.6 Determine the drift layer doping concentration and thickness required to obtain a breakdown voltage of 100 volts if 80 percent of the ideal parallel plane breakdown voltage is achieved at the edge termination.

7.7 A power MOSFET is fabricated using the drift region parameters obtained in Problem 7.6 with the DMOS process and a linear cell topology. The polysilicon gate length is 10 microns, and the polysilicon window is 16 microns. The N^+ source has a length of 5 microns and a junction depth of 1 micron. The P-base region has depth of 3 microns and is uniformly doped with a concentration of 1×10^{17} per cm^3. Determine the total specific on-resistance of the device. Compare the relative contributions of the channel, accumulation, and JFET regions with that for the drift region.

Chapter 7 : POWER MOSFET 425

7.8   Determine the ratio of the specific on-resistance for the power MOSFET obtained in Problem 7.7 with the ideal specific on-resistance determined in Problem 7.5.

7.9   What is the impact of the resistance of an antimony doped substrate with a resistivity of 0.01 ohm-cm and thickness of 500 microns on the specific on-resistance of the power MOSFET from Problem 7.7?

7.10 What is the impact of the resistance of an arsenic doped substrate with a resistivity of 0.002 ohm-cm and thickness of 250 microns on the specific on-resistance of the power MOSFET from Problem 7.7?

7.11 Determine the frequency response of the power MOSFET from problem 7.7 with an active area of 0.1 cm^2. Assume that the gate capacitance is due only to the gate oxide and that the gate drive circuit has an impedance of 50 ohms.

7.12 Calculate the transconductance of this MOSFET in the current saturation regime.

Chapter 8

INSULATED GATE BIPOLAR TRANSISTOR

The bipolar power transistors discussed in Chapter 5 were extensively used for applications operating at above 1 kHz where the switching speed of the gate turn-off thyristors is inadequate. These devices offer good on-state characteristics with low forward voltage drops. However, it was shown that the power bipolar transistor is a current controlled device and that the current gain for the high voltage power transistor is relatively small due to the wide base regions required to prevent reach-through breakdown during the blocking state and the influence of high level injection effects in the base and collector drift regions during current flow at higher current densities. The poor current gain of the power bipolar transistor complicates its gate drive circuit, making the control circuit bulky and expensive.

This stimulated the development of the power MOSFET discussed in Chapter 7. The power MOSFET is a voltage controlled device that can be controlled with relatively small input gate current flow during the switching transient. This makes its gate control circuit simple and amenable to integration. This is attractive in making the power electronics system compact and inexpensive. In addition, the on-resistance of power MOSFETs can be made extremely small if it is designed to block relatively low voltages (below 200 volts), and the device exhibits excellent fast switching capability and safe-operating-area. These features have led to the replacement of power bipolar transistors by power MOSFETs in applications where the DC power source voltage is less than 200 volts. However, the on-resistance of the power MOSFET increases very rapidly when its breakdown voltage is increased. This makes the on-state power losses unacceptable for applications where high DC supply voltages are used.

Based upon these features of the bipolar power transistors and power MOSFETs, it would be attractive to combine the best attributes of both devices. Since bipolar current conduction allows operation at high on-state current densities with a low on-state voltage drop and MOS-gate structures provide ease of gate control, it is advantageous to develop devices where bipolar current transport is controlled via an MOS-gate structure.

A simple approach to combining these features is to use a discrete bipolar transistor and power MOSFET connected in the Darlington configuration as shown in Fig. 8.1. Here, the power MOSFET provides the gate drive current to the bipolar power transistor when a gate drive voltage is applied to the power MOSFET. Thus, the composite device has the high input impedance of an MOS-gate structure making its gate control circuit simple. At the same time, most of the on-state current flow occurs in the bipolar transistor while the high voltage power

Chapter 8 : INSULATED GATE BIPOLAR TRANSISTOR

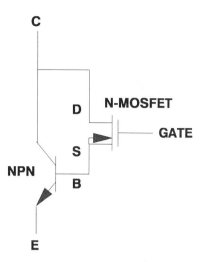

Fig. 8.1 Composite device with power MOSFET driving a power bipolar transistor.

MOSFET provides its base drive current. Since a combination of discrete high voltage bipolar and MOSFET elements is being used here, the overall (average) current density during forward conduction falls in between that observed in a bipolar transistor (typically 50 amperes per cm^2 for a 600 volt device) and a power MOSFET (typically 10 amperes per cm^2). The devices shown in Fig. 8.1 can be simultaneously fabricated on a single chip to provide the user with a three terminal composite device with characteristics superior to those of either the bipolar transistor or the power MOSFET. This is fundamentally a circuit solution to the problem of combining bipolar and MOS characteristics. Since no new device physics are involved, this approach will not be discussed any further.

A more powerful and innovative approach is based upon a combination of the physics of bipolar current conduction with the physics of MOS gated current control within the same semiconductor region. This concept is sometimes termed *functional integration of MOS and bipolar physics*. With this concept, a new class of power semiconductor devices has emerged. Among these devices, the *insulated gate bipolar transistor (IGBT)* has become commercially successful due to its superior on-state characteristics, reasonable switching speed, and excellent safe-operating-area. These devices have replaced bipolar power transistors in medium power applications, where the blocking voltages lie between 300 and 1500 volts.

The structure and physics of operation of the IGBT are discussed first in this chapter. This discussion is followed by an analysis of the blocking and on-state characteristics, as well as the switching behavior. An important aspect of the design of an IGBT is the prevention of latch-up of a parasitic thyristor within the structure. Methods for suppressing the latch-up of the parasitic thyristor are therefore described, followed by an analysis of its forward biased safe-operating-area. Some unique device structural designs that have led to significantly improved electrical characteristics for the IGBT are discussed at the end of the chapter.

8.1 DEVICE STRUCTURE AND OPERATION

At first sight, the structure of the insulated gate bipolar transistor appears to be identical to that of the MOS gated thyristor shown in Fig. 9.1. Its operation is, however, fundamentally different in that the IGBT structure is designed not to allow the regenerative turn-on inherent to the four layer thyristor structure. The conceptual breakthrough required to create the IGBT was the realization that the MOS gate can be used to create a (inversion layer) channel linking the N^+ emitter region to the N-drift layer without regenerative latch-up. Since the lower junction is forward biased in the on-state, current flow in the IGBT can now occur via this channel. This allows fully gate controlled output characteristics with forced gate turn-off capability.

A cross-section of the insulated gate bipolar transistor is shown in Fig. 8.2 based upon

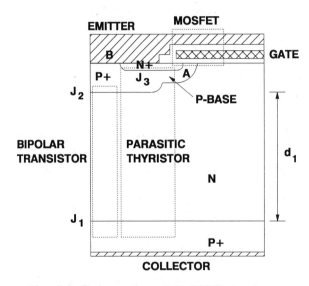

Fig. 8.2 Cross section of the IGBT structure.

the DMOS process. In this structure, current flow cannot occur when a negative voltage is applied to the collector with respect to the emitter because the lower junction (J_1) will become reverse biased. This provides the device with its reverse blocking capability. In this mode of operation, the depletion region extends in the N- drift region. When a positive voltage is applied to the collector with the gate shorted to the emitter, the upper junction (J_2) becomes reverse biased and the device operates in its forward blocking mode. In this mode of operation, the voltage is supported by a depletion region formed in the N- drift region. The forward and reverse blocking capability of the IGBT cell structure shown in Fig. 8.2 is approximately equal because it is determined by the thickness and resistivity of the same N- drift layer. However, considerable differences between these device parameters may arise due to differences in the edge terminations used for junction J_1 and junction J_2.

If a positive gate bias is applied of sufficient magnitude to invert the surface of the P-base region under the gate when the device is in its forward blocking mode, it can be switched into its on-state. In the forward conducting state, electrons flow from the N^+ emitter to the N- drift region. This provides the base drive current for the vertical P-N-P transistor in the IGBT structure. Since the emitter junction (J_1) for this bipolar transistor is forward biased, the P^+ region injects holes into the N-base region. (It is worth pointing out here that the IGBT terminals have been named from the point of view of replacing a power bipolar transistor with the IGBT in circuits. Unfortunately, this nomenclature is in conflict with the internal physics of operation of the device. Thus, the junction J_1 acts as an emitter of the internal P-N-P transistor even though the P^+ region is connected to the collector terminal.) When the positive bias on the collector terminal of the IGBT is increased, the injected hole concentration increases until it exceeds the background doping level of the N- drift region. In this regime of operation, the device characteristics are similar to those of a forward biased P-i-N diode. Consequently, these devices can be operated at high current densities even when designed to support high blocking voltages.

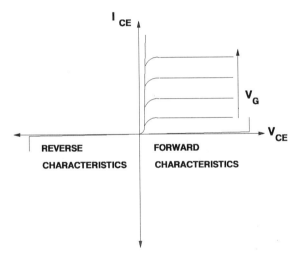

Fig. 8.3 Output characteristics of the IGBT.

As long as the gate bias is sufficiently large to produce enough inversion layer charge for providing electrons to the N-base region, the IGBT forward conduction characteristics look like those of a P-i-N diode. However, if the inversion layer conductivity is reduced by application of a gate bias close to the threshold voltage, a significant voltage drop occurs across this region due to the electron current flow in a manner similar to that observed in conventional MOSFETs. When this voltage drop becomes comparable to the difference between the gate bias and the threshold voltage, the channel becomes pinched-off. At this point, the electron current saturates. Since this limits the base drive current for the P-N-P transistor, the hole current flowing through this path is also limited. Consequently, the device operates with current saturation in its active region with a gate controlled output current. These characteristics of the

IGBT are shown in Fig. 8.3.

In order to switch the IGBT from its on-state to the off-state, it is necessary to discharge the gate by shorting it to the emitter. In the absence of a gate voltage, the inversion region at the surface of the P-base under the gate cannot be sustained. Removal of the gate bias cuts off the supply of electrons to the N- drift region and initiates the turn-off process. Due to the presence of the high concentration of minority carriers injected into the N- drift region during forward conduction, the turn-off does not occur abruptly. At first, an abrupt reduction in the anode current is observed because the electron current via the channel is terminated. The collector current then decays gradually with a characteristic time constant determined by the minority carrier lifetime.

The features of the insulated gate bipolar transistor structure are its high forward conduction current density, low drive power due its MOS gate structure, MOS gate controlled turn-off capability, fully gate controlled output characteristics with wide safe operating area, and high forward and reverse blocking capability. These characteristics approach those of an ideal power switch suitable for many DC and AC power control circuits. This has led to its extensive applications in power electronic systems immediately following its commercial introduction.

Since the IGBT structure is similar to that for the MOS-gated thyristor, it contains a parasitic P-N-P-N thyristor structure between the collector and the emitter terminals as indicated in Fig. 8.2. If this thyristor latches-up, the current can no longer be controlled by the MOS gate. It is important to design the device in such a manner that this thyristor action will be suppressed. This can be achieved by preventing the injection of electrons from the N$^+$ emitter region into the P-base during device operation. An equivalent circuit for the IGBT is shown in

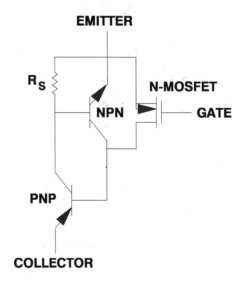

Fig. 8.4 Equivalent circuit for the IGBT.

Fig. 8.4 including the parasitic thyristor. The resistance R_s represents the resistance to hole current flow within the P-base region to the emitter terminal. If this resistance is sufficiently small, the thyristor will not be turned-on because of the low gain of the upper N-P-N transistor. The equivalent circuit for the IGBT then consists of a MOSFET driving the base of the P-N-P transistor in a Darlington configuration.

In the following sections, the operation of the IGBT in each of its operating modes is discussed in detail. For each mode of operation, the device structural parameters that dictate the performance are analyzed. Special process and design approaches used to improve the characteristics of the device are included in these sections. It is demonstrated that the IGBT has significantly better on-state characteristics than either the MOSFET or the bipolar transistor over a broad range of breakdown voltages. The characteristics of complementary devices are shown to be comparable to those for the n-channel structure shown in Fig. 8.2. This is attractive in many applications where complementary devices are used in various configurations to control power flow.

8.2 STATIC BLOCKING CHARACTERISTICS

The forward and reverse blocking voltage capability of the IGBT defines one of its important limits in performance for applications. It has been found the IGBT structure is particularly attractive for applications that require devices capable of blocking over 300 volts. For lower forward blocking voltages, the power MOSFET has a superior on-state characteristic. Thus, the IGBT structure must be designed to support relatively high voltages. In this section, the static blocking characteristics of the IGBT are analyzed based upon the assumption that the regenerative turn-on of the parasitic thyristor does not take place during device operation. During device design, it is necessary to ensure that this latch-up phenomenon is suppressed. Methods for suppression of the latch-up of the parasitic thyristor are discussed in a subsequent section.

8.2.1 Reverse Blocking Capability

When a negative collector voltage is applied, a large voltage can be supported by the IGBT structure shown in Fig. 8.2 because junction J_1 becomes reverse biased. When junction J_1 becomes reverse biased, its depletion layer extends primarily into the lightly doped N- drift region. It is important to note that the breakdown voltage during reverse blocking is determined by an open-base-transistor formed between the P^+ collector, N- drift region, and the P-base region. This structure is prone to punch-through breakdown if the N- drift region is too lightly-doped. The analysis of the breakdown voltage of an open-base-transistor was provided in Chapter 3. To obtain the desired reverse blocking capability, it is essential to optimally design the resistivity and thickness of the N- drift region. The design is also affected by the minority carrier diffusion length. As a general guideline, the width of the N- drift region is chosen so that its thickness is equal to the depletion width at the maximum operating voltage plus one diffusion length. Since the forward voltage drop increases with increasing N- drift region width, it is important to perform an optimization of the breakdown voltage with the objective of maintaining a narrow width for the N- drift region.

When the blocking voltage requirement increases, the N- drift region width must be correspondingly increased:

$$d_1 = \sqrt{\frac{2 \epsilon V_m}{q N_D}} + L_p \qquad (8.1)$$

where d_1 is the N- drift region width, V_m is the maximum blocking voltage, and L_p is the minority carrier diffusion length. At large blocking voltages, the depletion layer width becomes much larger than the diffusion length. The N- drift region width then increases approximately as the square-root of the blocking voltage.

The reverse blocking capability is also impacted by the method of terminating the reverse blocking junction (J_1). In a typical IGBT, the N- drift region width is in the range of 100 microns. Such devices require the use of a thick P^+ substrate (which forms the collector region) on which the N- drift region is epitaxially grown. Junction J_1 then extends across the entire plane of the wafer. When preparing individual devices, it is necessary to cut through the reverse blocking junction and passivate this surface to achieve reverse blocking capability. As discussed in Chapter 3, a positive-bevel angle at junction J_1 would be highly desirable to reduce the surface electric field. This can be achieved by using a V-shaped blade during sawing of the wafers, followed by chemical etching and surface passivation to suppress surface leakage.

8.2.2 Forward Blocking Capability

To operate the IGBT in the forward blocking mode, the gate must be shorted to the emitter. This prevents the formation of the surface inversion layer under the gate. When a positive collector bias is applied, the IGBT can then support a large voltage because the P-base/N- drift region junction (J_2) becomes reverse biased. A depletion layer extends from this junction on both sides. The breakdown voltage of this junction is limited by the considerations discussed in Chapter 7 for the power MOSFET with the additional impact of the presence of the lower junction (J_1) in the IGBT.

During analysis of the forward blocking capability, it is first necessary to examine the P-base doping profile. The P-base doping profile must be tailored with the objective of controlling the MOS channel threshold voltage while making sure that the depletion layer of junction J_2 in the P-base does not punch-through to the N^+ emitter region. This limits the minimum channel length that can be achieved as discussed earlier in Section 7.3.2 for power MOSFETs. The design considerations for the P-base and N^+ emitter depth for the IGBT are identical to those already discussed in that section for the power MOSFET.

In addition to the P-base doping profile, the spacing between the DMOS cells also affects the forward blocking capability. The effect of cell spacing upon the breakdown voltage of the DMOS structure was treated in Section 7.3.3.1 for power MOSFETs. As the DMOS cell spacing is increased, the breakdown voltage will decrease due to depletion layer curvature leading to field crowding at the edges of the P-base. In the case of the IGBT, the N- drift region doping level is lower than that for a power MOSFET with the same breakdown voltage. A larger DMOS cell spacing can consequently be tolerated in the IGBT.

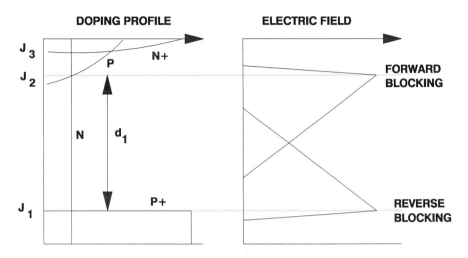

Fig. 8.5 Doping profile and electric field distribution for the symmetrical IGBT structure.

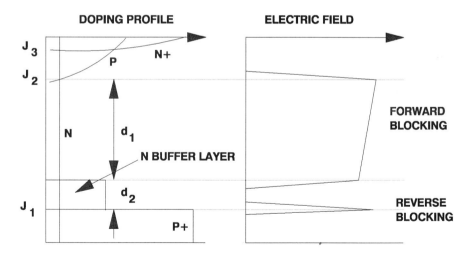

Fig. 8.6 Doping profile and electric field distribution for the asymmetrical IGBT structure.

In addition to these design considerations, the presence of the lower junction (J_1) must be taken into account. The IGBT breakdown voltage can be severely degraded by reach-through of the depletion layer of junction J_2 to the lower junction (J_1). In the case of symmetrical devices (i.e., devices with equal forward and reverse blocking capability that are used in AC circuits), the N- drift region width must be chosen in accordance with Eq. (8.1) used to design the reverse blocking capability. In DC circuit applications, where power MOSFETs and bipolar transistors are used, the IGBT is not required to support reverse voltage. This offers the opportunity to reconfigure the device structure to optimize the forward conduction characteristics for a given forward blocking capability without consideration for the reverse blocking capability. The doping profile and electric field distribution for such asymmetrical IGBTs can be compared with those of the symmetrical device with the aid of Fig. 8.5 and 8.6. In the asymmetrical IGBT structure, the uniformly doped N- drift region of the symmetrical IGBT is replaced by a two layer N- drift region. This alters the electric field distribution as illustrated on the right-hand side of the figures. If the critical electric field for breakdown is assumed to be independent of the N- drift region doping level and the N- drift region doping in layer 1 is very low, the electric field distribution changes from the triangular case in the symmetrical IGBT to a rectangular case in the asymmetrical structure. Under these circumstances, if the N-buffer layer thickness (d_2) is assumed to be approximately equal to one diffusion length (L_p), the forward blocking capability of the asymmetrical device will be twice as high as for the symmetrical device. In actual practice, the maximum electric field decreases with reduced N- drift region doping due to the field being distributed over a larger distance. This factor, in combination with the finite N- drift region doping level required to optimize the DMOS cell structure, results in an increase of forward blocking voltage by a factor of between 1.5 and 2.0 times for the same total N- drift region width for the symmetrical and asymmetrical structures.

During the design of the N- drift region doping profile for the asymmetrical IGBT structure, it is important to keep the thickness of the N-buffer layer as small as possible. Since the depletion layer of the forward blocking junction (J_2) must not punch-through to the collector junction (J_1), the charge (Q_B) in the buffer layer must be sufficient to allow the electric field to reduce to zero within it:

$$Q_B = (d_B N_B) \geq \epsilon_s E_c = 1.3 \times 10^{12} \tag{8.2}$$

where E_c is the critical electric field at breakdown and N_B is the buffer layer doping. This equation indicates that very narrow buffer layers can be used by arbitrarily raising the buffer layer doping concentration. An upper limit to the N-buffer layer doping concentration is imposed by a reduction of the injection efficiency of the junction (J_1), which degrades the forward conduction characteristic. The optimum doping concentration and thickness of the N-buffer layer are in the range of 10^{16} and 10^{17} per cm^3 and 10 to 15 microns, respectively.

8.3 FORWARD CONDUCTION CHARACTERISTICS

The IGBT can be operated in its forward conduction mode by the application of a positive

Chapter 8 : INSULATED GATE BIPOLAR TRANSISTOR   435

gate bias to create an inversion layer under the MOS gate. This forms a conducting channel which connects the N⁺ emitter to the N- drift region. As in the case of a power MOSFET, the gate voltage must be sufficiently above the threshold voltage to make the channel resistance small during current flow. Due to the tremendous reduction in the drift layer resistance in the IGBT because of the conductivity modulation arising from the injected carriers, the channel resistance must be designed to be much lower than that for a high voltage power MOSFET. A low channel resistance, similar to that in low voltage power MOSFETs is desirable to obtain the best IGBT performance. Once the channel is formed, forward current flow occurs by the injection of minority carriers across the forward biased collector junction (J_1). Over most of the N- drift region, the injected carrier density is typically 100 to 1,000 times greater than the N- drift region doping level resulting in a drastic reduction of its series resistance. This feature allows operation of the IGBT at very high current densities during forward conduction.

The analysis of the forward conduction characteristics can be performed by using two approaches. To understand these approaches, consider the equivalent circuit of the IGBT shown in Fig. 8.4. This equivalent circuit consists of a coupled P-N-P and N-P-N transistor pair representing the four layer thyristor structure with a MOSFET shunting the upper N-P-N transistor. Also note the shorting resistance (R_s) between the base and emitter of the N-P-N transistor. The magnitude of the shorting resistance (R_s) is determined by the sheet resistance of the P-base region and the distance between the edge of the N+ emitter at point A and its contact at point B (see Fig. 8.2). If the shorting resistance (R_s) is so small that the N⁺ emitter does not become forward biased above 0.7 volts during forward conduction, the upper N-P-N transistor can be assumed to be inactive for purposes of device analysis. The forward conduction characteristics of the IGBT can then be analyzed by using either of the two simplified

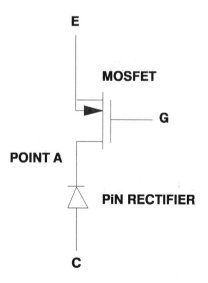

Fig. 8.7 MOSFET/P-i-N rectifier equivalent circuit for the IGBT.

equivalent circuits shown in Fig. 8.7 and Fig. 8.8.

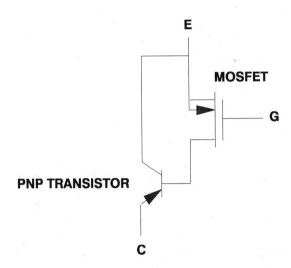

Fig. 8.8 Bipolar transistor/MOSFET equivalent circuit for the IGBT.

In the case of the circuit shown in Fig. 8.7, the IGBT is regarded as a P-i-N rectifier in series with a MOSFET. In the case of the circuit shown in Fig. 8.8, the IGBT is regarded as a wide-base P-N-P transistor driven by a MOSFET in a Darlington configuration. It is worth pointing out that these circuits are valid only for aiding device analysis and cannot be used to emulate IGBT characteristics by using discrete devices. The model shown in Fig. 8.8 based upon a bipolar transistor driven by a power MOSFET offers a more complete description of the IGBT, but the P-i-N rectifier/MOSFET model shown in Fig. 8.7 can be used to understand device behavior in many cases.

8.3.1 P-i-N Rectifier/MOSFET Model

In analyzing the forward conduction characteristics by using the P-i-N rectifier/MOSFET model, the device is treated as composed of two sections, shown by the dashed lines in Fig. 8.9. A single current flow path is assumed to exist through the P-i-N rectifier and MOSFET connected in series. The current-voltage relationship for the IGBT can be developed by coupling the equations derived earlier in Chapter 4 for the current conduction in the P-i-N rectifier and Chapter 7 for the MOSFET using their common potential at region C in the structure.

Using the analysis of the forward conduction characteristics of a P-i-N rectifier discussed in Section 4.2.1, the voltage drop across the P-i-N rectifier ($V_{F,PiN}$) is related to its forward conduction current density ($J_{F,PiN}$) by:

Chapter 8 : INSULATED GATE BIPOLAR TRANSISTOR

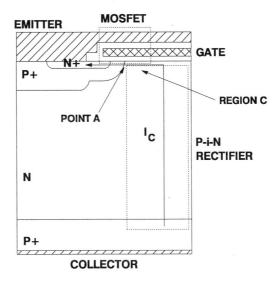

Fig. 8.9 P-i-N rectifier and MOSFET segments within the IGBT structure.

$$J_{F,PiN} = \frac{2 q D_a n_i}{d} F\left(\frac{d}{L_a}\right) e^{(qV_{F,PiN}/2kT)} \qquad (8.3)$$

The current density in the P-i-N rectifier can be assumed to be approximately equal to the collector current density because the current spreads from the bottom of the drift region and is uniformly distributed across the cross-section of the device cell over most of the distance between the collector and P-base region. Thus:

$$V_{F,PiN} = \frac{2 kT}{q} \ln\left[\frac{J_c d}{2 q D_a n_i F(d/L_a)}\right] \qquad (8.4)$$

Since the PiN rectifier current flows through the MOSFET channel, the MOSFET current is given by:

$$I_{MOSFET} = J_c W Z \qquad (8.5)$$

The voltage drop across the MOSFET is related to the current flowing through it and the gate bias voltage by the relationship derived in Section 7.4.3 on power MOSFETs:

$$I_c = \frac{\mu_{ns} C_{ox} Z}{2 L_{ch}} [2(V_G - V_T) V_{F,MOS} - V_{F,MOS}^2] \qquad (8.6)$$

where the term $V_{F,MOS}$ is used here in place of V_D to indicate that this expression applies to the MOSFET portion of the IGBT. Here, L_{CH} is the channel length. In the forward conduction mode, sufficient gate voltage is applied such that the forward voltage drop across the device is low. Under these conditions, $V_{F,MOS} << (V_G - V_T)$ and the MOSFET section of the devices is operating in its linear region. Thus:

$$I_{MOSFET} = \frac{\mu_{ns} C_{ox} Z}{L_{CH}} (V_G - V_T) V_{F,MOS} \qquad (8.7)$$

The voltage drop across the MOSFET section is, then, given by:

$$V_{F,MOS} = \frac{I_C L_{CH}}{\mu_{ns} C_{ox} Z (V_G - V_T)} \qquad (8.8)$$

Based upon this model for the IGBT, the forward voltage drop across the IGBT is the sum of the voltage drop across the MOSFET and the P-i-N rectifier :

$$V_F = \frac{2kT}{q} \ln\left[\frac{I_C d}{2 q W Z D_a n_i F(d/L_a)}\right] + \frac{I_C L_{CH}}{\mu_{ns} C_{ox} Z (V_G - V_T)} \qquad (8.9)$$

From this equation, the forward conduction characteristics can be computed as a function of the

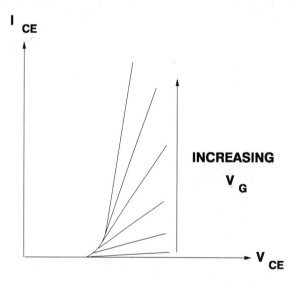

Fig. 8.10 On-state characteristics of the IGBT.

gate bias voltage. Typical forward conduction characteristics will take the form shown in Fig. 8.10. Note the presence of the diode knee below which very little current flow occurs because of lack of injection from the collector junction (J_1). Although the curves shown in Fig. 8.10 have a uniform spacing with increasing gate bias, in actual devices they tend to bunch together at high gate voltages due to a reduction in channel mobility with increasing gate field.

Several important conclusions about IGBT performance can be derived from the P-i-N rectifier/MOSFET model. Firstly, it becomes apparent that there will be a knee in the output conduction characteristic below which very little current flow will occur. Consequently, the IGBT is not suitable for applications requiring devices with forward voltage drops of less than 0.7 volts. From this model, it is also apparent that the IGBT forward conduction current density will rise exponentially as in a P-i-N rectifier. This behavior has been experimentally verified

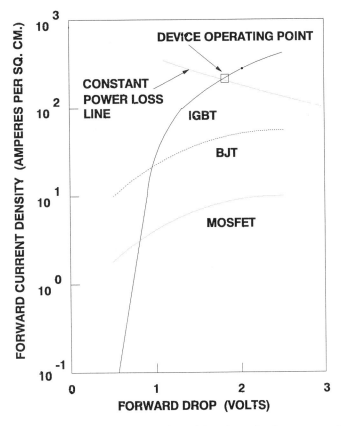

Fig. 8.11 Exponential increase in current density with voltage in the on-state for the IGBT.

in devices as shown in Fig. 8.11. In this figure, the forward conduction characteristics of a bipolar transistor operating at a current gain of 10 and a power MOSFET are also provided for comparison. Once the forward drop exceeds 1 volt, the current density of the IGBT surpasses that of the power MOSFET and bipolar transistor. At a typical operating forward voltage drop of 2 to 3 volts for devices with breakdown voltages of 600 volts, the IGBT current density is

20 times that of the power MOSFET and 5 times that of the bipolar transistor. The on-state voltage drop at which a device can be operated is usually determined by the ability to extract the heat generated by the current flow through the device without exceeding a junction temperature of about 200 °C. For a fixed thermal impedance, this implies that the operating point on the on-state characteristics can be obtained by drawing a constant power loss line as shown in Fig. 8.11. The intersection of this line with the on-state I-V characteristics of the three devices indicates that the on-state voltage drop of the IGBT is lower than that for the other devices. In addition, it can be seen that operating on-state current density for the IGBT is substantially greater than for the other devices. This indicates that a smaller chip size can be used, resulting in reduced cost for the IGBT.

Using the P-i-N rectifier/MOSFET model, it is also possible to derive the IGBT characteristics under current saturation. When the gate bias is reduced to a value close to the threshold voltage such that the voltage drop across the MOSFET channel becomes significant, the current becomes limited by the MOSFET. As in the case of a MOSFET, the IGBT collector current will saturate at a value given by:

$$I_{C,sat} = \frac{\mu_{ns} C_{ox} Z}{2 L_{CH}} (V_G - V_T)^2 \qquad (8.10)$$

and the transconductance of the IGBT will be identical to that for the MOSFET with the same cell size and channel length. In actual devices, the transconductance for the IGBT has been found to be larger than that for the MOSFET with the same cell size and channel length. The reason for this difference is discussed in the next section.

The P-i-N rectifier/MOSFET model can be used to understand the behavior of the forward conduction characteristics as a function of lifetime by taking into account the effect of changes in diffusion length upon the P-i-N rectifier. It also allows analysis of the impact of increasing breakdown voltage by accounting for the effect of a wider N- drift region upon the P-i-N rectifier portion. Further, it allows analysis of the change in forward conduction characteristics with temperature. These effects will be discussed further in subsequent sections. The major shortcoming of this model is that it omits the hole current component flowing into the P-base region. This component is included in the model based upon a MOSFET driving a bipolar transistor.

8.3.2 Bipolar Transistor/MOSFET Model

The bipolar transistor/MOSFET model is based upon the circuit shown in Fig. 8.8. Here, the MOSFET provides the base drive current for the wide base PNP bipolar transistor. These components of the device are illustrated in its cross-section shown in Fig. 8.12. In this figure, the electron current (I_e) flowing through the MOSFET channel and the hole current (I_h) flowing across the P-N-P bipolar transistor section are also indicated. These currents are related via the current gain of the wide-base P-N-P transistor:

Chapter 8 : INSULATED GATE BIPOLAR TRANSISTOR 441

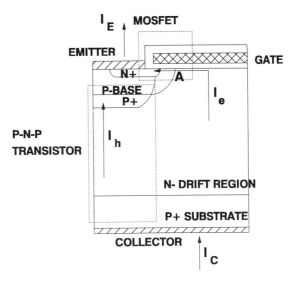

Fig. 8.12 IGBT cross-section indicating MOSFET and bipolar transistor current flow components.

$$I_h = \left(\frac{\alpha_{PNP}}{1 - \alpha_{PNP}}\right) I_e \qquad (8.11)$$

The emitter current is the sum of these components:

$$I_E = I_h + I_e = \frac{1}{(1 - \alpha_{PNP})} I_e \qquad (8.12)$$

This equation also gives the collector current because no gate current component exists due to the very high impedance of the MOS structure. From these equations, it can be concluded that the MOSFET channel current (I_e) is a fraction of the collector current (I_C). In contrast, the PiN rectifier/MOSFET model assumes that the entire collector current flows through the MOSFET channel.

The P-N-P transistor must have a large base width in order to support the forward and reverse blocking voltage. Its current gain (α_{PNP}) is primarily determined by the base transport factor (α_T) given by:

$$\alpha_T = \frac{1}{\cosh(l/L_a)} \qquad (8.13)$$

where l is the undepleted base width of the lower P-N-P transistor and L_a is the diffusion length.

The undepleted base width is essentially equal to the thickness of the N- drift region because the depletion width is small during forward conduction. Due to the high injection level conditions prevalent in the N- drift region, the ambipolar diffusion length (L_a) should be used during device analysis. This complicates the analysis because L_a is a function of injection level. The current gain (α_{PNP}) is typically about 0.5.

To compute the forward conduction characteristics, it is necessary to relate the forward drop across the device to the internal currents (I_e, I_h). This can be performed by analyzing the voltage drop along the MOSFET current path (I_e) in the same manner as described earlier for the P-i-N rectifier/MOSFET model. In the case of the bipolar transistor/MOSFET model, Eq. (8.9) must be modified by replacing I_C with I_e. The voltage drop in the N- drift region can still be approximated as that for a PiN rectifier with a current I_C flowing through it. Thus, the on-state voltage drop for the IGBT for this model is given by:

$$V_F = \frac{2kT}{q} \ln\left[\frac{I_C d}{2qWZD_a n_i F(d/L_a)}\right] + \frac{(1 - \alpha_{PNP}) I_C L_{CH}}{\mu_{ns} C_{ox} Z (V_G - V_T)} \qquad (8.14)$$

The forward voltage drop obtained by using this equation is smaller than that derived using the P-i-N rectifier/MOSFET model because all the collector current no longer flows through the MOSFET channel.

By using the bipolar transistor/MOSFET model for the IGBT, it can also be concluded that the bipolar transistor current component (I_h) alters the saturated current and consequently the transconductance of the device. When the gate bias is reduced so that the voltage drop in the MOSFET channel limits current flow, the electron current (I_e) is given by:

$$I_e = \frac{\mu_{ns} C_{ox} Z}{2 L_{CH}} (V_G - V_T)^2 \qquad (8.15)$$

Using Eq. (8.12), the saturated collector current is then given by:

$$I_{C,sat} = \frac{1}{(1 - \alpha_{PNP})} \frac{\mu_{ns} C_{ox} Z}{2 L_{CH}} (V_G - V_T)^2 \qquad (8.16)$$

From this equation, an expression for the transconductance of the IGBT in the active region can be obtained by differentiation with respect to V_G:

$$g_{ms} = \frac{1}{(1 - \alpha_{PNP})} \frac{\mu_{ns} C_{ox} Z}{L_{CH}} (V_G - V_T) \qquad (8.17)$$

It can be seen that the transconductance of the IGBT is substantially larger than for a power MOSFET with same channel aspect ratio (Z/L) because of the gain of the wide-base P-N-P bipolar transistor inherent in the IGBT structure. Since the current gain (α_{PNP}) of the P-N-P transistor is typically about 0.5, the transconductance of the IGBT can be a factor of 2 times larger than for a MOSFET with the same channel aspect ratio.

In the preceding analysis, it was assumed that the current will saturate and remain constant when the voltage drop across the MOSFET channel exceeds ($V_G - V_T$). Such an assumption would result in an infinite output resistance. The actual output characteristics of the symmetrical IGBT exhibit an increase in the collector current with increasing collector voltage with the rate of increase becoming greater as the collector bias is increased. The finite output resistance for the IGBT is due to two reasons. Firstly, as in the case of power MOSFETs, the effective channel length decreases when the collector voltage increases. Secondly, in the IGBT, an additional reduction in the drain output resistance arises from the presence of the bipolar transistor current flow. As the collector voltage increases, the current gain (α_{PNP}) of the bipolar

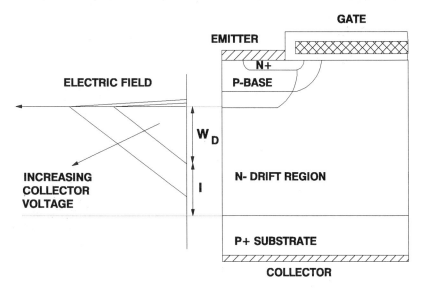

Fig. 8.13 Symmetrical IGBT structure and the electric field profiles at increasing collector bias.

transistor increases because its undepleted base width is reduced. The reduction in the undepleted N- drift region width with increasing collector bias is illustrated in Fig. 8.13 where the electric field profile is included to demonstrate an expansion in the depletion region. Since the base transport factor of this transistor controls its current gain:

$$\alpha_{PNP} = \alpha_T = \frac{1}{\cosh(l/L_a)} \qquad (8.18)$$

where the undepleted base width (l) is given by:

$$l = d_1 - W_D = d_1 - \sqrt{\frac{2\,\epsilon_s\,V_C}{q\,N_D}} \qquad (8.19)$$

The impact of the bipolar transistor current flow upon the collector output resistance becomes worse with increasing collector voltage. At low collector voltages, the undepleted base width is large and it does not change rapidly with increasing collector voltage. Consequently, the collector output resistance is large at low collector voltages. When the collector voltage is raised to the point at which the undepleted N- drift width becomes small, the current gain of the PNP transistor begins to grow rapidly with increasing collector voltage. The collector output resistance decreases rapidly with increasing collector voltage in this portion of the characteristics.

If the decrease in collector output resistance due to channel length reduction is neglected, it can be shown that the collector output resistance arising from an increase in the gain of the bipolar transistor will be given by:

$$\frac{1}{r_c} = \frac{dI_C}{dV_C} = \frac{\sinh(1/L_a)}{[\cosh(1/L_a)-2]^2}\sqrt{\frac{\epsilon_s}{2qN_DL_a^2V_C}} \qquad (8.20)$$

A typical set of output characteristics for an IGBT with the symmetrical structure are illustrated

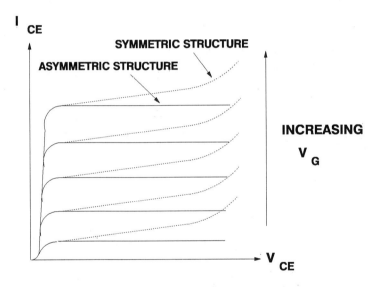

Fig. 8.14 Output characteristics of the IGBT.

in Fig. 8.14 by the dashed lines. It should be noted that there is a significant reduction in the collector output resistance with increasing collector voltage, and that the collector output resistance of an IGBT with uniformly doped N- drift region is much lower than for a power MOSFET. To obtain a higher collector output resistance, it is necessary to prevent the increase in the current gain of the bipolar transistor with increasing collector voltage.

One approach to increasing collector output resistance is by using the asymmetrical IGBT

Chapter 8 : INSULATED GATE BIPOLAR TRANSISTOR 445

structure illustrated in Fig. 8.6. Here, the depletion layer of the forward blocking junction (J_2) expands at low collector voltages to the width (d_1) of the lightly doped portion of the N- drift region. The voltage required for the depletion region to reach-through the lightly doped portion is given by:

$$V_{RT} = \frac{q N_D d_1^2}{2 \epsilon_s} \quad (8.21)$$

The depletion width does not increase significantly with further increase in collector voltage because of the relatively high doping concentration in the buffer layer. The depletion layer

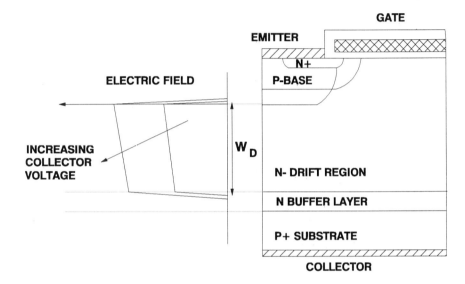

Fig. 8.15 Depletion layer extension for the asymmetrical IGBT structure.

extension for the asymmetrical structure is illustrated in Fig. 8.15 as a function of collector bias. It can be seen that the undepleted N- drift region width (l) does not change rapidly with increasing collector voltage for the asymmetrical IGBT structure because of the high buffer layer concentration. In contrast with the symmetrical IGBT structure, the undepleted base width (l) for the asymmetrical IGBT structure remains essentially equal to the thickness of the N-buffer layer (d_2) for all collector voltages. The current gain of the P-N-P transistor then remains constant at:

$$\alpha_{PNP} = \frac{1}{\cosh(d_2/L_a)} \quad (8.22)$$

for all collector voltages. This results in a higher collector output resistance as indicated by the solid lines in Fig. 8.14. The collector output resistance of the asymmetrical IGBT structure is therefore significantly superior to that for the symmetrical structure. It is worth pointing out that the buffer layer reduces the injection efficiency of the lower junction. This reduces the current gain of the P-N-P transistor, which also results in an improvement in the output resistance of the IGBT.

Another approach to increasing the collector output resistance of the IGBT is to utilize the dependence of the current gain of the P-N-P transistor up on the minority carrier diffusion length (L_a). The current gain (α_{PNP}) remains small until the undepleted N- drift region width becomes comparable to the diffusion length. For devices with smaller diffusion lengths, this will occur at higher collector voltages. The collector output resistance will therefore remain large until the collector voltage reaches close to the breakdown voltage. An increase in the collector output resistance can thus be achieved by reducing the diffusion length by decreasing the lifetime. This has been experimentally observed after electron irradiation of IGBTs with the symmetrical structure.

8.3.3 On-State Carrier Distribution

Although the on-state characteristics of the IGBT resemble those for a P-i-N rectifier, the carrier distribution within the N- drift region is not the same as that observed in a P-i-N rectifier. This difference occurs due to the presence of the reverse biased junction J_2 during the on-state. Since this junction is reverse biased, the free carrier density must become zero at this boundary. Thus, although the carrier distribution must follow the behavior obtained in a P-i-N rectifier due to high level injection, the boundary conditions for the IGBT are not identical to those for the P-i-N rectifier.

A one-dimensional analysis for the free carrier distribution in the IGBT along a line extending from junction J_1 to junction J_2 can be derived by solving the continuity equation under steady-state conditions:

$$\frac{d^2 p}{dx^2} - \frac{p}{L_{HL}^2} = 0 \qquad (8.23)$$

with the boundary conditions:

$$p(d_1) = 0 \qquad (8.24)$$

and

$$p(0) = p_0 \qquad (8.25)$$

where p_0 is the hole concentration in the N- drift region at junction J_1. The solution is given by:

$$p(x) = p_0 \frac{\sinh[(d_1 - x)/L_a]}{\sinh(d_1/L_a)} \qquad (8.26)$$

In order to determine p_0, it is necessary to use the following boundary conditions for the hole and electron currents:

$$J_p(x) = J \qquad at \quad x=0 \qquad (8.27)$$

and

$$J_n(x) = 0 \qquad at \quad x=0 \qquad (8.28)$$

where J is the collector current density. Using these relationships to solve for p_0, it can be shown that:

$$p_0 = \frac{J L_a}{2 q D_p} \tanh(d_1/L_a) \qquad (8.29)$$

Combining with Eq. (8.26), an expression for the carrier distribution in the N- drift region is obtained:

$$p(x) = \frac{J L_a}{2 q D_p} \frac{\sinh[(d_1 - x)/L_a]}{\cosh(d_1/L_a)} \qquad (8.30)$$

This charge distribution is illustrated on the right hand side of Fig. 8.16 together with the

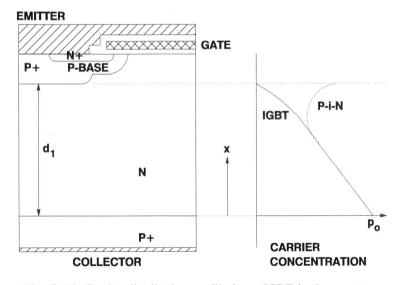

Fig. 8.16 Carrier distribution profile in an IGBT in the on-state.

catenary carrier distribution profile observed in the case of the P-i-N rectifier. It can be seen by making a comparison of these profiles, that the conductivity modulation of the N- drift region is quite similar near the collector junction J_1. However, there is a significant difference between them at junction J_2, where the conductivity modulation in the IGBT is much less than that observed in the P-i-N rectifier. From this difference, it can be concluded that, although the on-state characteristics of the IGBT resemble those for a P-i-N rectifier, the on-state voltage drop for the IGBT is larger than that for a P-i-N rectifier.

The stored charge (Q_s) in the N- drift region due to the on-state current flow can be obtained by integration of the carrier distribution:

$$Q_s = q \int_0^{d_1} p(x) \, dx = \frac{J L_a^2}{2 D_p} \left[\frac{1}{\cosh(d_1/L_a)} - 1 \right] \quad (8.31)$$

This charge must be removed from the drift region during the turn-off process. Since the diffusion length decreases when the lifetime is reduced, it can be concluded from this expression that the stored charge will also be smaller after electron irradiation. This enables faster turn-off of the IGBT because the rate of recombination is increased by the radiation and the stored charge that must be removed is also reduced.

8.3.4 On-State Voltage Drop

The on-state voltage drop for the IGBT can be derived from the carrier distribution profile obtained in the previous section. In the IGBT, the on-state voltage drop consists of three portions, namely, the voltage drop across the forward biased junction J_1, the voltage drop across the conductivity modulated N- drift region, and the voltage drop across the MOSFET. These components are analyzed below.

The voltage drop across the forward biased junction J_1 can be obtained by using the injected concentration of holes in the N- drift region at the junction (at x = 0):

$$V_{P^+N} = \frac{kT}{q} \ln\left(\frac{p_0}{p_{0N-}}\right) \quad (8.32)$$

where p_{0N-} is the equilibrium concentration of holes in the N- drift region, which is given by:

$$p_{0N-} = \frac{n_i^2}{N_D} \quad (8.33)$$

Combining these equations with Eq. (8.29):

$$V_{P^+N} = \frac{kT}{q} \ln\left[\frac{J L_a}{2 q D_p} \frac{N_D}{n_i^2} \tanh\left(\frac{d_1}{L_a}\right) \right] \quad (8.34)$$

This voltage drop is responsible for the "knee" in the forward conduction I-V curve. Its

Chapter 8 : INSULATED GATE BIPOLAR TRANSISTOR

magnitude is approximately 0.8 volts at room temperature. The knee voltage decreases with increasing temperature because of the very rapid increase in the intrinsic carrier concentration.

The voltage drop across the N- drift region can be obtained by integration of the electric field through the drift region. The electric field distribution can be derived by using the current transport equations:

$$J_p = q \mu_p \left(p E - \frac{kT}{q} \frac{dp}{dx} \right) \tag{8.35}$$

and

$$J_n = q \mu_n \left(n E - \frac{kT}{q} \frac{dn}{dx} \right) \tag{8.36}$$

with the high level injection condition:

$$n(x) = p(x) \tag{8.37}$$

For this one dimensional analysis, the total current density (J) flowing through the drift region is independent of position in the N- drift layer. Thus:

$$J(x) = J_n(x) + J_p(x) = constant \tag{8.38}$$

Combining these expressions, an equation for the electric field distribution is obtained:

$$E(x) = \frac{J}{q p (\mu_n + \mu_p)} - \frac{kT}{q} \frac{(\mu_n - \mu_p)}{(\mu_n + \mu_p)} \frac{1}{p} \frac{dp}{dx} \tag{8.39}$$

Based up on the distribution of the hole concentration given by Eq. (8.30), the electric field distribution is obtained:

$$E(x) = \frac{kT}{qL_a} \left\{ \frac{2\mu_p}{(\mu_n+\mu_p)} \frac{\cosh(d_1/L_a)}{\sinh[(d_1-x)/L_a]} + \frac{(\mu_n-\mu_p)}{(\mu_n+\mu_p)} \frac{1}{\tanh[(d_1-x)/L_a]} \right\} \tag{8.40}$$

An expression for the voltage drop across the N- drift (middle) region can be obtained by integration of the electric field distribution given by this equation between the limits $x = 0$ and $x = d_1$:

$$V_{M1} = \frac{kT}{q} \frac{2\mu_p}{(\mu_n+\mu_p)} \cosh(d_1/L_a) \ln[\tanh(d_1/2L_a)] \tag{8.41}$$

from the first part of the electric field expression and

$$V_{M2} = \frac{kT}{q} \frac{(\mu_n-\mu_p)}{(\mu_n+\mu_p)} \ln[\sinh(d_1/L_a)] \tag{8.42}$$

from the second part of the electric field expression. It is worth noting that the voltage drop across the N- drift (middle) region is usually less than 0.1 volt due to the strong conductivity modulation by the injected holes from the lower junction.

In the case of the voltage drop across the MOSFET, it is not sufficient to use only the voltage drop across the channel. In the IGBT, it can be seen from Fig. 8.16 that the conductivity modulation does not extend into the JFET region below the gate overlap region. Consequently, the voltage drop in this region must be added to the voltage drop across the channel. The voltage drop across the channel can be obtained by multiplying the electron current component of the IGBT collector current with the channel resistance. This voltage drop (V_{CH}) is given by:

$$V_{CH} = \frac{(1 - \alpha_{PNP}) \, J \, L_{CH} \, W_{Cell}}{\mu_{ns} \, C_{ox} \, (V_G - V_T)} \qquad (8.43)$$

The voltage drop in the JFET region can be obtained by using the same approach as in the case of the power MOSFET but with just the electron current component flowing through the channel:

$$V_{JFET} = \frac{\rho_{JFET} \, (1 - \alpha_{PNP}) \, J \, (x_P + W_0) \, W_{Cell}}{(L_G - 2 \, x_P - 2 \, W_0)} \qquad (8.44)$$

where ρ_{JFET} is the resistivity of the N- drift region below the gate overlap region, L_G is the gate length between the DMOS cell windows, and x_P is the depth of the P-base region. In deriving this expression, it has been assumed that the electron current component flows in the JFET region over a uniform cross section with width of ($L_G - 2 \, x_P - 2 \, W_0$) and length along the current flow path of ($x_P + W_0$). Since this expression is obtained under the assumption that the current is flowing uniformly down from the top surface, it is also appropriate to include a resistance in the accumulation layer with the correction factor for current spreading. Using the same approach as taken for the power MOSFET in Chapter 7 but with just the electron current component flowing through this region, the voltage drop contributed by the accumulation layer resistance is given by:

$$V_{ACC} = \frac{K \, (1 - \alpha_{PNP}) \, J \, (L_G - 2 \, x_P - 2 \, W_0) \, W_{Cell}}{2 \, q \, \mu_{nA} \, C_{ox} \, V_G} \qquad (8.45)$$

where K is the correction factor to account for current spreading from the accumulation layer into the JFET region. As in the case of the power MOSFET, this factor can be assumed to be 0.6 indicating that the effective accumulation resistance is 60 percent of the total resistance of the accumulation layer between the P-base regions. The total voltage drop ascribed to the MOSFET within the IGBT structure is then:

$$V_{MOSFET} = V_{CH} + V_{JFET} + V_{ACC} \qquad (8.46)$$

It is important to recognize that the doping concentration in the N- drift region in the

Chapter 8 : INSULATED GATE BIPOLAR TRANSISTOR

IGBT is typically far smaller that in power MOSFETs. This is because modulation of the conductivity of the drift layer by high level injection allows the use of low doping concentrations without encountering a very high resistance within the N- drift region. However, the low doping concentration in the N- drift region can result in a high voltage drop in the JFET region because the zero bias depletion region (W_0) is large. This not only leads to a large on-state voltage drop but can be sometimes manifested as a snap-back in the IGBT on-state characteristics. In order to obtain a low on-state voltage drop, it is important to increase the doping concentration in the JFET region. When this is performed, the voltage drop in the JFET region must be calculated by using the appropriate resistivity ρ_{JFET} in Eq. (8.44). It should be noted that the voltage drop from the JFET segment is worse for devices with lower carrier lifetime in the N- drift region. This occurs because the shorter diffusion length for the free carriers results in less conductivity modulation in the upper portion of the IGBT structure in the vicinity of the JFET region. For this reason, it is necessary to optimize the JFET region carefully to get the lowest on-state voltage drop for devices designed for higher switching frequency.

8.4 PARASITIC THYRISTOR LATCH-UP

The maximum operating current for the IGBT is limited by the presence of the parasitic thyristor within the structure. As shown in Fig. 8.17, this thyristor is formed between the N+ source region of the MOSFET acting as the cathode of the thyristor, the P-base region, the N-

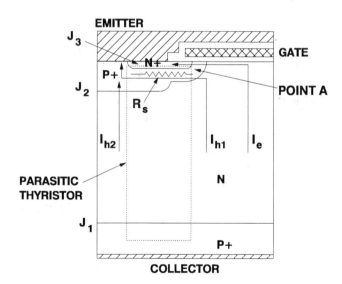

Fig. 8.17 Latch-up of parasitic thyristor in an IGBT.

drift region, and the P⁺ collector region. During operation of the IGBT in the on-state, electrons are supplied via the MOSFET channel and holes injected from junction J_1 are collected at junction J_2. The hole current shown as I_{h1} in the figure represents holes collected from the right hand side of the device. These holes flow to the emitter terminal via the resistance of the P-base region, indicated in the figure as R_s. This shunting resistance is also shown in the equivalent circuit for the IGBT in Fig. 8.4. This current flow produces a voltage drop across R_s that forward biases the junction J_3. At normal operating current levels for the IGBT, this voltage drop can be made much smaller than that for a forward biased diode (V_{bi}) by making the shunting resistance small. Under these conditions, the current gain of the N-P-N transistor is very small and the thyristor cannot latch-up. However, when the on-state current density is increased, the forward bias on junction J_3 can become sufficiently large to increase the current gain of the N-P-N transistor. If the sum of the current gains of the N-P-N and P-N-P transistor exceeds unity, the thyristor can latch-up. The collector current can now flow directly to the emitter terminal, bypassing the MOSFET channel. This implies that the IGBT current is no longer controlled by the gate bias.

From the above discussion, it can be concluded that it is essential to suppress the turn-on of the parasitic thyristor in order to maintain gate control over the collector current. This can be achieved by reducing the gain of either the N-P-N transistor, the P-N-P transistor, or of both transistors. Since the on-state current flows partially via the P-N-P transistor, a reduction in its gain produces an increase in the on-state voltage drop. It is therefore preferable to reduce the gain of the N-P-N transistor. However, methods employed to improve the turn-off time for the IGBT tend to reduce the gain of the P-N-P transistor, which then leads to an increase in the current density at which the parasitic thyristor will latch-up. In this section, the methods that have been explored to suppress the latch-up of the thyristor are described and their influence on other IGBT characteristics is discussed.

8.4.1 Parasitic Thyristor Latch-up Suppression via P-N-P Transistor

The latching current density of the parasitic thyristor can be increased by reducing the gain of the P-N-P transistor. This results in an increase in the electron current supplied via the MOSFET channel and a reduction in the hole current flowing via the collector junction J_2 of the P-N-P transistor. A decrease in the hole current component I_{h1} in the IGBT structure reduces the voltage drop across the shunt resistance R_s in the P-base region, which suppresses the latch-up of the parasitic thyristor. This implies that latch-up will occur at a higher total collector current when the gain of the P-N-P transistor is reduced.

There are two basic techniques for reducing the gain of the P-N-P transistor. The first method is based up on reducing the base transport factor. This can be achieved by electron irradiation, which reduces the minority carrier lifetime and diffusion length in the N- drift region (base of the P-N-P transistor). The base transport factor for the P-N-P transistor in the symmetrical IGBT structure depends upon the width (l) of the undepleted N- drift region. It is given by:

Chapter 8 : INSULATED GATE BIPOLAR TRANSISTOR

$$\alpha_T = \frac{1}{\cosh(l/L_a)} \tag{8.47}$$

where L_a is the ambipolar diffusion length. The diffusion length can be reduced by the electron irradiation. This is also beneficial from the point of view of reducing the turn-off time as discussed later in this chapter. However, it is accompanied by an increase in the on-state voltage drop.

The second method for reducing the gain of the P-N-P transistor is based upon decreasing the injection efficiency of junction J_1. The injection efficiency for the emitter of the P-N-P transistor is given by an expression similar to that derived in Chapter 5 in Section 5.2.1 for the N-P-N transistor:

$$\beta_E = \left(\frac{D_{pB}}{D_{nE}}\right)\left(\frac{L_{nE}}{W_{N-}}\right)\left(\frac{N_{AE}}{N_{DB}}\right) \tag{8.48}$$

where D_{pB} and D_{nE} are the diffusion coefficients for holes in the base (N- drift region) and electrons in the emitter (P+ region), respectively; L_{nE} is the diffusion length for electrons in the emitter (P+ region); W_{N-} is the width of the drift region; and N_{AE} and N_{DB} are the acceptor concentration in the emitter (P+ region) and the donor concentration in the base (N- drift region), respectively. From this expression, it can be concluded that the injection efficiency can be reduced by increasing the doping concentration in the N- drift region. However, if the doping in the N- drift region is uniformly increased, a decrease in the forward and reverse blocking voltage results. It is possible to increase the effective doping in the base region of the P-N-P transistor by incorporation of a buffer layer as shown in Fig. 8.6. This method allows a reduction in the injection efficiency while maintaining a high forward blocking capability. However, as discussed earlier, this structure has very low reverse blocking capability. It is typical to use a doping concentration of between 1×10^{16} and 1×10^{17} per cm^3 for the buffer layer with a thickness of between 10 and 20 microns. Although a higher doping concentration in the buffer layer can be used to improve the latch-up current level, this has been found to result in a very high on-state voltage drop for the IGBT. It is worth noting that the incorporation of the buffer layer allows a reduction in the total drift region thickness, which can offset the effect of the reduced injection efficiency. A decrease in the turn-off time can also be expected for higher buffer layer doping concentrations because of the decrease in the current gain as discussed in the section on switching characteristics.

8.4.2 Parasitic Thyristor Latch-up Suppression via N-P-N Transistor

The suppression of the latch-up of the parasitic thyristor can also be achieved by reduction in the gain of the N-P-N transistor. As mentioned earlier, this is preferable to reducing the gain of the P-N-P transistor because the hole current flow in the P-N-P transistor is essential to IGBT operation with low on-state voltage drop. In contrast, the IGBT structure can function with the N-P-N transistor being inactive. For this reason, many methods for reducing the gain of the N-P-N transistor have been conceived and experimentally verified. As in the case of the P-N-P transistor, these methods fall into two basic categories. The first is a reduction in the injection efficiency of junction J_3 and the second is a reduction in the base transport factor of the N-P-N transistor. The method for reducing the gain of the N-P-N transistor are described below.

8.4.2.1 Deep P$^+$ diffusion: One of the first and most effective techniques developed to suppress latch-up of the parasitic thyristor is based up on the addition of a deep, highly doped P$^+$ region within the IGBT cell structure. As discussed earlier with reference to Fig. 8.17, the N+ emitter of the N-P-N transistor is short-circuited to the P-base region by the emitter metal. In spite of this, the emitter junction J_3 can become forward biased at point A when the hole current I_{h1} flows through the P-base region. It is not possible to arbitrarily increase the doping concentration in the P-base region because this will result in a very high threshold voltage. However, an additional diffusion can be added to the IGBT process to reduce the resistance for the hole current without altering the threshold voltage. In order to accomplish this, the P$^+$ diffusion must be located as far as possible under the N$^+$ emitter but it must not encroach up on the MOSFET channel. It is important to recognize that the peak channel doping, which determines the threshold voltage, occurs at junction J_3 under the gate electrode. For a typical DMOS process with a threshold voltage of between 2 to 3 volts, the peak doping concentration is about 1×10^{17} per cm^3. The depth and concentration of the P$^+$ diffusion can therefore be extended until its contribution to the channel doping approaches 1×10^{17} per cm^3.

The introduction of the deep P$^+$ diffusion into the IGBT structure creates two regions

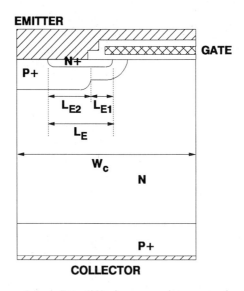

Fig. 8.18 Influence of deep P+ diffusion on resistance under N+ emitter.

under the N$^+$ emitter with different sheet resistances as illustrated in Fig. 8.18. The segment of length L_{E1}, which contains the N$^+$ emitter over the P-base diffusion, has a relatively high sheet resistance of about 3000 ohms per square. In contrast, the segment of length L_{E2}, which contains the N$^+$ emitter over the P$^+$ diffusion, has a relatively low sheet resistance of about 40 ohms per square. The resulting value for the shunting resistance R_s is given by:

$$R_s = \frac{1}{Z} (\rho_{SB} L_{E1} + \rho_{SP+} L_{E2}) \quad (8.49)$$

where Z is the width of IGBT cell orthogonal to the cross-section in the figure; and ρ_{SB} and ρ_{SP+} are the sheet resistances of the P-base and P$^+$ regions under the N$^+$ emitter, respectively. The criterion for latch-up of the parasitic thyristor can be written as:

$$I_{h1} R_s = V_{bi} \quad (8.50)$$

If the hole current component I_{h2} is neglected, then the collector current is related to the hole current by:

$$I_C = \frac{I_{h1}}{\alpha_{PNP}} \quad (8.51)$$

Using these equations, the collector current at which latch-up of the parasitic thyristor takes place is given by:

$$I_{CL} = \frac{V_{bi}}{\alpha_{PNP} R_s} = \frac{V_{bi} Z}{\alpha_{PNP} (\rho_{SB} L_{E1} + \rho_{SP+} L_{E2})} \quad (8.52)$$

The corresponding collector current density at which the latch-up of the parasitic thyristor takes place is given by:

$$J_{CL} = \frac{I_{CL}}{Z W_c} = \frac{V_{bi}}{W_c \alpha_{PNP} (\rho_{SB} L_{E1} + \rho_{SP+} L_{E2})} \quad (8.53)$$

An assessment of the impact of the addition of the deep P$^+$ diffusion into the IGBT cell upon the latch-up of the parasitic thyristor can be derived by considering an N$^+$ emitter with a total length of 8 microns with the P$^+$ diffusion extending to within 1 micron of the channel. Without the P$^+$ diffusion, the shunting resistance is that for the P-base region under the entire N$^+$ emitter length. The ratio of the shunting resistances without the deep P$^+$ diffusion to that with the P$^+$ diffusion is then given by:

$$\frac{R_{SP}}{R_{SP+}} = \frac{L_{N+} \rho_{SB}}{(\rho_{SB} L_{E1} + \rho_{SP+} L_{E2})} \quad (8.54)$$

For the case of the parameters given above, this ratio is approximately 8. Thus, the latching current density for the parasitic thyristor can be expected to increase by eight-fold by the incorporation of the deep P⁺ region.

The position of the deep P⁺ region plays an important role in determining the latching current density of the parasitic thyristor. From Eq. (8.53), it can be concluded that the latching current density is strongly dependent upon the length L_{E1} of the N⁺ emitter over the P-base region. Thus, if the position of the P⁺ diffusion varies either due to alignment tolerances during device fabrication or due to the design rules used to position the edge of the P⁺ diffusion window relative to the polysilicon edge, the latching current density of the parasitic thyristor will also vary. Measurements of the dependence of the latching current density upon the N⁺ emitter length L_{E1} have been performed by fabrication of IGBTs with different positions for the edge of the P⁺ diffusion relative to the polysilicon edge. The results are shown in Fig. 8.19 for the case

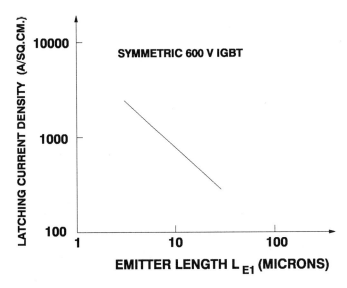

Fig. 8.19 Effect of position of P⁺ diffusion on the latching current density of the symmetrical IGBT structure.

of the symmetrical blocking structure with blocking voltage capability of 600 volts. The latching current density is found to decrease inversely with increasing N⁺ emitter length L_{E1}. This is consistent with the expression given in Eq. (8.53) because the contribution to the shunting resistance R_s from the sheet resistance of the P-base region is much greater than that from the P⁺ region. Based upon these results, it can be concluded that it is important to obtain good alignment tolerances between the P⁺ diffusion and the polysilicon edge during IGBT fabrication. It is also important to use the largest possible depth for the P⁺ diffusion without affecting the threshold voltage.

8.4.2.2 Shallow P⁺ Diffusion: In the previous section, it was demonstrated that the latching current density for the IGBT can be increased by the addition of a deep P⁺ diffusion to the IGBT cell. During device fabrication, the P⁺ diffusion must be aligned to the polysilicon edge. This

Chapter 8 : INSULATED GATE BIPOLAR TRANSISTOR

places a limitation on the smallest possible N$^+$ emitter length L_{E1} that can be obtained during device fabrication. For this reason, an alternate approach to reducing the sheet resistance of the P-base region under the N$^+$ emitter has been developed based upon a shallow P$^+$ diffusion.

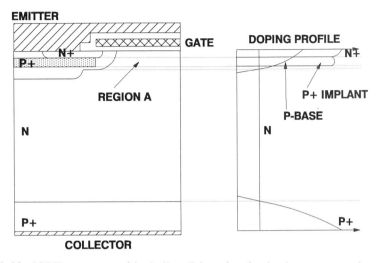

Fig. 8.20 IGBT structure with shallow P$^+$ region for latch-up suppression.

The IGBT structure with a shallow P$^+$ region is shown in Fig. 8.20. It is important to note that the shallow P$^+$ region is self-aligned to the edge of the polysilicon gate and is located deeper than the N$^+$ emitter. This can be accomplished by high energy ion implantation of boron. When boron is implanted into silicon, it has a range of 1 micron for an energy of 450 keV. It is therefore difficult to obtain a depth much greater than 1 micron for the center of the P$^+$ implanted region. In order to achieve a low sheet resistance for the P$^+$ region, it is necessary to obtain a high doping concentration comparable to that for the N$^+$ emitter region. This creates two problems. First, the dose for the ion implantation must be relatively large, which complicates the processing. Second, the P$^+$ region compensates the N$^+$ emitter doping. This can lead to a very high resistance for the N$^+$ emitter if the energy for the boron implant is insufficient to make its depth larger than the N$^+$ emitter depth, resulting in a very high on-state voltage drop. It should also be noted that the maximum energy for the boron implantation is limited by the ability of the polysilicon and underlying gate oxide to mask it. At higher implant energies, the boron can penetrate the polysilicon and gate oxide. This would destroy the IGBT structure because of the formation of a P layer at point A in the structure that would interrupt the flow of electrons from the MOSFET channel to the N- drift region.

Experiments have been conducted on the asymmetrical IGBT structure with a combination of a deep P$^+$ region and a shallow P$^+$ region. The shallow P$^+$ region was formed with boron implantation at an energy of 120 keV and doses of up to 1×10^{14} per cm^2. Due to the low implantation energy for the boron, the N$^+$ emitter depth was scaled down to 0.2 microns. It was found that the latching current density for the IGBT could be increased by a factor of 2 when the shallow P$^+$ region was included without altering the threshold voltage.

8.4.2.3 Reduced N⁺ Doping Concentration:

Another approach to reducing the gain of the N-P-N transistor is by reducing the doping concentration of the N⁺ emitter region. A reduced N⁺ emitter concentration leads to a smaller injection efficiency. Although this method can be used in principle to obtain a smaller current gain for the N-P-N transistor, it is difficult to implement because of several reasons. Firstly, the surface concentration of the P-base diffusion is typically about 1×10^{18} per cm³. Consequently, it is not possible to control the depth of the N⁺ emitter, and hence the MOSFET channel length, unless its doping concentration at the surface is at least one-order of magnitude larger than the surface concentration of the P-base region. This restricts the minimum N⁺ emitter doping concentration at the surface to above 1×10^{19} per cm³ which is sufficient to result in a high emitter injection efficiency. The second reason is an increase in the sheet resistance of the N⁺ emitter when its doping concentration is reduced which can lead to a high on-state voltage drop. For these reasons, this method is not commonly used to improve the latching current level for IGBTs.

8.4.2.4 Minority Carrier By-Pass:

As discussed earlier, the latch-up of the parasitic thyristor occurs due to the hole current that flows under the N⁺ emitter region. This hole current flow forward biases the junction J_3 leading to latch-up of the parasitic thyristor. It can therefore be expected that an improvement in the latching current density of the parasitic thyristor can be achieved by reducing the hole current component I_{h1} (see Fig. 8.17). This has been accomplished by using an IGBT cell design referred to as the *minority carrier by-pass design*.

A cross-section of the IGBT cell with the minority carrier by-pass is shown in Fig. 8.21 where two adjacent cells have been included to clarify the current flow paths. Unlike the IGBT cell shown in Fig. 8.2, this design contains an N⁺ emitter region only on' one side of the polysilicon window. In this design, the hole current can flow via two paths. In the first path,

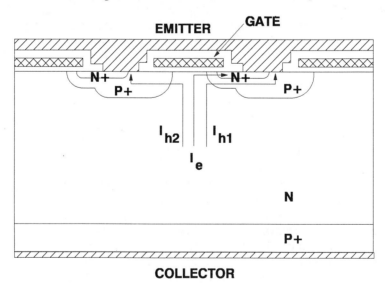

Fig. 8.21 IGBT structure with minority carrier by-pass.

indicated by I_{h1}, the hole current flows via the P-base region under the N$^+$ emitter in a manner similar to that for the earlier IGBT cell. In the second path, indicated by I_{h2}, the hole current flows to the emitter contact via the P$^+$ region without flowing under the N$^+$ emitter. As a consequence, the hole current flow under the N$^+$ emitter is reduced in half. This results in an improvement in the latching current density by a factor of two times.

A disadvantage of the IGBT cell structure with the minority carrier bypass is a reduction in the MOSFET channel density by a factor of two. This leads to an increase in the on-state voltage drop. For this reason, IGBT cells with minority carrier bypass have also been fabricated where the N$^+$ emitter is periodically interrupted along the length of the cell orthogonal to the cross-section shown in Fig. 8.2. The degree of minority carrier current flow that bypasses the N$^+$ emitter can then be traded off against an increase in the on-state voltage drop. Although this concept has been used to enhance the latch-up performance of the IGBT cell with a linear topology, it can be applied to other cell topologies. In the case of the linear cell topology with hole by-pass over half the cell edge, the on-state voltage drop has been found to be increased by about 0.5 volts over the cell without hole bypass. The reduction in the channel width for these devices has also been found to result in a decrease in the transconductance. This can result in the device current becoming saturated before the current density is sufficient to latch-up the parasitic thyristor. In spite of its adverse impact upon the on-state characteristics, the IGBT cell with minority carrier bypass has been popular because it enables design of IGBTs that can operate at higher currents and temperatures without the latch-up of the parasitic thyristor.

8.4.2.5 Counter-Doped Channel: From the discussion in the previous sections, it can be concluded that the latching current density of the parasitic thyristor is strongly dependent upon the sheet resistance of the P-base region. From Eq. (8.53), it can be seen that a decrease in the P-base sheet resistance will result in a proportionate increase in the latching current density. The sheet resistance of the P-base region can be reduced by increasing its surface concentration by using a higher ion implantation dose for the P-type impurity (boron). However, this is accompanied by an increase in the peak doping concentration of the P-base region under the gate oxide of the MOSFET. This results in an unacceptable increase in the threshold voltage for the IGBT.

It is possible to increase the surface concentration for the P-base diffusion while obtaining an acceptable threshold voltage if an additional N-type ion implantation is performed in the channel of the MOSFET to compensate for the higher boron concentration. This method is therefore referred to as *counter-doping* of the channel region. An IGBT structure with a counter-doped channel is illustrated in Fig. 8.22. Although the counter-doping implant is shown to extend only in the channel region in this figure, it can be performed over the entire cell area to provide greater processing convenience and eliminate an additional mask. It is important that the N-type ion implanted region remain shallow. This can be achieved by using arsenic as the N-type dopant. Although this method has been experimentally demonstrated to allow fabrication of IGBTs with higher latching current densities with acceptable threshold voltages, it is difficult to perform the counter-doping because this requires an accurate knowledge of the boron distribution in the channel. If the counter-doping is excessive, a channel can be formed at zero gate bias because the threshold voltage can become less than zero. This results in a high leakage

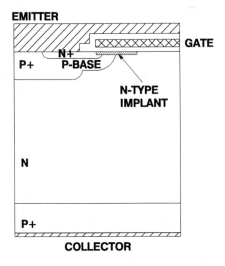

Fig. 8.22 IGBT structure with an N-type ion implant used to counter-dope the channel region.

current flow during forward blocking. Another disadvantage of this method is that the net dopant concentration in the channel is the sum of the P-base and the N-type counter doping concentrations. This higher concentration reduces the channel inversion layer mobility which tends to raise the on-state voltage drop.

8.4.2.6 Thinner Gate Oxide: One of most attractive and powerful methods for increasing the latching current density for the IGBT is by reducing the gate oxide thickness. It was shown in chapter 7 on power MOSFETs that the threshold voltage is given by the expression:

$$V_T = \frac{t_{ox}}{\epsilon_{ox}} \sqrt{4 \epsilon_s k T N_{AP} \ln\left(\frac{N_{AP}}{n_i}\right)} + \frac{2 k T}{q} \ln\left(\frac{N_{AP}}{n_i}\right) \qquad (8.55)$$

At the relatively high doping concentrations in the MOSFET channel, the first term of this expression becomes dominant. Based up on this, it can be concluded that the threshold voltage is linearly proportional to the gate oxide thickness (t_{ox}) and varies as the square root of the peak channel doping (N_{AP}). Note that the logarithmic term within the first term in the equation makes a very small contribution to the dependence of threshold voltage up on the peak channel doping which can be neglected. Thus, if the gate oxide thickness is reduced in half, the peak channel doping must be increased by a factor of four times to maintain the same threshold voltage. This increase in the P-base concentration can be utilized to raise the latching current density for the parasitic thyristor.

Experimental results have verified the ability to improve the latching current density for the IGBT by reducing the gate oxide thickness. In these experiments, the dose for the P-base

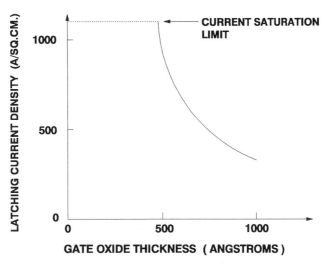

Fig. 8.23 Impact of reducing gate oxide thickness on the latching current density for the IGBT.

ion implantation was increased in proportion to the inverse square of the gate oxide thickness. It was found that the threshold voltage for the IGBT was approximately the same for all the devices. As shown in Fig. 8.23, the latching current density increased in proportion to the inverse of the square of the oxide thickness until it was no longer possible to latch-up the parasitic thyristor even at 150°C when a gate oxide of less than 500 angstroms was used.

An advantage of this method for increasing the latching current density is that it enables reduction in the gate drive voltage. This occurs because the channel resistance is given by:

$$R_{CH} = \frac{L_{CH}\, t_{ox}}{Z\, \mu_{ns}\, \varepsilon_{ox}\, (V_G - V_T)} \qquad (8.56)$$

From this expression, it can be seen that when the gate oxide thickness is reduced, the gate drive voltage can be proportionately reduced (assuming the threshold voltage is small compared with the gate voltage). This behavior has been verified for the IGBTs fabricated with gate oxide thickness ranging from 1000 angstroms to 280 angstroms. It was found that the gate drive voltage could be reduced from 25 volts to less than 12 volts to obtain the same on-state voltage drop. The reduction in gate drive voltage is not exactly proportional to the gate oxide thickness because these devices had a relatively high threshold voltage of 6 volts. In addition, there is a slight reduction in the inversion layer mobility as the peak channel doping is increased (see Chapter 3), which must be compensated for by an increase in gate voltage.

8.4.2.7 IGBT Cell Topology: As indicated in Fig. 8.17, there are two basic paths for the hole current flow during on-state operation of the IGBT. These paths are shown by the arrows marked I_{h1} and I_{h2} in the figure. The hole current flow via path I_{h2} does not forward bias the N$^+$

462 Chapter 8 : INSULATED GATE BIPOLAR TRANSISTOR

emitter/P-base junction while the hole current path I_{h1} which flows through the P-base region under the N$^+$ emitter produces the forward bias across junction J_3 leading to latch-up of the parasitic thyristor. The partition of the total collector current via these two paths depends up on the IGBT DMOS cell topology. In all the previous discussion in this chapter, it has been assumed that the IGBT cell has a linear topology. Although this design is often used, IGBTs with other cell topologies have been fabricated and characterized.

In all the IGBT cell topologies, the hole current component (I_{h1}) responsible for latch-up of the parasitic thyristor flows from the area under the polysilicon gate area into the P-base region. In the case of the linear cell topology, this occurs uniformly along its length as shown

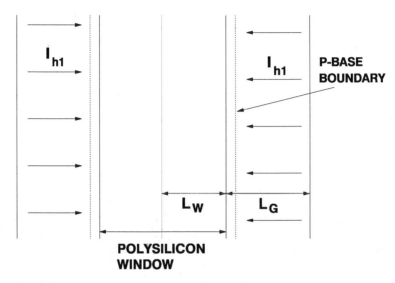

Fig. 8.24 Top view of linear IGBT cell with hole current distribution.

in Fig. 8.24 where a top view of the cell is shown with the hole current vectors. The hole current responsible for determining the latch-up of the parasitic thyristor is therefore given by:

$$I_{h1} = I_C \alpha_{PNP} \left(\frac{L_G}{L_G + L_W} \right) \quad (8.57)$$

When this current flows via the P-base region under the N$^+$ emitter region, it produces a voltage drop across the shunting resistance (R_s). As shown earlier, the magnitude of the shunting resistance is primarily determined by the length L_{E1} of the N$^+$ emitter over the P-base region. For the linear cell, the shunting resistance is therefore given by:

$$R_s = \frac{\rho_{SB} L_{E1}}{Z} \quad (8.58)$$

Chapter 8 : INSULATED GATE BIPOLAR TRANSISTOR   463

where Z is the length of the IGBT cell. The parasitic thyristor latches up when the voltage drop caused by the hole current (I_{h1}) across the shunting resistance (R_s) becomes equal to that for a forward biased diode (V_{bi}). Combining these equations:

$$I_{CL,LIN} = \frac{V_{bi} \, Z}{\alpha_{PNP} \, \rho_{SB} \, L_{E1}} \left(\frac{L_G + L_W}{L_G} \right) \quad (8.59)$$

The collector current density at which the parasitic thyristor latches up in the linear cell design is then given by:

$$J_{L,LIN} = \frac{I_{CL}}{Z \, (L_W + L_G)} = \frac{V_{bi}}{\alpha_{PNP} \, \rho_{SB} \, L_{E1} \, L_G} \quad (8.60)$$

In the case of the circular IGBT cell topology shown in Fig. 8.25, the hole current I_{h1} responsible for causing latch-up of the parasitic thyristor is concentrated into the P-base region from the surrounding region below the polysilicon gate as indicated by the hole current vectors.

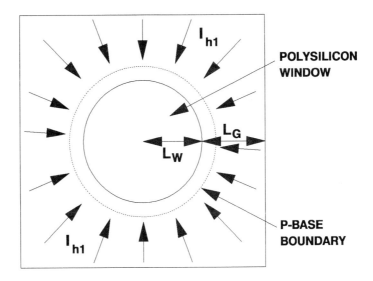

Fig. 8.25 Top view of circular IGBT cell with hole current distribution.

The total hole current responsible for determining the latch-up of the parasitic thyristor is therefore given by:

$$I_{h1} = I_C \, \alpha_{PNP} \, \frac{[4 \, (L_W + L_G)^2 - \pi \, L_W^2]}{4 \, (L_G + L_W)^2} \quad (8.61)$$

When this current flows via the P-base region under the N$^+$ emitter region, it produces a voltage drop across the shunting resistance (R_s). As in the case of the linear cell, the magnitude of the shunting resistance is primarily determined by the length L_{E1} of the N$^+$ emitter over the P-base region. For the circular cell, the shunting resistance is then given by:

$$R_s = \int_{L_W-L_{E1}}^{L_W} \frac{\rho_{SB} \, dr}{2 \pi r} = \frac{\rho_{SB}}{2 \pi} \ln\left(\frac{L_W}{L_W - L_{E1}}\right) \quad (8.62)$$

The parasitic thyristor latches up when the voltage drop caused by the hole current (I_{h1}) across the shunting resistance (R_s) becomes equal to that for a forward biased diode (V_{bi}). Combining these equations:

$$I_{CL,CIR} = \frac{V_{bi}}{\alpha_{PNP} \rho_{SB}} \frac{8 \pi (L_W + L_G)^2}{[4(L_G + L_W)^2 - \pi L_W^2] \ln[L_W/(L_W - L_{E1})]} \quad (8.63)$$

The collector current density at which the parasitic thyristor latches up for the circular cell design is then given by:

$$J_{CL,CIR} = \frac{V_{bi}}{\alpha_{PNP} \rho_{SB}} \frac{2 \pi}{[4(L_G + L_W)^2 - \pi L_W^2] \ln[L_W/(L_W - L_{E1})]} \quad (8.64)$$

In the case of the square cell IGBT design shown in Fig. 8.26, the hole current concentrates into the diffusion window from the surrounding region under the polysilicon gate

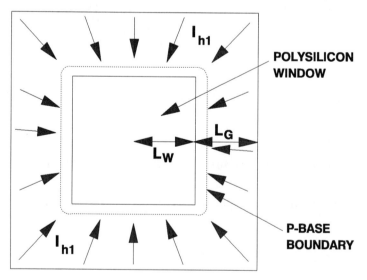

Fig. 8.26 Top view of square IGBT cell with hole current distribution.

Chapter 8 : INSULATED GATE BIPOLAR TRANSISTOR

in a manner similar to that for the circular cell design. However, the latch-up current density for the parasitic thyristor in the square cell design is lower than that for the circular cell. This is because the length of the N$^+$ emitter over the P-base region is larger at the corners of the square cell by a factor of $\sqrt{2}$. The IGBTs with square cell design are therefore susceptible to latch-up at relatively low current densities at the corners of the DMOS cell.

In the case of the circular and square cell designs, it was shown that the latch-up current density will be lower than that for the linear cell design because the hole current concentrates into the diffusion window. This effect can be reversed with the atomic lattice layout (A-L-L)

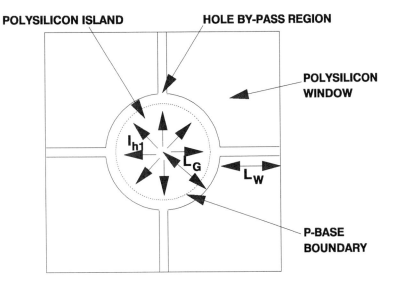

Fig. 8.27 Top view of A-L-L IGBT cell with hole current distribution.

IGBT cell topology shown in Fig. 8.27. In this case, the hole current I_{h1} responsible for causing latch-up of the parasitic thyristor spreads outward from the area below the polysilicon gate as indicated by the hole current vectors. The total hole current responsible for determining the latch-up of the parasitic thyristor is therefore given by:

$$I_{h1} = I_C \, \alpha_{PNP} \, \frac{\pi \, L_G^2}{4 \, (L_G + L_W)^2} \qquad (8.65)$$

When this current flows via the P-base region under the N$^+$ emitter region, it produces a voltage drop across the shunting resistance (R_s). As in the case of the linear cell, the magnitude of the shunting resistance is primarily determined by the length L_{E1} of the N$^+$ emitter over the P-base region. For the A-L-L cell, the shunting resistance is then given by:

$$R_s = \int_{L_G}^{L_G+L_{E1}} \frac{\rho_{SB} \, dr}{2\pi r} = \frac{\rho_{SB}}{2\pi} \ln\left(\frac{L_G + L_{E1}}{L_G}\right) \tag{8.66}$$

The parasitic thyristor latches up when the voltage drop caused by the hole current (I_{h1}) across the shunting resistance (R_s) becomes equal to that for a forward biased diode (V_{bi}). Combining these equations:

$$I_{CL,ALL} = \frac{V_{bi}}{\alpha_{PNP} \rho_{SB}} \frac{8(L_W + L_G)^2}{L_G^2 \ln[(L_G + L_{E1})/L_G]} \tag{8.67}$$

The collector current density at which the parasitic thyristor latches up for the A-L-L cell design is then given by:

$$J_{CL,ALL} = \frac{V_{bi}}{\alpha_{PNP} \rho_{SB}} \frac{2}{L_G^2 \ln[(L_G + L_{E1})/L_G]} \tag{8.68}$$

A comparison between the linear cell, the circular cell, and the A-L-L cell can be made by using a typical set of device parameters. Since all the devices are assumed to have the same current gain (α_{PNP}) and P-base sheet resistance (ρ_{SB}), their latching current densities can be compared by using typical values for the geometrical parameters: L_G = 8 microns, L_W = 8 microns, and L_{E1} = 1 micron. Substitution of these values into Eqs. (8.60), (8.64), and (8.68) results in a latching current density for the circular cell that is half that for the linear cell, while the latching current density for the A-L-L cell is twice that for the linear cell topology. Consequently, the A-L-L cell design has a much superior immunity against the latch-up of the parasitic thyristor. The latching current density for the A-L-L design can be improved even further by forming hole bypass paths at the locations where the polysilicon bars are used to interconnect the circular polysilicon pads. Unlike the case of the linear cell design, in the A-L-L cell design there is no significant loss of N+ emitter (or MOSFET channel) width caused by the incorporation of the hole bypass paths because the polysilicon bars are required to enable fabrication of the A-L-L geometry with a single polysilicon layer. As discussed in subsequent sections, the A-L-L cell design also has a superior forward and reverse biased safe operating area.

8.4.2.8 IGBT Cell with Diverter: From the previous sections, it can be surmised that an increase in the latching current density for the IGBT can be achieved by reducing the hole current component I_{h1} flowing into the P-base region. An IGBT structure that is specifically designed to perform this function is shown in Fig. 8.28. In this structure, holes are collected not only across the reverse biased junction J_2 but also across the junction J_4 formed between the P+ diverter diffusion and the N- drift region. It should be noted that the diverter is connected to the emitter metal. Consequently, the current flowing via the diverter comprises an additional component of the total emitter current.

The improvement in the latching current density of the IGBT can be analyzed for the

Chapter 8 : INSULATED GATE BIPOLAR TRANSISTOR 467

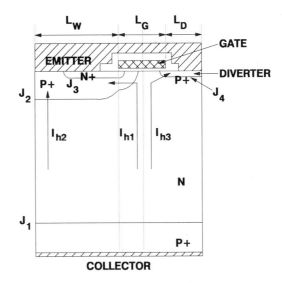

Fig. 8.28 IGBT cell structure with a diverter region.

IGBT cell containing the diverter by using the same approach used to analyze the various IGBT cell topologies. Although the diverter can be incorporated into any of the IGBT cell topologies, only the case of the linear cell design will be considered here. The impact of including the diverter region for other cell topologies can be analyzed in a similar manner. In the case of the linear cell topology with the diverter region it will be assumed that the hole current component I_{h1} responsible for causing latch-up of the parasitic thyristor flows from half the area under the gate electrode as shown in Fig. 8.28. The hole current from the rest of the area under the gate electrode is assumed to flow into the diverter region. Under these assumptions, the hole current responsible for determining the latch-up of the parasitic thyristor is given by:

$$I_{h1} = I_C \, \alpha_{PNP} \, \frac{L_G}{2(L_G + L_W + L_D)} \quad (8.69)$$

When this current flows via the P-base region under the N⁺ emitter region, it produces a voltage drop across the shunting resistance (R_s). The shunting resistance for the IGBT cell with the diverter region is identical to that derived for the linear cell design. Thus:

$$R_s = \frac{\rho_{SB} \, L_{E1}}{Z} \quad (8.70)$$

where Z is the length of the IGBT cell. The parasitic thyristor latches up when the voltage drop caused by the hole current (I_{h1}) across the shunting resistance (R_s) becomes equal to that for a forward biased diode (V_{bi}). Combining these equations:

$$I_{CL,D} = \frac{V_{bi} Z}{\alpha_{PNP} \rho_{SB}} \left(\frac{2 (L_G + L_W + L_D)}{L_G L_{E1}} \right) \tag{8.71}$$

The collector current density at which the parasitic thyristor latches up in the linear cell design is then given by:

$$J_{CL,D} = \frac{I_{CL,D}}{Z (L_W + L_G + L_D)} = \frac{2 V_{bi}}{\alpha_{PNP} \rho_{SB} L_{E1} L_G} \tag{8.72}$$

Thus, the incorporation of the diverter region increases the latching current density by a factor of two times for the linear cell topology. In addition, as discussed in a subsequent section, the diverter region acts as a guard ring for the junction J_2. This results in an improvement in the forward biased safe operating area. However, it is important to point out that the diverter region increases the resistance to electron current flow from the channel due to the JFET action. This results in an increase in the on-state voltage drop if the diverter junction is located very close to the P-base junction and when its depth is made comparable to the P-base region. By using a shallow diverter region that is located at least several microns from the edge of the P-base region, the increase in the on-state voltage drop of the IGBT can be made to be less than 200 millivolts.

8.5 SAFE-OPERATING-AREA

The *safe-operating-area (S-O-A)* is defined as the area within the output characteristics of the IGBT where it can be operated without destructive failure as long as the power dissipation is kept within the thermal constraints of the device package. For the IGBT, it is relevant to consider the S-O-A only for the first quadrant in its I-V characteristics. In general, there are three distinct boundaries for the safe-operating-area. With high applied voltages at low current levels, the maximum voltage that can be supported is limited by the breakdown voltage of the edge termination. As discussed in earlier sections, this breakdown voltage is determined by the open base transistor formed between the P-base region of the IGBT and the P^+ collector region. At high current levels with small collector voltages, the maximum collector current for the IGBT is limited by the on-set of latch-up of the parasitic thyristor. This limitation is observed at high gate bias voltages especially when the device is operating at higher temperatures. The suppression of the latch-up of the parasitic thyristor in the IGBT has been a challenging technological achievement that has led to devices with acceptable latch-up current levels for the parasitic thyristor. This phenomenon is sometimes referred to as *current induced latch-up* because it occurs when the collector current exceeds a certain current level irrespective of the collector bias as long as the collector voltage is relatively small.

In addition to these boundaries for the safe-operating-area, there is a boundary where the current and voltage become simultaneously large. Due to the high power dissipation within the device under these conditions, one limitation to the maximum current-voltage product is the temperature rise in the device. This thermal limit is determined by the die mount down, the package, and the heat sink. If it is assumed that the time duration during which the device is subjected to the simultaneous high current and voltage stress is short, then the power dissipation is no longer the limiting factor. The S-O-A is then dictated by a phenomenon referred to as *avalanche induced second breakdown*. This phenomenon can occur during two modes of IGBT operation.

These modes are encountered during the switching of the IGBT with an inductive load

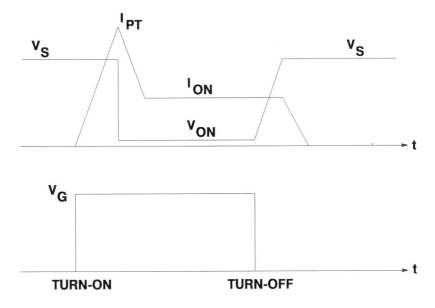

Fig. 8.29 Device switching waveforms with an inductive load.

as shown by the waveforms shown in Fig. 8.29 for the collector current and voltage. During each switching cycle, the device must withstand the presence of a high current flow simultaneously with a high voltage across it during both the turn-on and turn-off transients. During turn-on, the current exceeds the on-state value due to the reverse recovery current flowing in the flyback diodes as discussed in detail in chapter 10. During the turn-on transient, a positive gate bias is applied to the n-channel IGBT resulting in both electron and hole current flow through the structure with a high voltage across the device. This condition is called *forward biased safe operating area (FBSOA)*. During the turn-off transient, the gate bias is switched from positive to either zero or a negative value, resulting in only hole current transport with a high voltage across the device. This condition is called *reverse biased safe operating area (RBSOA)*. The physics that limits the safe operating area in these two modes is discussed below.

8.5.1 Forward Biased Safe-Operating-Area

The forward biased safe operating area (FBSOA) of the IGBT is an important characteristic for applications with inductive loads. It has been found that IGBTs have excellent FBSOA without resorting to expensive snubber networks. The FBSOA of the IGBT is defined by the maximum voltage that the device can withstand without destructive failure while the collector current is saturated. During this mode of operation, both electrons and holes are transported through the drift region, which is supporting a high collector voltage. The electric field in the drift region is sufficiently large to result in velocity saturation for the carriers. Consequently, the electron and hole concentrations in the drift region are related to the corresponding current densities by:

$$n = \frac{J_n}{q \, v_{sat,n}} \quad (8.73)$$

and

$$p = \frac{J_p}{q \, v_{sat,p}} \quad (8.74)$$

where $v_{sat,n}$ and $v_{sat,p}$ are the saturated drift velocities for electrons and holes, respectively. The net positive charge in the drift region is then given by:

$$N^+ = N_D + \frac{J_p}{q \, v_{sat,p}} - \frac{J_n}{q \, v_{sat,n}} \quad (8.75)$$

The electric field distribution in the drift region is determined by this charge. Unlike the steady state forward blocking condition where the drift region charge is equal to the doping concentration (N_D), under FBSOA conditions the net charge is usually much larger because the hole current density is significantly larger than the electron current density. An increase in the charge in the drift region results in a change in the electric field profile as shown in Fig. 8.30. The greater slope for the electric field profile results in the collector voltage being supported across a narrower "depletion" region with a higher electric field at the P-base/N- drift region junction than observed under the forward blocking conditions. This leads to breakdown in the IGBT cell at voltages lower than the breakdown voltage of the edge termination.

For a simple one-dimensional analysis, Poisson's equation can be solved with a net positive charge in the drift region given by N^+. Using Fulop's formula for the impact ionization coefficient, the breakdown voltage is then given by:

$$BV_{SOA} = \frac{5.34 \times 10^{13}}{(N^+)^{3/4}} \quad (8.76)$$

As in the case of the forward blocking voltage, it is necessary to account for the current gain of the open-base transistor in determining the maximum voltage that can be supported.

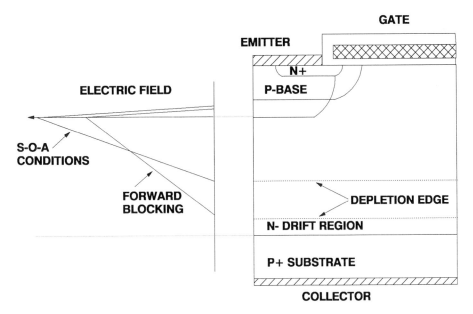

Fig. 8.30 Alteration in electric field profile under S-O-A conditions.

Thus, the FBSOA limit is given by the criterion:

$$\alpha_{PNP} M = 1 \qquad (8.77)$$

with

$$\alpha_{PNP} = \frac{1}{\cosh(1/L_a)} \qquad (8.78)$$

and

$$M = \left[1 - \left(\frac{V}{BV_{SOA}}\right)^n\right]^{-1} \qquad (8.79)$$

where l is the undepleted N-base width. These equations indicate that the on-set of avalanche breakdown will occur in the IGBT cell at a lower collector bias when the collector current is increased. It should be noted that a higher breakdown voltage (BV_{SOA}) can be obtained by reducing the doping concentration (N_D) in the drift region. This cannot be done for a symmetrical IGBT structure because it would lead to reach-through breakdown problems. However, this is possible in the asymmetrical structure because the buffer layer prevents reach-through problems.

8.5.2 Reverse Biased Safe-Operating-Area

The reverse biased safe operating area (RBSOA) is of importance during the turn-off transient. Since the gate bias is at zero or at a negative value under these conditions, the current transport in the drift region occurs exclusively via holes for an n-channel IGBT under these conditions. The presence of the holes adds charge to the drift region, resulting in an increase in the electric field at the P-base/N drift region junction as illustrated in Fig. 8.30. Since there are no electrons present in the space charge region, the electric field enhancement during the RBSOA conditions can be expected to be worse than for the FBSOA conditions.

The net charge in the space charge region under the RBSOA conditions is given by:

$$N^+ = N_D + \frac{J_C}{q\, v_{sat,p}} \qquad (8.80)$$

where J_C is the total collector current. As in the case of the FBSOA analysis, the avalanche breakdown limit under RBSOA conditions is given by Eq. (8.77) with the breakdown voltage determined by the above net charge. In the case of IGBTs fabricated using the asymmetrical structure, the doping concentration in the drift region can be made much lower than the free carrier concentration. In this case, the avalanche breakdown voltage is given by:

$$BV_{SOA} = 5.34 \times 10^{13} \left(\frac{q\, v_{sat,p}}{J_C} \right)^{3/4} \qquad (8.81)$$

When combined with the current gain of the P-N-P transistor, the RBSOA limit defined by this equation is as indicated in Fig. 8.31. A wider RBSOA is obtained when the current gain of the P-N-P transistor is reduced.

8.5.3 p-Channel Versus n-Channel IGBT

It is important to compare the characteristics of p- and n-channel IGBTs because of the interest in using these devices when complementary switches are needed in applications. An example of such an application would be the formation of an AC switch by the connection of an n-channel and p-channel IGBT in parallel. This allows control of the combination with gate signals referenced to a common terminal. A comparison of the RBSOA of the devices can be performed under the assumption that both structures have identical doping profiles and cell structures. The RBSOA obtained under these conditions is shown in Fig. 8.32.

From the figure, it can be seen that the p-channel IGBT has a superior current induced latching current density but that its avalanche induced S-O-A limit is much worse than that of the n-channel IGBT. This behavior can be understood by examining the physics underlying the operation of the two types of devices in each mode. The current induced latch-up limit is determined by the current density at which the emitter-base junction J_1 of the N-P-N transistor becomes forward biased due to the current collected in the base region. For the same doping profile for the base region, the sheet resistance of the P-base region in the n-channel IGBT is about 2.5 times larger than the sheet resistance of the N-base region in the p-channel IGBT

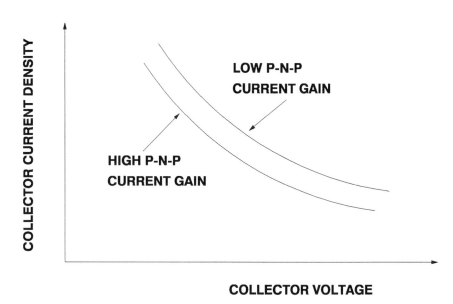

Fig. 8.31 Reverse biased safe operating area for an IGBT.

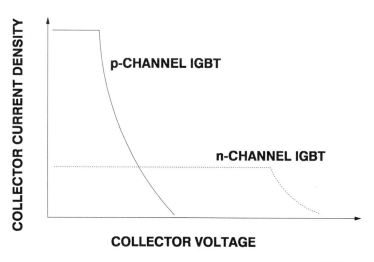

Fig. 8.32 Comparison of RBSOA of p- and n-channel IGBT.

because of the higher mobility for electrons. This difference in the sheet resistance results in a proportionate increase in the current induced latch-up limit for the p-channel IGBT over the n-channel IGBT. In the case of the avalanche induced SOA limit, the difference between the devices arises from the differences in the impact ionization coefficients for holes and electrons in silicon. As discussed in chapter 3, the impact ionization coefficient for electrons is significantly larger than that for holes. Under RBSOA conditions, electrons are transported through the depletion layer for the p-channel IGBT whiles holes are transported through the depletion layer for the n-channel IGBT. Consequently, the lower impact ionization coefficient for holes in silicon results in a wider SOA for n-channel devices when compared with p-channel devices. The net result is a cross-over of the RBSOA curves for the n- and p-channel devices.

8.5.4 DMOS Cell Design

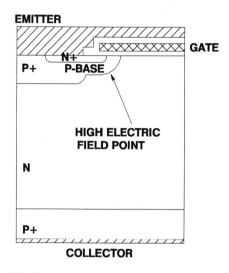

Fig. 8.33 IGBT DMOS cell indicating point with enhanced electric field due to junction curvature.

In the previous sections, the avalanche induced SOA limit was analyzed for the case of a one-dimensional device structure. In this case, it was assumed that the peak electric field at the P-base/N- drift region junction was uniform. In the case of the actual DMOS structure, this assumption is not applicable because of junction curvature effects. As indicated in Fig. 8.33, a higher electric field occurs at the edges of the P-base region due to junction curvature, in spite of the presence of the polysilicon gate which acts as a field plate. This enhanced electric field results in a reduction in the voltage at which avalanche breakdown is induced in the presence of the hole current flow through the drift region.

The magnitude of the electric field enhancement in the DMOS structure depends up on the topology of the cell design. The cell topologies were discussed in section 8.4.2.7. In the case of the linear IGBT cell topology shown in Fig. 8.24, the junction has a cylindrical shape. However, it is common to end the linear cell window with sharp corners. This creates a

Chapter 8 : INSULATED GATE BIPOLAR TRANSISTOR 475

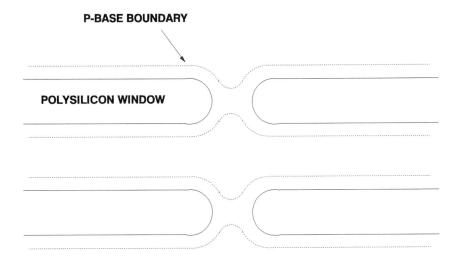

Fig. 8.34 Linear cell topology with rounded ends.

spherical junction at the corners where the electric field enhancement is greater than in the linear region. This can degrade the RBSOA. One method that has been found to be effective in preventing this problem is by rounding the ends of the polysilicon windows as shown in Fig. 8.34. Although it is not essential, it is preferable to make the distance between the polysilicon windows sufficiently narrow so that the P-base diffusions merge as shown by the dashed lines. This makes the electric field enhancement close to that for a cylindrical junction.

In the case of the circular cell topology shown in Fig. 8.25, the electric field enhancement is worse than that for the linear cell with rounded ends but less than that for the linear cell with the sharp corners. This is because, in the circular cell, the junction curvature is enhanced by the cylindrical junction being rotated with a radius equal to the radius of the polysilicon window. As extreme cases, the circular cell topology becomes equivalent to the linear cell with the sharp corners when the radius of the polysilicon window is zero leading to the formation of spherical junction curvature, while it becomes equivalent to a linear cell when the radius of the polysilicon window is infinity leading to the formation of cylindrical junction curvature.

An electric field enhancement that is less than that observed with the linear cell topology can be obtained with the Atomic-Lattice-Layout (A-L-L) shown in Fig. 8.27. This cell topology is equivalent to a cylindrical junction rotated in the opposite manner to that for the circular cell. This creates a saddle junction under the polysilicon gate which reduces the electric field enhancement. For the p-channel IGBT whose avalanche induced RBSOA limit is poor, numerical simulations have indicated that this leads to an increase in the collector current density required to induce avalanche breakdown at the junction by a factor of three times when compared with the linear cell. This has been experimentally confirmed.

8.5.5 Switching Locus

The current-voltage locus during the turn-on and turn-off transients are shown in

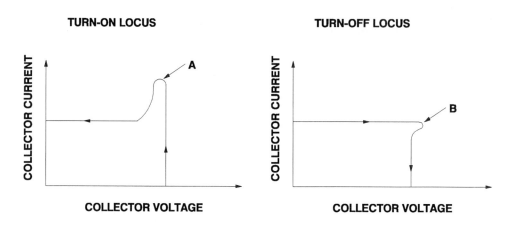

Fig. 8.35 Switching locus during turn-on and turn-off with an inductive load.

Fig. 8.35. During turn-on, the collector current first rises while the device is supporting the supply voltage, goes through an overshoot due to the reverse recovery current of the fly-back rectifier, and then settles down to its on-state value. The highest stress on the device occurs at point marked A. During this time interval, a gate bias is applied to the IGBT to induce current conduction. Consequently, in order to avoid destructive failure of the IGBT, it is essential that the turn-on locus be located within the forward biased safe operating area. This locus indicates that the use of improved rectifiers with reduced stored charge and smaller reverse recovery currents can greatly reduce the stress on the IGBT.

During turn-off, the collector voltage first rises with a constant collector current until it exceeds the supply voltage. This is necessary in order to transfer the current to the flyback diode. When the collector current falls to zero, an overshoot in the collector voltage can occur because of the finite stray inductance between the IGBT and the flyback diode. It is typical to observe an overshoot of 30 percent of the supply voltage when the switching speed of the IGBT is increased to reduce switching losses. During this time interval, the gate bias is either zero or at a negative value for an n-channel IGBT. Consequently, the switching locus during turn-off should be kept within the reverse biased safe operating area of the device.

8.6 SWITCHING CHARACTERISTICS

An important characteristic of the insulated gate bipolar transistor is its gate turn-off capability. Since the current flowing through the MOS channel controls the output characteristics, the collector current flow can be interrupted by removing the gate drive voltage. To perform the transition from the on-state to the forward blocking state, the gate is connected to the emitter terminal via an external circuit which allows discharging the gate capacitance. When the gate voltage falls below the threshold voltage of the MOS gate structure, the channel inversion layer can no longer exist. At this point, the electron current (I_e) ceases. If the gate turn-off is performed using a low external resistance in the gate drive circuit so as to abruptly reduce the gate voltage to zero, the collector current will drop abruptly because the channel current (I_e) is suddenly discontinued. Even after this occurs, the collector current continues to flow because the hole current (I_h) does not cease abruptly. The high concentration of minority carriers stored in the N- drift region during the on-state supports the hole current flow. As the minority carrier density decays due to recombination, it leads to a gradual reduction in the collector current. This current flow is sometimes referred to as a *current tail*.

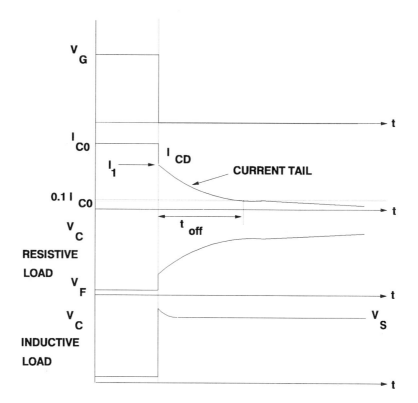

Fig. 8.36 Typical gate controlled turn-off waveforms observed for IGBTs.

Typical examples of the IGBT voltage and current waveforms during device turn-off are provided in Fig. 8.36 for gate controlled turn-off. In the case of a resistive load, the collector voltage rises during turn-off at the same rate at which the collector current decays. In contrast, in the case of an inductive load, the collector voltage abruptly rises to the supply voltage when the gate drive voltage is reduced to zero. In either case, it is important to note that a large power dissipation occurs during the time that the collector current decays to zero because of the relatively high voltage across the IGBT during this time. It is therefore important to minimize the duration of the current tail.

8.6.1 Turn-Off Time

An analysis of the turn-off time for the IGBT can be performed based up on the on-state current flow discussed in earlier sections. It is customary to define the *turn-off time* as the time taken for the collector current to decrease from its on-state value (I_{CO}) to 10 percent of this value. In order to determine this time interval, it is necessary to obtain the magnitude of the initial abrupt fall in the collector current. The magnitude of the abrupt drop in the collector current (I_{CD}) at time t_1 is due to the cessation of the electron current provided during the on-state via the MOSFET channel. Its magnitude is determined by the current gain of the P-N-P transistor:

$$I_{CD} = I_e = (1 - \alpha_{PNP}) I_{CO} \qquad (8.82)$$

This relationship has been experimentally verified by using devices with different N- drift region widths processed with a broad range of minority carrier lifetimes.

After the abrupt drop in the collector current, collector current is sustained by the hole current flow due to the presence of the stored charge in the N- drift region. At time t_1, the hole current flow is equal to its value during on-state conduction just prior to turn-off. Thus, the magnitude of I_1 is given by:

$$I_1 = I_{CO} - I_{CD} = I_h = \alpha_{PNP} I_{CO} \qquad (8.83)$$

After this, the collector current decays exponentially at a rate determined by the lifetime. Since significant current flow occurs during this time with a large density of free carriers in the N-drift region, it is appropriate to use the high level lifetime to characterize the rate of decay of the current. Thus:

$$I_C(t) = I_1 e^{-t/\tau_{HL}} = \alpha_{PNP} I_{CO} e^{-t/\tau_{HL}} \qquad (8.84)$$

Since the turn-off time (t_{off}) is defined as the time taken for the collector current to decay to 10 percent of its on-state value, an expression for the turn-off time can be derived from Eq. (8.84):

$$t_{off} = \tau_{HL} \ln(10 \, \alpha_{PNP}) \qquad (8.85)$$

In deriving this expression, it has been assumed that the current decays while high level injection conditions prevail within the N- drift region, and that the lifetime remains independent of the injection level during the decay. In actual practice, the decay rate can change during the current tail because of the reduction in injection level leading to a non-exponential tail. This

Chapter 8 : INSULATED GATE BIPOLAR TRANSISTOR 479

expression takes into account an increase in the initial abrupt reduction of the collector current (I_{CD}) as well as the change in the rate of the exponential decay in the collector current during the turn-off when the lifetime in the N- drift region is altered.

8.6.2 Lifetime Control

In principle, it is possible to use the diffusion of deep level impurities, such as gold or platinum, to reduce the lifetime in the N- drift region. However, this method has not found favor because it produces an undesirable shift in the threshold voltage and poor MOS interface characteristics due to the accumulation of the impurity at the oxide-silicon interface. For these reasons, the use of high energy particle bombardment has been found to be more favorable for controlling the switching speed in IGBTs. This can be achieved by using either electrons or neutrons to produce a uniform concentration of deep levels within the wafer or by using protons to create a localized plane with a low lifetime deep within the N- drift region. Since the proton implantation technique requires very specialized high voltage equipment, it has not been utilized for the manufacturing of IGBTs. The neutron irradiation technique has the drawback of producing radio-activity which makes the processing and handling inconvenient.

8.6.2.1 Turn-off time: Electron irradiation has therefore been found to be the most widely accepted method for controlling lifetime in the drift region for IGBTs. Although this process also results in the introduction of positive charge in the gate oxide that leads to a threshold voltage shift, fortunately this shift can be recovered by low temperature annealing without

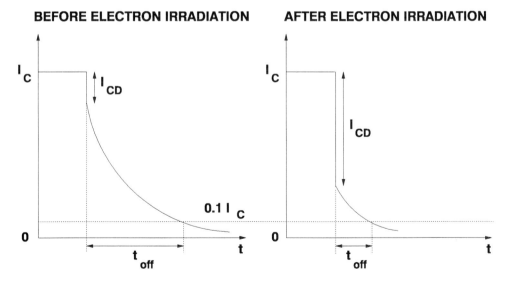

Fig. 8.37 Comparison of the collector current turn-off waveform before and after electron irradiation.

removal of the defects in the N- drift region that are responsible for reducing the lifetime. When electron irradiation of the IGBT is performed, the turn-off time decreases not only because of a reduction in the lifetime (τ_{HL}) but also due to a decrease in the current gain (α_{PNP}). The changes in the turn-off waveform for the collector current after electron irradiation can be seen in Fig. 8.37 where waveforms for pre- and post electron irradiated devices are compared. After electron irradiation, there is an increase in the magnitude of the abrupt reduction of collector current (I_{CD}) and a decrease in the time taken for the current tail to decay to zero.

The measured reduction in the turn-off time with increasing electron irradiation dose is

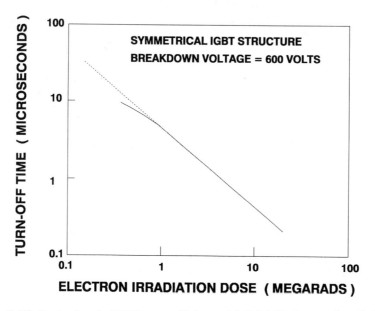

Fig. 8.38 Reduction in IGBT turn-off time with 3 MeV electron irradiation.

shown in Fig. 8.38 for the case of 3 MeV energy. It can be seen that the turn-off time decreases with increasing radiation dose. Since the turn-off time is dominated by the recombination tail, it can be anticipated that the turn-off time should decrease in proportion to the minority carrier lifetime. The minority carrier lifetime after irradiation (τ_f) is dependent upon the pre-irradiation lifetime (τ_i) and the electron irradiation dose (ϕ):

$$\frac{1}{\tau_f} = \frac{1}{\tau_i} + K\phi \qquad (8.86)$$

where K is the radiation damage coefficient. At high doses, the first term in Eq. (8.86) is negligible if the initial lifetime is large, and the lifetime after irradiation decreases inversely proportional to the radiation dose. In Fig. 8.38, the turn-off time can be observed to decrease inversely with increasing radiation dose at levels above 1 Megarads. Since the turn-off time before radiation typically ranges from 15 to 30 microseconds, at doses below 1 Megarads the

Chapter 8 : INSULATED GATE BIPOLAR TRANSISTOR   481

turn-off time is determined by a combination of both the initial lifetime and the radiation dose, so that the curves exhibit a departure from the inverse relationship observed at the higher doses. These results demonstrate that electron irradiation is effective in reducing the turn-off time in IGBTs. A reduction in the turn-off time from over 20 microseconds to less than 200 nanoseconds can be achieved by using a radiation dose of 16 Megarads.

The turn-off time is also a function of the collector voltage and current. The turn-off time increases with increasing collector voltage. This phenomenon can be ascribed to the need to establish larger depletion layer widths with increasing collector voltage as discussed for the gate turn-off thyristor in section 6.7.2.4. An unusual phenomenon observed for IGBTs is the reduction in the turn-off time with increasing collector current. This effect occurs both before and after electron irradiation. The reason for this behavior is that as the collector current increases, the current gain of the PNP transistor decreases due to a reduction in the diffusion length caused by carrier scattering. This results in an increase in the channel current component

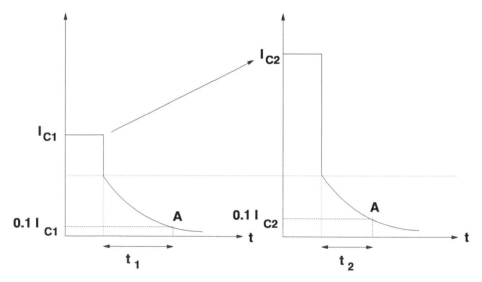

Fig. 8.39 Effect of increasing the on-state collector current on the turn-off waveform for the IGBT.

with increasing collector current. The waveforms in Fig. 8.39 illustrate the impact of increasing collector current upon the turn-off waveform under these circumstances assuming that the current tail remains invariant. Since the turn-off time is defined as the time taken for the collector current to decrease from its on-state value to 10 percent of this value, as I_C increases, the turn-off time decreases because the 10 percent point (A) moves up the decay tail in the collector current waveform. Note that even though the turn-off time decreases, the switching losses due to the current tail will be independent of the collector current. In this regard, it can be misleading to use the turn-off time for power loss analysis. A more accurate assessment of the turn-off losses can be obtained by integration of the product of the instantaneous collector current and voltage during the turn-off transient. This power loss is referred to as the *switching energy per cycle*.

8.6.2.2 Forward Voltage Drop: In bipolar power devices, an improvement in switching speed is generally accompanied by a degradation in current conduction capability. In the case of the IGBT, the devices behave like P-i-N rectifiers during forward current conduction with high level injection of minority carriers into the drift region which drastically lowers its resistance. When the diffusion length is reduced by the electron irradiation, the conductivity modulation of the drift region is also reduced and the IGBT forward current density decreases in a manner similar to that observed with P-i-N rectifiers.

The measured reduction in the forward current with increasing electron irradiation dose

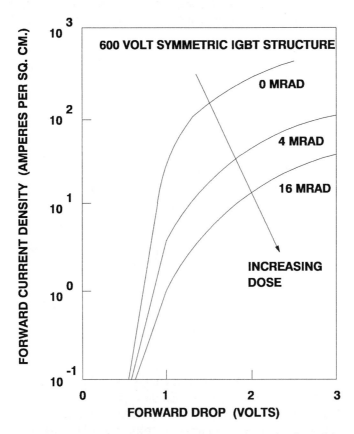

Fig. 8.40 Impact of electron irradiation on the forward conduction characteristics of the symmetrical IGBT structure.

is shown in Fig. 8.40 for the symmetrical structure. It can be seen that at a fixed forward drop of 2 volts, the current density decreases from about 150 amperes per cm^2 before radiation down to 15 amperes per cm² after radiation at a dose of 16 Megarads. Another more commonly used measure of the impact of lifetime reduction is in terms of an increase in forward voltage drop at a fixed collector current density. For example, at a fixed collector current density of 150 amperes per cm², the forward drop can be seen to increase from 1.8 volts before radiation to about 5 volts after radiation at a dose of 16 Megarads. This can be analyzed using Eq. (8.9) by taking into account the effect of the lifetime reduction due to electron irradiation upon the diffusion length L_a.

Thus, the increase in switching speed of the IGBTs is accompanied by a loss in current handling capability. Although this is undesirable, the IGBT performance is still substantially superior to that of a power MOSFET with the same forward blocking voltage capability of 600 volts. If a power MOSFET is operated at a current density of 150 amperes per cm², it would have a forward drop of 35 volts. Based upon this, it can be concluded that even after increasing the speed of the IGBT to obtain a turn-off time of 200 nanoseconds, its forward conduction characteristics remain superior to those of the power MOSFET.

8.6.2.3 Trade-Off Curve: The results described above demonstrate that electron irradiation can be utilized to control the switching speed of IGBTs. The radiation has been shown to be capable of reducing the gate turn-off time from over 20 microseconds down to 200 nanoseconds. However, this increase in the switching speed is accompanied by an increase in the forward voltage drop. Since a short gate turn-off time is desirable in order to reduce switching losses and a low forward voltage drop is desirable in order to reduce the conduction losses, it becomes necessary to perform a trade-off between these device characteristics.

This can be most conveniently performed by using a plot of the forward voltage drop versus the gate controlled turn-off time as shown in Fig. 8.41 for the symmetrical and asymmetrical IGBT structures. Depending upon the application, the appropriate device characteristics can be selected by choosing the irradiation dose. In the case of circuits operating at low frequencies with large duty cycles, where the conduction losses dominate the switching losses, IGBTs with turn-off times in the range of 5 to 20 microseconds would be the best. An example of this is line-operated phase control circuits. For higher frequency circuits with shorter duty cycles, where the switching losses are comparable to the conduction losses, IGBTs with turn-off times in the range of 0.5 to 2.0 microseconds would be appropriate. An example of this type of application is in AC motor drives operating at frequencies ranging from 1 kHz to 10 kHz. For high frequency circuits, the switching losses would become dominant and IGBTs with gate turn-off times ranging from 100 to 500 nanoseconds would be necessary. An example of such circuits is in un-interruptible power supplies (UPS) operating at between 20 kHz and 100 kHz. Thus, the IGBTs can be tailored to match the power switching requirements of a broad range of applications.

It is worth pointing out that the trade-off curve for the asymmetric IGBT structure has been found to be superior to that for the symmetric IGBT structure as shown in the figure. This is because of the smaller thickness for the drift region in the asymmetrical structure for obtaining the same forward blocking voltage as discussed earlier in Section 8.2.2. Due to the smaller drift region thickness, the (d/L_a) ratio is also smaller for the asymmetrical structure, which results in a lower on-state voltage drop for the same lifetime in the drift region. It has also been found that the trade-off curve can be improved upon by increasing the doping concentration in the

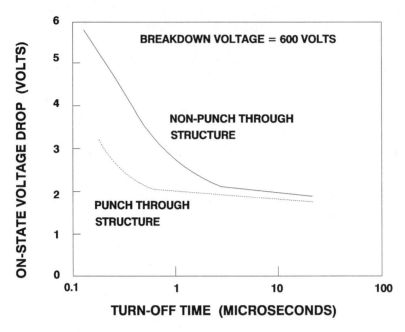

Fig. 8.41 Comparison of the trade-off curve between on-state voltage drop and turn-off time for the symmetrical and asymmetrical IGBT structures.

buffer layer. This results in a reduction in the injection efficiency of the lower junction (J_1), which reduces the current gain of the PNP transistor. As explained in the context of the turn-off waveform, a lower PNP current gain will enhance the initial rapid fall in the collector current during turn-off, which results in a shorter turn-off time.

The choice of the power device for any application is often decided by the maximization of the system efficiency that is equivalent to minimization of the power loss in the devices. In the case of the IGBT, it is possible to optimize the switching speed to obtain the lowest possible power loss. This provides a unique ability to tailor the device characteristics for a wide range of operating frequencies, as compared with the power MOSFET, which is best suited for very high frequency operation. In order to illustrate this, consider the power loss as a function of frequency as shown in Fig. 8.42 for three IGBTs and the power MOSFET. The IGBT (A) in the figure has a slow switching speed with a turn-off time of 15 microseconds; the IGBT (B) has a moderate switching speed with a turn-off time of 1 microsecond; and IGBT (C) has the fastest switching speed with a turn-off time of 0.25 microseconds. The on-state voltage drop for the devices increases as the switching speed is increased in accordance with the trade-off curve shown in Fig. 8.41 for the symmetric blocking device. The power MOSFET is assumed to be operated with a turn-off time of 0.1 microseconds. From Fig. 8.42, it can be seen that the power loss for the IGBT can be minimized by altering its switching speed. This results in a net minimum power loss curve for the IGBT as indicated by the dotted line. This curve intersects with the power loss curve for the power MOSFET at a frequency of 200 kHz. Thus, the use

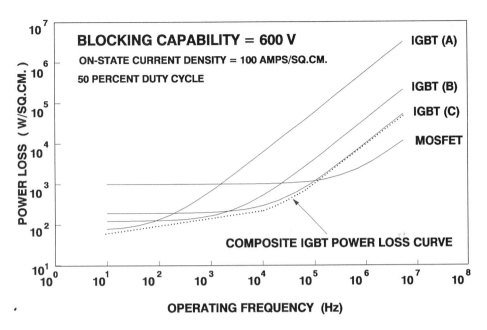

Fig. 8.42 Optimization of IGBT for wide range of operating frequencies.

of the IGBT will allow lower power losses in systems operating at up to 200 kHz. This conclusion is valid for the case of the devices with 600 volt blocking capability. When similar plots and analyses are performed with lower and higher blocking voltages, it is found that the power losses in the case of the power MOSFET are less than those for the case of the IGBT at a frequency of 50 kHz for a blocking voltage of 300 volts and at a frequency of 500 kHz for a blocking voltage of 1200 volts. When the blocking voltage is less than 200 volts, the power losses in the power MOSFET become less than for the case of the IGBT at all operating frequencies. Thus, the IGBT is particularly well suited for high blocking voltages and lower operating frequencies, while the power MOSFET is well suited for low blocking voltages and high operating frequencies.

8.7 COMPLEMENTARY DEVICES

Complementary devices are often needed in power circuits. An example of an application is an AC switch for numerical and appliance controls. It is preferable to use one n-channel device and one p-channel device in parallel to form a composite AC switch as shown in Fig. 8.43. This allows control of both IGBTs with a common reference terminal. If a composite AC switch is formed using two n-channel IGBTs in anti-parallel, as shown in

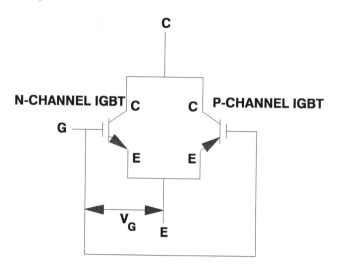

Fig. 8.43 AC switch formed by using an n-channel IGBT in parallel with a p-channel IGBT.

Fig. 8.44, it is necessary to drive one of the devices by level shifting the gate signal. This is undesirable because of the added complexity and cost of the control circuit. For such applications, p-channel power MOSFETs have been developed to complement the n-channel devices. Due to the lower mobility for holes in silicon, p-channel power MOSFETs have a higher specific on-resistance than n-channel devices. Consequently, these devices have to be made about three times larger in area than n-channel devices to handle the same power rating. It has been demonstrated that this penalty in size is not experienced with IGBTs. In this section, the characteristics of p-channel IGBTs are compared with those of n-channel devices.

In the IGBT, the drift region is flooded with minority carriers during forward conduction. Since the concentration of the free carriers greatly exceeds the doping level, the carrier transport is determined by ambipolar diffusion and drift, which is similar for both n-channel and p-channel devices. These devices differ in terms of the contribution of the channel resistance and the gain of the wide base transistor through which the bipolar current flow takes place. For devices with long turn-off times, the lifetime in the drift region is large and the gain of the lower transistor (P-N-P for n-channel and N-P-N for p-channel IGBTs) is high. The channel contribution is then negligible. Consequently, complementary IGBTs with slow switching speed are found to exhibit nearly identical forward conduction characteristics. It is worth pointing out that, for the same lifetime in the drift region, the current gain of the NPN transistor in the p-channel IGBT is greater than that of the PNP transistor in the n-channel IGBT because of the longer diffusion length. This reduces the channel current for the p-channel IGBT, which compensates for the lower channel mobility, resulting in its on-state voltage drop becoming very close to that of the n-channel IGBT.

When the switching speed is increased by lifetime reduction, the current gain of the lower transistor is reduced. The current contribution from the channel then grows with increasing switching speed. Since the mobility for holes in the channel is smaller than for electrons, the

Chapter 8 : INSULATED GATE BIPOLAR TRANSISTOR

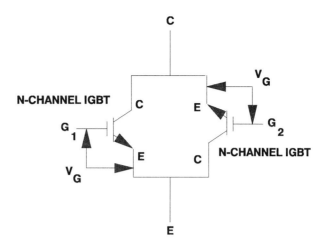

Fig. 8.44 AC switch formed by using two n-channel IGBTs connected in anti-parallel.

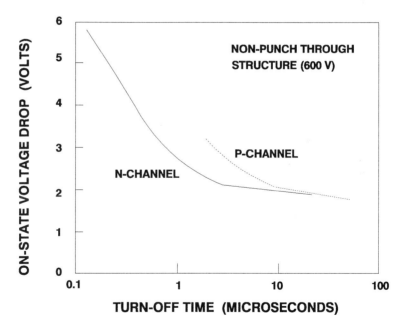

Fig. 8.45 Comparison of the trade-off curve for p-channel and n-channel symmetrical IGBT structures.

rate of increase in forward voltage drop with decreasing turn-off time will be worse for p-channel IGBTs. The results of experimental measurements performed on 600 volt n- and

488   Chapter 8 : INSULATED GATE BIPOLAR TRANSISTOR

p-channel IGBTs over a broad range of switching speeds by using electron irradiation are shown in Fig. 8.45. As expected, the n- and p-channel trade-off curves merge for turn-off times over 10 microseconds. At shorter turn-off times, the trade-off curve for the p-channel devices lies above that for the n-channel IGBTs. It should be noted that, even for turn-off times of 1 microsecond, the difference in the forward voltage drop is less than 50 percent compared with 300 percent for power MOSFETs. When the turn-off time is reduced to the point at which the channel current becomes a significant proportion of the total collector current, the difference in forward voltage drop will approach a factor of 3, as in the case of power MOSFETs. Despite this factor, p-channel IGBTs continue to be attractive because their absolute forward drop is still far lower than for p-channel power MOSFETs.

8.8 HIGH VOLTAGE DEVICES

The IGBT is ideally suited for scaling up the blocking voltage capability. In the case of the power MOSFET, it was shown that the on-resistance increases sharply with breakdown voltage. This has limited the development of high current power MOSFETs with high blocking voltage ratings. For the power MOSFET, the increase in on-resistance with breakdown voltage arises from an increase in the resistivity and thickness of the drift region required to support the operating voltage. In contrast, for the IGBT, the drift region resistance is drastically reduced by the high concentration of injected minority carriers during on-state current conduction. The contribution to the forward drop from the drift region then becomes dependent upon its thickness but independent of its original resistivity.

In the IGBT, the drift region thickness determines the current gain (α_{PNP}) of the integral wide base bipolar transistor. When the blocking voltage capability of the IGBT is increased by increasing the drift region width, the current gain (α_{PNP}) is reduced. More channel current is then required for these devices with higher blocking voltages. This raises their forward voltage drop. Since the voltage drop in the MOSFET portion of the IGBT is a small fraction of the total voltage drop, there is a relatively small increase in the on-state voltage drop with increasing blocking voltage capability when compared with power MOSFETs.

A comparison of the forward conduction characteristic of 300-, 600- and 1200-volt IGBTs has been experimentally performed. These devices had the symmetrical blocking structure. For the symmetrical blocking structure, the drift layer thickness and resistivity have to be increased with increasing blocking voltage rating to satisfy the open-base transistor breakdown criteria. The drift layer for the 300-volt devices was 60 microns with a resistivity of 10 ohm-cm, that for the 600 volt devices was 120 microns with a resistivity of 30 ohm-cm, and that for the 1200-volt devices was 200 microns with a resistivity of 90 ohm-cm. The devices were simultaneously fabricated with the DMOS cell structure. The impact of the increase in drift region thickness on the forward conduction characteristics is shown in Fig. 8.46 for the case of devices with identical turn-off times. As expected, the forward voltage drop increases with increase in voltage rating. An important point to note is that this effect is much greater for the power MOSFETs, as indicated by the dashed lines. As a consequence, the ratio of on-state current density for the

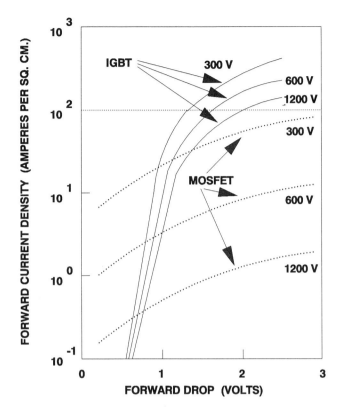

Fig. 8.46 Comparison of on-state characteristics of symmetrical IGBTs with different blocking voltage ratings.

IGBT to the power MOSFET at a forward drop of 2 volts grows from 10 for the 300 volt devices to 150 for the 1200 volt devices. A similar effect has been reported for the asymmetrical IGBT structure. These results demonstrate that the forward conduction current density of the IGBT will decrease approximately as the square root of the breakdown voltage. This moderate rate of reduction in current density has allowed the rapid development of devices with both high current and voltage capability.

The trade-off curve between forward voltage drop and turn-off speed changes when the blocking voltage capability, and hence the drift region width, is altered. A typical set of curves for 300V, 600V and 1200V n-channel symmetrical IGBTs is provided in Fig. 8.47 for the case when electron irradiation is used as a lifetime control method. The forward voltage drop increases more sharply with decreasing turn-off time for the higher voltage devices. This behavior is due to the wider drift region in the higher voltage structures. This results in a larger value for the ratio (d_1/L_a) in Eq. (8.41) and Eq. (8.42) for the drift region voltage drop for the higher blocking voltage devices for the same lifetime in the drift layer. Thus, the drift region voltage drop increases more rapidly with reduction in the lifetime in the drift region for the

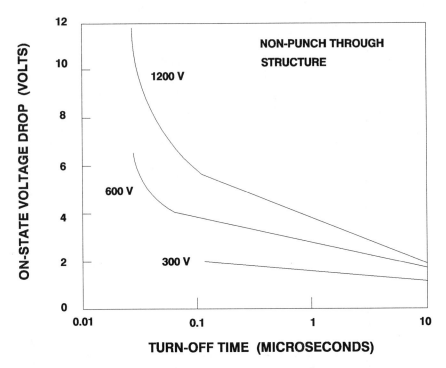

Fig. 8.47 Trade-off curve for symmetric blocking IGBTs with different blocking voltage capabilities.

higher blocking voltage devices. These trade-off curves demonstrate that IGBTs with symmetric blocking voltage capability of over 1000 volts can be developed with turn-off speeds of well below 1 microsecond. Asymmetrical IGBTs have been developed with blocking voltage ratings of over 1500 volts with very high current handling capability. It is likely that the blocking voltage rating for the IGBT can be extended to over 2000 volts.

8.9 HIGH TEMPERATURE CHARACTERISTICS

One of the important characteristics of the insulated gate bipolar transistor is its excellent high temperature forward conduction characteristic. This feature makes the device attractive for applications in which high ambient temperatures may be encountered. An example of such an application is the regulation of the temperature of the base-plate in a steam iron. Devices have been operated successfully with heat sink temperatures approaching 200°C. To take advantage of this feature, it is imperative to raise the latching current density even higher than described previously in order to maintain fully gate controlled operation.

8.9.1 On-State Characteristics

When a sufficient gate bias voltage is applied to the IGBT, the conductivity of the inversion region in the channel under the MOS gate becomes very high, and the device operates without current saturation. Under these conditions, the forward I-V characteristics are similar to those of a P-i-N rectifier with a MOSFET in series. The on-state voltage drop for the IGBT can then be modeled as that across the diode plus that across the MOSFET.

The forward conduction characteristics measured at an elevated temperature is shown in

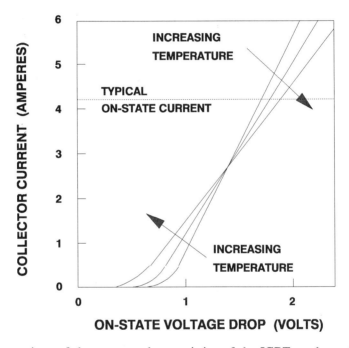

Fig. 8.48 Comparison of the on-state characteristics of the IGBT at elevated temperature with that at room temperature.

Fig. 8.48 for comparison with the room temperature characteristic. The on-state characteristics can be viewed to consist of two segments: a diode drop portion followed by a resistive portion. The diode voltage drop can be seen to decrease when the temperature is increased. This behavior is typical for a P-i-N diode, where the injection across the P-N junction becomes more favorable with increasing temperature. At the same time, the resistance of the second segment increases when the temperature increases. It is important to note that the decrease in the diode forward drop compensates for the increase in channel resistance. This results in a relatively small increase in the on-state voltage drop for the IGBT with increasing temperature. In contrast, for the power MOSFET, the on-resistance increases rapidly, resulting in the need to derate the current handling capability more severely than for the IGBT.

When the IGBT forward voltage drop is examined as a function of temperature at various fixed forward currents, as shown in Fig. 8.49, a decrease in the forward drop is observed at current densities below 100 amperes per cm^2 while an increase is observed at current densities

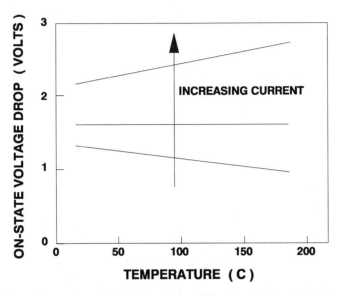

Fig. 8.49 Variation of on-state voltage drop of the 600 volt symmetric IGBT structure with temperature.

above 100 amperes per cm^2. The forward drop at an anode current density of 100 amperes per cm^2 is found to remain independent of temperature for this example. The current density at which the on-state voltage drop of the IGBT becomes independent of temperature is a function of the device switching speed and blocking voltage rating.

The small temperature coefficient for the on-state voltage drop is a unique feature of the IGBT in contrast to other power switching devices. As an example, in the power MOSFET, the decrease in electron mobility with increasing temperature results in an increase in the forward voltage drop by a factor of 3 between room temperature and 200°C. This requires a large derating of the current handling capability of power MOSFETs with increasing temperature, which is not required for the IGBT. Thus, the 20 fold increase in forward conduction current density of a 600-volt IGBT compared to a MOSFET at room temperature is increased to a 60 fold improvement in operating current density at 200°C. This feature makes the IGBT extraordinarily well suited for applications in which a high temperature ambient is encountered. It is worth pointing out that the small positive temperature coefficient of the forward drop at higher current levels observed in IGBTs is beneficial in ensuring homogeneous current

Chapter 8 : INSULATED GATE BIPOLAR TRANSISTOR   493

distribution within chips, and for achieving good current sharing when paralleling devices. The paralleling of multiple IGBTs without any special need for matching devices or providing emitter ballasting has been successfully used to achieve high current circuit performance.

8.9.2 Switching Characteristics

In the IGBT, the gate turn-off time is dominated by the tail in the collector current waveform. As the temperature increases the minority carrier lifetime in the drift region has been found to increase. This not only slows down the recombination process but it increases the P-N-P transistor gain. The latter effect produces a smaller initial drop (I_{CD}) in the collector current during turn-off. The resulting change in the collector current waveforms with

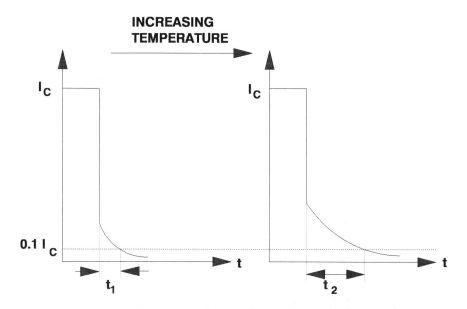

Fig. 8.50 Changes in collector current turn-off waveform with increase in temperature.

temperature is illustrated in Fig. 8.50. Both these phenomena cause an increase in the turn-off time with increasing temperature.

A typical example of the measured change in turn-off time with temperature is provided in Fig. 8.51 for the symmetric blocking IGBT structure. In this case, the turn-off time shows an approximately linear change with temperature with a 50 percent increase between room temperature and 200°C. A relatively smaller change in the turn-off time is observed with temperature for the asymmetrical IGBT structure. This is believed to be due to the recombination of the carriers in the buffer layer which has a higher concentration than the N drift region for the symmetric blocking structure.

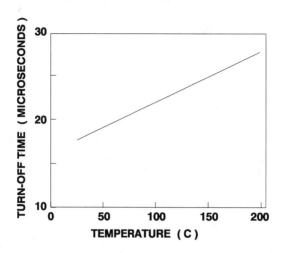

Fig. 8.51 Increase in turn-off time with temperature for the symmetric blocking IGBT structure.

8.9.3 Latching Current Density

One of the problems encountered when operating the IGBT at high current levels has been found to be latch-up of the parasitic P-N-P-N thyristor structure inherent in the device structure. Latch-up of this thyristor can occur causing loss of gate controlled current conduction. At room temperature, it has been found that, in the early developmental devices, the latching current was about 6 times greater than the average current level at which the device is expected to operate. Since the current gains of the N-P-N and P-N-P transistors increase with increasing temperature, the latching current was found to decrease with increasing temperature. This effect is also aggravated by an increase in the resistance (R_p) of the P-base with temperature due to a decrease in the mobility for holes.

An example of the measured reduction in the latching current with temperature is shown in Fig. 8.52 for the symmetrical blocking structure. In this figure, the latching current was measured under switching conditions with a resistive load. The latching current is observed to decrease by a factor of 2 between room temperature and 150°C. A similar reduction of the latching current with temperature has been observed under dynamic switching with an inductive load. For an operating current of 10 amperes, these devices can be operated with a peak current higher by a factor of 2 for junction temperatures of up to 175°C. By using the asymmetric device structure with hole by-pass regions, the latching current density has been raised significantly to provide a current margin to over a factor of 10, making the IGBTs useful in applications with ambient temperatures of even 200°C. Devices that limit the collector current by saturation prior to latch-up have been demonstrated to operate at up to 125°C.

The above decrease in the latching current density with increasing temperature can be

Chapter 8 : INSULATED GATE BIPOLAR TRANSISTOR 495

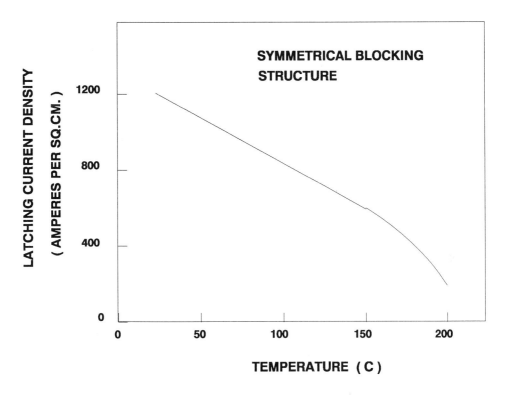

Fig. 8.52 Reduction in latching current with increasing temperature for symmetric blocking IGBT structure.

understood by examination of those terms in Eq. (8.53) that vary significantly with temperature. These terms are highlighted in the following equation:

$$J_{CL} \propto \frac{V_{bi}(T)}{\alpha_{PNP}(T) \, \rho_{SB}(T)} \propto \frac{V_{bi}(T) \, \mu_{PB}(T)}{\alpha_{PNP}(T)} \qquad (8.87)$$

The built-in potential of a diode decreases relatively slowly with increasing temperature. The current gain of the PNP transistor (α_{PNP}) increases slowly with temperature due an increase in the minority carrier lifetime. The most strongly temperature dependent term is the mobility for holes in the P-base region, which decreases with increasing temperature. As discussed in chapter 2, the rate of decrease in the mobility with temperature depends up on the doping concentration. For the relatively high doping concentration in the P-base and P$^+$ regions, the mobility for holes decreases by less than a factor of two over the temperature range discussed above. Therefore, it is a combination of all three factors that governs the observed reduction in the latching current level for the IGBT with increasing temperature.

8.10 TRENCH GATE IGBT STRUCTURE

It was shown in Chapter 7 that the performance of the power MOSFET can be improved by using the UMOS gate structure in place of the DMOS gate structure. The reason for the reduced on-resistance for the power MOSFET in the UMOS structure was shown to be due to the elimination of the JFET resistance contribution and due to an increase in the channel density. In the case of the IGBT, the voltage drop from the MOSFET portion is a small percentage of the total on-state voltage when the lifetime in the N- drift region is large. In these devices, the bipolar current flow via the P-N-P transistor is much larger than the MOS channel current component because of the high gain of the bipolar transistor. However, when the switching speed is increased by lifetime control, the current gain of the P-N-P transistor is reduced and a greater percentage of the collector current flows via the MOSFET. Consequently, a reduction in the resistance of the MOS current path becomes important to obtain a lower on-state voltage drop.

One of the methods of improving the resistance for the MOS current path is to replace the DMOS structure with the UMOS structure. The UMOS-IGBT structure is illustrated in

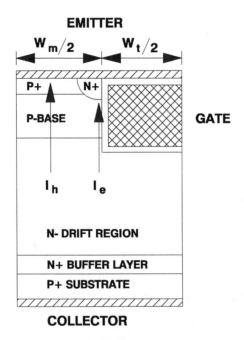

Fig. 8.53 The UMOS IGBT structure.

Fig. 8.53. As in the case of the power MOSFET, the trench gate must extend below the junction between the P-base region and the N- drift region to form a gate bias induced channel

Chapter 8 : INSULATED GATE BIPOLAR TRANSISTOR   497

between the N⁺ emitter and the N- drift region. The electron current path indicated in the figure illustrates that there is no JFET or accumulation layer resistance for the UMOS structure. This reduces the resistance for the MOS current flow. In addition, the cell pitch can be made relatively small when compared with the DMOS structure, allowing a five-fold increase in channel density. The improvement in the MOS path resistance for the UMOS structure has been shown to result in a superior on-state characteristic. Even for devices with high lifetime in the drift region, the on-state voltage drop at a current density of 200 amperes per cm² was found to be 1.2 volts for the UMOS structure compared with 1.8 volts for the DMOS structure. A larger difference can be expected to occur when the lifetime is reduced to increase the switching speed.

The latching current density for the UMOS IGBT structure has been found to be superior to that for the DMOS structure. This is attributed to the improved hole current flow path in the UMOS structure. As shown in Fig. 8.53, the hole current flow can take place along a vertical trajectory in the UMOS structure, while in the DMOS structure hole current flow occurs below the N⁺ emitter in the lateral direction. As a consequence, the resistance for the hole current flow is determined only by the depth of the N⁺ emitter region. A shallow P⁺ layer can be used as shown in the figure to reduce this resistance. This P⁺ region is similar to the deep P⁺ region required for the DMOS structure to suppress latch-up of the parasitic thyristor.

The latching current density for the UMOS structure can be analyzed in a manner similar to that used for the DMOS structure in Section 8.4.2.1. In the UMOS structure, the hole current I_{h1} that forward biases the N⁺ emitter flows through a resistance R_s determined by the depth of the N⁺ emitter diffusion (x_{N+}) and the resistivity of the P⁺ region (ρ_{P+}). It is worth pointing out that the sheet resistance of the P⁺ region must not be used here because the hole current flows vertically through it and not laterally as in the DMOS structure. Using the geometrical parameters given in Fig. 8.53, the resistance R_s is given by:

$$R_s = \frac{\rho_{P+}}{Z} \frac{2 \, x_{N+}}{(W_m - x_{N+})} \tag{8.88}$$

As in the case of the DMOS structure, the latch-up criteria is:

$$I_h R_s = V_{bi} \tag{8.89}$$

The hole current can be related to the collector current density (J_C) by:

$$I_h = \alpha_{PNP} Z \frac{(W_m + W_t)}{2} J_C \tag{8.90}$$

Combining these expressions, the latching current density ($J_{CL,UMOS}$) for the UMOS structure is given by:

$$J_{CL,UMOS} = \frac{V_{bi}}{\alpha_{PNP} \rho_{P+} x_{N+}} \left(\frac{W_m - 2 \, x_{N+}}{W_m + W_t} \right) \tag{8.91}$$

In comparison with the DMOS structure, the latching current density for the UMOS structure is larger because of the smaller resistance of the current path and because of the smaller cell size. This can be demonstrated by taking the ratio of the latching current density for the UMOS and DMOS structures:

$$\frac{J_{CL,UMOS}}{J_{CL,DMOS}} = \frac{(\rho_{SB} L_{E1} + \rho_{SP} + L_{E2})}{\rho_{P+}} \left(\frac{W_m - 2 X_{N+}}{X_{N+}} \right) \left(\frac{W_{cell,DMOS}}{W_{cell,UMOS}} \right) \quad (8.92)$$

Using the same parameters as used before for the DMOS structure and a resistivity of 0.1 ohm-cm for the P^+ region in the UMOS structure, the latching current density for the UMOS structure is found to be about 8 times larger than for the DMOS structure. The measured latching current density for the UMOS structure has been reported to be about 4 times larger than for the DMOS structure.

8.11 TRENDS

The IGBT was the first commercially successful device based upon combining the physics of MOS-gate control with bipolar current conduction. The MOS gate provides an easy interface that can be addressed using low cost integrated circuits. The bipolar current conduction makes high current, high voltage devices available to circuit designers at a relatively low cost. Further, the IGBT structure is very similar to that for the power MOSFET from a fabrication standpoint. This has made its manufacturing relatively easy immediately after conception and the suppression of the parasitic thyristor. After commercial introduction of IGBTs in the mid 1980s, the ratings of devices have grown at a very rapid pace due to the ability to scale up both the current and blocking voltage ratings. In addition, the differences in cost for production of these MOS technology based devices and the bipolar power transistor have diminished. This has made the IGBT very attractive from an applications point of view, resulting in rapid displacement of power bipolar transistors. Some of the applications for which the IGBT is used are adjustable speed motor control for air-conditioning, numerical controls for factory automation and robotics, and appliance controls. The device has also been chosen for the development of the electric car. As the ratings of the devices are scaled up, it is anticipated that the devices may find application in traction (electric street cars) as well.

REFERENCES

1. N. Zommer, "The monolithic HV BIPMOS," IEEE Int. Electron Devices Meeting Digest, Abs. 11.5, pp. 263-266 (1981).

2. D.Y. Chen and S. Chin, "Design considerations for FET-gated power transistors," IEEE Trans. Electron Devices, Vol. ED-31, pp. 1834-1837 (1984).

3. B.J. Baliga, "Enhancement and depletion mode vertical channel MOS gated thyristors," Electronics Letters, Vol. 15, pp. 645-647 (1979).

4 B.J. Baliga, M.S. Adler, P.V. Gray, R. Love, and N. Zommer, "The insulated gate rectifier (IGR) : A new power switching device," IEEE Int. Electron Devices Meeting Digest, Abs. 10.6, pp. 264-267 (1982).

4 J.P. Russell, A.M. Goodman, L.A. Goodman, and J.M. Nielson," The COMFET: a new high conductance MOS gated device," IEEE Electron Device Letters, Vol.EDL-4, pp.63-65 (1983)

5. J.D. Plummer and B.W. Scharf,"Insulated gate planar thyristors," IEEE Trans. Electron Devices, Vol. ED-27, pp. 380-394 (1980).

6. M.F. Chang, G.C. Pifer, B.J. Baliga, M.S. Adler, and P.V. Gray, "25 Amp, 500 volt insulated gate transistors," IEEE Int. Electron Devices Meeting Digest, Abs. 4.4, pp. 83-86 (1983).

7. M.F. Chang, G.C. Pifer, H. Yilmaz, R.F. Dyer, B.J. Baliga, T.P. Chow, and M.S. Adler, "Comparison of N and P Channel IGTs," IEEE Int. Electron Devices Meeting Digest, Abs. 10.6, pp. 278-281 (1984).

8. B.J. Baliga, M.S. Adler, R.P. Love, P.V. Gray, and N. Zommer, "The insulated gate transistor : A new three terminal MOS controlled bipolar power device," IEEE Trans. Electron Devices, Vol. ED-31, pp. 821-828 (1984).

9. B.J. Baliga, "Analysis of insulated gate transistor turn-off characteristics," IEEE Electron Device Letters, Vol. EDL-6, pp.74-77 (1985).

10. B.J. Baliga, M. Chang, P. Shafer, and M.W. Smith, "The insulated gate transistor (IGT) - A new power switching device," IEEE Industrial Applications Society Meeting Digest, pp. 794-803 (1983).

11. B.J. Baliga, M.S. Adler, P.V. Gray, and R.P. Love, "Suppressing latch-up in insulated gate transistors," IEEE Electron Device Letters, Vol. EDL-5, pp. 323-325 (1984).

12. A.M. Goodman, J.P. Russell, L.A. Goodman, J.C. Neuse, and J.M. Neilson, "Improved COMFETs with fast switching speed and high current capability," IEEE Int. Electron Devices Meeting Digest, Abs. 4.3, pp. 79-82 (1983).

13. B.J. Baliga, "Analysis of the output conductance of insulated gate transistors," IEEE Electron Device Letters, Vol. EDL-7, pp. 686-688 (1986).

14. B.J. Baliga, "Switching speed enhancement in insulated gate transistors by electron irradiation," IEEE Trans. Electron Devices, Vol. ED-31, pp. 1790-1795 (1984).

15. T.P. Chow and B.J. Baliga, "Comparison of 300, 600, 1200 Volt n-channel insulated gate transistors," IEEE Electron Device Letters, Vol. EDL-6, pp. 161-163 (1985).

16. A. Nakagawa, H. Ohashi, M. Kurata, H. Yamaguchi, and K. Watanabe, "Non-latch-up, 1200 volt bipolar mode MOSFET with large SOA," IEEE Int. Electron Devices Meeting Digest, Abs. 16.8, pp. 860-861 (1984).

17. B.J. Baliga, "Temperature behavior of insulated gate transistor characteristics," Solid State Electronics, Vol. 28, pp. 289-297 (1985).

18. T.P. Chow, B.J. Baliga, P.V. Gray, M.F. Chang, G.C. Pifer, and H. Yilmaz,"A self-aligned short process for insulated gate transistors," IEEE Int. Electron Devices Meeting Digest, Abs. 6.2, pp. 146-149 (1985).

19. T.P. Chow and B.J. Baliga,"The effect of MOS channel length on the performance of insulated gate transistors," IEEE Electron Device Letters, Vol. EDL-6, pp. 413-415 (1985).

20. D.S. Kuo, J.Y. Choi, D. Giandomenico, C. Hu, S.P. Sapp, K.A. Sassaman, and R. Bregar,"Modelling the turn-off characteristics of the bipolar-MOS transistor," IEEE Electron Device Letters, Vol.EDL-6, pp. 211-214 (1985).

21. A. Mogro-Campero, R.P. Love, M.F. Chang, and R.F. Dyer,"Localized lifetime control in insulated gate transistors by proton implantation," IEEE Trans. Electron Devices, Vol. ED-33, pp. 1667-1671 (1986).

22. A. Nakagawa and H. Ohashi, "600 and 1200V bipolar mode MOSFETs with high current capability," IEEE Electron Device Letters, Vol.EDL-6, pp. 378-380 (1985).

23. T.P. Chow and B.J. Baliga, "Counter-doping of MOS channel (CDC) - new technique of improving suppression of latching in insulated gate bipolar transistors," IEEE Electron Device Letters, Vol.EDL-6, pp. 29-31 (1988).

24. B.J. Baliga, S.R. Chang, P.V. Gray, and T.P. Chow, "New cell designs for improving IGBT safe-operating-area," IEEE Int. Electron Devices Meeting Digest, Abs. 34.5, pp. 809-812 (1988).

25. A.R. Heffner and D.L. Blackburn, "A performance trade-off for the insulated gate bipolar transistor: buffer layer versus base lifetime reduction," IEEE Trans. Power Electronics, Vol. PE-2, pp. 194-207 (1987).

26. A.R. Heffner and D.L. Blackburn, "An analytical model for the steady-state and transient characteristics of the power insulated gate bipolar transistor," Solid State Electronics, Vol. 31, pp. 1513-1532 (1988).

27. H.R. Chang and B.J. Baliga,"500V n-channel IGBT with trench gate structure," IEEE Trans. Electron Devices, Vol. ED-36, pp. 1824-1829 (1989).

28. N. Iwamuro, A. Okamoto, S. Tagami, and H. Motoyama, "Numerical analysis of short-circuit safe operating area for p-channel and n-channel IGBTs," IEEE Trans. Electron Devices, Vol. ED-38, pp. 303-309 (1991).

29. H. Shigekane, H. Kirihata, and Y. Uchida, "Developments in modern high power semiconductor devices," IEEE Symp. on Power Semiconductor Devices and ICs, Abstr. 1.3, pp. 16-21 (1993).

30. N. Thapar and B.J. Baliga, "A new IGBT structure with wider safe operating-area (SOA)," IEEE Symp. on Power Semiconductor Devices and ICs, Abstr. 4.3 (1994).

PROBLEMS

8.1 Calculate the resistivity and width of the N- drift region for a symmetric IGBT structure designed to support 600 volts if the minority carrier lifetime in the drift region is 1 microsecond.

8.2 What is the width of the N- drift region for an asymmetric IGBT structure designed to support 600 volts if the lightly doped portion of the drift region has a doping concentration of 1×10^{13} per cm^3?

8.3 Determine the total width of the N-base region for the asymmetric IGBT structure defined in Problem 8.2 if the N-buffer layer has a doping concentration of 1×10^{17} per cm^3 and a minority carrier lifetime of 0.01 microsecond.

8.4 Compare the on-state voltage drop of the symmetric IGBT structure defined in Problem 8.1 with that for a 600-volt MOSFET at a current density of 200 amperes per cm^2 if the effective high level diffusion length in the N-base region is one-quarter of the N-base thickness. Ignore the channel voltage drop contribution for both devices.

Chapter 8 : INSULATED GATE BIPOLAR TRANSISTOR

8.5 For the same effective diffusion length, what is the on-state voltage drop for the asymmetric IGBT structure defined in Problem 8.3?

8.6 Determine the ratio of the transconductance of the symmetrical IGBT structure defined in Problem 8.1 with that for a power MOSFET with identical channel parameters if the diffusion length in the N-base region is 40 microns. Assume that the collector bias is 100 volts.

8.7 Determine the ratio of the transconductance of the asymmetrical IGBT structure defined in Problem 8.3 with that for a power MOSFET with identical channel parameters if the diffusion length in the N-base region is 40 microns. Assume that the collector bias is 100 volts.

8.8 The symmetrical IGBT defined in Problem 8.1 is fabricated with a homogeneously doped P-base region with a doping concentration of 1×10^{17} per cm^3 and thickness of 5 microns. What is the latching current density for the device if the N$^+$ emitter length is 5 microns. Assume that there is no deep P$^+$ region in the cell.

8.9 Symmetric p- and n-channel IGBT structures are fabricated with a drift region doping concentration of 1×10^{14} per cm^3 and thickness of 100 microns. The lifetime in the drift region is 1 microsecond. The DMOS cell parameters are a channel length of 2 microns, a gate oxide thickness of 1000 °A, a polysilicon window of 20 microns, a gate length between polysilicon windows of 15 microns, and a gate bias of 15 volts. The inversion and accumulation layer mobilities are 400 and 800 cm^2/V.s, respectively. Compare the on-state voltage drop for the two devices at a current density of 200 amperes per cm^2 by using the P-i-N/MOSFET model.

8.10 Calculate and compare the latching current density for the two devices in Problem 8.9 if the highly doped base has a doping concentration of 1×10^{17} per cm^3.

Chapter 9

MOS-GATED THYRISTORS

The insulated gate bipolar transistor (IGBT) discussed in the previous chapter has many excellent characteristics that make it an attractive device from an applications point of view. In particular, its high input impedance and good safe-operating-area have led to a rapid replacement of the power bipolar transistor with the IGBT in motor control applications. Although the on-state voltage drop for the IGBT is much superior to the power MOSFET, the on-state voltage drop is substantially larger than that observed in thyristors. The investigation and development of MOS-gated thyristor structures have been pursued with the goal of utilizing the superior on-state characteristics of thyristors in combination with the ease of control obtained using an MOS-gate structure. These structures are particularly of interest for very high voltage power switching applications because the on-state voltage drop of the IGBT increases with increasing blocking voltage capability.

In order to control the current flowing in a thyristor structure, it is necessary both to trigger the device from its off-state to its on-state and to turn-off the device once it is carry the on-state current. The turn-on of a thyristor by means of an MOS-gate structure integrated within the thyristor structure is relatively easier than the turn-off of a thyristor once it has entered into its regenerative current conduction mode of operation. Historically, the MOS-gated turn-on of a vertical thyristor structure was demonstrated first and it was shown that the MOS-gated structure has a much superior turn-on characteristic when compared with the conventional structure. These devices have become commercially available. In contrast, the development of thyristor structures that can be turned-off by means of an integrated MOS-gate region has been slow because of the difficulties with stopping the regenerative action within the thyristor when it is in its on-state.

The first part of this chapter discusses the MOS-gated turn-on of a thyristor and compares this MOS-gated structure with the conventional thyristor. This MOS-gate structure for turning on the thyristor is incorporated within all the MOS-gated thyristors with turn-off capability. The rest of the chapter deals with a variety of MOS-gate structures that have been proposed for turning-off the thyristor. For these devices, an important parameter is the *maximum controllable current density*, which is defined as the highest current density that can be switched off under MOS-gate control. A large maximum controllable current density is highly desirable in order to obtain a large operating current range for the device. Most of the MOS-gated thyristors do not exhibit current saturation once the thyristor is triggered into its on-state even though the device can be switched off. Since the current saturation capability is a very useful characteristic in many applications, a thyristor structure called the *emitter switched thyristor*, has been developed with this capability at the expense of a slight increase in the on-state voltage drop. This structure is also promising because it promotes more uniform current distribution within a multi-cellular device, which is of importance in scaling the current handling capability of these devices.

9.1 MOS-GATED THYRISTOR TURN-ON

The turn-on of a thyristor by using a control signal applied to an MOS-gate structure was achieved prior to the ability to turn-off the thyristor with an integrated MOS-gate structure. The basic concept used is to supply the base drive current for one of the coupled transistors within the thyristor structure via a channel formed under the MOS-gate region. Although this was first demonstrated in a device by formation of the MOS-gate region in a V-shaped groove formed between the cathode and N- drift region, it is more commonly achieved by using the DMOS process. A cross-section of the basic cell for an MOS-gated thyristor is illustrated in Fig. 9.1

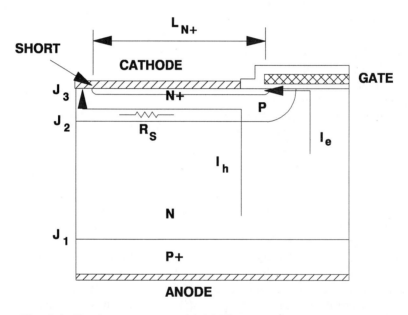

Fig. 9.1 Thyristor structure with MOS-gate region for turn on.

with the current flow paths for electrons and holes during the turn-on process.

When the gate bias is zero and a positive voltage is applied to the anode, the device exhibits a high forward blocking voltage by supporting the voltage across the reverse biased junction J_2. When a positive bias is applied to the gate electrode, electrons are supplied to the

Chapter 9 : MOS-GATED THYRISTORS 505

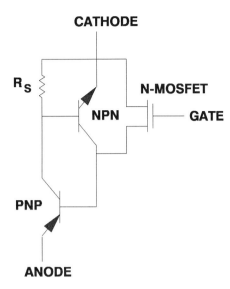

Fig. 9.2 Equivalent circuit for the thyristor structure with MOS-gated turn-on region.

base region of the P-N-P transistor. This results in the injection of holes from the anode into the N- drift region. These holes diffuse across the N- drift region and are collected at the reverse biased junction J_2. As in the case of the conventional thyristor structure, the P-base region of the thyristor is shorted to the N^+ emitter region at a location along the emitter finger as shown in Fig. 9.1. The current in the P-base region created by the collection of holes across junction J_2 flows into the short. However, this current flow produces a voltage drop across the pinch resistance of the P-base region. This resistance is represented by R_S in this figure and in the equivalent circuit for the device shown in Fig. 9.2. The voltage drop across this resistance becomes more than that of a forward biased diode at a sufficiently large hole current level. At this point, the N^+ emitter/P-base junction J_1 becomes sufficiently forward biased to begin injection of electrons from the N^+ emitter into the P-base region. This triggers the regenerative feedback mechanism between the two coupled transistors within the thyristor structure. The thyristor can therefore be turned-on by the application of the gate voltage to the MOS electrode.

The [dV/dt] capability of the MOS-gated thyristor structure is dependent up on the shunting resistance R_S as in the case of the conventional thyristor structure analyzed in Chapter 6. A high [dV/dt] capability can be achieved by reducing the resistance R_S by locating the short-circuit between the N^+ emitter and P-base region at a shorter spacing, i.e. by making L_{N+} smaller, or by reducing the sheet resistance of the P-base region. However, the sheet resistance of the P-base region must be selected to obtain an acceptable threshold voltage. As discussed in the chapter on the power MOSFET, the threshold voltage depends up on the peak P-base doping concentration (N_{AP}) and the gate oxide thickness (t_{ox}):

$$V_T = \frac{t_{ox}}{\epsilon_{ox}} \sqrt{4 \epsilon_s k T N_{AP} \ln\left(\frac{N_{AP}}{n_i}\right)} \qquad (9.1)$$

It is typical to design the devices to obtain a threshold voltage in the range of 2 to 3 volts. For a gate oxide thickness of 1000 angstroms, this results in a sheet resistance for the P-base region (under the N$^+$ emitter) of about 2000 ohms per square when its depth is about 3 microns. The depth of the P-base region and the N$^+$ emitter region must be selected to prevent reach-through breakdown as discussed in Chapter 7 on power MOSFETs. Since the P-base sheet resistance is dictated by the threshold voltage of the MOS-gate region, the [dV/dt] capability must be designed by choosing the distance L_{N+}.

An important advantage of the MOS-gate structure for turning on the thyristor is that the current used to trigger the device is not provided by the gate circuit. This current is derived from the anode circuit as shown in Fig. 9.1. Consequently, a very high [dV/dt] capability can be obtained without influencing the gate drive requirements. The cathode shorting density is limited only by its influence on the on-state voltage drop. In addition, a very large turn-on gate length can be provided to the thyristor without influencing the gate drive requirements. This enables the design of devices with very high [dI/dt] capability without encountering the severe gate drive problems observed in the conventional thyristor triggering method.

9.2 MOS-CONTROLLED THYRISTOR OR MOS-GTO

The control of power flow from a DC power source is of importance for a large variety of applications. This requires the use of power switches that can be turned-off under gate control. As indicated earlier, the ability to turn-off a thyristor under MOS-gate control is desirable in order to obtain a device with lower on-state voltage drop than the IGBT while retaining the advantages of the high input impedance offered by the MOS-gate structure when compared with the GTO. The first thyristor structure which had the ability to turn-off the thyristor on-state current flow under the control of an MOS-gate structure was called either the *MOS-Controlled Thyristor (MCT) or the MOS-Controlled Gate Turn-Off Thyristor (MOS-GTO)*. The basic concept proposed for this structure is to introduce an MOS-controlled short-circuit between the N$^+$ emitter and the P-base region by the formation of a lateral MOSFET located within the P-base region. In general, the MOSFET can be either an enhancement-mode device or a depletion-mode device, and it can be configured as either an n-channel device or a p-channel device. When the MOSFET gate is biased so that this short-circuit does not occur, the thyristor can be triggered into its on-state because the injection efficiency of the N$^+$ emitter/P-base junction is large. However, if the gate of the MOSFET is biased so that it forms a conductive path between the N$^+$ emitter and the P-base region, the injection efficiency of the emitter is greatly reduced. If the resulting current gain of the N-P-N transistor is sufficiently

Chapter 9 : MOS-GATED THYRISTORS   507

low, the thyristor can be turned-off. This method for turning-off the thyristor can be considered to consist of raising the holding current of the thyristor to above its on-state current density.

9.2.1 Device Structure and Operation

There are two basic structures for the MCT. In the structure illustrated in Fig. 9.3, a

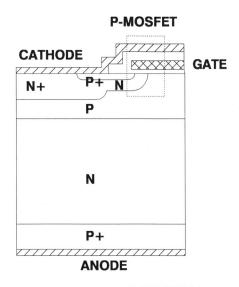

Fig. 9.3 MCT structure with p-channel turn-off MOSFET integrated within the P-base region.

p-channel MOSFET is integrated within the P-base region. The p-channel MOSFET is formed by diffusion of an additional N-region (N-well) into the P-base region to form the substrate for the p-channel MOSFET, followed by the diffusion of a P^+ region that acts as the source of the p-channel MOSFET. This P^+ region is shorted to the N^+ cathode region of the thyristor by the cathode metallization. In order to obtain a low resistance for the p-channel MOSFET, the N-well and the P^+ source diffusions are self-aligned by using the DMOS process with the polysilicon gate as a common masking boundary. However, unlike in the case of the power MOSFET and IGBT, these diffusions must be performed into a relatively highly doped P-base region. This creates difficulty in obtaining a low threshold voltage for the p-channel MOSFET.

The equivalent circuit for the MCT structure shown in Fig. 9.3 is given in Fig. 9.4. Note that both the source and the substrate regions for the p-channel MOSFET are connected to the cathode of the thyristor. Since the p-channel MOSFET is used to short-circuit the N^+ emitter to P-base junction, it is connected across the base-emitter junction of the upper NPN transistor of the thyristor. In the case of zero gate bias, the p-channel MOSFET has a high impedance and the gain of the NPN transistor is large. This allows turning-on the thyristor

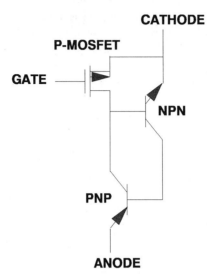

Fig. 9.4 Equivalent circuit for MCT with p-channel MOSFET formed within the P-base region.

either by a base drive current applied to the P-base region as in the case of conventional thyristors or by means of an MOS turn-on gate as discussed in the previous section. The lowest on-state voltage drop for the thyristor is obtained when the injection efficiency of the N$^+$ emitter is high. This requires that the p-channel MOSFET have good sub-threshold characteristics (i.e. that its channel impedance be very large when the gate bias is zero). This imposes a limitation to the reduction in the channel length of the p-channel MOSFET.

When a negative bias is applied to the gate of the p-channel MOSFET, a channel is formed by the inversion of the N-well surface. This provides a path for the flow of holes from the P-base region into the cathode contact that bypasses the N$^+$ emitter/P-base junction. The holes that are flowing into the P-base region when the thyristor was operating in its on-state can then be diverted via the p-channel MOSFET into the cathode electrode. If the voltage drop for the hole current flow via the MOSFET is significantly below that of the on-state diode drop, the voltage across the N$^+$ emitter/P-base junction will become too small for significant injection of electrons into the P-base region. This is equivalent to a large reduction in the current gain of the N-P-N transistor or to a large increase in the holding current for the thyristor. The thyristor regenerative action is then terminated. The stored charge in the N-base region is now removed by recombination of the holes and electrons, leading to a decay in the anode current.

An alternative MCT structure, shown in Fig. 9.5, has been proposed with an n-channel MOSFET integrated within the P-base region. In this case, an additional N$^+$ region is formed adjacent to the N$^+$ cathode region of the thyristor to form the drain of the n-channel MOSFET. This region is then shorted to the P-base region by a metal strap that is not connected to the cathode metallization. The n-channel MOSFET can be formed without the additional processing steps required in the case of the MCT with the p-channel MOSFET. However, the channel length of the n-channel MOSFET is controlled by the gate length and the lateral diffusion of the

Chapter 9 : MOS-GATED THYRISTORS 509

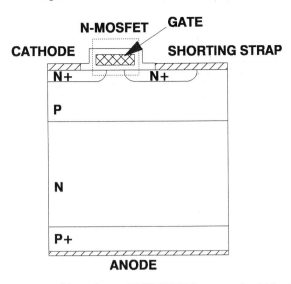

Fig. 9.5 MCT structure with n-channel MOSFET integrated within the P-base region.

N$^+$ regions and the formation of a floating metal strap requires fine patterning of the cathode metallization which is undesirable due the large thickness of the cathode metal.

An equivalent circuit is shown in Fig. 9.6 for the MCT with the n-channel MOSFET

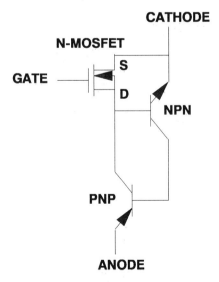

Fig. 9.6 Equivalent circuit for MCT with n-channel MOSFET integrated within the P-base region.

integrated within the P-base region. As in the case of the device with the p-channel MOSFET, the drain of the n-channel MOSFET is connected to the P-base region. However, note that in this case the substrate of the MOSFET is shorted to the drain and not to the source. Consequently, the MOSFET substrate potential is not fixed during device operation.

In general, both of the above structures can be modified to replace the enhancement mode MOSFETs with depletion mode devices. This would be preferable because it would ensure that the N$^+$ emitter is shorted to the P-base region via the MOSFET when the gate bias is zero. This results in a good forward blocking characteristic for the MCT with a high [dV/dt] capability even if the gate supply malfunctions. In contrast, with the enhancement mode MOSFETs, a gate bias must be applied at all times during forward blocking to prevent inadvertent turn-on of the MCT due to either the leakage current flow or the [dV/dt] induced current flow. In spite of this, all of the MCTs that have been developed have been based upon enhancement mode MOSFETs for control of the turn-off process. In addition, the MCT with the p-channel MOSFET has found favor because a single gate electrode can be used with a positive gate bias to turn on the device via an IGBT-like region while a negative gate bias can be used to turn off the device via the p-channel MOSFETs to short-circuit the emitter-base junction.

9.2.2 Maximum Controllable Current Density

For utilization in applications, it is essential that the MCT be able to turn off a high current density when the gate bias is applied to the turn-off MOSFET while having a low holding current density in the absence of the gate bias. This provides a large operating range for control of the anode current. The highest current density that can be turned off with the application of the gate bias to the MOSFET is referred to as the *maximum controllable current density*. Its magnitude depends not only upon the MOSFET channel resistance but also upon the design of the MCT cell because of additional resistances in the turn-off path.

Consider the cross-section of the MCT shown in Fig. 9.7 with the p-channel MOSFET integrated in the P-base region. The MCT is initially assumed to be operating in the on-state with the current flowing via the N$^+$ emitter. When the negative gate bias is applied to turn off the MCT, the holes entering the P-base region are diverted to the cathode metal as indicated by the arrow. The resistances in the path for the hole current can be lumped into three components, namely, the resistance below the N$^+$ emitter (R_P), the resistance between the channel and the P-base region (R_{SP}), and the resistance of the p-channel MOSFET (R_{CH}). The second component is termed a *spreading resistance* because the current flows through a region similar to the JFET region with the power MOSFET where the cross-sectional area for current flow increases from under the gate towards the P-base region.

A similar conclusion can be made with reference to the MCT structure with the n-channel MOSFET. The three lumped resistances for this structure are shown in Fig. 9.8. It is worth noting that in both structures the forward bias across the N$^+$ emitter/P-base junction is the largest at the left-hand side of the cell structure at the center of the N$^+$ emitter. This voltage drop is determined by the sum of the voltage drop across all three resistances due to the hole current flow. Although the hole current (I_h) increases when proceeding from the left-hand side to the right-hand side of the cell, a worst-case analysis can be performed by assuming that all

Chapter 9 : MOS-GATED THYRISTORS

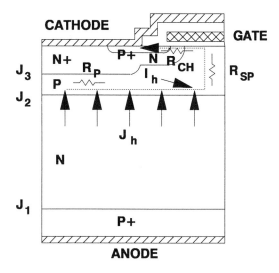

Fig. 9.7 Resistances in turn-off path for MCT with p-channel MOSFET.

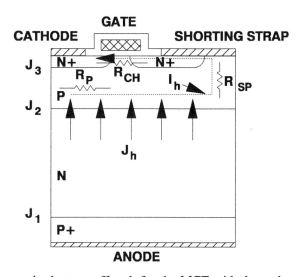

Fig. 9.8 Resistances in the turn-off path for the MCT with the n-channel MOSFET.

of the current collected by junction J_2 flows through all three resistances. Since the hole current density collected by junction J_2 is equal to the anode current density (J_A) multiplied by the current gain of the P-N-P transistor:

$$I_h = J_h \, W \, Z = \alpha_{PNP} \, J_A \, W \, Z \qquad (9.2)$$

where W is the MCT cell width and Z is the cell length orthogonal to the cross-section shown in the figures. The voltage drop in the turn-off path due to the hole current flow is given by:

$$V_R = I_h (R_P + R_{SP} + R_{CH}) \qquad (9.3)$$

If the channel resistance is assumed to be dominant:

$$V_R = \alpha_{PNP} J_A W Z R_{CH} \qquad (9.4)$$

The resistance of the MOSFET channel is dependent upon the gate bias:

$$R_{CH} = \frac{L_{CH}}{Z \mu_{is} C_{ox} (V_G - V_T)} \qquad (9.5)$$

where L_{CH} is the MOSFET channel length, μ_{is} is the inversion layer mobility, C_{ox} is the specific capacitance of the gate oxide, V_G is the applied gate bias, and V_T is the threshold voltage. Substituting this expression into Eq. (9.4):

$$V_R = \frac{\alpha_{PNP} L_{CH} t_{ox} W}{\mu_{is} \epsilon_{ox} (V_G - V_T)} J_A \qquad (9.6)$$

Thus, the voltage drop along the turn-off path increases with increasing anode current density. The MCT can be turned off as long as this voltage drop is less than that required to forward bias the N⁺ emitter/P-base junction to produce electron injection into the P-base region. If the forward biased diode potential is termed V_{bi}, the MCT will fail to turn off when the voltage drop V_R in the turn-off path becomes equal to V_{bi}. Thus, the voltage drive available to extract the holes from the P-base region is limited to the magnitude of the forward biased junction potential. This is substantially smaller than that for the case of a GTO, which is limited by the breakdown voltage of the N⁺ emitter/P-base junction as discussed in Chapter 6. In order to obtain turn-off with such a small driving voltage, it is necessary to make the MCT cell size very small.

Using the criterion $V_R = V_{bi}$ in Eq. (9.6), an expression for the maximum controllable current density (J_{mcc}) is obtained:

$$J_{mcc} = \frac{\mu_{is} \epsilon_{ox} (V_G - V_T) V_{bi}}{\alpha_{PNP} L_{CH} t_{ox} W} \qquad (9.7)$$

From this expression, it can be concluded that a large maximum controllable current density can be obtained by (a) using the n-channel MOSFET structure due to a higher channel mobility for electrons, (b) a shorter channel length L_{CH}, (c) a smaller gate oxide thickness t_{ox}, and (d) a smaller cell pitch (W). The maximum controllable current density can also be increased by application of a larger gate bias.

In practice, all the methods cannot be simultaneously utilized because of limitations on the maximum gate oxide electric field, which is determined by the ratio of the gate bias (V_G) to the gate oxide thickness (t_{ox}). Further, the channel mobility decreases with increasing gate bias

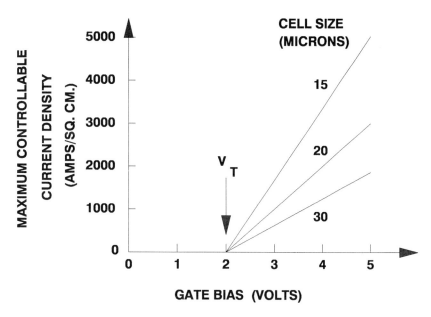

Fig. 9.9 Maximum controllable current density for a MCT with integrated p-channel MOSFET.

due to the increase in the electric field normal to the surface as discussed in chapter 2. Devices with the p-channel turn-off MOSFET have been fabricated with sub-micron channel length using the DMOS process with gate oxide thickness of 500 Angstroms. The maximum controllable current density increases with gate bias once it exceeds the threshold voltage as shown in Fig. 9.9. The data in this figure also demonstrates that a higher maximum controllable current can be obtained by reduction of the cell size as predicted by Eq. (9.7).

The data in Fig. 9.9 was obtained by measurement of the maximum controllable current with small anode voltages. Under these conditions, a very high maximum controllable current density is observed. However, when the device is used in power circuits, the anode voltage rises to a high value during turn-off. It has been found that the maximum controllable current decreases very rapidly with increasing anode voltage if no snubber is used. An example of data obtained on MCTs is provided in Fig. 9.10 with the p-channel and the n-channel structures. At small anode voltages, the maximum controllable current is over 1000 amperes per cm^2 for the device with the p-channel turn-off MOSFET. Without snubbers, the maximum controllable current density decreases drastically with increasing anode voltage until it reaches a value of about 50 amperes per cm^2. It is possible to maintain a high value for the controllable current density by the use of a snubber capacitor across the MCT. This helps by not only reducing the [dV/dt] during turn-off but it also diverts the load current from the device into the snubber capacitor allowing safe turn-off of the MCT. Although effective in improving the turn-off performance of the MCT, the addition of the snubber capacitor is highly undesirable in power circuits due to the additional cost and power losses.

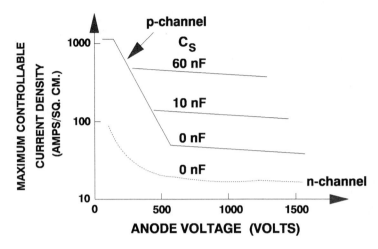

Fig. 9.10 Maximum controllable current measured for MCTs with p-channel and n-channel structure.

9.2.3 Complementary MCT

The MCT structure with the n-channel turn-off MOSFET integrated into the P-base region has been found to be inferior in performance to the MCT structure with the p-channel MOSFET integrated into the P-base region. Since the mobility for electrons is larger than that for holes, an improvement in the maximum controllable current density can be expected by

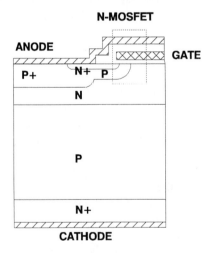

Fig. 9.11 Complementary MCT structure with n-channel MOSFET integrated in the N-base region.

fabrication of the device structure shown in Fig. 9.3 with all the regions replaced by their complementary conductivity type. This device structure, shown in Fig. 9.11, is referred to as the *Complementary MCT or CMCT structure*. It has been found that a high maximum controllable current density can be obtained with this structure. However, this structure has the drawback that the gate is referenced to the anode terminal, which is inconvenient for applications operating from a positive power bus. Another problem with this structure is that the uniformity of high resistivity P-type drift regions is inferior to that for N-type drift regions because the neutron transmutation doping process can be used to produce only uniformly doped N-type silicon. Further, it has been found that the passivation of high resistivity P-type silicon is more difficult than N-type silicon, which leads to problems in manufacturing the CMCT structure.

9.2.4 On-State Voltage Drop

From the equivalent circuit for the MCT shown in Fig. 9.4 and Fig 9.6, it may be concluded that the MCT behaves like a thyristor in the on-state because the turn-off MOSFET has a high impedance in this operating condition. However, the on-state voltage drop of the actual devices have been observed to exceed those for a thyristor. The reason for the higher on-state voltage drop is the formation of parasitic bipolar transistors when the turn-off MOSFET is integrated into the P-base region. These parasitic transistors produce a degradation in the injection efficiency of the N^+ emitter/P-base junction in a different manner for the p-channel and n-channel MOSFET structures as discussed below.

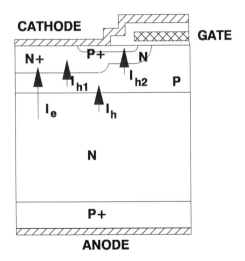

Fig. 9.12 MCT structure with parasitic hole current flow through the p-channel MOSFET.

Consider the MCT structure shown in Fig. 9.12 with the p-channel MOSFET. During operation of the MCT in its on-state, there is injection of electrons from the N^+ emitter into the P-base region and the injection of holes from the P-base region into the N^+ emitter. This hole

516 Chapter 9 : MOS-GATED THYRISTORS

current flow is indicated by the component I_{h1} in Fig. 9.12. A high injection efficiency for the N^+ emitter/P-base junction is needed for good thyristor action with a low on-state voltage drop. This can be achieved by optimum doping of the N^+ emitter region as discussed previously for the P-i-N rectifiers and power bipolar transistors.

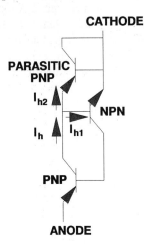

Fig. 9.13 MCT structure with parasitic P-N-P transistor formed with the integration of the p-channel MOSFET.

However, in the MCT structure shown in Fig. 9.12, an additional hole current component I_{h2} is observed because the formation of the integrated p-channel MOSFET also creates a parasitic P-N-P bipolar transistor. The P-base region acts as the emitter of this P-N-P transistor and the P^+ source region of the p-channel MOSFET acts as the collector of the parasitic P-N-P transistor. The additional hole current component I_{h2} flowing through this parasitic P-N-P transistor reduces the injection efficiency of the N^+ emitter/P-base junction. An equivalent circuit for the MCT is shown in Fig. 9.13 with this parasitic P-N-P transistor included. The parasitic transistor acts as a shunt to the N^+ emitter/P-base junction and degrades its injection efficiency.

The degradation in the injection efficiency of the N^+ emitter/P-base junction has been found to be sufficient to cause a snap back in the on-state characteristics, and to result in a relatively high on-state voltage drop of about 4 volts for an MCT designed with a forward blocking voltage of 2000 volts. The parasitic transistor action can be suppressed by the addition of a deep, highly doped, N-type, sub-surface region to the p-channel MOSFET structure as shown in Fig. 9.14. The goal of this region is to increase the doping concentration in the base region of the parasitic P-N-P transistor and thus reduce the injection efficiency for holes from the P-base region into the N-well region. This region must be formed below the surface because a high doping concentration in the N-well at the surface would result in a very high threshold voltage for the p-channel MOSFET. Such a deep, sub-surface, N^+ region was obtained by using ion-implantation of arsenic into the P-base region followed by epitaxial growth of a lightly doped N-type layer. In addition to the process complexity associated with these steps, the epitaxial

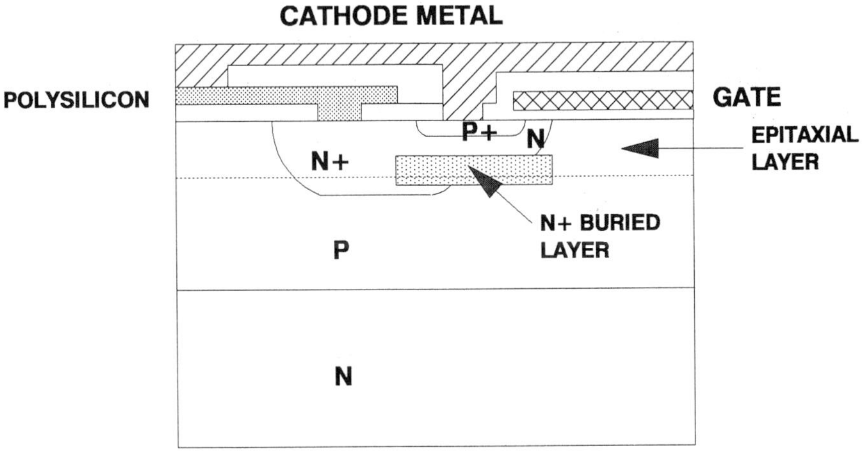

Chapter 9 : MOS-GATED THYRISTORS

Fig. 9.14 MCT cell structure with p-channel turn-off MOSFET containing a N+ buried layer and polysilicon series resistance.

layer must be removed from the edge termination region of the device to prevent its adverse effect on the forward blocking voltage.

A parasitic bipolar transistor is also formed during the integration of the n-channel turn-

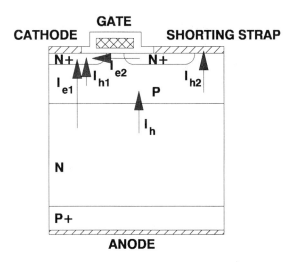

Fig. 9.15 MCT structure with n-channel turn-off MOSFET indicating parasitic lateral N-P-N transistor current flow.

off MOSFET as shown in the MCT structure in Fig. 9.15. In this case, the N^+ drain region of the n-channel MOSFET acts as the collector of a parasitic N-P-N transistor. The floating metal strap serves to short the base of this parasitic N-P-N transistor to the collector. The emitter current I_{e2} flows laterally and recombines with the hole current I_{h2} at the ohmic contact formed by the floating metal strap. Thus, all the emitter current no longer serves to drive the thyristor.

The equivalent circuit for the MCT with integrated n-channel turn-off MOSFET is shown

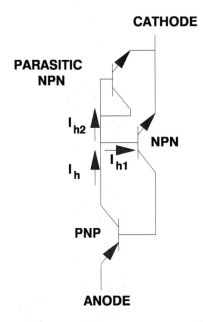

Fig. 9.16 Equivalent circuit for the MCT structure with the parasitic N-P-N transistor created by integration of the n-channel turn-off MOSFET.

in Fig. 9.16 with the parasitic N-P-N transistor. It can be seen that the parasitic N-P-N transistor shunts the N$^+$ emitter/P-base junction and reduces its injection efficiency. This is because of the additional hole current component I_{h2} created by the activation of the lateral parasitic N-P-N transistor.

Based upon the above discussion, it can be concluded that the MCT structure does not operate like a simple thyristor structure in its on-state. The integration of the turn-off MOSFET creates parasitic transistors that reduce the injection efficiency of the N$^+$ emitter/P-base junction leading to an increase in the forward voltage drop. When the size of the N$^+$ emitter diffusion within the MCT cell is reduced in order to increase the MOSFET channel density, it has been found that the effect of the parasitic transistor becomes worse. For this reason, the forward voltage drop has been found to increase with decreasing cell size. Thus, there is a trade-off between improvement of the maximum controllable current density and the on-state voltage drop. Further, as discussed in the next section, it has been found that the addition of emitter ballasting resistances becomes necessary in multi-cellular structures to prevent the formation of a current filament. The emitter ballasting resistance produces a significant increase in the on-state voltage drop, typically in the range of 1 volt, which can double the on-state voltage drop of the MCT when compared with a thyristor.

9.2.5 Multi-Cellular Devices

Multi-cellular devices are necessary to scale the current handling capability of the MCT from the order of milliamperes for a single cell to the order of amperes as needed for applications. The maximum controllable current in the MCT has been found to be strongly affected by inhomogeneous current distribution within a multi-cellular device. This inhomogeneous distribution of current can occur in the steady-state current conduction mode due to the presence of local turn-on gate regions or it can be created during turn-off by a redistribution of the current.

The most attractive method to turn-on the MCT is by the incorporation of a MOS-turn-on gate region. This region is similar to that discussed earlier in section 9.1. In the MCT with the p-channel turn-off MOSFET, the MOS-turn-on gate region can be formed by bringing the P-base region to the surface. This can be performed as shown in Fig. 9.17 by using the polysilicon

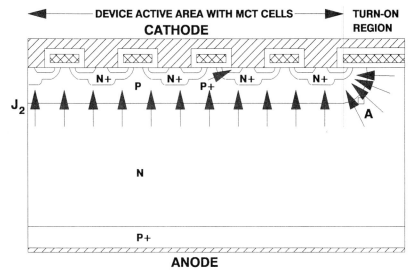

Fig. 9.17 Multi-cell MCT device with MOS turn-on region.

gate electrode as a common mask for the P-base, N-well, and the P$^+$ source regions. When a positive bias is applied to the gate, the N$^+$ emitter can supply electrons to the N-drift region via the channel formed at the surface of the P-base region under the gate in the turn-on region. This allows holes to be injected from the P$^+$ anode which are collected at junction J_2 triggering the turn-on of the thyristor. Since the turn-on occurs at a local region, the thyristor current spreads from the turn-on region to the rest of the device as in the case of conventional thyristors. The density and distribution of the turn-on gates requires a trade-off between good turn-on performance and good turn-off performance for the MCT. For a low density of turn-on gates per unit area, a snap back in the on-state characteristics is observed because the anode voltage must rise well above the on-state voltage drop of the MCT before sufficient current is provided via the gate region to trigger the thyristor into its on-state. In addition, the turn-on di/dt will be small because of the small turn-on gate edge. As the density of turn-on gates per unit area is increased, the snap back no longer occurs and a higher [di/dt] capability can be obtained. However, this reduces the density of the turn-off gates which reduces the maximum controllable current density.

The presence of the turn-on gate region also creates an inhomogeneous distribution in the current density when the MCT is operating in its on-state. A higher current density flows through the MCT cell adjacent to the turn-on region because of the hole current from the turn-on region being collected by the P-base region at point A as indicated by the arrows. During turn-off, the internal MCT cells will turn-off more easily than the cell adjacent to the turn-on region. Thus, the maximum controllable current of the multi-cell MCT can be considerably smaller than for the basic MCT cell discussed in earlier sections.

The degradation of the maximum controllable current by the additional current flowing into the MCT cell adjacent to the turn-on gate region can be avoided by using the structure

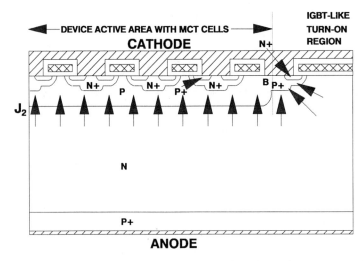

Fig. 9.18 Multicellular MCT structure with IGBT-like turn-on cell.

shown in Fig. 9.18 with an IGBT-like turn-on cell structure. An important aspect of this design is that the hole current in the turn-on region is now removed via the contact between the P$^+$ region and the cathode metal at point B. This allows the MCT cells to be operated with a relatively homogeneous current density.

The maximum controllable current density for this multicellular MCT design is given in Fig. 9.19 as a function of the anode voltage for un-snubbered turn-off. For this design, at lower

Chapter 9 : MOS-GATED THYRISTORS

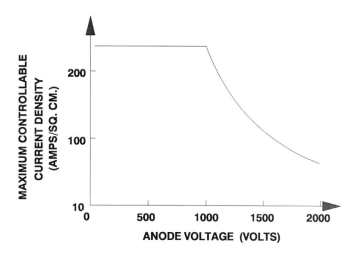

Fig. 9.19 Maximum controllable current density for multi-cellular MCT design with IGBT-like turn-on cells.

anode voltages (below 1000 volts), the maximum controllable current density is determined by the resistance of the turn-off path in the MCT cell as described by the equation derived in section 9.2.2. When the anode voltage exceeds 1000 volts, the maximum controllable currents becomes limited by the onset of the current induced avalanche breakdown phenomenon discussed in the chapter on IGBTs. As in the case of the IGBT, during turn-off, a high concentration of holes traverses the depletion region formed in the N-drift region at saturated drift velocity. The presence of the holes increases the space charge density within the depletion region leading to enhancement of the electric field at the junction J_2. A higher voltage can be sustained by the device at lower current densities because of a reduced hole concentration in the depletion region. As in the case of the IGBT, the highest electric field occurs at the junction curvature in the turn-on region. Thus, the reverse biased safe-operating-area for the MCT is limited by the same phenomenon observed in the IGBT. In general, it can be expected that the RBSOA for the MCT will be worse than that for the IGBT because the turn-on region for the MCT has a relatively small area.

Another important design consideration in a multi-cellular MCT is the inhomogeneous current distribution at the boundary of the device where the active area and the edge termination are merged. As shown in Fig. 9.20, the holes injected from the P^+ anode in the edge termination area are also collected by the junction J_2. This increases the current density in the MCT cell A that is adjacent to the boundary. During turn-off, this cell must shunt not only the hole current collected from the junction under its N^+ emitter but an additional very large hole current that is collected from the edge termination region because the edge termination area is considerably larger than the MCT cell size. The maximum controllable current for the device is then limited by the performance of the edge cells and it does not scale with increase in the active area of the device.

The above problem can be solved as shown in Fig. 9.21 by the addition of a boundary

522   Chapter 9 : MOS-GATED THYRISTORS

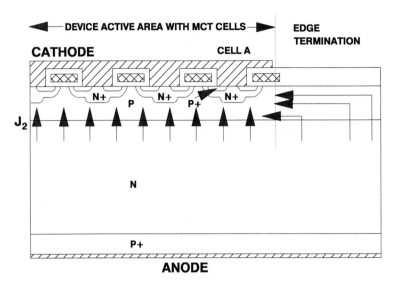

Fig. 9.20 Effect of device boundary on current flow in a multi-cellular MCT.

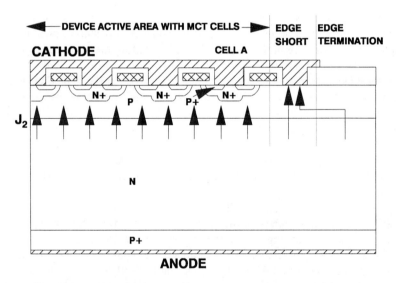

Fig. 9.21 Multi-cellular MCT structure with boundary short.

short. In this design, the hole current flowing from the edge termination region is collected at the boundary short by the cathode electrode. Thus, the MCT cell A no longer has to handle this current during the turn-off process. In fact, the boundary short provides the benefit of a permanent cathode short within the MCT structure, which ensures good forward blocking and [dV/dt] capability. Further, it is important to note that the presence of this cathode short can even prevent the turn-on of MCT cell A. In spite of this, there is no loss in the conduction area because hole current is still collected under MCT cell A. The RBSOA data given in Fig. 9.19 was obtained with the boundary short. Without the boundary short, it was found that the maximum controllable current density was limited to less than 50 amperes per cm^2 independent of the anode voltage because the cells at the boundary limited the performance of the device.

9.2.6 Turn-Off Time

The on-state current conduction in the MCT is accompanied by the injection of a high concentration of holes and electrons into the N- drift region. When the gate bias is applied to the turn-off MOSFET in the MCT, injection from the N$^+$ emitter almost immediately ceases. The high concentration of holes at junction J_2 must then be removed before the device can support voltage. Once the depletion region is established and the device is supporting a large anode voltage, holes that are left in the N- drift region near the P$^+$ anode junction J_1 are removed by recombination. This produces a current tail as observed in the case of IGBTs. The turn-off time is usually determined by the time constant for the tail current. A faster turn-off time can be obtained by using lifetime control processes to increase the recombination rate or by using anode shorts to extract the electrons in the N- drift region.

Electron irradiation has been used to reduce the turn-off time for the MCT with the p-channel turn-off MOSFET. These devices were fabricated from 200 ohm-cm, 500 micron thick N- starting material, which is capable of provided a blocking voltage of 3500 volts. By using an electron irradiation dose of 1 x 10^{14} per cm^2 at an energy of 1.5 MeV, a turn-off time of about 2 microseconds was observed. However, the on-state voltage drop measured at a current density of 100 amperes per cm^3 increased from 1.9 volts without the electron irradiation to 3.35 volts. These results indicate that the trade-off curve between the on-state voltage drop and turn-off time for the MCT is not much superior to that obtained for IGBTs.

An improvement in the turn-off time can also be obtained by the introduction of anode shorts in the MCT structure. This allows extraction of electrons from the anode side during the turn-off process, which greatly reduces the tail current. However, the anode short degrades the injection efficiency of the P$^+$ anode/N- drift junction, which results in an increase in the on-state voltage drop. MCTs with the p-channel turn-off MOSFET structure have been fabricated using 200 ohm-cm, 500 micron thick, N- drift regions with anode shorting densities of 0, 50 and 70 percent. IGBTs were simultaneously fabricated for comparison. These devices had a forward blocking voltage of 3500 volts. Without the anode short, the on-state voltage drop of the MCT was reported to be 1.3 volts at a current density of 100 amperes per cm^2 while that for the IGBT was 2.3 volts. For the anode-shorted devices, the on-state voltage drop was 1.8 volts for the MCT versus 3.3 volts for the IGBT. Thus, the power loss for the MCT is substantially smaller than that for the IGBT. However, when devices with the same anode shorting density are compared with each other, the turn-off power loss for the IGBT was found to be about one-third that observed for the MCT at lower anode voltages and nearly equal at high anode voltages.

9.2.7 Temperature Dependence

It is important to consider the operation of the MCT at above room temperature because of self-heating during operation. In addition, these devices have been considered for applications with high ambient temperatures. The most important temperature dependent parameters for the MCT are its on-state voltage drop, the turn-off time, and the maximum controllable current density.

The temperature dependence of the on-state voltage drop of the MCT is similar to that for a thyristor. As the temperature is increased, the voltage required at the junctions for the injection of holes and electrons into the N- drift region reduces. This results in a decrease in the on-state voltage drop with increasing temperature. Although this is advantageous from the point of view of reduced power losses, it can lead to catastrophic failure in the devices because it promotes localization of the current flow within multi-cellular devices. The formation of current filaments in an MCT structure can be prevented by the incorporation of emitter ballasting resistance. This prevents the current from increasing in a local region because of the increased voltage drop across the ballast resistance. The emitter ballasting resistance must be integrated into the MCT cell structure so that it is distributed among all the cells. This has been achieved by using polysilicon resistors as shown in the cross-section of the MCT in Fig. 9.14 on the left-hand side. Note that the contact to the cathode metal is displaced from the N^+ emitter region. Although the emitter ballasting resistances have been successfully used to promote uniform current distribution during turn-off, they produce a significant increase in the on-state voltage drop (about 1 volt). Further, their incorporation within the MCT cell structure results in a reduced channel density for the turn-off MOSFET, which degrades the maximum controllable current density.

The turn-off time for the MCT is also a function of temperature. As in the case of the

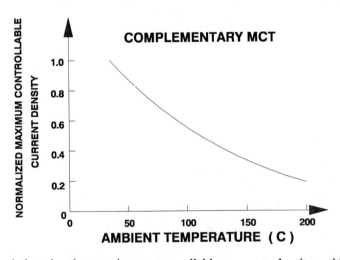

Fig. 9.22 Degradation in the maximum controllable current density with increasing temperature for the complementary MCT (CMCT).

IGBT, the turn-off time increases with increasing temperature. This is attributed to an increase in the minority carrier lifetime with increasing temperature. The increase in the turn-off time is undesirable due to an increase in the switching power losses. The degradation of the maximum controllable current density for the MCT is the most significant limitation to its high temperature operation. The maximum controllable current density was given by the expression in Eq. (9.7). There are many temperature-dependent terms in this expression. With increasing temperature: (a) the channel mobility decreases, (b) the threshold voltage decreases, (c) the diode potential V_{bi} decreases, and (d) the current gain of the P-N-P transistor increases. Among these effects, the dominant ones are the decrease in the channel mobility and the decrease in the diode potential. When the temperature is raised from room temperature to 200°C, the inversion layer mobility decreases by a factor of 1.8x while the bulk mobility decreases by a factor of 2x. At the same time, the diode potential required for injection decreases by 20 percent and the current gain increases by a factor of two. These effects combine together to produce a reduction in the maximum controllable current density by a factor of five, as shown in Fig. 9.22 for the case of the complementary MCT structure.

9.2.8 MCT Characteristics

In this section, the important electrical characteristics of the MCT will be summarized. The MCT is a thyristor structure in which a turn-off MOSFET has been integrated within the P-base region. During the on-state, the current flows via the thyristor structure with an on-state voltage drop that is smaller than that for the IGBT. However, the incorporation of the turn-off gate structure has been found to result in an increase in the on-state voltage drop from that observed in a simple thyristor. The regenerative action in the thyristor is turned-off by the application of a gate bias to the turn-off MOSFET. This produces a resistive shunt path that bypasses the N^+ emitter junction. If the voltage drop in the shunt path is less than that of a forward biased diode, the thyristor will turn off. In order to obtain a uniform turn-off, it has been found that the incorporation of emitter ballasting resistance may be essential. However, this resistance will increase the on-state voltage drop substantially.

The main advantages of the MCT is that it has an on-state voltage drop lower than that of the IGBT and that it can be operated at high surge current levels (without gate controlled turn-off). However, it lacks the current saturation feature and the excellent forward-biased safe-operating-area (FBSOA) of the IGBT. This does not allow control of the inrush current during turn-on by the gate waveform. It is therefore essential to use a small series inductance to limit the rate of rise of current. This not only increases the components and cost in the power circuit but the presence of this inductance also increases the turn-off power losses. The absence of any FBSOA for the MCT is also a problem because this feature is useful for short-circuit protection in power electronic circuits. For these reasons, it is likely that the MCT will be used in those applications where the GTO is now being used because it has similar output characteristics with the advantage of a much reduced gate drive current for its turn-off.

9.3 BASE RESISTANCE CONTROLLED THYRISTOR

The base resistance controlled thyristor (BRT) is an MOS-gated thyristor structure that can be turned-on and turned-off by the voltage applied to its MOS gate electrode. The turn-on is performed using the same physical process described in section 9.1. The principle of operation for turn-off under MOS gate control is based up on diverting the holes from the base region of the thyristor to an alternative route. For this purpose, a diverter region is formed adjacent to the P-base region of the thyristor. This device structure differs fundamentally from the MCT structure because the turn-off MOSFET is not integrated within the P-base region but is formed within the N-base region. This has important implications not only from the point of view of device characteristics but also for simplification of the fabrication process.

9.3.1 Device Structure and Operation

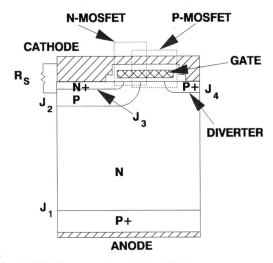

Fig. 9.23 Cross-section of the BRT structure.

A cross-section of the BRT structure is shown in Fig. 9.23. The diverter region is a shallow P-type junction formed adjacent to the P-base region of the thyristor. This diverter region is connected to the cathode electrode. The P-base region and the N^+ emitter region of the thyristor are formed by using the polysilicon gate as the mask. This results in the formation of a DMOS structure at the polysilicon edges. It is also important to short the P-base region to the N+ emitter at a location orthogonal to the cross-section to create a shunting resistance R_S. This ensures good forward blocking capability and a high [dV/dt] capability even when no gate bias is applied. The length of the N+ emitter between the shorts determines the holding current of the thyristor as discussed in section 9.1. A longer distance can be used to obtain a lower holding current. This will reduce the [dV/dt] capability. However, this problem can be overcome if a negative bias is applied whenever the BRT is in its forward blocking mode as

Chapter 9 : MOS-GATED THYRISTORS

discussed below. It is possible to even exclude the short between the N^+ emitter and the P-base region but this requires the application of a negative gate bias during initialization of device operation during start-up of power circuits.

When the gate bias is zero and a positive bias is applied to the anode, the BRT operates in its forward blocking mode. In this case, the junctions J_2 and J_4 are both simultaneously reverse biased. These regions act as guard rings for each other which results in reducing the effects of junction curvature up on the breakdown voltage. The forward blocking voltage is determined by the open-base P-N-P transistor breakdown as in the case of the IGBT. Although a symmetric blocking device structure with homogeneously doped N-base (drift region) is shown in Fig. 9.23, the asymmetric device structure with an N^+ buffer layer adjacent to the P^+ anode can be used to improve the forward blocking and on-state voltage drop, as described in the case of the IGBT in chapter 8, if the device is to be used in DC power circuits.

The BRT can be switched from its forward blocking state to the on-state by the application of a positive gate bias. The positive gate bias creates an inversion layer at the surface of the P-base region under the gate. The electrons supplied from the N^+ emitter into the N-base region then provide the base drive current to the P-N-P transistor. Holes are injected from the P^+ anode and collected at junction J_2 and J_4. Thus, at low current levels, the device behaves like an IGBT. In this mode of operation, the anode current can be saturated by reduction of the gate bias to a value close to the threshold voltage of the n-channel MOSFET. The maximum current for the IGBT mode of operation is determined by the density of the short between the N^+ emitter and the P-base region. This IGBT mode of operation can be used to control the turn-on of the BRT.

When the anode current increases, sufficient hole current is collected at junction J_2 to produce a voltage drop across the shunt resistance R_S that forward biases the N^+ emitter/P-base junction to allow the N^+ emitter to inject electrons. This initiates the turn-on of the thyristor. When the BRT switches from the IGBT mode to the thyristor mode, its on-state voltage drop

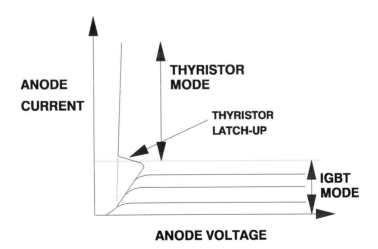

Fig. 9.24 Output characteristics of the BRT.

decreases. This produces a small snap-back in the I-V characteristics as shown in Fig. 9.24. After the thyristor in the BRT latches-up, the on-state voltage drop of the device becomes close to that of a thyristor. It is worth pointing out that, even after the latch-up of the thyristor, hole current flow continues to occur via the diverter region. When the BRT switches from the IGBT mode to the thyristor mode, the junction J_2 goes from reverse biased operation to forward biased operation, and the N-drift region under it becomes highly conductivity modulated. This produces the reduction in the on-state voltage drop. In contrast, the junction J_4 always remains in the reverse based condition. The thyristor-like on-state characteristics of the BRT are attractive for reducing the on-state power dissipation and for obtaining a high surge current handling capability.

The MOS-gated turn-off of the BRT structure is achieved by application of a negative gate bias. This creates an inversion layer at the surface of the N- drift region under the gate electrode between the P-base region and the P+ diverter region. The hole current flowing into the P-base region can now flow via an alternate path from that provided by the emitter shunt resistance R_S. If the resistance for this path for hole current flow is much smaller than the

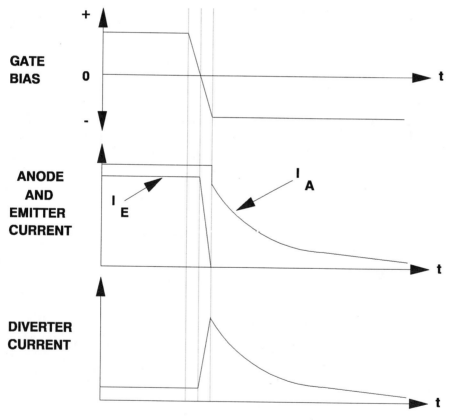

Fig. 9.25 Current waveforms during turn-off of the BRT.

Chapter 9 : MOS-GATED THYRISTORS   529

shunting resistance, it is possible to turn-off the thyristor current flow. As in the case of the MCT, the turn-off is accomplished by raising the holding current of the thyristor to above its on-state operating current level. Thus, the application of negative gate bias reduces the bias across the N+ emitter/P-base junction J_1 so that the N^+ emitter stops injecting electrons. Consequently, during the turn-off process, the anode current is being diverted from the N^+ emitter to the diverter. This process is illustrated with the current waveforms for the anode (I_A), the N^+ emitter (I_E), and the diverter (I_D) in Fig. 9.25. Note that prior to turn-off, there is a substantial current flowing through the diverter. When the gate bias is ramped from its positive value to its negative value, the resistance of the p-channel MOSFET decreases once the gate bias exceeds its threshold voltage. The emitter current falls to zero during this time, while the diverter current increases from its on-state value, keeping the anode current relatively constant. Once the thyristor has turned-off at the end of the gate ramp, the current decays due to the recombination of the holes that were injected into the N- drift region during the on-state. This produces a current tail as in the case of the IGBT and the MCT. It is worth noting that, since the diverter is connected to the cathode metal, although the current takes an alternate path internally within the BRT structure, the external current flows only between the anode and the cathode terminals. Thus, large anode currents can be controlled with a voltage signal applied to the gate electrode resulting in a device with a high input impedance.

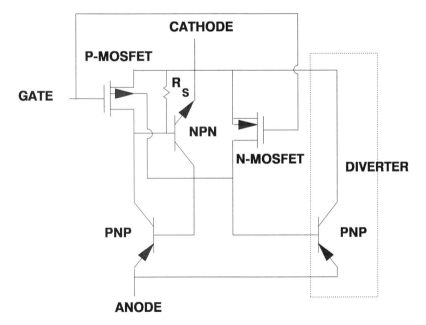

Fig. 9.26 Equivalent circuit for the BRT.

The equivalent circuit for the BRT is shown in Fig. 9.26. In addition to the coupled NPN and PNP transistors that form the thyristor, it contains two MOSFETs and the shunting resistance that represents the short between the N^+ emitter and the P-base region, as well as a

P-N-P transistor that is created between the P⁺ anode and the P⁺ diverter regions. The n-channel MOSFET is formed between the N⁺ emitter as the source and the N- drift region as the drain. It is used to turn-on the BRT and operate it in the IGBT mode. The p-channel MOSFET is formed between the P-base region as the drain and the P+ diverter region as the source. It should be noted that the substrate for the p-channel MOSFET is connected to the collector of the NPN transistor because this MOSFET is formed within the N- drift region. From the equivalent circuit, it can be clearly seen that the p-channel MOSFET and the shunting resistance are connected in parallel. The relative values of the resistance of the p-channel MOSFET and the shunt resistance determines the operating range of the BRT in the thyristor mode. A large shunting resistance and small p-channel MOSFET resistance is required to obtain a operating range in the thyristor mode for the BRT.

9.3.2 Maximum Controllable Current Density

As in the case of the MCT, the maximum controllable current density is defined as the highest current density that can be turned-off by the application of the negative gate bias. Since the turn-off of the thyristor regenerative action requires diverting the hole current via the p-channel MOSFET with a voltage drop below that of a forward biased diode, it is determined by the resistances in the turn-off path. These resistances are shown in Fig. 9.27 together with the

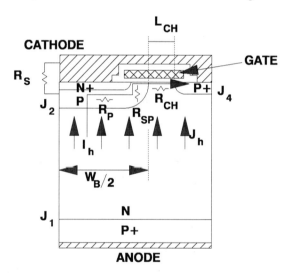

Fig. 9.27 Resistances in the hole current flow path in the BRT during turn-off.

hole current flow during the turn-off process. As in the case of the MCT, the hole current path includes resistances in the P-base region under the N⁺ emitter and the spreading resistance in the P-base region in the DMOS portion, as well as the resistance of the p-channel MOSFET.

The maximum controllable current density for the BRT can be analyzed by considering the hole current collected by the P-base region. The hole current can be assumed to be approximately uniform across the cross-section of the device. Then, the hole current collected by the P-base region across junction J_2 is:

$$I_h = J_h \frac{W_B}{2} Z = \alpha_{PNP} J_A \frac{W_B}{2} Z \qquad (9.8)$$

where J_A is the anode current density, J_h is the hole current near junction J_2, α_{PNP} is the current gain of the wide base P-N-P transistor, and Z is the width of the cell orthogonal to the cross-section. The voltage drop along the turn-off path is then given by:

$$V_R = I_h (R_P + R_{SP} + R_{CH}) \qquad (9.9)$$

If the channel resistance is assumed to be dominant:

$$V_R = \alpha_{PNP} J_A \frac{W_B}{2} Z R_{CH} \qquad (9.10)$$

The resistance of the MOSFET channel is determined by the gate bias and its threshold voltage. Since the p-channel MOSFET in the BRT is formed with a relatively low substrate doping, its threshold voltage is approximately zero. In this case, the channel resistance is given by:

$$R_{CH} = \frac{L_{CH}}{Z \mu_{pi} C_{ox} V_G} \qquad (9.11)$$

where L_{CH} is the MOSFET channel length, μ_{pi} is the inversion layer mobility for holes, C_{ox} is the specific capacitance of the gate oxide, and V_G is the applied gate bias. Substituting this expression into Eq. (9.10):

$$V_R = \frac{\alpha_{PNP} L_{CH} t_{ox} W_B}{2 \mu_{pi} e_{ox} V_G} J_A \qquad (9.12)$$

As in the case of the MCT, the voltage drop along the turn-off path increases with increasing anode current density. However, in the BRT only a fraction of the hole current proportional to the width of the P-base region flows through the turn-off path and a significant fraction of the hole current flows directly into the diverter. The BRT can be turned off as long as this voltage drop is less than that required to forward bias the N$^+$ emitter/P-base junction to produce electron injection into the P-base region. If the forward biased diode potential is termed V_{bi}, the BRT will fail to turn-off when the voltage drop (V_R) in the turn-off path becomes equal to V_{bi}. Thus, the voltage drive available to extract the holes from the P-base region is limited to the magnitude of the forward biased junction potential.

Using the criterion $V_R = V_{bi}$ in Eq. (9.12), an expression for the maximum controllable current density (J_{mcc}) is obtained:

$$J_{mcc} = \frac{2 \mu_{pi} \epsilon_{ox} V_G V_{bi}}{\alpha_{PNP} L_{CH} t_{ox} W_B} \qquad (9.13)$$

From this expression, it can be concluded that a large maximum controllable current can be obtained by (a) a shorter channel length L_{CH}, (b) a smaller gate oxide thickness t_{ox}, and (c) a smaller base width (W_B). The maximum controllable current can also be increased by application of a larger gate bias. In comparison with the MCT with the p-channel turn-off MOSFET, the BRT has a much higher maximum controllable current density because although their cell pitch is nearly equal for the same lithographic design rules, the width of the P-base region (W_B) is only about one-half to one-third of the cell pitch (W) in the BRT. From the above equation, it can also be concluded that the complementary BRT structure with a wide base N-P-N transistor would have a higher maximum controllable current due to the higher mobility in the inversion layer of the n-channel turn-off MOSFET. However, such devices have not been pursued due to the difficulties associated with passivation of P-type silicon as discussed earlier in section 9.2.3 for the complementary MCT.

The maximum controllable current density has been measured on asymmetrical BRT structures (with N^+ buffer layer at the anode junction) with 600 volt forward blocking capability. The maximum controllable current density was found to increase with increasing negative gate

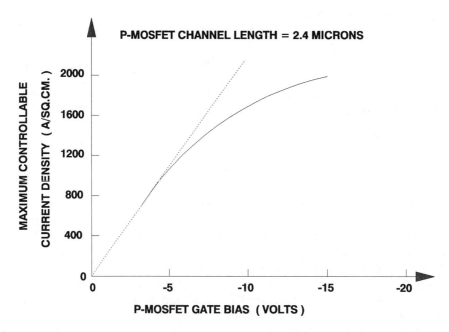

Fig. 9.28 Maximum controllable current for the BRT as a function of the turn-off gate bias.

bias as shown in Fig. 9.28 as described by Eq. (9.13). Note that the threshold voltage is close to zero for the p-channel MOSFET. Although Eq. (9.13) predicts a linear increase in the maximum controllable current density with increasing gate voltage, this is observed only at lower gate biases. As the gate bias increases beyond 10 volts, the maximum controllable current density rises sub-linearly because of a reduction in the inversion layer mobility as discussed in chapter 2. It has been found that, under resistive load switching conditions, the maximum controllable current density is independent of the anode voltage even when the voltage was increased to 500 volts for these devices rated for a forward blocking capability of 600 volts.

It has also been verified that the maximum controllable current density can be increased by reducing the P-base and N$^+$ emitter widths within the BRT cell while keeping the channel length of the P-MOSFET and the size of the diverter region constant. A plot of the measured maximum controllable current density as a function of the P-base width (W_B) is provided in

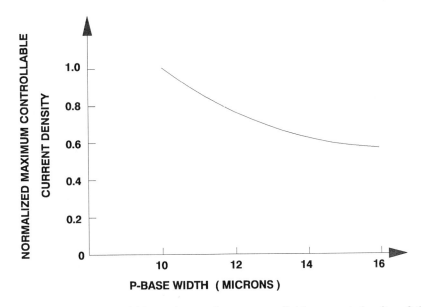

Fig. 9.29 Influence of P-base width on the maximum controllable current density of the BRT.

Fig. 9.29. The maximum controllable current nearly doubles when the P-base width is reduced from 16 microns to 10 microns. This is consistent with the inverse relationship between the maximum controllable current density and the P-base width predicted by Eq. (9.13).

The equation derived earlier also indicates that the maximum controllable current can be increased by reducing the channel length of the p-channel MOSFET. However, it is difficult to reduce the channel length below 2 microns in the BRT because this enhances the pinch-off between the P-base regions due to the JFET action, which makes it difficult to turn-on the device. It also creates a more pronounced snap-back in the I-V characteristics shown in Fig. 9.24, which is undesirable for operation of the device in power circuits. This fundamental limitation of the BRT structure is compensated for by the improved maximum controllable current resulting from the smaller hole current that needs to be extracted via the turn-off MOSFET.

9.3.3 On-state Voltage Drop

In the BRT, the on-state current flow occurs via a thyristor operating in its self-sustaining mode. Its on-state characteristics can therefore be expected to be similar to those of a thyristor. As discussed in chapter 6, the on-state voltage drop for a thyristor can be modelled like that for a P-i-N rectifier. The forward voltage drop is then given by:

$$V_F = \frac{2kT}{q} \ln\left[\frac{J_F d}{2 q D_a n_i F(d/L_a)}\right] \tag{9.14}$$

However, the on-state voltage drop for the BRT has been observed to be slightly higher than that for the simple thyristor structure with the same emitter and base region parameters. One reason for the higher forward voltage drop is due to the smaller area of the N^+ emitter when compared with the cell area. Further, the N^+ emitter must supply electrons not only to the thyristor region but also to drive the vertical P-N-P transistor formed under the P^+ diverter diffusion. The second reason for the increase in the on-state voltage drop is associated with the formation of a parasitic lateral P-N-P transistor due to the introduction of the diverter region. The lateral parasitic P-N-P transistor is formed between the P-base region as its emitter and the P^+ diverter region as its collector. As already pointed out, when the BRT is operating in its on-state after latch-up of the thyristor, the junction J_2 is forward biased while junction J_4 is reverse biased. This allows the injection of holes from the P-base region towards the diverter region.

A cross-section of the BRT is shown in Fig. 9.30 with the hole and electron current

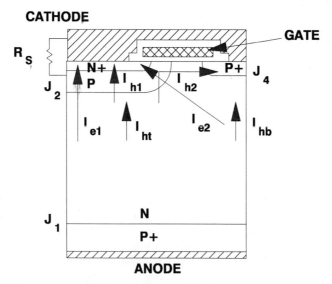

Fig. 9.30 BRT structure with the electron and hole current components shown.

components indicated. The electron current component I_{e1} serves to provide the base drive

Chapter 9 : MOS-GATED THYRISTORS   535

current for the vertical P-N-P transistor in the thyristor region below the N$^+$ emitter. An additional electron current component I_{e2} must be supplied by the N$^+$ emitter to provide the base drive current for the vertical P-N-P transistor formed between the P$^+$ anode and the P$^+$ diverter regions. This implies that the N$^+$ emitter/P-base junction must be operated at a higher forward bias to supply electrons to the thyristor region because some of the electrons are being diverted to the vertical P-N-P transistor region.

In the BRT structure, all of the hole current collected at the junction J_2 is not available

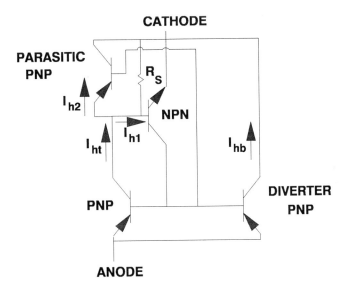

Fig. 9.31 Equivalent circuit for the BRT including the parasitic lateral P-N-P transistor.

to sustain the thyristor regenerative action. A fraction of the hole current I_{h2} is lost because of the presence of the lateral parasitic P-N-P transistor. Fortunately, this lateral P-N-P transistor has a poor current gain because its emitter injection efficiency is low. This is due to the relatively low doping concentration and small thickness of the P-base region which is acting as the emitter. Typically, the peak doping concentration in the P-base region is only about 1 x 10^{17} per cm^3 and its thickness is only 2 microns. Further, the doping concentration of the base region of the lateral P-N-P transistor is relatively high (about 1 x 10^{16} per cm^3) because it is customary to use an ion-implantation to enhance this doping to reduce the JFET effect between the P-base and P$^+$ diverter regions to produce good turn-on characteristics. The equivalent circuit for the BRT is shown in Fig. 9.31 with the lateral parasitic thyristor included. The lateral parasitic thyristor creates an additional shunting path for the hole current even when the p-channel MOSFET is turned-off. This reduces the injection efficiency of the N$^+$ emitter/P-base junction for the thyristor resulting in an increase in the on-state voltage drop.

The effect of changes in the BRT structural dimensions on the on-state voltage drop have been studied for the asymmetrical blocking devices designed to support 600 volts. The impact of reducing the N$^+$ emitter width on the on-state voltage drop is given in Fig. 9.32, for the case

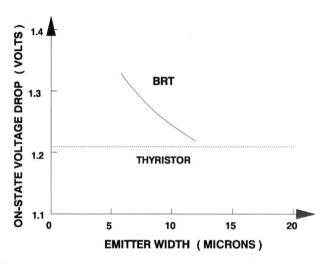

Fig. 9.32 Impact of emitter size on the on-state voltage drop for the BRT.

of a on-state current density of 300 amperes per cm^2. When the emitter width is reduced from 12 microns to 6 microns, the on-state voltage drop increases from 1.21 volts, which is almost equal to that for the simple thyristor structure, to about 1.35 volts. This trend is observed because the relative area of the N$^+$ emitter area to the cell area is becoming smaller as its width is reduced. Consequently, a larger fraction of the N$^+$ emitter injection is being used to support the current flow via the vertical P-N-P transistor under the diverter.

The impact of changing the length of the polysilicon gate electrode is shown in Fig. 9.33 for the case of an on-state current density of 300 amperes per cm^2. When the gate electrode length is decreased from 8 microns to about 6 microns, the on-state voltage drop increases from 1.21 volts to 1.3 volts. This trend is due to an increase in the gain of the parasitic lateral P-N-P transistor because its base width (and effective base charge) becomes smaller when the gate length is reduced. As discussed earlier, a smaller gate length is desirable to reduce the channel resistance of the p-channel turn-off MOSFET in order to obtain a higher maximum controllable current density. Thus, there is a trade off between obtaining the lowest possible on-state voltage drop and the highest possible maximum controllable current density.

From the above analysis, it can be concluded that the on-state voltage drop for the BRT will be slightly larger than that for the thyristor because of a parasitic lateral transistor. Although this increases the on-state power losses, it also creates an emitter ballasting effect within the structure that is beneficial for promoting more uniform current distribution within multi-cellular devices. A comparison of the on-state characteristics of the BRT with the thyristor and the IGBT is provided in Fig. 9.34. The BRT on-state characteristic lies close to that for the thyristor making it much superior to that for the IGBT. As in the case of the MCT, the BRT has a good surge current handling capability.

The on-state operating current density for these devices is dictated by the ability to remove the heat generated within the devices due to the on-state and switching power losses.

Chapter 9 : MOS-GATED THYRISTORS

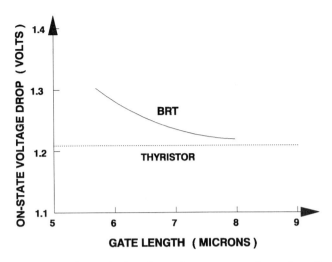

Fig. 9.33 Impact of gate electrode length on the on-state voltage drop for the BRT.

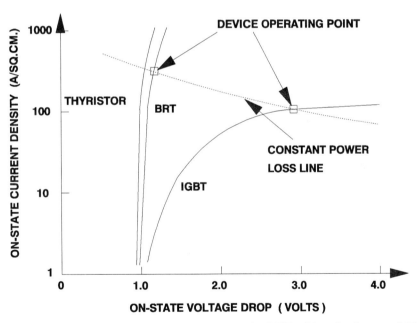

Fig. 9.34 Comparison of on-state characteristics of the BRT with a thyristor and IGBT.

The net power loss depends up on the frequency of operation and the duty cycle. As an example, consider the dashed line in the figure which represents a power dissipation of 300 watts per cm² in the on-state. The intersection of this line with the on-state characteristic defines the operating point for each device. Based up on this criterion, the BRT can be operated at an on-state current density of 300 amperes per cm² with an on-state voltage drop of 1.3 volts. In comparison, the IGBT must be operated at a current density of 100 amperes per cm² with an on-state voltage drop of 3 volts. Thus, the superior on-state characteristics of the BRT can be beneficial in reducing the area (and hence cost) of the device required to satisfy the current required for an application.

9.3.4 Turn-Off Time

As discussed earlier, the turn-off process in the BRT consists initially of the transfer of the hole current path from the thyristor region to the diverter. During this time, the N^+ emitter stops injecting electrons into the N- drift region. This results in an abrupt decrease in the anode current with a rate determined by the ramp rate for the gate drive voltage. After the initial fall in anode current, there is a long current tail created by the hole current flow. This results in a tail current with a long time constant. As in the case of other bipolar power devices, the turn-off time for the BRT can be reduced by the introduction of recombination centers into the N-drift region. The recombination centers remove the holes and electrons stored in the N- drift region reducing the duration of the current tail.

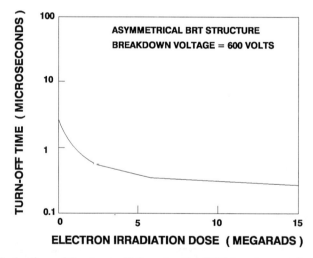

Fig. 9.35 Reduction of the turn-off time for the BRT by electron irradiation.

The most convenient method for the introduction of the recombination centers is by using electron irradiation. As discussed in chapter 2, electron irradiation creates deep level centers uniformly throughout the drift region whose density can be accurately controlled by the radiation dose. As in the case of the MOSFET and IGBT, the charge induced in the gate oxide by the electron irradiation must be removed by a low temperature annealing step to recover the threshold voltage of the MOSFET. Electron irradiation at an energy of 3 MeV has been used to reduce the turn-off time for BRTs fabricated with the asymmetrical structure with a blocking voltage of 600 volts. The turn-off time could be reduced from about 3 microseconds to 0.2 microseconds by using a dose of 16 Megarads, as shown in Fig. 9.35. The change in the turn-off time with increasing dose is very similar to that obtained for the IGBT.

As in the case of other devices with bipolar current transport, a reduction in the lifetime in the drift region causes an increase in the on-state voltage drop. The on-state voltage drop for the BRT is approximately given by Eq. (9.14) in spite of the influence of the parasitic lateral P-N-P transistor. When the lifetime, and hence the diffusion length (L_a) is reduced by electron irradiation, the function $F(d/L_a)$ decreases (see Fig. 4.25) because (d/L_a) becomes larger than

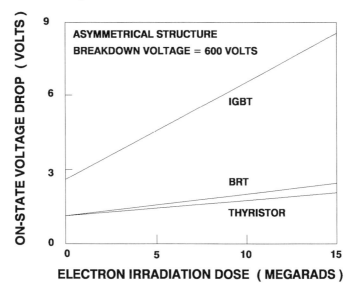

Fig. 9.36 Impact of electron irradiation on the on-state voltage drop of the BRT.

one. This results in an increase in the on-state voltage drop. The increase in the on-state voltage drop is not severe with increase in radiation dose because of the functional dependence given in Eq. (9.14) between V_F and $F(d/L_a)$. This has been experimentally verified. In Fig. 9.36, the change in the on-state voltage drop with electron radiation dose for the BRT is compared with that for the thyristor and the IGBT. The increase in the on-state voltage drop for the BRT is similar to that for the thyristor and not as severe as in the case of the IGBT.

Since the turn-off time for the BRT can be reduced without a severe increase in its on-state voltage drop, it can be expected that the trade-off curve between these parameters will be superior to those obtained for the IGBT. This can be seen in Fig. 9.37 for the case of the asymmetrical structures with a blocking voltage of 600 volts. From these results, it can be concluded that the BRT's characteristics become more attractive for applications at higher frequencies.

The reduction in the lifetime by electron irradiation can also be expected to result in a

540 Chapter 9 : MOS-GATED THYRISTORS

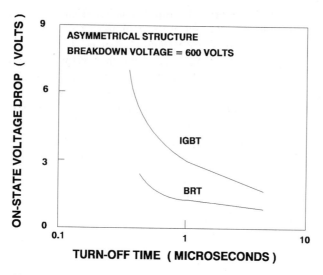

Fig. 9.37 Trade-off curve between on-state voltage drop and turn-off time for the BRT.

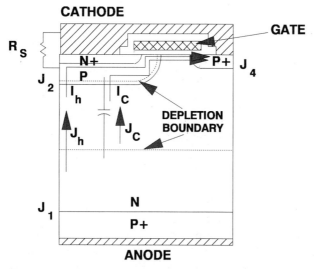

Fig. 9.38 Displacement current flow in BRT during turn-off.

change in the maximum controllable current density. It can be seen from Eq. (9.13) that the maximum controllable current density is inversely proportional to the current gain of the PNP transistor. When the lifetime (and hence the diffusion length) is reduced by the electron irradiation, the base transport factor for the PNP transistor is reduced. Thus, an increase in the maximum controllable current density can be expected after electron irradiation. Although such an increase in the maximum controllable current with electron irradiation dose has been observed in devices, it is much smaller in magnitude than that predicted by the reduction in the current gain of the P-N-P transistor. The reason is that the devices switch at a faster rate after electron irradiation. Consequently, an increasing displacement current flows through the device junction capacitance as the electron irradiation dose is increased. When turning-off the BRT, this displacement current also flows through the p-channel MOSFET as illustrated in Fig. 9.38. Thus, although a reduction in the lifetime decreases the hole current component J_h due to the decrease in the base transport factor for the P-N-P transistor, there is an increase in the displacement current J_C across the junction capacitance due to the higher [dV/dt] resulting from the faster turn-off time. The net result is an increase in the maximum controllable current density by a factor of about 2.

9.3.5 Temperature Dependence

The temperature dependence of the BRT characteristics is of importance because the device normally operates above room temperature due to self-heating and may be used in applications with high ambient temperatures. The most important performance limitation with increasing temperature is a decrease in its maximum controllable current density. Other parameters of interest are the on-state voltage drop and the turn-off time. These parameters are discussed in this section.

The maximum controllable current density for the BRT was given by Eq. (9.13). In this expression, the temperature dependent terms are the inversion layer mobility (μ_{pi}), the forward biased diode potential (V_{bi}), and the current gain of the P-N-P transistor. The inversion layer

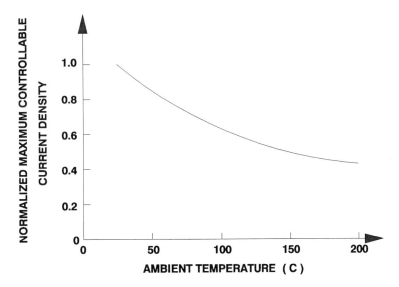

Fig. 9.39 Reduction in the maximum controllable current density with increasing temperature.

mobility decreases with increasing temperature. Its rate of change with temperature is less than that for the bulk mobility. From room temperature to 200° C, the inversion layer mobility decreases by a factor of 1.8 as compared with a three-fold decrease in the bulk mobility. The forward biased diode potential decreases with increasing temperature by about 20 percent between room temperature and 200° C. Since the lifetime in the drift region increases with increasing temperature, the current gain of the P-N-P transistor also increases with temperature. Thus, all the temperature dependent parameters conspire to reduce the maximum controllable current density as shown in Fig. 9.39. The combined effect is a reduction in the maximum controllable current by a factor of 2.5. This decrease is less than that observed for the MCT where the maximum controllable current decreases by a factor of 5. This is because the bulk resistance components in the turn-off path for the hole current play a large role in the MCT and this resistance increases much more than the channel resistance.

Since the on-state current in the BRT flows via a thyristor region, the temperature dependence of its on-state voltage drop can be expected to behave like that for a thyristor. The variation of the on-state voltage drop for the BRT is compared with that for a thyristor in

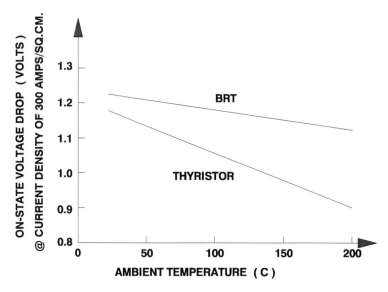

Fig. 9.40 Reduction of on-state voltage drop in the BRT with increasing temperature.

Fig. 9.40 for temperatures up to 200° C. There is a decrease in the on-state voltage drop with increasing temperature for both devices. This behavior is undesirable because it can lead to current filamentation. However, it can be clearly seen that the on-state voltage drop for the BRT changes by less than 10 percent when compared with a 30 percent decrease for the thyristor. This difference is related to the emitter ballasting effect associated with the parasitic lateral P-N-P transistor in the BRT. Since the gain of this transistor increases with increasing temperature, its shunting action on the N^+ emitter/P-base junction becomes greater at higher temperatures. This tends to increase the on-state voltage drop offsetting the effect of temperature on the on-state voltage drop of a thyristor. The emitter ballasting inherent in the BRT structure is therefore beneficial in promoting a homogeneous current distribution within multi-cellular devices.

It should be noted that there is an increase in the turn-off time with increasing temperature as in the case of other bipolar devices due to an increase in the lifetime. It is typical to observe an increase in the turn-off time by a factor of two between room temperature and 200°C. This produces an increase in the switching losses when the device is used in power circuits. Thus, a greater degree of lifetime reduction must be performed if the device is operated at high temperatures.

9.3.6 BRT Characteristics

In this section, the important characteristics of the BRT will be summarized. The BRT is an MOS-gated thyristor structure in which a turn-off MOSFET has been integrated into its N-base region by including a P^+ diverter region adjacent to the P-base region. During the on-state, the device operates like an IGBT at lower current levels with gate controlled current saturation capability. At higher currents, the current flows through a thyristor region, providing the device with a relatively low on-state voltage drop when compared with an IGBT. The on-state voltage drop has been found to be greater than that for a thyristor because the emitter injection efficiency is reduced by the formation of a parasitic lateral P-N-P transistor. This produces an emitter ballasting effect that is beneficial for promoting homogeneous current distribution.

In comparison with the IGBT, the BRT has a lower on-state voltage drop and superior surge current handling capability. However, it does not have the current saturation capability at higher current levels. Although its current saturation capability at lower current levels may be sufficient to control the inrush current during turn-on, the device does not have a sufficient forward biased safe operating area (FBSOA) for short-circuit protection in power circuits. In comparison with the MCT, the BRT has a similar on-state and surge current capability and superior maximum controllable current capability. An equally important feature of the BRT is that it can be fabricated with a DMOS process similar to that used to manufacture MOSFETs and IGBTs. This process requires only 8 masking steps when compared with typically 12 masking steps required to build the MCT. Due to its limited FBSOA, it is unlikely that the BRT will be used in place of the IGBT in applications. It is more likely that the BRT will be used to replace the GTO because it offers a superior on-state voltage drop and a much reduced gate drive current during turn-off.

9.4 EMITTER SWITCHED THYRISTOR

The MCT and BRT are devices in which the on-state current flows via a thyristor region whose regenerative self-sustaining action can be interrupted by the voltage applied to a gate electrode. Although these devices offer a lower on-state voltage drop and a superior surge current capability when compared with an IGBT, they lack the controlled turn-on and current

544   Chapter 9 : MOS-GATED THYRISTORS

saturation capability. These attributes are often indispensable in power circuits for regulating the in-rush current during device turn-on and for short-circuit protection. This makes the MCT and BRT structure more suitable for replacement of GTOs than as a substitute for the IGBT. Even for these circuits a device with controlled turn-on and current limiting capability would be preferable. The emitter switched thyristor offers these features.

9.4.1 Device Structure and Operation

The emitter switched thyristor (EST) is a device in which the on-state current also flows via a thyristor region. However, in this device, the N^+ emitter of the thyristor is not directly connected to the cathode metallization. Instead, the thyristor emitter forms the drain region of a lateral n-channel MOSFET which is integrated into the P-base region of the thyristor as

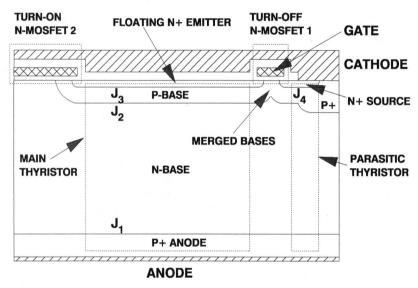

Fig. 9.41 Emitter switched thyristor structure.

illustrated in Fig. 9.41. In the structure shown in this figure, an n-channel MOSFET has been included on the left-hand side. This MOSFET, formed by using the DMOS process with the polysilicon gate as a mask, is used for turning-on the EST. Since there is no external contact to the N^+ emitter region of the thyristor, it is referred to as a *floating N^+ emitter region*. The vertical thyristor formed between the N+ emitter, the P-base region, the N- drift region, and the P^+ anode region is called the *main thyristor*. It should be noted that the EST structure shown in the figure is formed by merging the lateral diffusion of the P-base region formed using a DMOS process with the gate electrode of MOSFET-1 as the mask. This allows fabrication of the turn-on MOSFET-1 and turn-off MOSFET-2 simultaneously with a single DMOS process, which greatly simplifies the fabrication process.

With zero gate bias, the EST can support a large forward blocking voltage because junction J_2 becomes reverse biased and a depletion layer extends into the N- drift region. The forward blocking voltage capability is determined by the open-base P-N-P transistor breakdown as in the case of the IGBT and thyristor structures discussed previously. The structure shown in Fig. 9.41 has a high reverse blocking capability. As in the case of the IGBT, asymmetrical structures can be built with the N^+ buffer layer if the device is to be optimized for DC applications.

When the EST is operated with a positive anode voltage, it can be switched from its forward blocking mode to the on-state by the application of a positive gate bias. This results in the formation of an inversion layer at the surface of the P-base region under the gate electrode for both MOSFET-1 and MOSFET-2. This provides a path for electron transport from the N^+ cathode region on the right hand side via the channel of MOSFET-1, the floating N^+ emitter region, and the channel of MOSFET-2 to the N- drift region. These electrons provide the base drive current for the vertical P-N-P transistor formed between the P^+ anode as the emitter, the N- drift region as the base, and the P-base region as the collector. In response to the base drive current, holes are injected from the P^+ anode into the N- drift region. These holes diffuse through the N- drift region and are collected by the reverse biased junction J_2. The hole current collected by the P-base region flows to its contact with the cathode electrode on the right hand side. Under these conditions, the device operates like an IGBT. In the IGBT mode of operation, the EST can support a high voltage with current limiting capability.

The hole current flow in the P-base region produces a voltage drop that forward biases junction J_1 between the floating N^+ emitter and the P-base region. When the anode current is increased, this voltage drop becomes sufficient to allow injection of electrons from the N^+ emitter. This triggers the vertical main thyristor into its regenerative self-sustaining mode of operation. In the EST structure, the anode current flow through the main thyristor is constrained to flow through the lateral MOSFET-1. This allows control of the thyristor current flow by the gate bias even after the thyristor has latched up. Not only can the thyristor be switched off by reducing the gate bias to zero, but also the anode current can be saturated or limited by reducing the gate bias to a value close to the threshold voltage of the lateral MOSFET-1. The current saturation is achieved by biasing the MOSFET-1 gate such that its drain (the N^+ floating emitter) potential can pinch-off the channel. This results in most of the anode voltage being supported across the lateral MOSFET-1 because the vertical thyristor is still in its on-state and has a voltage drop of only about 1 volt. Since the lateral MOSFET-1 must have a short channel length in order to obtain a low on-state voltage drop, the breakdown voltage of MOSFET-1 is relatively low (typically about 15 to 20 volts). Consequently, the current saturation capability of the EST structure shown in the figure extends only up to the breakdown voltage of the lateral MOSFET-1. Although this voltage is much smaller than the breakdown voltage of the device, it can be useful for performing controlled turn-on of the device.

The integration of the lateral MOSFET-1 into the P-base region of the thyristor creates a parasitic thyristor within the EST structure. This thyristor consists of the P^+ anode, the N-drift region, the P-base/P^+ region, and the N^+ source region of the lateral MOSFET-1. It has been found that the parasitic thyristor can also latch-up at high current densities. Since the current in the parasitic thyristor flows directly between the anode and cathode terminals, its regenerative self-sustaining action cannot be controlled by the gate bias to the lateral

MOSFET 1. Thus, the maximum operating current density for the EST becomes limited by the latching current density of the parasitic thyristor. A large latching current density for the parasitic thyristor can be obtained by the inclusion of the deep, highly doped P^+ region shown on the right hand of the figure. This region serves to reduce the gain of the N-P-N transistor and the forward bias across the N^+ source/P-base junction when the device is operating in the EST mode. This suppresses latch-up of the parasitic thyristor in a manner similar to that used in the IGBT.

The output characteristics of the EST are shown in Fig. 9.42 with the three operating

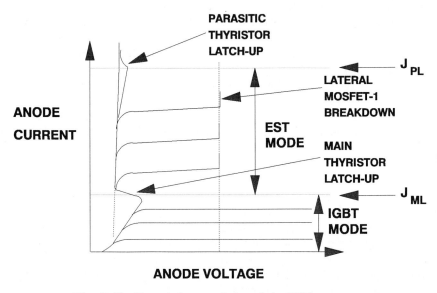

Fig. 9.42 Output characteristics of the EST structure.

modes indicated. At low anode current densities, the device operates in the IGBT mode. In this mode, it is capable of supporting a high voltage close to the breakdown voltage in the forward blocking mode. When the anode current density exceeds J_{ML}, the main thyristor latches-up and the device switches to the EST mode of operation. If the gate bias is large, the device can carry a high current density with a low on-state voltage drop. The on-state voltage drop of the EST can be considered to be the sum of the voltage drop across a high voltage vertical thyristor and a lateral low breakdown voltage MOSFET. Both these devices have a low on-state voltage drop which provide the EST with good on-state characteristics. In a typical device, the on-state voltage drop is approximately 0.5 volts greater than that for the thyristor.

As shown in Fig. 9.42, current saturation occurs in the EST mode of operation up to the breakdown voltage of the lateral MOSFET-1. When either the lateral MOSFET-1 breaks down resulting in a very rapid rise in the current with increasing anode voltage, or when the anode current exceeds a current limit indicated as J_{PL} in the figure, the parasitic thyristor latches-up. Beyond this current density, gate control over the thyristor current is no longer possible. Although the EST can be operated at current levels above J_{PL} to obtain a large surge current

Chapter 9 : MOS-GATED THYRISTORS 547

handling capability, the current density must be less than J_{PL} in order to obtain gate controlled turn-off.

The equivalent circuit for the EST provided in Fig. 9.43 includes the parasitic thyristor.

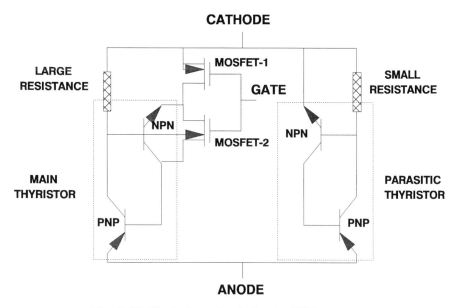

Fig. 9.43 Equivalent circuit for the EST structure.

The large resistance R_P on the left hand side represents the resistance for the hole current flow through the P-base region under the N^+ emitter and the region under the gate of the lateral MOSFET-1 where the P-base regions have been merged. The small resistance R_{P+} on the right hand side represents the resistance for hole current flow below the N^+ source region of the lateral MOSFET-1. This resistance determines the latching current level for the parasitic thyristor. It should be noted that the current through the main thyristor flows via MOSFET-1 to the cathode electrode while that through the parasitic thyristor flows directly between the anode and cathode terminals. In the IGBT mode of operation, the base drive current for the PNP transistor in the main thyristor flows via both MOSFET-1 and MOSFET-2, and the collector current of the P-N-P transistor flows via the large resistance R_P to the cathode terminal. The voltage drop produced by this current flow across R_P forward biases the emitter-base junction of the N-P-N transistor in the main thyristor leading to its turn-on.

The EST can be switched from the on-state to the forward blocking state by switching the gate bias from its positive value to zero. When the gate bias crosses the threshold voltage for MOSFET-1, the N^+ floating emitter is disconnected from the cathode terminal and it stops injecting electrons. Although this stops the regenerative thyristor action, anode current flow continues due to the presence of a high density of holes in the N- drift region that were injected during the on-state. These holes must be removed by recombination leading to a current tail as observed in other bipolar devices. It is typical to observe a sharp drop in the anode current followed by the current tail. The initial sharp drop in current is due to the removal of the electron current component immediately following the gate ramp. Unlike the MCT and BRT, the EST can be gated on and off by using a single polarity for the gate drive voltage.

9.4.2 Dual Channel Structure

Although the EST structure discussed in the previous section (which will be referred to as the single-channel EST structure) offers the ability to turn-off the thyristor current flow after latch-up of the main thyristor, the current saturation capability is limited by the breakdown voltage of the lateral n-channel MOSFET. This restricts the forward biased safe-operating-area (FB-SOA) to relatively small (about 20 volts) anode voltages when the device enters the thyristor mode. The FB-SOA of the EST can be extended to high voltages by modifying the structure as described below. This structure is called the *Dual Channel EST structure* because the N^+ floating emitter region is connected to the N+ source region via two channels. In this structure, the turn-on MOSFET-2 and the turn-off MOSFET-1 of the single-channel EST structure are merged into a single region.

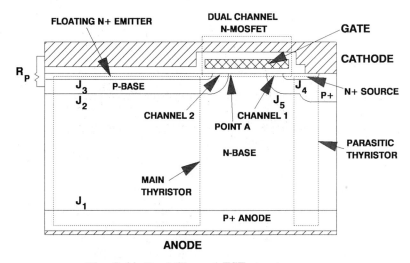

Fig. 9.44 Dual-Channel EST structure.

A cross-section of the dual channel EST structure is shown in Fig. 9.44. In this structure, the separation between the P-base regions of the turn-off MOSFET-1 has been increased so that the P-base regions no longer merge under the gate electrode. This results in the formation of a lateral MOSFET with two channel regions at the surface of each P-base region under the gate and an accumulation region formed at the surface of the N- drift region under the gate electrode between the two P-base regions when the positive gate bias is applied to turn-on the device. It should be noted that, since the P-base regions are no longer merged under the gate electrode, the P-base region under the floating N^+ emitter is no longer connected

Chapter 9 : MOS-GATED THYRISTORS

to the cathode metal on the right hand side of the structure. Although it is possible to leave this P-base region open-circuited, it has been found that a superior FBSOA can be obtained by connecting it to the cathode metal at a point located orthogonal to the device cross-section shown in Fig. 9.44. The resulting equivalent resistance is labeled R_P in the figure. The value of this resistance is determined by the sheet resistance of the P-base region under the floating N^+ emitter (pinch-resistance) and the distance to the contact with the cathode metal.

When a positive bias is applied to the anode with zero gate bias, the junctions J_2 and J_5 become simultaneously reverse biased. These junctions act as guard rings for each other, greatly ameliorating the effect of junction curvature under the gate electrode. The depletion region spreads into the N- drift region, allowing the device to support a high voltage in this forward blocking mode. When a positive gate bias is applied, an inversion layer is formed at the surface of the P-base region to create the channels 1 and 2. At the same time, an accumulation layer is formed at the surface of the N- drift region under the gate electrode. Upon application of a positive bias to the anode, electrons flow from the N^+ source region into the N- drift region via channel 1. This provides base drive current for the vertical P-N-P transistor formed between junctions J_1 and J_5, and junctions J_1 and J_2. Holes injected from the anode are then collected by both junction J_5 and junction J_2. The holes collected by junction J_5 flow to the cathode contact via the P^+ region on the right hand side. The holes collected by junction J_2 flow to the cathode contact via the resistance R_P. The dual-channel EST operates in the IGBT mode under these

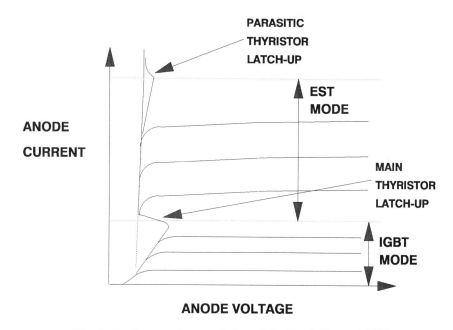

Fig. 9.45 Output characteristics of the Dual-Channel EST.

conditions as shown in Fig. 9.45, which provides its output characteristics.

When the hole current flow through the resistance R_P produces a sufficient voltage drop to forward bias junction J_3 by the potential for a forward biased diode (V_{bi}), it begins to inject electrons and the main thyristor latches-up. In the on-state, the thyristor current flows from the N^+ floating emitter through this lateral MOSFET to the cathode electrode. Thus, the thyristor current can be turned off by reducing the gate bias to zero as in the case of the previous EST structure. If the gate bias is large, the lateral MOSFET operates in its linear region and the total on-state voltage drop of the device is the sum of the voltage drop across the main thyristor and the dual-channel MOSFET. Since the resistance of the dual-channel MOSFET is larger than that of the lateral MOSFET-1 in the single-channel EST structure, the on-state voltage drop of the dual-channel EST will also be larger if it has the same length (L_{N+}) for the floating N^+ emitter region. However, the length of the floating N^+ emitter can be made much smaller in the dual-channel EST than in the single-channel EST structure because this does not alter the latching current, whose value is determined by R_P. Consequently, by using a small floating N^+ emitter length, the on-state voltage drop for the dual-channel EST can be made comparable to the single-channel EST structure.

An important feature of the dual-channel EST structure is that the current can be saturated to very high voltages even after latch-up of the main thyristor. This occurs because it is possible to pinch-off the lateral MOSFET when higher anode voltages are applied. With increasing anode voltage, the potential of the N- drift region under the gate electrode increases. When this voltage becomes comparable to the gate bias voltage, the accumulation region can no longer form and a depletion region begins to extend from junction J_2 at point A (see Fig. 9.44). This cuts off the main thyristor current from flowing into the cathode electrode via the lateral MOSFET. However, since the gate bias exceeds the threshold voltage, electrons can be supplied via channel-1 to the N- drift region. This allows the device to operate like an IGBT, with holes being collected by junction J_2 and junction J_5. The dual-channel EST can support a high voltage in this mode of operation as discussed in subsequent sections.

9.4.3 Maximum Controllable Current Density

The maximum controllable current density is an important parameter that defines the upper bound for the operating current level for the EST. This limit is set by the turn-on of the parasitic thyristor within the EST structure. Although the device can be operated beyond the current level at which the parasitic thyristor latches up to obtain a high surge current handling capability, the device cannot be turned off by reducing the gate bias to zero under these conditions. The operating range for the anode current is then set by the latching current density of the main thyristor and the maximum controllable current density. The latching current density for the main thyristor will be analyzed in a subsequent section.

In order to analyze the maximum controllable current density for the single channel EST structure, consider the cross-section of the device shown in Fig. 9.46 with various dimensions indicated. Since the main thyristor is in its on-state, most of the anode current flows via the floating N^+ emitter through the lateral MOSFET-1 to the cathode terminal. However, the base drive current (I_B) for the N-P-N transistor flows via the merged base regions and the resistance of the P-base region under the N^+ source region (which is also the N^+ emitter of the parasitic

Chapter 9 : MOS-GATED THYRISTORS 551

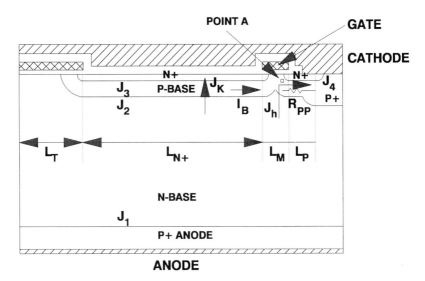

Fig. 9.46 EST structure with current flow pattern prior to latch-up of the parasitic thyristor.

thyristor) to the cathode terminal. In addition, the hole current collected by junction J_2 under the merged base region (segment L_M) and under the N^+ source region (segment L_P) flows through the resistance R_{PP}. This current flow produces a voltage drop that forward biases the junction J_4 at point A. When this forward bias becomes sufficient for junction J_4 to inject electrons, the parasitic thyristor latches-up.

The condition for latch-up of the parasitic thyristor can be analyzed by integration of the hole current flowing via the resistance R_{PP}. This current is given by:

$$I_{PP} = (1 - \alpha_{NPN})\, J_K\, (L_T + L_{N+})\, Z + \alpha_{PNP}\, J_A\, (L_M + L_P)\, Z \quad (9.15)$$

where J_K is the cathode current density, J_A is the anode current density, α_{NPN} is the current gain of the N-P-N transistor, α_{PNP} is the current gain of the P-N-P transistor, L_T is the length of the turn-on region, L_{N+} is the length of the floating N^+ emitter region, L_M is the length of the merged segment, and L_P is the length of the P-base region under the N^+ source region. The first term in this equation is due to the base current of the N-P-N transistor within the main thyristor segment, while the second term is due to the collection of holes at junction J_2 in segments L_M and L_P. The resistance R_{PP} of the P-base region under the N^+ source region depends up on the sheet resistance (ρ_{SP}) of the base region:

$$R_{PP} = \rho_{SP}\, \frac{L_P}{Z} \quad (9.16)$$

Using these expressions, the voltage drop at point A is obtained:

$$V_{PP} = R_{PP}\, I_{PP} = J_A\, \rho_{SP}\, L_P\, [(1-\alpha_{NPN})(L_T+L_{N+}) + \alpha_{PNP}(L_M+L_P)] \quad (9.17)$$

where it has been assumed that the anode and cathode current densities are approximately equal. In addition, the hole current collected from under the N⁺ source region has been neglected. The latch-up of the parasitic thyristor takes place when this voltage drop becomes equal to that for a forward biased diode (V_{bi}). Using this criterion, the parasitic thyristor latch-up current density is obtained:

$$J_{PL} = \frac{V_{bi}}{\rho_{SP} L_P [(1-\alpha_{NPN})(L_T+L_{N+}) + \alpha_{PNP}(L_M+L_P)]} \quad (9.18)$$

Among the terms in this expression, the length of the floating N⁺ region (L_{N+}) is much larger than either the turn-on segment length or the merged segment length. Consequently, the latching current density of the parasitic thyristor varies approximately inversely with increasing floating N⁺ emitter length. This behavior has been observed in experimental devices as shown in

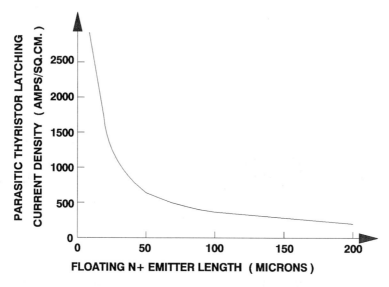

Fig. 9.47 Effect of increasing the length of the floating N⁺ emitter on the latching current density of the parasitic thyristor in the EST.

Fig. 9.47. It is desirable to reduce the floating N⁺ emitter length to avoid the latch-up of the parasitic thyristor and obtain a large maximum controllable current density.

The above analysis is valid when the anode current is increased with a large gate bias applied. The maximum controllable current observed when the single channel EST is switched from the on-state to the forward blocking state is observed to be smaller than the parasitic thyristor latching current density (J_{PL}). When the gate bias is reduced to zero during turn-off, the current flow via the floating N⁺ emitter stops almost instantaneously. If the device is connected to an inductive load, the anode current cannot immediately decrease. This results in a rapid increase in the anode voltage while the device is carrying current via the holes that are

Chapter 9 : MOS-GATED THYRISTORS 553

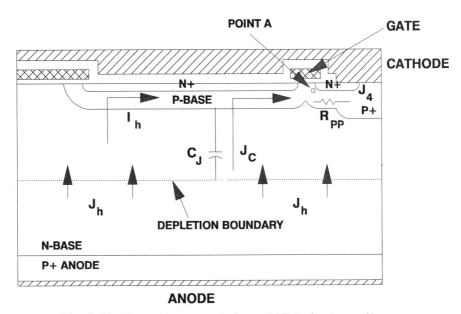

Fig. 9.48 Current components in an EST during turn-off.

are stored in the N- drift region. This produces a displacement current that flows via the resistance R_{PP} as shown in Fig. 9.48. The current flowing through the resistance R_{PP} then consists of an enhanced hole current flow (I_h) because the current flow path via MOSFET-1 no longer exists together with the additional displacement current (J_C). The current flowing via the resistance R_{PP} during the turn-off transient is therefore greater than that flowing under steady state conditions. This results in a lower *maximum controllable current density*, which is defined as the maximum on-state current density that can be switched off under gate control, than the latching current density of the parasitic thyristor.

It has been found the maximum controllable current density of the single-channel EST structure varies with the gate bias voltage as shown in Fig. 9.49. The trend observed in this figure is very similar to that observed for the MCT and the BRT. It can be seen that the maximum controllable current density increases with gate bias until it approaches the parasitic thyristor latching current density. However, the reason for the lower maximum controllable current density observed at small gate biases is related to the current saturation in the EST mode of operation. At small gate bias voltages, when the anode voltage is increased, the EST current becomes limited by current saturation until the breakdown voltage of the lateral MOSFET-1 is reached. At this point, the current rises abruptly as shown in Fig. 9.42, leading to latch-up of the parasitic thyristor. If the gate bias is increased until current saturation does not occur, the maximum controllable current density approaches the parasitic thyristor latching current density. Thus, although the increase in the maximum controllable current density in the single-channel EST appears to be very similar to that for the MCT and BRT, the reason for the variation is quite different.

554   Chapter 9 : MOS-GATED THYRISTORS

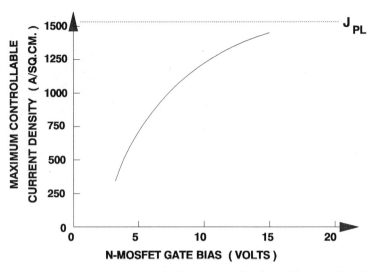

Fig. 9.49 Increase in maximum controllable current density with gate bias for the EST.

9.4.4 Latching Current Density

The latching current density defines the transition from the IGBT mode of operation to the EST mode of operation. This value together with the maximum controllable current density defines the operating current range in the thyristor mode where gate control is maintained. In

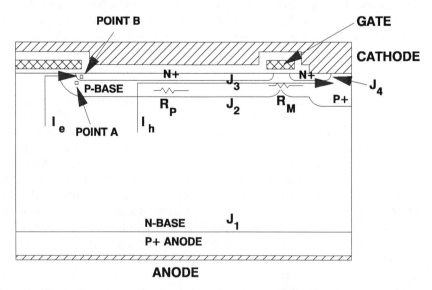

Fig. 9.50 Current flow components in the EST in the IGBT mode of operation.

Chapter 9 : MOS-GATED THYRISTORS 555

order to analyze the latching current density of the main thyristor consider the cross-section of the single channel EST shown in Fig. 9.50 with the electron and hole current components indicated. In this mode of operation, the electron current (I_e) flows via both the MOSFETs and the floating N^+ emitter to the N- drift region. The holes injected from the anode region are collected at junction J_2 and flow laterally through the P-base region under the floating N^+ emitter via the merged base region to the cathode contact on the right hand side. The hole current flow produces a voltage drop across the resistance R_P of the P-base region under the floating N^+ emitter and the resistance R_M of the merged base region, which forward biases junction J_3. The largest forward bias occurs at point A. When the voltage in the base at point A becomes equal to the potential of a forward biased diode (V_{bi}) with respect to the potential of the floating N^+ emitter at point B, junction J_3 begins to inject electrons. These electrons provide the base drive current for the P-N-P transistor leading to the turn-on of the main thyristor.

The current density at which the main thyristor turns on is defined as the *latching current density*. It can be related to the device parameters by considering the current distribution shown

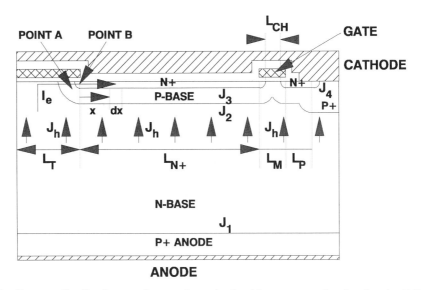

Fig. 9.51 Current distribution used to analyze the latching current density for the EST.

in Fig. 9.51. In performing this analysis, it will be assumed that the hole current density is uniformly distributed across the cross-section of the device. The voltage drop at point A in the P-base region under the floating N^+ emitter can be obtained by addition of the voltage drop produced in the P-base region under the floating N^+ emitter and the voltage drop produced in the merged base region. The voltage drop V_{AP} in the P-base region under the floating N^+ emitter is given by:

$$V_{AP} = \int_0^{L_{N^+}} J_h \, \rho_{SP} \, L_T \, dx + \int_0^{L_{N^+}} J_h \, \rho_{SP} \, x \, dx \qquad (9.19)$$

In this expression, the first term on the right hand side is due to the hole current collected from the turn-on region with length L_T flowing through the P-base resistance, while the second term on the right hand side is due to the collection of holes in the P-base region under the floating N^+ emitter. Here, the hole current density (J_h) at junction J_2 has been related to the anode current density (J_A) by the current gain of the P-N-P transistor. After performing the integration:

$$V_{AP} = \alpha_{PNP} J_A \rho_{SP} \left[L_T L_{N+} + \frac{L_{N+}^2}{2} \right] \quad (9.20)$$

Similarly, the voltage drop (V_{AM}) produced at point A due to the hole current flow through the merged base region is given by:

$$V_{AM} = \int_{L_{N+}}^{L_{N+}+L_M} J_h \, \rho_{SM} \, (L_T + L_{N+}) \, dx + \int_{L_{N+}}^{L_{N+}+L_M} J_h \, \rho_{SM} \, x \, dx \quad (9.21)$$

In this expression, the first term on the right hand side is due to the hole current collected from the turn-on region and the region under the floating N^+ region flowing through the merged P-base resistance, while the second term on the right hand side is due to the collection of holes in the merged P-base region. After performing the integration:

$$V_{AM} = \alpha_{PNP} J_A \rho_{SM} \left[(L_T + L_{N+}) L_M + \frac{L_M^2}{2} \right] \quad (9.22)$$

As indicated earlier, the forward bias across junction J_3 is determined by the difference in potential between points A and B. The potential at point B in the floating N^+ emitter is not equal to zero because the electron current component produces a voltage drop in the floating N^+ emitter and in the channel of the lateral MOSFET-1. It will be assumed that this electron current provides the base drive current for a P-N-P transistor equal in width to the entire width (W) of the EST. Then, the electron current is given by:

$$I_e = (1 - \alpha_{PNP}) J_A W Z \quad (9.23)$$

where Z is the length of the cell orthogonal to the cross-section. The voltage drop at point B produced by this current flowing through the floating N^+ emitter is given by:

$$V_{BN+} = \rho_{SN+} \frac{L_{N+}}{Z} I_e \quad (9.24)$$

and the voltage drop at point B due to this current flowing through the channel of MOSFET-1 is given by:

$$V_{BMOS} = \frac{L_{CH} I_e}{\mu_{ns} C_{ox} Z (V_G - V_T)} \quad (9.25)$$

The voltage at point A is then given by:

$$V_A = V_{AP} + V_{AM} \quad (9.26)$$

while the voltage at point B is given by:

$$V_B = V_{BN+} + V_{BMOS} \quad (9.27)$$

The latching current density (J_{LM}) of the main thyristor is determined by the criterion:

$$V_A - V_B = V_{bi} \quad (9.28)$$

Measurements of the latching current density of the main thyristor in the EST have been performed for asymmetrical devices designed with a forward blocking voltage of 600 volts. It has been found that the latching current density is most strongly dependent upon the length of

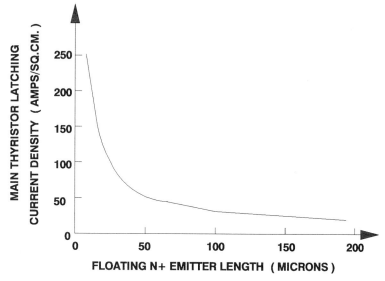

Fig. 9.52 Effect of floating N^+ emitter length on the latching current density for the EST.

the floating N^+ emitter. This variation, provided in Fig. 9.52, is consistent with the above analysis.

9.4.5 On-State Voltage Drop

A simplified equivalent circuit for analysis of the on-state voltage drop of the EST consists of a thyristor in series with a MOSFET. Since the on-state voltage drop of a thyristor is similar to that for a P-i-N rectifier, this equivalent circuit is similar to the simplified equivalent circuit for the IGBT. However, this equivalent circuit is more valid for the EST than the IGBT because the current flows via a thyristor in the EST in contrast with a bipolar transistor in the IGBT. The on-state voltage drop for the EST can therefore be calculated by adding the voltage drop across the thyristor and the lateral MOSFET:

$$V_{F,EST} = V_{PiN} + V_{MOS} \tag{9.29}$$

In the EST, the area of the cathode region of the thyristor is equal to the area of the floating N^+ emitter region. This area is smaller than the cell area because of the space taken by the turn-on region and the control MOSFET-1 region. Consequently, the current density at the cathode of the thyristor is given by:

$$J_K = \frac{W}{L_{N+}} J_A \tag{9.30}$$

If this current density is used to model the voltage drop in the P-i-N diode:

$$V_{PiN} = \frac{2kT}{q} \ln\left[\frac{J_A\, d}{2 q D_a n_{ie} F(d/L_a)} \frac{W}{L_{N+}}\right] \tag{9.31}$$

If the entire anode current is assumed to flow through the channel of the MOSFET, the voltage drop in the MOSFET is given by:

$$V_{MOS} = \frac{L_{CH} J_A W}{\mu_{ns} C_{ox} (V_G - V_T)} \tag{9.32}$$

Using these expressions, the on-state voltage drop of the EST can be analyzed. In order to obtain a small on-state voltage drop, a short channel length for the lateral MOSFET-1 is essential.

It can be shown that there is an optimum length for the floating N^+ emitter from the point of view of minimizing the on-state voltage drop. This occurs because the MOSFET voltage drop decreases when the floating N^+ emitter length is reduced because less current flows through the channel for any given anode current density. However, as the length of the floating N^+ emitter is reduced, the area of the cathode in the thyristor decreases as a fraction of the total cell area. This results in an increase in the voltage drop across the thyristor with a decrease in the floating N^+ emitter length. Since the thyristor voltage drop and the MOSFET voltage drop components have opposing trends with respect to the floating N^+ emitter length, it can be expected that there will be an optimum length for the floating N^+ emitter at which the on-state voltage drop of the EST will have a minimum value. This behavior of the total on-state voltage drop for the EST is shown in Fig. 9.53, where the voltage drop for the thyristor and MOSFET components have been included. It should be noted that the optimum floating N^+ emitter length has a relatively small value because the on-state voltage drop of the thyristor increases rapidly only when the cathode current density becomes large. This optimization was performed for the case of a

Chapter 9 : MOS-GATED THYRISTORS

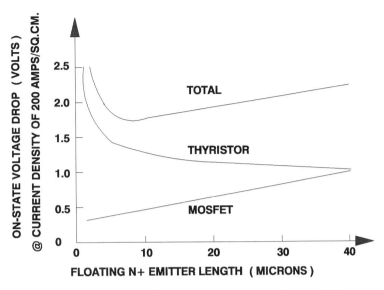

Fig. 9.53 Optimization of on-state voltage drop for the EST.

lifetime of 1 microsecond in the base region of the thyristor. If the lifetime in the N-base region is reduced, the thyristor voltage drop increases and the optimum floating N^+ emitter length becomes larger.

While performing the optimization of the on-state voltage drop for the EST, it is important to recognize that the latching current density for the main thyristor increases when the floating N^+ emitter length is reduced. As a consequence, the latching current density may exceed the on-state operating current density for small values of the floating $N+$ emitter length. In this case, the device will not operate in the EST mode and the on-state voltage drop will be that of the IGBT mode. Since this voltage drop is large, it is important to make sure that the floating N^+ emitter length is sufficiently large to ensure latch-up of the main thyristor.

The on-state characteristics of the EST are compared with those of a thyristor and an IGBT in Fig. 9.54. The on-state voltage drop for the EST can be observed to lie in between that for the thyristor and IGBT. The increase in on-state voltage drop over that for the thyristor is due to the voltage drop in the MOSFET channel. This increases the power losses during operation of the EST in the forward conduction mode and reduces its on-state operating current density when compared with the thyristor as indicated by the operating point obtained by using a constant power loss line in the figure. However, the MOSFET resistance also acts as an emitter ballast resistance, which has the beneficial impact of preventing current filamentation in the EST. As discussed earlier, a resistance has been integrated into the MCT cell by using polysilicon resistors to provide emitter ballasting. This is unnecessary in the EST because the emitter ballasting resistance is inherent within the device structure because the thyristor current flows through the MOSFET that controls the device operation.

560   Chapter 9 : MOS-GATED THYRISTORS

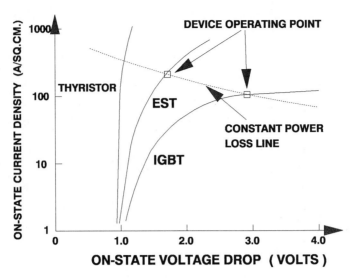

Fig. 9.54 Comparison of on-state characteristics of the EST with a thyristor and IGBT.

9.4.6 Current Saturation

The EST has the unique ability among the thyristor based structures of exhibiting current saturation even after latch-up of the thyristor region used to carry the on-state current. Current saturation capability is useful in power circuits to enable control of the rate of rise of the anode current during device turn-on. It has also been used to limit the current during fault conditions to enable *short-circuit protection*. For short-circuit protection, the device is required to withstand a high voltage (approximately equal to the DC supply voltage) for a short duration while limiting the anode current. This capability is determined by the *forward biased safe operating area (FB-SOA)* of the device. These characteristics are discussed for the EST in this section.

In the EST, current saturation occurs when the gate bias voltage is reduced to values close to the threshold voltage of the turn-off MOSFET-1. Under these conditions, the current flow through the channel produces a sufficient voltage drop to produce channel pinch-off on the drain side as discussed in chapter 7 on the power MOSFET. The drain current then becomes saturated. The relationship between the MOSFET-1 current I_{MOS} and the channel parameters is given by:

$$I_{MOS} = \frac{\mu_{ns} C_{ox}}{2} \frac{Z}{L_{CH}} (V_G - V_T)^2 \qquad (9.33)$$

If all the anode current is assumed to flow through the MOSFET:

$$I_{MOS} = J_A W Z \qquad (9.34)$$

Combining these equations, the saturated anode current density for the EST is obtained as a function of the device parameters:

$$J_{A,sat} = \frac{\mu_{ns} C_{ox}}{2 L_{CH} W} (V_G - V_T)^2 \qquad (9.35)$$

Using this expression, the transconductance for the EST in the current saturation regime of operation can be obtained:

$$g_m = \frac{\mu_{ns} C_{ox}}{L_{CH} W} (V_G - V_T) = \frac{\mu_{ns}}{L_{CH} W} \frac{\epsilon_{ox}}{t_{ox}} (V_G - V_T) \qquad (9.36)$$

From this expression, it can be concluded that a higher transconductance can be obtained by reducing the gate oxide thickness for the lateral turn-off MOSFET-1 and by reducing its channel length. The single-channel EST has a higher transconductance than the dual-channel EST because of its shorter channel length. However, the maximum voltage that can be supported during current saturation is limited to the breakdown voltage of the lateral MOSFET in the single-channel EST. In contrast to this, the dual-channel EST can support high voltages in the current saturation mode.

In the dual-channel EST, under current saturation conditions, pinch-off occurs at point A between channel 1 and channel 2. When this occurs, the main thyristor stops conducting current and the device shifts into IGBT-like operation with electron current supplied via channel 1. In this mode of operation, the hole current flows, as shown in Fig. 9.55, into both junction

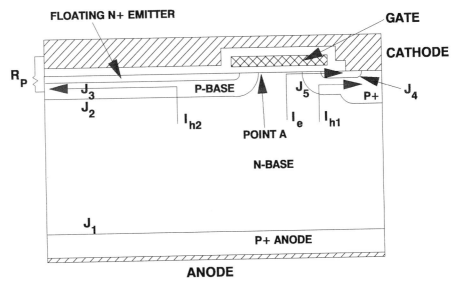

Fig. 9.55 Current flow pattern in the dual channel EST after current saturation.

J_2 and junction J_5. This makes the hole current density under the gate electrode much smaller than that observed in the case of an IGBT. At high anode voltages, the holes are transported through the drift region at their saturated drift velocity. The presence of these holes in the depletion region alters the space charge as discussed earlier for the IGBT. Since the charge due to the holes is positive, it adds to the donor charge, which enhances the electric field at the reverse biased junctions. This mechanism was discussed in Chapter 8 as the current induced avalanche breakdown mechanism. Current induced avalanche breakdown is strongly dependent upon the hole concentration. In the EST, the combination of the reduced hole concentration by removal of holes via the P-base region (junction J_2) and the shunting resistance (R_P), and the reduced electric field at junction J_5 due to junction J_2 acting as a guard ring results in a substantial increase in the FBSOA when compared with the IGBT.

The FBSOA of the EST is shown in Fig. 9.56. There are three boundaries defined in this figure. The highest current level for operation at low anode voltages is determined by the latch-up of the parasitic thyristor. The highest voltage of operation at low current levels is determined by the breakdown voltage of the cell or the edge termination. The region between these limits is determined by the current induced avalanche breakdown phenomenon discussed above. It has been found that this limit can be enlarged by reduction of the P-base shunting resistance (R_P) and by reduction of the current gain of the P-N-P transistor using either lifetime control or an increase in the doping concentration of the buffer layer. It should be noted that a reduction in R_P will result in an increase in the latching current density of the main thyristor while a reduction in the gain of the P-N-P transistor results in a larger on-state voltage drop.

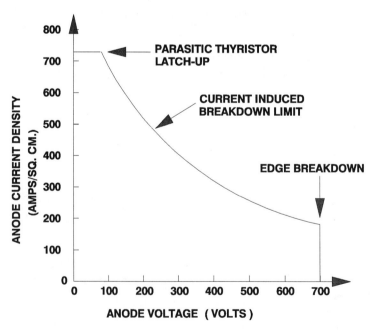

Fig. 9.56 Forward biased safe operating area for the EST.

9.4.7 Turn-Off Time

The turn-off process for the EST occurs in a manner similar to that for the IGBT. When the gate bias is reduced to zero, the floating N^+ emitter immediately stops injecting electrons because it is disconnected from the cathode terminal. The anode current continues to flow via the holes that are present within the N- drift region. This produces a current tail whose duration depends upon the lifetime for the carriers in the N- drift region. The duration of the tail can be reduced by the introduction of recombination centers.

A reduction in the turn-off time for the EST has been demonstrated by using electron irradiation to reduce the lifetime. As in the case of other MOS-gated devices, the positive charge introduced in the gate oxide by the radiation process must be removed by annealing the devices at about 200° C to recover the threshold voltage of the MOSFETs. The measured turn-off time for asymmetrical devices with 600-volt forward blocking capability is shown in

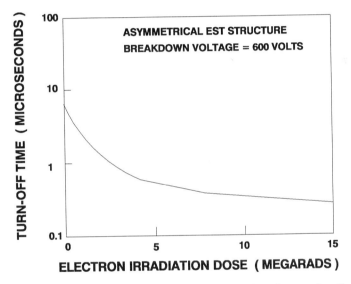

Fig. 9.57 Reduction of turn-off time for the EST by electron irradiation.

Fig. 9.57. The turn-off time can be reduced from about 3 microseconds for the as-fabricated devices to about 0.2 microsecond by using a radiation dose of 16 megarads.

The electron irradiation has been found to produce an increase in the on-state voltage drop of the EST as shown in Fig. 9.58. This is due to an increase in the voltage drop of the thyristor because of the decrease in the diffusion length. The observed increase in the on-state voltage drop for the EST is larger than the observed increase in voltage drop across a thyristor. This is not due to an increase in the MOSFET channel resistance but is due to a redistribution of the electron and hole current components in the EST. As the electron irradiation dose is increased, the gain of the P-N-P transistor increases. As a consequence, the floating N^+ emitter must supply a larger electron current to satisfy recombination in the base region of the P-N-P

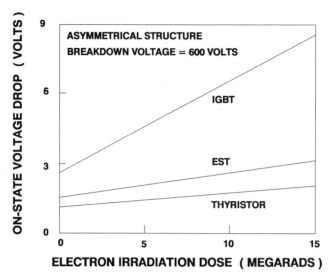

Fig. 9.58 Increase in on-state voltage drop for the EST after electron irradiation.

transistor. This results in an increase in the electron current flow via the channel of the MOSFET with increasing radiation dose. Thus, the electron irradiation results in an increase in the voltage drop across the thyristor and the MOSFET.

The electron irradiation can also be expected to have an effect upon the latching current density of the main and parasitic thyristors. A small increase in the latching current density of the main thyristor has been measured on devices. This effect can be explained with the aid of the equations derived in Section 9.4.4. Since the electron irradiation produces a decrease in the current gain of the P-N-P transistor, an increase in J_{LM} occurs because this latching current density is determined mainly by the voltage drop in the P-base and merged base regions. In contrast, it has been found that the latching current density of the parasitic thyristor remains essentially independent of the radiation dose. This can be explained by considering the expression for the parasitic thyristor latching current density (J_{PL}) given in Eq. (9.18). In this equation, the first term in the denominator is dominant because the lengths of the turn-on region (L_T) and the floating N^+ emitter region (L_{N+}) are much greater than that of the merged base region (L_M). This term in the expression is not sensitive to the electron irradiation dose because the current gain of the N-P-N transistor is governed mainly by the injection efficiency of junction J_3 and does not vary with electron radiation.

9.4.8 Temperature Dependence

As in the case of the other MOS-gated thyristors, the temperature dependence of the EST characteristics is of interest due to self-heating during operation of the device in circuits and for applications with high ambient temperatures. Among the MOS-gated thyristors, the on-state characteristics of the EST are the most favorable from the point of view of preventing current filamentation. Its other characteristics behave in a manner similar to those of the previously discussed thyristor-based structures as discussed below.

The changes in the on-state characteristics of the EST with an increase in operating

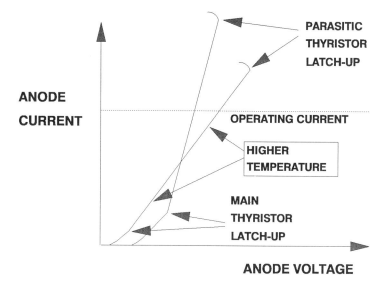

Fig. 9.59 Changes in the forward conduction characteristics of the EST with temperature.

temperature are shown in Fig. 9.59. There are many distinct changes in the characteristics that can be identified. First, there is a decrease in the knee voltage at which the current flow begins to increase rapidly. This is due to the larger thermal energy for injection of carriers across the junctions. Second, the slope of the I-V curve becomes less steep at higher temperatures. This is due to an increase in the resistance of the lateral MOSFET-1 because the channel mobility decreases. These two opposing trends cause an intersection of the two curves at a current density of about 150 amperes per cm^2.

The temperature dependence of the on-state voltage drop is provided in Fig. 9.60 for three values of the on-state current density. It can be seen that, at low current densities, the on-state voltage drop decreases with increasing temperature. At a current density of 167 amperes per cm^2, the on-state voltage drop for this EST has a zero temperature coefficient. When the current density is increased further to 230 amperes per cm^2, the on-state voltage drop has a small positive temperature coefficient. Consequently, at typical on-state current densities, the on-state voltage drop of the EST has a positive temperature coefficient. This behavior, similar to that observed in the IGBT, promotes uniform current distribution during the on-state and during turn-off.

It can be observed from the I-V characteristics in Fig. 9.59 that the latching current densities of the main and parasitic thyristor are also changing with temperature. Both of these parameters decrease with increasing temperature as shown in Fig. 9.61, where the curves are drawn by normalizing to the room temperature value. In the case of the latching current density

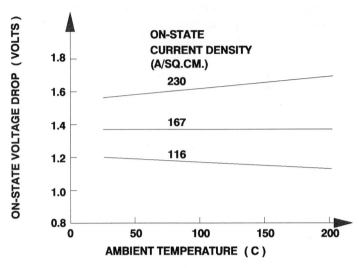

Fig. 9.60 Temperature dependence of the on-state voltage drop for the EST.

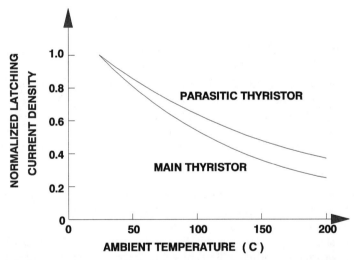

Fig. 9.61 Variation of latching current density for the main and parasitic thyristors with temperature.

of the main thyristor, the decrease in latching current density is due to an increase in the sheet resistance of the P-base region (ρ_{SB}) due to a reduction in the hole mobility. In the case of the latching current density of the parasitic thyristor, the decrease in latching current density is due to an increase in the sheet resistance of the merged P-base region (ρ_{SM}) due to a reduction in the hole mobility. The temperature dependence of these resistances are not equal because of the difference in doping concentration in the P-base and the merged base regions. Due to its lower doping concentration, the mobility decreases more for the merged base region as compared with the P-base region.

9.4.9 EST Characteristics

In this section, the important characteristics of the EST are summarized. The EST is an MOS-gated thyristor structure in which the thyristor current is constrained to flow through a turn-off MOSFET that has been integrated into its N-base region. During the on-state, the device operates like an IGBT at lower current levels with gate controlled current saturation capability until the thyristor region latches up. At higher currents, the current flows through the main thyristor region and via the turn-off MOSFET to the cathode terminal. This allows control of the thyristor current flow by the bias applied to the gate electrode, providing the device with a unique current saturating capability even after latch-up of the main thyristor. The on-state voltage drop of the EST is larger than that of the thyristor because of the additional voltage drop produced by the current flowing via the control MOSFET, but its on-state voltage drop is lower than that for the IGBT. An important benefit of the voltage drop in the control MOSFET is that it provides a built-in emitter ballasting effect that ensures uniform current distribution within the EST in the on-state and during turn-off.

In comparison with the IGBT, the EST has a lower on-state voltage drop and superior

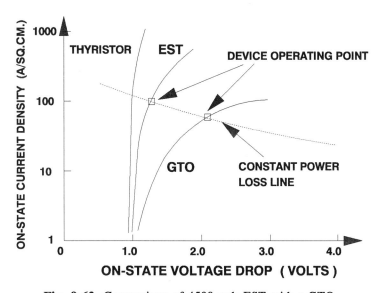

Fig. 9.62 Comparison of 4500 volt EST with a GTO.

forward biased safe operating area. In comparison with the MCT and BRT, the EST has a higher on-state voltage drop and similar surge current capability. An important feature of the

EST is that it can be fabricated with a DMOS process similar to that used to manufacture MOSFETs and IGBTs. This process requires only 8 masks when compared with 12 masks required to build the MCT. Due to its excellent FBSOA and lower on-state voltage drop, it is possible that the EST will be used in place of the IGBT in applications. However, it is more likely that the EST will be used to replace the GTO because it offers superior on-state voltage drop, reduced gate drive current, and much superior output characteristics including current saturation with good FBSOA. The on-state characteristics of a 4500-volt EST have been obtained by numerical simulations for comparison with the GTO and thyristor. From these characteristics, shown in Fig. 9.62, it can be seen that the EST is expected to operate at a lower on-state voltage drop and higher on-state current density than the GTO. This is attractive from the point of view of improving the efficiency and reducing the cost of the power electronics systems.

9.5 TRENDS

The development of MOS-gated thyristors has been motivated by the desire to utilize the low on-state voltage drop of thyristors and the high input impedance of MOS gate structures. In principle, this approach could lead to the replacement of the IGBT in power electronics applications. However, the excellent fully gate-controlled output characteristics of the IGBT with its wide safe-operating-area have been difficult to match while obtaining a lower on-state voltage drop in the thyristor-based devices. With continual improvements in the performance of the IGBT, it appears unlikely that the MOS-gated thyristors will have sufficiently superior characteristics to justify replacement of the IGBT.

In contrast, there is a strong motivation to replace the GTO in high voltage, high power applications. The relatively high on-state voltage drop of the GTO and its enormous gate drive current requirements call for its replacement with other devices. All of the MOS-gated thyristor structures offer the features of a lower on-state voltage drop and much reduced gate drive current when compared with the GTO, while providing an equivalent surge current handling capability. Further, the EST offers the possibility for development of an MOS-gated thyristor with fully gate controlled output characteristics for the high voltage, high power applications for the first time. Consequently, its commercial introduction has the potential for radically altering the way in which power systems are designed in the future.

REFERENCES

1. B.J. Baliga, "Enhancement and depletion mode vertical channel MOS-gated thyristors," Electronic Letters, Vol. 15, pp. 645-647 (1979).

2. V.A.K. Temple, "MOS controlled thyristors (MCTs)," IEEE Int. Electron Devices Meeting, Abstr. 10.7, pp. 282-285 (1984).

3. M. Stoisiek and H. Strack, "MOS GTO - a turn-off thyristor with MOS controlled emitter shorts," IEEE Int. Electron Devices Meeting, Abstr. 6.5, pp. 158-161 (1985).

4. V.A.K. Temple, "MOS controlled thyristors - a new class of power devices," IEEE Trans. Electron Devices, Vol. ED-33, pp. 1609-1618 (1986).

5. V.A.K. Temple and W. Tantraporn, "Effect of temperature and load on MCT turn-off capability," IEEE Int. Electron Devices Meeting, Abstr. 5.5, pp. 118-121 (1986).

6. M. Stoisiek, M. Beyer, W. Kiffe, H.J. Schultz, H. Schmid, H. Schwartzbauer, R. Stengle, P. Turkes, and D. Theis, "A large area MOS-GTO with wafer repair technique," IEEE Int. Electron Devices Meeting, Abstr. 29.3, pp. 666-669 (1987).

7. F. Bauer, P. Roggwiler, A. Aemmer, W. Fichtner, R. Vuilleumier, and J.M. Moret, "Design aspects of MOS controlled thyristor elements," IEEE Int. Electron Devices Meeting, Abstr. 11.6.1, pp. 297-300 (1989).

8. F. Bauer, H. Hollenbeck, T. Stockmeier, and W. Fichtner, "Current handling and switching performance of MOS controlled thyristor (MCT) structures," IEEE Electron Device Letters, Vol. EDL-12, pp. 297-299 (1991).

9. H. Lendenmann, H. Dettmer, W. Fichtner, B.J. Baliga, F. Bauer, and T. Stockmeier, "Switching behavior and current handling performance of MCT-IGBT cell ensembles," IEEE Int. Electron Devices Meeting, Abstr. 6.3.1, pp. 149-152 (1991).

10. F. Bauer, T. Stockmeier, H. Lendenmann, H. Dettmer, and W. Fichtner, "Static and dynamic characteristics of high voltage (3kV) IGBT and MCT devices," IEEE Int. Symp. on Power Semiconductor Devices and ICs, Abstr. 2.1, pp. 22-27 (1992).

11. Q. Huang, A.J. Amartunga, E.M.S Narayanan, and W.I. Milne, "Analysis of n-channel MOS controlled thyristors," IEEE Trans. Electron Devices, Vol. ED-38, pp. 1612-1618 (1991).

12. M. Stoisiek, K.G. Oppermann, and R. Stengle, "A 400A/2000V MOS-GTO with improved cell design," IEEE Trans. Electron Devices, Vol. ED-39, pp. 1521-1528 (1992).

13. M. Nandakumar, B.J. Baliga, M.S. Shekar, S. Tandon, and A. Reisman, "A new MOS-gated power thyristor structure with turn-off achieved by controlling the base resistance," IEEE Electron Device Letters, Vol. EDL-12, pp. 227-229 (1991).

14. M. Nandakumar, B.J. Baliga, M.S. Shekar, S. Tandon, and A. Reisman, "Theoretical and experimental characteristics of the base resistance controlled thyristor (BRT)," IEEE Trans. Electron Devices, Vol. ED-39, pp. 1938-1945 (1992).

15. B.J. Baliga, "Trench gate base resistance controlled thyristors (UMOS-BRTs)," IEEE Electron Device Letters, Vol. EDL-12, pp. 597-599 (1991).

16. B.J. Baliga, "The MOS-gated emitter switched thyristor," IEEE Electron Device Letters, Vol. EDL-11, pp. 75-77 (1990).

17. M.S. Shekar, B.J. Baliga, M. Nandakumar, S. Tandon, and A. Reisman, "Characteristics of the emitter switched thyristor," IEEE Trans. Electron Devices, Vol. ED-38, pp. 1619-1623 (1991).

18. M.S. Shekar, B.J. Baliga, M. Nandakumar, S. Tandon, and A. Reisman, "High voltage current saturation in emitter switched thyristors," IEEE Electron Device Letters, Vol. EDL-12, pp. 387-389 (1991).

19. M. Nandakumar, M.S. Shekar, and B.J. Baliga, "Fast switching power MOS-gated (EST and BRT) thyristors," IEEE Int. Symp. on Power Semiconductor Devices and ICs, Abstr. 10.1, pp. 256-260 (1992).

20. N. Iwamuro, M.S. Shekar, and B.J. Baliga, "A study of ESTs short-circuit SOA," IEEE Int. Symp. on Power Semiconductor Devices and ICs, pp. 71-76 (1993).

21. M.S. Shekar, M. Nandakumar, and B.J. Baliga, "An emitter switched thyristor with base resistance control," IEEE Electron Device Letters, Vol. EDL-14, pp. 280-282 (1993).

22. M.S. Shekar, J. Korec, and B.J. Baliga, "Trench gate emitter switched thyristors," IEEE Int. Symp. on Power Semiconductor Devices and ICs, Abstr. 5.1, pp. 189-194 (1994).

23. N. Iwamuro, B.J. Baliga, R. Kurlagunda, G. Mann, and A.W. Kelley, "Comparison of RBSOAs of ESTs with IGBTs and MCTs," IEEE Int. Symp. on Power Semiconductor Devices and ICs, Abstr. 5.2, pp. 195-200 (1994).

Chapter 9 : MOS-GATED THYRISTORS

PROBLEMS

9.1 An MOS-gated thyristor is fabricated with a P-base region that is homogeneously doped at a concentration of 1×10^{17} per cm^3 and has a thickness of 5 microns. The N-drift region has a doping concentration of 1×10^{14} per cm^3. Determine the [dV/dt] capability at an anode bias of 1000 volts assuming that the N$^+$ cathode has a length of 200 microns between the shorts. Assume a linear cathode short topology.

9.2 An MCT structure is fabricated using an n-channel turn-off MOSFET with a channel length of 1 micron and a gate oxide thickness of 500 °A. Calculate the maximum controllable current density for the device at a gate bias of 15 volts if the cell pitch is 20 microns. Assume that the bulk resistance components can be neglected.

9.3 A BRT structure is fabricated using an n-channel turn-off MOSFET with a channel length of 1 micron and a gate oxide thickness of 500 °A. Calculate the maximum controllable current density for the device at a gate bias of 15 volts if the cell pitch is 20 microns. Assume that the bulk resistance components can be neglected.

9.4 A BRT structure is fabricated using a p-channel turn-off MOSFET with a channel length of 1 micron and a gate oxide thickness of 500 °A. Calculate the maximum controllable current density for the device at a gate bias of 15 volts if the cell pitch is 20 microns. Assume that the bulk resistance components can be neglected.

9.5 Determine the maximum controllable current density for the basic EST structure with a floating N$^+$ emitter length of 50 microns, a turn-on gate length of 10 microns, a turn-off gate length of 3 microns, and an N$^+$ source length (L_p) of 1 micron. Assume a sheet resistance of 3000 ohms per square for the P-base region and a current gain of 0.8 and 0.4 for the N-P-N and P-N-P transistors.

9.6 Determine the main thyristor latching current density for the EST structure described in Problem 9.5 if the voltage drop in the floating N$^+$ emitter and the turn-off MOSFET can be neglected. Assume that the merged P-base region has a sheet resistance of 10,000 ohms per square.

9.7 Calculate the on-state voltage drop for the EST structure described in Problem 9.5 at an anode current density of 200 amperes per cm^2. The device has an N-drift region with thickness of 100 microns, a P-base region a with thickness of 2 microns, and a high level lifetime of 1 microsecond. The n-channel turn-off MOSFET has a channel length of 1 micron and a gate oxide thickness of 500 °A.

9.8 What is the saturated current density of the EST structure described in Problem 9.7 at a gate bias of 5 volts if the threshold voltage of the turn-off MOSFET is 2 volts?

Chapter 10

SYNOPSIS

This book is intended to provide the reader with a basic knowledge of the physics of operation of power semiconductor devices. Using the information provided in each chapter, it is possible to perform the analysis and design of each of the devices. Since a variety of devices are available, it becomes important to consider their relative merit for each application. In this chapter, a comparison of the characteristics of power rectifiers and three terminal (active) switching devices is provided. Based upon this method of comparison, it is possible to select the device that will result in the highest system operating efficiency. This analysis assumes that high voltages are encountered in the applications. A similar procedure can be used to compare the characteristics of devices operating in systems at other (lower or higher) voltages.

10.1 COMPARISON OF GATE-CONTROLLED DEVICES

With the development of many alternative devices to the power bipolar transistors, the device designer is faced with the task of making a judicious choice between these devices. As an aid to device selection, the gate controlled devices discussed in the previous chapters are compared here on the basis of several criteria. To begin with, normally-on devices have been found to be undesirable for power switching because of the need to ensure that a negative gate drive is available during circuit power-up. Since the power JFET (SIT) and FCD (SITH) exhibit this characteristic, their use is relegated to those applications in which some of their unique characteristics (such as the very high frequency response of the JFET or the high [dV/dt] and radiation tolerance of the FCD) are necessary. Since these requirements do not exist in most applications, the circuit designer is left with a choice between the remaining normally-off devices, namely, the bipolar transistor, the GTO, the power MOSFET, the IGBT, and the MOS-gated thyristor. For very high power levels, such as traction, only GTOs with sufficient current and voltage ratings are available, thus making the choice quite limited. However, at lower power levels, where gate controlled operation of the devices is highly desirable, the circuit designer must choose between the bipolar transistor, the power MOSFET, the IGBT, and the MOS-gated thyristor. The relative merits of these three devices are discussed here.

10.1.1 Gate Control Requirements

One criterion in selecting a power device is its gate drive power requirement. Since the power bipolar transistor is a current controlled device with a typical current gain of 10, it requires a relatively high gate drive power during steady state current conduction as well as during turn-off. The gate drive circuitry for the power bipolar transistor becomes complex and expensive. For this reason, the power loss analysis in subsequent sections does not include the bipolar transistor. In contrast, the power MOSFET, the power IGBT, and the MOS-gated thyristor are voltage controlled devices with very high input impedance. The gate drive power required to control these devices is relatively small. This eliminates complexity in the gate drive and often allows control of these devices directly from an integrated circuit since the gate circuit must merely provide enough current to charge and discharge the input capacitance of these devices. In this regard, the IGBT is even superior to an equivalent power MOSFET because its input capacitance is an order of magnitude smaller for the same power rating. Further, since the technology for the fabrication of the power MOSFET and IGBT is similar, the IGBT offers a lower cost to the circuit designer because the chip area is an order of magnitude smaller than for the power MOSFET.

10.1.2 Simplified Power Loss Analysis

The performance of a power device is ultimately limited by the power dissipation within its structure, which determines the temperature rise:

$$\Delta T = T_J - T_A = P_D R_\theta \tag{10.1}$$

where P_D is the power dissipation and R_θ is the thermal resistance. From reliability considerations, the maximum junction temperature (T_J) of a power device is generally maintained below 125°C. For an ambient temperature (T_A) of 25°C and a typical thermal resistance of 1°C per watt, the maximum power dissipation in power devices must therefore be kept below 100 watts per cm^2.

The power dissipation in a power device can arise during steady state current conduction, during steady-state voltage blocking, and during switching between these conditions. The leakage currents of the power devices are generally so low that the power losses incurred in their steady state forward blocking modes can usually be neglected. In addition, if the turn-off time is much longer than the turn-on time, the switching losses during device turn-on can be neglected. The device analysis can then be performed by considering the sum of the power loss during the on-state and the power loss during switching:

$$P_D = V_F I_F \frac{t_{on}}{P} + \frac{1}{2} I_F V_S \frac{t_f}{P} \tag{10.2}$$

where V_F is the forward voltage drop at a current I_F and V_S is the blocking voltage. In order to obtain this equation, the device is assumed to be operated with a constant current I_F in the on-state for a fraction t_{on} of the total period P, and the current is assumed to fall to zero in a switching time t_f from the on-state to the off-state while the voltage supported by the device remains at the supply voltage V_S. This situation is commonly encountered when switching currents with inductive loads.

To aid in choosing devices for applications operating at different frequencies, a plot of the power dissipation versus frequency is useful. As an example, the case of devices operated from a 400 volt DC bus will be considered. In this case, it is typical to use devices with a forward blocking voltage rating of 600 volts to allow for voltage overshoots during the switching transients. In calculating the power loss curves, it will be assumed that all the devices are operating at a current density of 100 amperes per cm^2 with a duty cycle of 50 percent. Other cases can be analyzed using the same approach. The forward voltage drop and turn-off time of the devices used for the calculations are provided in Table 10.1.

Table 10.1 Characteristics of power switches used for power loss calculations.

Device	Forward Drop	Current Density	Turn-off Time
Si MOSFET	20 V	100 A/cm^2	0.01 μs
IGBT	3.0 V	100 A/cm^2	0.30 μs
BRT/MCT	1.1 V	100 A/cm^2	0.30 μs
EST	1.5 V	100 A/cm^2	0.30 μs
SiC MOSFET	0.1 V	100 A/cm^2	0.01 μs

From both a device and system point of view, the best power device is the one that provides the lowest power dissipation. For the case of an on-state current of 15 amperes, the power loss calculated using Eq. (10.2) is provided in Fig. 10.1 for frequencies up to 30 kHz. This range has been chosen because it is preferable to increase the switching frequency to above the acoustic range for commercial applications. From this plot, it can be seen that the power loss in the silicon MOSFET is much larger than for the other devices because of its high on-state voltage drop. For this reason, silicon power MOSFETs are not good candidates for high voltage (above 300 volts) power conditioning applications. It is important to note that the thyristor based silicon devices (BRT and EST) offer significantly lower losses than the IGBT at low operating frequencies (below 5 kHz) because the on-state power dissipation dominates over the switching power loss. However, as the frequency increases, the power losses among the MOS-gated bipolar silicon devices become comparable. In contrast, the power losses in silicon carbide power MOSFETs are much smaller than for the silicon devices at all operating frequencies. This makes it a very attractive candidates for a wide range of power conditioning applications.

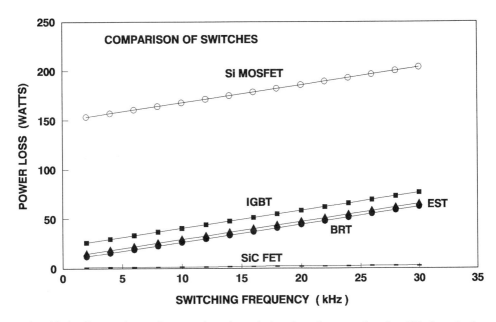

Fig. 10.1 Comparison of power loss in switches based up on the simplified analysis.

Unfortunately, this method for comparison grossly underestimates the power losses in the applications because it does not take into account the turn-on losses and the power dissipation incurred in the rectifiers. It is, therefore, important to perform the power loss analysis by considering the entire waveform for the transistor and rectifier for the complete switching period. The power losses obtained with this method are discussed in the next section for the case of a variable speed motor control application to illustrate the significance of the power rectifier reverse recovery behavior upon the power dissipation.

10.2 BASIC VARIABLE SPEED MOTOR CONTROL CIRCUIT

The most commonly used adjustable speed motor drive technology is based upon the PWM invertor in which the input AC line power is first rectified to form a DC bus. The variable frequency AC power to the motor is then provided by using six switches and flyback rectifiers. The switches are connected in a totem-pole configuration as illustrated in Fig. 10.2. The power delivered to the motor is regulated by adjusting the time duration for the on and off states for the power switch. By using a sinusoidal reference waveform, a variable frequency output current can be synthesized by using a switching frequency well above the motor operating frequency. In order to reduce the acoustic noise from the motor, it is desirable to increase the

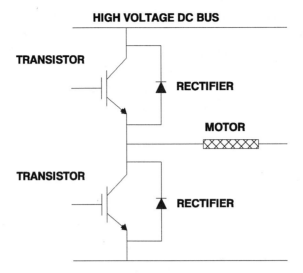

Fig. 10.2 Totem pole switch configuration used in PWM motor control circuits.

switching frequency for the transistors to above the acoustic range (preferably above 15 kHz).

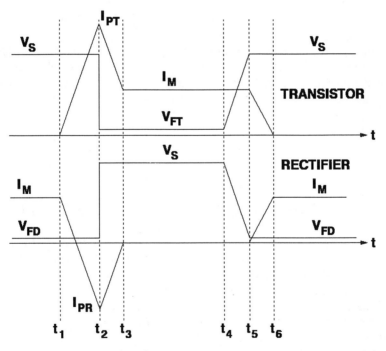

Fig. 10.3 Linearized current and voltage waveforms for switch and rectifier.

Typical current and voltage waveforms for the transistor and rectifier during one switching cycle are shown in Fig. 10.3. These waveforms have been linearized to simplify the power loss analysis. It is assumed that the motor current is initially flowing through the flyback rectifier at the bottom of the totem pole circuit. At time t_1, the upper transistor is switched on with a controlled rate of rise of current. During the time interval from t_1 to t_2, the motor current transfers from the rectifier to the transistor. Unfortunately, the P-i-N rectifiers that are used in these circuits are unable to recover instantaneously from their forward conduction state to their reverse blocking state. Instead, a large reverse current flow occurs with a peak value (I_{PR}) before the rectifier becomes capable of supporting reverse voltage. This reverse recovery current flows through the transistor. It is worth pointing out that the net transistor current during this time interval is the sum of the motor current (I_M) and the diode reverse recovery current. Further, during this time, the transistor must support the entire DC bus voltage because the rectifier is not yet able to support voltage. This produces a significant power dissipation in the power switch when it is being turned on. This will be referred to as the *turn-on power loss* for the transistor. Further, the transistor is subjected to a high stress due to the presence of a high current (I_{PT}) and a high voltage (V_S) simultaneously. This can place the transistor in a destructive failure mode if the stress exceeds its safe-operating-area limit. At time t_2, the rectifier begins to support reverse voltage and its reverse current decreases to zero during the time interval from t_2 to t_3. During this time, the transistor current falls to the motor current. During this time interval, a large power dissipation occurs in the diode because it is supporting a high voltage while conducting a large reverse current.

The other important switching interval is from t_4 to t_6, during which the transistor is turned off and the motor current is transferred to the rectifier. During the first part of this time interval, from t_4 to t_5, the voltage across the transistor rises to the bus voltage while its current remains essentially equal to the motor current because of the large inductance in the motor winding. In the second portion of this time interval, from t_5 to t_6, the current in the transistor decreases to zero while it is supporting the bus voltage. Since a large current and voltage are impressed on the transistor during both of these time intervals, there is significant power dissipation in the transistor during its turn-off.

Due to a relatively long turn-off time for bipolar power transistors and the first IGBTs

Table 10.2 Characteristics of power rectifiers used for power loss calculations.

Device	Forward Drop	Current Density	Reverse Recovery Time	Peak Recovery Current
P-i-N	2.0 V	100 A/cm²	0.50 μsec	45 Amps
MPS/SSD	1.0 V	100 A/cm²	0.25 μsec	10 Amps
SiC Schottky	1.1 V	100 A/cm²	0.01 μsec	0 Amps

introduced into the marketplace, the emphasis has been on reducing the power loss in the switches during their turn-off. It has been found that methods employed to reduce the turn-off time for the switches is usually accompanied by an increase in their on-state voltage drop which increases their on-state power dissipation during the time interval t_2 to t_4. It has been customary to compare power switches by calculating the sum of the on-state and turn-off power losses as a function of frequency as given by Eq. (10.2). As pointed out earlier, this results in an under estimation of the total power loss. Since the power losses are a strong function of the characteristics of the power rectifier, it is important to consider the impact of using different rectifiers in the circuit. The characteristics of power rectifiers are summarized in Table 10.2. These characteristics are used in the subsequent sections to obtain a more accurate evaluation of the total power loss.

10.3 IGBT POWER LOSS COMPONENTS

As shown in Fig. 10.1, the power losses in the power MOSFET are very large due to its high on-state voltage drop. For this reason, the availability of high performance IGBTs has made them the device of choice for variable speed drives. Power MOSFETs will therefore not be considered for further analysis. In this section, the power losses obtained when the IGBT is used with the P-i-N and MPS rectifiers will be analyzed. These calculations of the power losses were performed for the case of device operation with a 50 percent duty cycle, a device on-state current of 15 amperes, and a DC bus voltage of 400 volts.

10.3.1 IGBT with PiN Rectifier

The power loss in the IGBT and the PiN rectifier are compared with the total power loss in Fig. 10.4. The electrical characteristics of the IGBT and PiN rectifier used for these calculations are given in Tables 10.1 and 10.2. It can be seen that the total power loss increases rapidly with an increase in operating frequency, indicating that the switching losses are more important than the conduction losses. The power loss in the rectifier is about one half that in the IGBT. This may lead to the conclusion that the switch characteristics are more important than those of the rectifier in determining the power losses. However, this is misleading because the power losses in the IGBT are strongly dependent on the rectifier characteristics.

In order to demonstrate this, it is important to analyze the power losses in the IGBT during various phases of operation. The power losses in the IGBT during the on-state, turn-off, and turn-on are shown in Fig. 10.5 together with the total power loss in the IGBT. It is interesting to note that, although the on-state power loss in the IGBT is dominant at switching frequencies below 5 kHz, the switching power losses become dominant at higher frequencies. More importantly, the turn-on power loss is observed to be larger than the turn-off power loss. This occurs because of the large reverse recovery current for the PiN rectifier that the IGBT

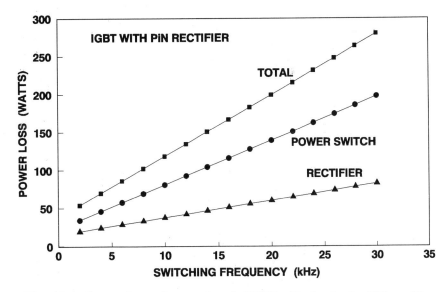

Fig. 10.4 Comparison of power loss in IGBT with that in the PiN rectifier.

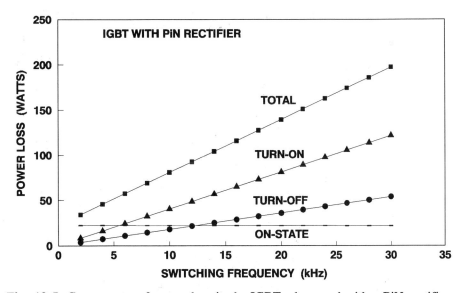

Fig. 10.5 Components of power loss in the IGBT when used with a PiN rectifier.

switch must conduct during its turn-on. From these figures, it can be concluded that a reduction in the power losses can be achieved by improving the reverse recovery performance of the power rectifier. Thus, progress in power rectifier technology has been essential to obtaining high performance variable frequency drives.

10.3.2 IGBT with MPS Rectifier

The impact of replacing the PiN rectifier with the MPS rectifier is illustrated in Fig. 10.6 which provides the total power losses in the drive together with the power losses in the IGBT

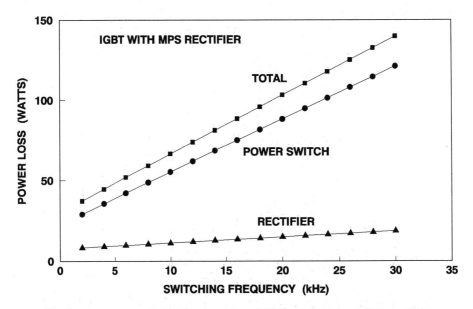

Fig. 10.6 Power loss in the case of an IGBT used with a MPS rectifier.

and power rectifier. By comparison with Fig. 10.4, it can be seen that the total power loss has been reduced by nearly a factor of two by replacement of the PiN rectifier with the MPS rectifier. This is due to not only a smaller power loss in the rectifier but also in the IGBT. This can be seen more clearly from Fig. 10.7 where the power loss components in the IGBT are provided. By comparison of the power loss components in this case to those obtained for operation of the IGBT with the PiN rectifier (see Fig. 10.5), it is evident that a significant reduction in the turn-on losses for the IGBT is responsible for the improved performance. These charts demonstrate the importance of the rectifier characteristics on the system performance.

10.4 MOS-GATED THYRISTOR POWER LOSS COMPONENTS

In comparison with the IGBT, the MCT, BRT, and EST have a lower on-state voltage

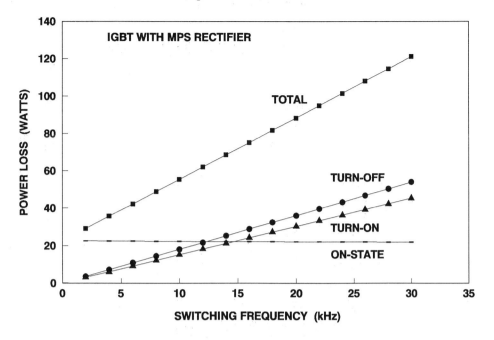

Fig. 10.7 Components of power loss in the IGBT when used with a MPS rectifier.

drop for the same turn-off switching time (see Table 10.1). The impact of replacing the IGBT with these MOS-gated thyristor structures is analyzed in this section. As expected, these devices exhibit lower on-state power losses which has its strongest effect at low operating frequencies. In this section, the power losses obtained when these devices are used with the PiN and MPS rectifiers will be analyzed. As in the case of the IGBT, these calculations of the power losses were performed for the case of device operation with a 50 percent duty cycle, a device on-state current of 15 amperes, and a DC bus voltage of 400 volts.

10.4.1 BRT with PiN Rectifier

The on-state characteristics of the MCT and BRT are very similar. The power losses obtained for both of these devices are therefore expected to be identical. The impact of the lower on-state voltage drop for these devices upon the power losses for the motor drive circuit when these devices are used with a PiN rectifier is shown in Fig. 10.8. In comparison with the IGBT, although a reduction in the total power loss is observed at low operating frequencies due to the reduced on-state voltage drop, the impact is small at the higher operating frequencies. This indicates that the choice of the power switch is less critical than the choice of the power rectifier from the point of view of reducing the power losses in the drive.

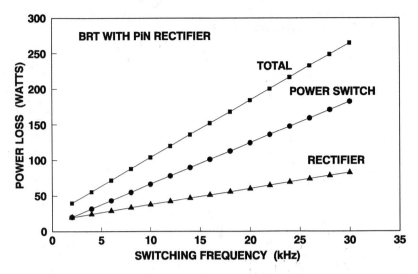

Fig. 10.8 Comparison of power loss in BRT with that in the PiN rectifier.

10.4.2 BRT with MPS Rectifier

The impact of replacing the PiN rectifier with the MPS rectifier is illustrated in Fig. 10.9

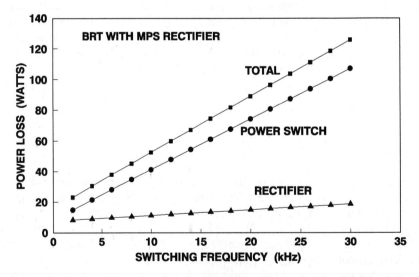

Fig. 10.9 Comparison of power loss in BRT with that in the MPS rectifier.

which provides the total power losses in the drive together with the power losses in the BRT and

MPS rectifier. By comparison with Fig. 10.8, it can be seen that the total power loss has been reduced by nearly a factor of two by replacement of the PiN rectifier with the MPS rectifier. This is due to a smaller power loss not only in the rectifier but also in the BRT. These figures demonstrate the importance of the rectifier characteristics on the system performance.

10.4.3 EST with PiN Rectifier

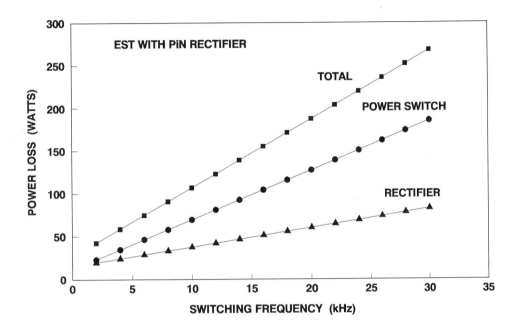

Fig. 10.10 Power loss for drive with EST and PiN rectifier.

The emitter switched thyristor (EST) structure has a slightly higher forward voltage drop than the MCT and BRT because the thyristor current flows through a MOSFET, as indicated in its equivalent circuit in Section 9.4, which produces an additional voltage drop. The typical on-state voltage drop of the EST is about 1.5 volts for a turn-off switching time of 0.3 microseconds as given in Table 10.1. The power losses in the motor drive circuit for the case of an EST were calculated for the case of operation with a PiN rectifier for the same load and bus voltage as the IGBT. These results are plotted in Fig. 10.10 for comparison with the IGBT and MCT/BRT cases. Although the power loss is slightly greater than that for the MCT/BRT case at low operating frequencies, the difference becomes small at higher operating frequencies. It can therefore be concluded that the EST may be a better choice than the BRT or MCT because of its excellent FBSOA.

10.4.4 EST with MPS Rectifier

The impact of replacing the PiN rectifier with the MPS rectifier is illustrated in Fig. 10.11 which provides the total power losses in the drive together with the power losses in

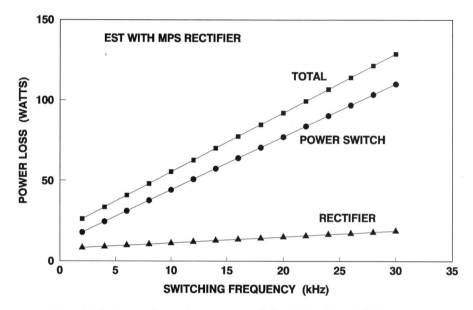

Fig. 10.11 Power losses for the case of the EST with a MPS rectifier.

the EST and MPS rectifier. By comparison with Fig. 10.10, it can be seen that the total power loss has been reduced by nearly a factor of two by replacement of the PiN rectifier with the MPS rectifier. This is due to not only a smaller power loss in the rectifier but also in the EST. Once again, these charts demonstrate the importance of the rectifier characteristics on the system performance.

10.5 SILICON CARBIDE DEVICES

From the above discussion, it can be surmised that further improvement in the power losses in the drive can be accomplished only by a reduction in the switching time for the power switch and the rectifier, as well as a reduction in the peak reverse recovery current for the rectifier. Recent theoretical analysis has shown that very high performance FETs and Schottky rectifiers can be obtained by replacing silicon with gallium arsenide, silicon carbide or semiconducting diamond. Among these, silicon carbide is the most promising material because its technology is more mature than for diamond and the performance of silicon carbide devices is expected to be an order of magnitude better than that of gallium arsenide devices. In these devices, the high breakdown electric field strength of silicon carbide leads to a 200 fold reduction in the resistance of the drift region. As a consequence of this low drift region resistance, the on-state voltage drop for even the high voltage FET is much smaller than for any unipolar or bipolar silicon device as shown in Table 10.1. These switches can be expected to switch off in less than 10 nanoseconds and have superb FBSOA. The analysis also indicates that high voltage Schottky barrier rectifiers with on-state voltage drops close to 1 volt may be feasible with no reverse recovery transient. Recently, these theoretical predictions have been experimentally confirmed by the fabrication of Schottky barrier rectifiers with breakdown voltages of 400 volts.

If the silicon power switch and rectifier are replaced with the silicon carbide devices, it becomes possible to reduce the turn-off time to 10 nanoseconds because both the switch and the rectifier behave as nearly ideal devices. The power losses calculated for this case are shown in

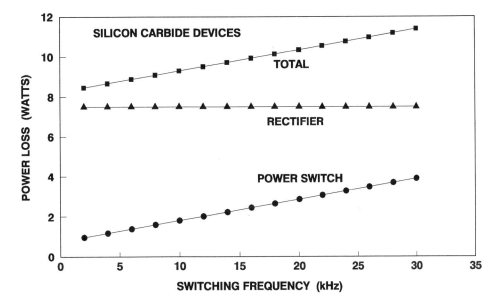

Fig. 10.12 Power losses for the case of silicon carbide based switch and rectifier.

Fig. 10.12 for comparison with the silicon devices. It is obvious that the power losses have been drastically reduced at all switching frequencies. Note that the power loss in the SiC rectifier is higher than that in the SiC FET because of its larger on-state voltage drop. This calculation assumes that the short switching time of the SiC devices can be utilized without encountering severe dI/dt and dV/dt problems in the system. If this becomes a problem, it may be necessary to increase the switching time by adjusting the input gate waveform driving the SiC FETs. It is quite obvious that the power loss obtained with the silicon carbide devices is much smaller than that for the silicon devices. In addition, it is anticipated that these devices can be operated at higher junction temperatures due to the large band gap of SiC, making the heat sink requirement less stringent.

10.6 COMPARISON OF DEVICES

A comparison of the power losses for all the devices considered in the previous sections of this chapter can be performed with the aid of Fig. 10.13 where the total loss for each case

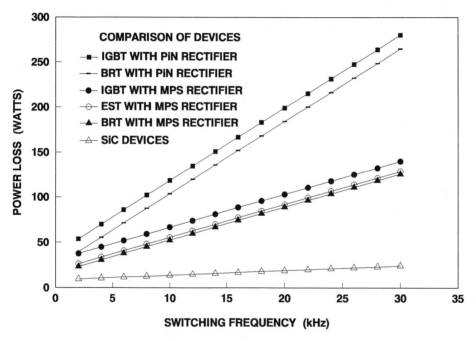

Fig. 10.13 Comparison of total power losses obtained with the different device technologies.

has been plotted. From this figure, it is clear that the impact of replacing the PiN rectifier with an MPS rectifier (or any other rectifier with faster reverse recovery time, reduced reverse recovery current, and comparable on-state voltage drop) is much greater than replacement of the IGBT with the BRT (or any other MOS-gated thyristor based device with on-state voltage drop of about 1 volt) unless the operating frequency is relatively low. It is also obvious that the silicon carbide devices are an extremely attractive choice for reducing power losses in variable speed motor drives.

10.7 SMART POWER INTEGRATED CIRCUITS

The development of the power MOSFET and IGBT resulted in greatly simplifying the gate drive circuits for motor drive applications. Since the IGBT gate drive circuit has fewer components than the base drive circuit for bipolar transistors and only relatively small currents are needed to control the IGBT, it became possible to integrate the gate drive circuit into a monolithic chip for the first time. This in turn created the opportunity to add other functions, such as protection against adverse operating conditions and logic circuits to interface with microprocessors.

In a three-phase-leg motor drive circuit, it is possible to partition the drive in two basic ways. In one case, all the drive circuits for the lower switches in the totem pole configuration are integrated together on one chip while all the drive circuits for the upper switches in the totem pole configuration are integrated on a second chip. This avoids the need to integrate level shifting capability within the chip but requires a separate high voltage chip for providing this feature. Alternately, the drive can be partitioned with the drive circuits for both the upper and lower switch in each phase leg integrated on a monolithic chip. In order to achieve this capability, technology has been developed to integrate the level-shifting circuits for the upper switch. This also requires the ability for the drive circuit to rise in potential to above the DC bus voltage in order to turn on the upper switch. Three such control chips would be required in a three phase system.

In the smart power control chip, the sensing and protection circuits are usually implemented using analog circuits with high speed bipolar transistors. These circuits must sense the following adverse operating conditions: over-temperature, over-current, over-voltage, and under-voltage. It is obvious that an over-temperature and over-current condition can cause thermal runaway, leading to destructive failure, while an over-voltage condition can lead to avalanche injection induced failure. The under-voltage lockout feature is also necessary because sufficient gate drive voltages are not generated at low bus voltages, leading to very high power dissipation in the output transistors. An example of this condition is during system start-up. The bipolar transistors used in the analog portion must have a high frequency response because of the high dI/dt during short-circuit conditions. When the current exceeds a threshold value, the feedback loop must react in a short duration to prevent the current from rising to destructive levels.

Today's smart power chips are manufactured using a junction isolation technology. In these chips, the high voltage level-shifting transistors are usually lateral structures made using the RESURF principle to obtain a high breakdown voltage with thin epitaxial layers. It is anticipated that dielectric isolation (DI) technology will replace the junction isolation (JI) technology that is now being used for most smart power ICs. Dielectric isolation offers reduced parasitics, a more compact isolation area, and the prospects for integrating MOS-gated bipolar devices that occupy less space than the lateral MOSFETs. A detailed discussion of power integrated circuits is beyond the scope of this book.

10.8 TRENDS

This chapter has been written with the point of view of illustrating the impact of the power switch and rectifier technology on variable speed motor drive applications. By performing power loss calculations during a typical switching cycle, it has been shown that improvement in the reverse recovery behavior of power rectifiers is critical for reducing the total power loss. Although MOS-gated power switches with lower on-state voltage drop are useful for reducing the power loss in systems operating at low switching frequencies, they become of less importance for systems operating at higher frequencies of interest for motor control operation above the acoustic range. It has been shown that, in the future, power switches and rectifiers fabricated from silicon carbide offer tremendous promise for reduction of power losses in the drive. Thus, it can be concluded that advances in power semiconductor technology continue to look very promising for improving the performance of motor drives. A similar conclusion can be derived by performing the analysis of other applications.

REFERENCES

1. B.K. Bose, "Power electronics and motion control - technology status and recent trends," IEEE PESC Conference Record, pp. 3-10 (1992).

2. B. Mokrytzki, "Survey of adjustable frequency drive technology," IEEE IAS Conference Record, pp. 1118-1126 (1991).

3. M.S. Adler and S.R. Westbrook, "Power semiconductor switching devices - a comparison based on inductive switching," IEEE Trans. Electron Devices, Vol. ED-29, pp. 947-952 (1982).

4. M.S. Adler, K.L. Owyang, B.J. Baliga, and R.A. Kokosa, "The evolution of power device technology," IEEE Trans. Electron Devices, Vol. ED-31, pp. 1570-1591 (1984).

5. B.J. Baliga, "Evolution of MOS-bipolar power semiconductor technology," Proc. IEEE, Vol. 76, pp. 408-418 (1988).

6. P.L. Hower, "Power semiconductor devices - an overview," Proc. IEEE, Vol. 76, pp. 335-342 (1988).

INDEX

AC circuit, 258
AC power, 291, 322
AC switch, 322, 472, 485
Accumulation, 26, 351
Acid, 110, 120, 152
Acoustic range, 574
Air-conditioning, 8, 498
Alpha, 200
Aluminum, 27, 33, 40, 103, 110, 152, 263, 320
Ambipolar, 16
Anisotropy, 16
Annealing, 36, 40, 414
Anode-shorted:
   GTO, 316
   MCT, 523
Antimony, 27, 33, 148, 368, 410
Appliance control, 322, 485
Applications, 1
Arsenic, 27, 33, 368, 410, 459
Asymmetrical IGBT, 433
Atomic-lattice-layout, 416, 465, 475
Auger recombination, 166
Autodoping, 410
Automotive electronics, 2, 420
Avalanche breakdown, 40, 66
Avalanche multiplication, 140, 202

Baliga's figure-of-merit, 150, 417
Baliga-pair, 418
Band gap, 150
Band gap narrowing, 29, 166, 205
Band structure, 130
Band tail, 29
Base-resistance-controlled thyristor, 526

Base transport factor, 115, 202, 211
Beta, 199
Bevel, 103
   double-positive, 117
   negative-positive, 116
   negative, 107
   positive, 105
Bipolar transistor, 198, 572
   base widening, 222
   blocking characteristics, 232
   carrier distribution, 219, 238, 242
   control circuit, 198
   current crowding, 248
   current gain, 201, 226
   current source drive, 246
   current tail, 243
   Darlington configuration, 252
   doping profile, 229
   drift resistance, 217
   emitter ballast, 246
   emitter current crowding, 228
   emitter periphery, 231
   forward active region, 216
   interdigitation, 228, 232
   leakage current, 235
   on-state, 217
   output characteristics, 213
   power dissipation, 217, 236
   quasi-saturation, 218
   saturation region, 216
   second breakdown, 243
   storage time, 241
   structure, 202
   switching characteristics, 236
   turn-off, 240
   turn-on, 237

Blocking characteristics:
   bipolar transistor, 232
   IGBT, 431
   MOSFET, 342
   thyristor, 262
Body-bias effect, 400
Boltzmann, 28, 156, 207
Bonding pad, 361
Boron, 27, 33, 38, 411, 457
Breakdown, 66
   avalanche, 66
   bulk, 105
   open-base transistor, 113
   punch-through, 431
   reach-through, 343, 471
   soft, 148
   surface, 103
Breakdown voltage, 35, 66
   cylindrical junction, 86
   floating field ring, 94
   linearly-graded junction, 79
   open-base, 235
   open-emitter, 233
   parallel-plane, 73
   shorted-emitter, 233
   spherical junction, 90
BRT, 526
   channel resistance, 531
   current saturation, 543
   current tail, 538
   emitter ballasting, 536, 542
   equivalent circuit, 529, 535
   holding current, 526, 529
   IGBT mode, 527
   JFET action, 533
   maximum controllable current, 530
   on-state, 534
   output characteristics, 527
   parasitic transistor, 534
   power dissipation, 537
   structure, 526
   surge current, 528
   temperature, 541
   thyristor mode, 527
   trade-off curve, 540
   turn-off, 538
Buffer layer, 434, 445, 484, 545
   bipolar transistor, 251
   GTO, 304
Built-in potential, 130

Capacitance:
   input, 381, 573
   junction, 280
   MOS, 382
   overlap, 382
Capture cross-section, 43
Capture rate, 43
Caro's solution, 152
Carrier distribution:
   bipolar transistor, 219, 238, 242
   IGBT, 446
   P-i-N rectifier, 160
   thyristor, 273
Carrier-carrier scattering, 166
Cascade, 66
Cathode short, 300
   array, 270
   linear, 269, 278
   thyristor, 266
Cell pitch, 373
Cell topology, 461
   IGBT, 461
   MOSFET, 415
Channel density, 371
Channel resistance:
   BRT, 531
   IGBT, 435
   MOSFET, 362
Charge balance, 106
Charge control analysis, 239, 306
Charge neutrality, 158
Chemical etch, 110
Chlorine, 415
Circuit inductance, 154
Cleaning, 120
Collector efficiency, 202
Common base configuration, 200

Common emitter configuration, 199
Compensation, 54, 344, 405
Complementary devices, 472, 485
Computers, 7, 420
Conduction band, 10, 16
Conductivity modulation, 159, 187, 435
Contact, 129, 412
Contact potential, 130
Continuity equation, 159, 189, 204, 219, 446
Control circuit bipolar transistor, 198
Counter-doped channel, 459
Crystal growth, 36
Current constriction, 307
Current filamentation, 542
    EST, 559
    MCT, 524
Current gain, 199, 200
    bipolar transistor, 226
    common base, 200
    common emitter, 199
    shunting resistance, 267
Current-induced base, 224
Current saturation, 6
    BRT, 543
    EST, 545, 560
    IGBT, 439
    MCT, 525
    MOSFET, 341, 365
Current tail, 243, 293
    BRT, 538
    EST, 547
    GTO, 315
    IGBT, 477
    MCT, 523
Czochralski, 36, 258

Darlington configuration, 252, 426
    equivalent circuit, 254
    monolithic, 254
Deep level, 42, 57, 155
Delay time, 282
Density of states, 28

Depletion, 351
Depletion layer, 69, 97, 100, 116, 138
    maximum, 357
Depletion region, 46, 212
Depletion width, 72
    critical, 73, 79
Device:
    comparison, 586
    normally-on, 572
    silicon carbide, 584
Device testing, 56
Dielectric isolation, 587
Diffusion, 211
    lateral, 82
    window, 82, 90, 91
Diffusion coefficient, 33
    ambipolar, 16, 159
Diffusion current, 45, 157
Diffusion length, 115, 156
    ambipolar, 159
Diffusion temperature, 55
Diode:
    abrupt-junction, 71
    anti-parallel, 244, 253, 402
    diffused, 79
    linearly-graded, 77
    punch-through, 75
Disorder, 29
Displacement current, 294, 301, 540
    EST, 553
Display drives, 2
Divacancy, 55
Diverter, 466, 526
DMOS cell:
    IGBT, 474
    optimization, 373
DMOSFET, 337
Doping profile:
    bipolar transistor, 229
    IGBT, 433
    MOSFET, 343
    thyristor, 260
Double-diffusion process, 336

Drift current, 157
Drift region resistance
  bipolar transistor, 217
  MOSFET, 371, 373
Drift velocity, 13
  saturation, 13
Duty cycle, 144, 537, 574
dV/dt capability:
  MOSFET, 395
  Thyristor, 280
  Triac, 329

Early voltage, 213
Edge termination, 81, 340, 346
  bevel, 103
  comparison, 121
  etch contour, 110
  floating field ring, 91
  planar, 82, 148
  surface implantation, 111
Einstein relationship, 16, 220
Electric car, 498
Electric field:
  critical, 74, 85, 89, 150
  crowding, 82
  surface, 107
Electron affinity, 130, 146, 350
Electron irradiation, 56, 57, 175, 316, 330, 404, 446, 479, 523, 538, 563
Emitter ballast, 246
  bipolar transistor, 246
  BRT, 536, 542
  EST, 559
  MCT, 518, 524
Emitter-switched thyristor, 543
Equivalent circuit:
  BRT, 529, 535
  EST, 547
  IGBT, 430, 436
  MCT, 508
EST, 543
  current filamentation, 559
  current saturation, 545, 560
  current tail, 547

displacement current, 553
dual-channel, 548
emitter ballast, 559
equivalent circuit, 547
floating emitter, 544
high voltage, 567
latching current, 547, 554
maximum controllable current, 550
on-state, 557
optimization, 559
output characteristics, 546, 549
parasitic thyristor, 545, 551
safe-operating-area, 548, 560
structure, 544
surge current, 546
temperature, 564
transconductance, 561
turn-off, 563
Etch, 110, 336
  anisotropic, 24
Ethylene diamine, 24

Factory automation, 8
FCD (field-controlled diode), 572
Fermi level, 129, 350, 356
Field oxide, 361, 410
Field plate, 100, 148, 346, 474
Fixed charge, 359
Flatband, 350
Float zone, 36, 258
Floating emitter (EST), 544
Floating field ring, 91, 98
  optimum spacing, 93, 95
Fluence, 56
Flux, 56
Flyback diode, 420
Forward blocking, 6
  IGBT, 432
  Thyristor, 264
Forward voltage overshoot, 153
Frequency response (MOSFET), 381
Fulop's Formula, 470

Gallium, 27, 33, 103, 320

Gallium arsenide, 132, 143, 150, 417
Gamma, 201
Gamma irradiation, 56
Gamma recoil, 40
Gate drive, 573
    thyristor, 274
Gate oxide, 359, 411, 414
    IGBT, 460
Gate turn-off thyristor, 302
Gauss's law, 102, 354
Gaussian, 18, 22, 79
Generation, 27
    space-charge, 45
    thermal, 28
Germanium, 9
Gettering, 56
Gold, 55, 57, 175, 316, 405
Grade constant, 77
Grit blasting, 107, 118
Groove, 336
    circular, 118
GTO, 302
    anode short, 304, 316
    buffer layer, 304
    current constriction, 307
    current tail, 315
    fall-time, 312
    gate breakdown, 320
    layout, 321
    maximum controllable current, 317
    ratings, 322
    recessed gate, 321
    turn-off gain, 304, 311
    turn-off time, 305
Guard ring, 149, 386, 549, 562

Heat sink, 38, 145
High-level injection, 50, 158, 206
High voltage:
    EST, 567
    IGBT, 488
High-voltage DC transmission, 2, 7
Historical perspective, 7
Holding current, 272

BRT, 526, 529
    thyristor, 272, 277
Hydrogen peroxide, 152

Ideal ohmic contact, 178
Ideal rectifier, 4
Ideal switch, 3, 5
IGBT, 426, 572
    asymmetrical, 433
    avalanche-induced latch-up, 469
    blocking characteristics, 431
    carrier distribution, 446
    cell topology, 461
    channel resistance, 435
    complementary devices, 485
    current saturation, 439
    current tail, 477
    current-induced latch-up, 468
    DMOS cell, 474
    doping profile, 433
    equivalent circuit, 430, 436
    forward blocking, 432
    gate oxide, 460
    JFET region, 450
    latch-up, 495
    latch-up suppression, 451
    lifetime control, 479
    on-state, 434
    output characteristics, 429, 444
    output resistance, 430, 443, 451
    reverse blocking, 431
    safe-operating-area, 468
    stored charge, 448, 478
    structure, 428
    switching characteristics, 476
    symmetrical, 433
    temperature, 490
    trade-off curve, 483
    transconductance, 442
    trench-gate, 496
    turn-off, 478
    UMOS structure, 496
IGFET, 336
Image force lowering, 138

Impact ionization, 66, 140, 233
　coefficient, 67
Impurity band, 29
Impurity level, 29
　splitting, 29
Inductive load, 244, 552
Ingot, 37
　rotation, 38
Injection
　high-level, 158, 206
　low-level, 156
Injection level, 16, 42
Input capacitance, 573
　MOSFET, 384
Insulated gate bipolar transistor, 426
Integral diode (MOSFET), 402
Integrated circuits, 36, 335, 587
Integration, 253
Interdigitation, 287
Intermetallic compounds, 34
Intrinsic concentration, 28
　effective, 32
Inversion, 20, 351
　strong, 353
　weak, 20, 353
Inversion layer, 18
Inverters, 149
Involute gate, 287
Ion implantation, 111, 147
　proton, 177
Ionization integral, 70

JBS rectifier, 182
　leakage current, 186
　on-state, 184
　reverse blocking, 185
　structure, 183
JFET (junction field effect transistor), 572
JFET action
　BRT, 533
　MOSFET, 370
Junction, 71
　abrupt, 71
　back-to-back, 116
　capacitance, 301
　cylindrical, 82, 83, 321, 346, 475
　diffused, 109
　grid, 183
　linearly-graded, 77
　negative bevel, 71, 104, 107
　planar, 82
　positive bevel, 103, 105
　saddle, 475
　saturation current, 165, 314
　shallow, 83
　spherical, 82, 86, 346, 475
Junction isolation, 587

Kirk current density, 224
Kirchhoff's law, 114, 199, 200, 264, 273, 285, 303

Lamp ballasts, 2, 336
Lapping, 110, 118
Laser, 296
Latch-up (IGBT), 495
Latch-up suppression (IGBT), 451
Latching current (EST), 547, 554
Lattice damage, 40, 55
Lattice vibration, 10
Leakage current, 28, 45, 114, 138, 153, 169
　bipolar transistor, 235
　diffusion, 170
　leakage current, 186
　P-i-N rectifier, 169
　Schottky rectifier, 140
　space-charge-generation, 169
　thyristor, 265
Lifetime, 41
　ambipolar, 16
　Auger, 60
　high-level, 44, 159
　low-level, 44
　space-charge generation, 46
Lifetime control, 55, 175
　BRT, 538

Lifetime control, (*Continued*)
  EST, 563
  IGBT, 479
  P-i-N rectifier, 175
Load (inductive), 248
Load-line, 237
Locomotive, 322
LOCOS, 149
Low-level injection, 156

Mask, 82, 96, 113, 330
  shadow, 152
Maximum controllable current
  BRT, 530
  EST, 550
  GTO, 317
  MCT, 510
MCT, 506
  anode short, 523
  boundary current, 522
  boundary short, 522
  complementary, 514
  current filament, 524
  current saturation, 525
  current tail, 523
  edge termination, 521
  emitter ballast, 518, 524
  equivalent circuit, 508
  maximum controllable current, 510, 520
  multicell, 519
  parasitic transistor, 515
  safe-operating-area, 521
  structure, 507
  temperature, 524
  turn-off, 523
  turn-on gate, 520
Mesa, 321, 378
MESFET, 418
Mesoplasma, 28
Metal-semiconductor contact, 129
Miller effect, 385
Mobile ions, 359, 414
Mobility, 9, 137, 150, 152, 166, 220, 350, 542
  accumulation, 26
  average, 13
  effective, 19
  fixed charge, 21
  surface, 19
  surface orientation, 23
Monte Carlo method, 15
MOS physics, 350
MOS-controlled thyristor, 506
MOS-gated thyristors, 503
MOS-GTO, 506
MOSFET, 18, 335, 572
  accumulation resistance, 370
  blocking characteristics, 342
  body bias effect, 400
  cell structure, 345
  cell topology, 415
  channel density, 371
  channel length, 344
  channel pinch-off, 364
  channel resistance, 362, 369
  contact resistance, 372
  current saturation, 341, 365
  DMOS cell, 346
  doping profile, 343
  drift resistance, 371
  [dV/dt] capability, 395
  frequency response, 348, 381
  gate oxide, 414
  gate resistance, 386
  input capacitance, 384
  integral diode, 402
  JFET resistance, 370
  lateral, 363
  on-resistance, 340, 367
  on-state, 349
  optimum profile, 380
  output characteristics, 340
  parasitic transistor, 339, 342, 396, 406
  power dissipation, 367
  process, 410
  safe-operating-area, 397

MOSFET, (Continued)
  silicon carbide, 417
  source resistance, 368
  structure, 336
  substrate resistance, 368
  switching performance, 387
  temperature, 406
  terraced gate structure, 395
  threshold voltage, 357
  transconductance, 341, 366
  turn-off, 392
  turn-on, 388
  UMOS cell, 348
  VMOS cell, 347
Motor control, 2, 7, 128, 336, 402, 483, 498, 575
MPS rectifier, 187
  carrier profile, 190
  on-state, 188
  reverse blocking, 192
  reverse recovery, 192
  stored charge, 191
  structure, 187
Multicell MCT, 519
Multiplication coefficient, 68, 115, 265

Negative bevel, 107
Negative resistance, 233
Neutron:
  absorption, 37
  decay length, 37
  flux, 37
Neutron transmutation doping, 36
Neutrons:
  fast, 40
  thermal, 36
Noise, 410
Normally-off device, 6
Normally-on device, 6
Numerical analysis, 68
Numerical controls, 8, 485, 498

Off-state, 3

Ohm's law, 313
On-resistance
  DMOSFET, 367
  ideal, 373
  specific, 152
  UMOSFET, 377
On-state, 3
  bipolar transistor, 216
  BRT, 534
  EST, 557
  IGBT, 434
  JBS rectifier, 184
  MOSFET, 349
  MPS rectifier, 188
  P-i-N rectifier, 155
  Schottky rectifier, 131
  thyristor, 272
Open-base transistor, 36, 70, 113, 263
Optical fiber, 295
Optimization
  DMOS cell, 373
  EST, 559
Optocoupler, 296
Output characteristics, 6
  bipolar transistor, 213
  BRT, 527
  EST, 546, 549
  IGBT, 429, 444
  MOSFET, 340
  thyristor, 262
Output conductance
  bipolar transistor, 213
  MOSFET, 366
  IGBT, 443
Over-current, 587
Over-temperature, 587
Over-voltage, 587
Oxidation, 22, 120
  dry, 23
  wet, 23
Oxide:
  fixed charge, 21
  tapered, 149
Oxide charge, 405

Oxygen, 55
   precipitates, 35

p-channel
   IGBT, 472, 485
   MOSFET, 373
P-i-N rectifier, 153
   carrier distribution, 160
   doping profile, 177
   end-region recombination, 164
   leakage current, 169
   lifetime control, 175
   maximum temperature, 181
   ohmic drop, 161
   on-state, 155
   power dissipation, 171, 182
   recombination current, 155
   reverse blocking, 169
   reverse recovery, 171
   saturation current, 165
   structure, 158
   trade-off curve, 176
Parasitic thyristor
   EST, 545, 551
   IGBT, 430, 451
Parasitic transistor, 396
   BRT, 534
   MCT, 515
Passivation, 119
Phase control, 483
Phonon, 10
Phosphorus, 27, 33, 37, 55
Photo-current, 297
Pinch rectifier, 182
Pinch-resistance, 228
Planarization, 413
Platinum, 55, 57, 153, 175, 405
Poisson's equation, 30, 71, 77, 83, 87, 222, 353, 380, 470
Polyamide, 120
Polysilicon, 336, 362, 411
Polysilicon resistor, 524
Positive bevel, 105
Potassium hydroxide, 24

Potential distribution, 72, 78
Power dissipation, 4, 171, 573
   BRT, 537
   MOSFET, 367
Power loss analysis, 573
Power supplies, 2, 7, 128, 193, 336, 342, 402, 420, 483
Process (MOSFET), 410, 412
Protection, 587
PWM (Pulse-width-modulation), 576
Punch-through, 305

Quadrant:
   first, 4
   third, 4
Quantum levels, 18
Quasi-saturation, 237

Radius (screening), 30
Reach-through, 96, 114, 117, 203
Reactive ion etching, 25, 111, 338, 412
Recombination, 41, 211, 239, 314, 477, 538
   Auger, 41, 59, 205, 273
   multiphonon, 41
   radiative, 41
   Shockley-Read-Hall, 42
Recombination level, 42
   optimization, 47
Rectifier, 128, 578
   blocking voltage, 264
   epitaxial, 178
   forward recovery, 154
   JBS, 182
   MPS, 187
   P-i-N, 153
   reverse recovery, 154
   Schottky, 129
   synchronous, 341
Regenerative action, 261
Relaxation time, 383
Remote emitter, 273
Resistive load, 312, 533

Resistivity, 27
  extrinsic, 33
  intrinsic, 28
  variation, 36, 38
Resonant circuit, 146
RESURF, 587
Reverse blocking, 6
  IGBT, 431
  MPS rectifier, 192
  P-i-N rectifier, 169
  Schottky rectifier, 137
  thyristor, 263
Reverse recovery, 153, 154, 171, 404
  MPS rectifier, 192
  snappy, 176
  soft, 175, 176
  thyristor, 291
  time, 174
  triac, 329
Reactive-Ion-Etching, 25, 412
Rise time, 283
Rittner current density, 210
Rittner effect, 206
Robotics, 8, 498

Safe-operating-area, 243, 391, 577
  EST, 548, 560
  IGBT, 468
  MCT, 521
  MOSFET, 397
Saturated drift velocity, 222
Saturated velocity, 470
Saturation current, 133
Scattering, 11
  acoustical phonon, 10
  carrier-carrier, 17, 273
  inter-valley, 10
  ionized impurity, 11
  lattice, 10
  optical phonon, 10
  surface, 18
Schottky barrier
  lowering, 138, 140
  optimization, 147

  shielding, 183
Schottky barrier height, 131, 146
Schottky rectifier, 129
  drift resistance, 135
  edge termination, 148
  fabrication, 146
  gallium arsenide, 151
  high voltage, 150
  leakage current, 138, 140
  maximum temperature, 141
  on-state, 131
  power dissipation, 142, 144
  reverse blocking, 137
  reverse characteristics, 141
  saturation current, 133
  series resistance, 133
  silicon carbide, 152
  structure, 133
  substrate resistance, 133
  switching, 145
  trade-off curve, 143
Screening, 29
  radius, 30
Scribing, 103
Second breakdown, 243, 398
  forward-biased, 244
  reverse-biased, 247
Segregation, 112
Segregation coefficient, 36, 38
Self-aligned contacts, 413
Self-heating, 4
Self-protection, 587
Shockley, 42
Shockley-Read-Hall recombination, 156
Short-circuit protection, 525, 544, 560
Shunting resistance, 267, 395, 526, 530
Silicate glass, 121
Silicide, 362, 372
Silicon carbide, 9, 143, 150, 417, 574, 584
  MESFET, 419
  poly-types, 152

Silicon dioxide, 120
Silicon nitride, 120, 412
SIPOS, 120
SIT (static induction transistor), 572
SITH (static induction thyristor), 572
Smart power technology, 2, 587
Snap-back, 533
Snubber, 513
Sodium, 360
Solid solubility, 55
Space charge, 155, 354
Space-charge-generation, 169
Specific resistance, 133
Spreading resistance, 26, 371
Spreading time, 285
Spreading velocity, 285
Steam iron, 490
Steel mills, 7
Storage time:
   bipolar transistor, 241
   GTO, 306
Stored charge, 154, 172, 189, 191, 222, 240, 306
   IGBT, 448, 478
   thyristor, 284
   triac, 330
Stray inductance, 244
Street car, 322
Striation, 36
Structure:
   bipolar transistor, 202
   BRT, 526
   EST, 544
   IGBT, 428
   JBS rectifier, 183
   MCT, 507
   MOSFET, 336
   MPS rectifier, 187
   P-i-N rectifier, 158
   Schottky rectifier, 133
   thyristor, 260
   triac, 325
Surface charge, 18, 96, 353
Surface passivation, 119
Surface polish, 24
Surface potential, 354
Surface roughness, 18, 24
Surface states, 20
Surge current, 40, 60, 167, 189
   BRT, 528
   EST, 546
Switching:
   bipolar transistor, 236
   IGBT, 476
   MOSFET, 387
   MPS rectifier, 192
   P-i-N rectifier, 171
   Schottky rectifier, 145
   thyristor, 279
Switching energy, 482
Switching locus, 476
Switching speed, 50, 58
Symmetrical IGBT, 433

Telecommunications, 128
Temperature, 10, 15, 136, 263
   BRT, 541
   EST, 564
   IGBT, 490
   MCT, 524
   MOSFET, 406
Terraced gate structure, 395
Thermal impedance, 145, 368, 573
Thermal runaway, 4, 28, 181, 407
Threshold voltage, 357, 409
Thyristor, 258
   amplifying gate, 287
   base-resistance-controlled, 526
   blocking characteristics, 262
   carrier distribution, 273
   cathode short, 266
   design, 276
   dI/dt capability, 287
   doping profile, 260
   [dV/dt] capability, 280
   [dV/dt] protection, 300
   emitter-switched, 543
   equivalent circuit, 261

Thyristor, (Continued)
  forward blocking, 264
  gate design, 287
  gate drive, 274
  gate turn-off, 302
  holding current, 272, 277
  inverter grade, 286
  leakage current, 265
  light-triggered, 295
  MOS-gated, 503, 572
  MOS-gated turn-on, 504
  on-state, 272
  output characteristics, 262
  ratings, 259
  regenerative action, 261
  reverse blocking, 263
  reverse recovery, 291
  self-protection, 298
  self-sustaining mode, 272
  stored charge, 284
  structure, 260
  switching, 279
  triggering current, 276
  turn-off, 291, 303
  turn-on, 273, 282
Titanium, 153
Totem-pole, 575
Traction, 2, 8, 322, 498, 572
Trade-off curve:
  BRT, 540
  IGBT, 483
Transconductance, 20, 408
  EST, 561
  IGBT, 442
  MOSFET, 341, 366
Transistor, 261, 505
  bipolar, 198, 395, 427, 440, 572
  open-base, 36, 70, 113, 263
Transit time, 282, 381
Trench, 298, 338, 378
Trench-gate, 412
  IGBT, 496
  MOSFET, 338, 377, 413
Trends, 192, 255, 330, 420, 498, 568, 588
Triac, 322
  area, 325
  [dV/dt] capability, 329
  gate, 326
  remote gate, 328
  reverse recovery, 329
  stored charge, 330
  structure, 325
  triggering, 328
Triode, 322
Turn-off:
  bipolar transistor, 240
  BRT, 538
  EST, 563
  IGBT, 478
  MCT, 523
  MOSFET, 392
  thyristor, 291
Turn-off locus, 244
Turn-on:
  bipolar transistor, 237
  MOSFET, 388
  thyristor, 273, 282
Turn-on locus, 245
TV deflection, 7

UMOSFET, 338
  on-resistance, 377
  process, 413
Unipolar, 335
Unipolar device, 47

V-groove, 24
Vacancy, 55
Valency band, 10
Velocity, 9, 13, 18
  saturation, 27
VLSI, 128
VMOSFET, 337
Voltage overshoot, 244

Wave function, 18
Work function, 130, 350